An Algorithmic Approach To Hemostasis Testing

Kandice Kottke-Marchant, MD, PhD

Editor

Section Editors

Hemostasis Physiology
Jerry B. Lefkowitz, MD

Basics of Laboratory Testing in Hemostasis
Kandice Kottke-Marchant, MD, PhD

Algorithmic Approach to Bleeding Disorders
Elizabeth M. Van Cott, MD

Algorithmic Approach to Thrombophilic Disorders
Charles Eby, MD
John D. Olson, MD, PhD

Antiplatelet and Anticoagulant Drugs
John T. Brandt, MD

Advancing Excellence

Library of Congress Control Number: 2008936747
ISBN: 978-0-930304-93-5
Printed in the United States of America

Advancing Excellence

College of American Pathologists
325 Waukegan Road
Northfield, Illinois 60093
800-323-4040

Contributors

Kandice Kottke-Marchant, MD, PhD
Chair, Pathology and Laboratory Medicine
  Institute
Section Head, Hemostasis and Thrombosis
Cleveland Clinic
Chair, Department of Pathology
Cleveland Clinic Lerner College of Medicine
Cleveland, Ohio

Charissa A. Bailey, MD
Department of Pathology
University of Texas Health Science Center and
  University Health System
San Antonio, Texas

John T. Brandt, MD
Eli Lilly and Company
Lilly Corporate Center
Indianapolis, Indiana

Wayne L. Chandler, MD
Professor and Vice Chair
Department of Laboratory Medicine
Harborview Medical Center
Seattle, Washington

Mark T. Cunningham, MD
Associate Professor
Director, Hematology Laboratory
Department of Pathology and Laboratory
  Medicine
University of Kansas Medical Center
Kansas City, Kansas

Charles Eby, MD
Associate Professor
Department of Pathology and Immunology
Washington University School of Medicine
St. Louis, Missouri

Timothy Hayes, MD, DVM
Department of Pathology
Maine Medical Center
Portland, Maine

Jayashree Krishnan, MD
Medical Director, Hematopathology and
  Automated Services
Department of Pathology
Washington Hospital Center
Washington, DC

Jerry B. Lefkowitz, MD
Professor of Pathology and Laboratory Medicine
Medical Director, Coagulation and Special
  Hematology
Weill Cornell College of Medicine of Cornell
  University
New York Presbyterian Hospital
New York, New York

John D. Olson, MD, PhD
Professor and Director of Clinical Laboratories
Department of Pathology
University of Texas Health Science Center and
  University Health System
San Antonio, Texas

Chad R. Rund, DO
Fellow
Department of Pathology
University of Pittsburgh
School of Medicine
Pittsburgh, Pennsylvania

Elizabeth M. Van Cott, MD
Associate Professor of Pathology
Harvard Medical School
Director, Coagulation Laboratory
Massachusetts General Hospital
Boston, Massachusetts

This book is dedicated to our friend, colleague, and mentor, Douglas A. Triplett, MD. Dr. Triplett is well known for his original contributions to the field of hemostasis and thrombosis, particularly with respect to lupus anticoagulants and the antiphospholipid syndrome, and for his ability to communicate and teach the principles of hemostasis and thrombosis. Through these efforts, he has had a profound influence on a whole generation of "clotters."

Dr. Triplett was the first Chair of the Coagulation Resource Committee of the College of American Pathologists. In this role, he provided a vision for the Coagulation Resource Committee that included supervision of meaningful proficiency exercises for clinical laboratories, careful analysis of the results of these surveys, and communication of the results of the surveys to the medical community to educate and improve laboratory assessment of hemostasis.

Dr. Triplett truly regarded the Coagulation Resource Committee as a resource to the College of American Pathologists, its members, and the wider medical community. His commitment to serving the medical community is well illustrated by the series of published consensus conferences on critical topics in laboratory hemostasis. This volume is another step in implementing his vision for the Coagulation Resource Committee.

Contents

Preface   xvii

Section 1. Hemostasis Physiology   1

CHAPTER 1. Coagulation Pathway and Physiology   3
Introduction   3
Constituents of the Hemostatic System   3
    Endothelium   3
    Platelets   5
    Coagulation Proteins   5
    The Extrinsic Pathway and the PT   6
    The Intrinsic Pathway and the aPTT   7
    The PT and aPTT Pathways   7
    Newer Coagulation Model   7
    Regulatory Mechanisms   10
Operation of the Hemostasis and Thrombosis Pathway   11
Suggested Reading   11

CHAPTER 2. Platelet Structure and Function   13
Platelet Production   13
    Megakaryocyte Development and Maturation   13
    Platelet Production from Megakaryocytes   15
Platelet Structure   15
    Surface Glycoproteins and Plasma Membrane   15
        Von Willebrand Receptor (GPIb/IX/V)   15
        Collagen Receptors (GPVI and GPIa/IIa)   16
        Adenosine Diphosphate Receptors   17
        Thrombin Receptors (PAR-1 and PAR-4)   17
        Thromboxane Receptor and Prostaglandin Receptors 18
        Fibrinogen Receptor (GPIIb/IIIa)   18
    Platelet Cytoskeleton   19
    Platelet Membrane Systems   21
    Platelet Granules   21
        Alpha Granules   21
        Dense Granules   22
Platelet Activation   22
    Platelet Adhesion   23
        Role of GPIb/IX/V and von Willebrand Factor   23
        Role of Collagen Receptors   23
        Cytoskeletal Changes   23
    Platelet Activation and Granule Release   24
        Phospholipase Pathway   24
        Cyclooxygenase Pathway   25
        Platelet Granule Release   25

Platelet Aggregation   25
Platelet Procoagulant Activity   26
Suggested Reading   26

CHAPTER 3. Fibrinolytic System Physiology   **29**
History   29
Proteins and Function   29
Regulation of Fibrinolysis   32
Fibrinolysis at the Site of a Thrombus   33
Suggested Reading   34

Section 2. Basics of Laboratory Testing in Hemostasis   **35**

CHAPTER 4. Sample Collection and Processing   **37**
Introduction   37
Sample Collection Technique for Peripheral Venipuncture   37
    Needle Selection for Venipuncture   38
    Tourniquet Effect   38
    Collection Sets   38
    Specimen Labeling   38
Blood Collection from Indwelling Lines and Catheters   39
Blood Collection Tubes and Anticoagulant Considerations   39
    Blood Collection Tubes   39
    Blood Tube Draw Order   40
    Anticoagulant   40
    Volume Effect   40
    Hematocrit Effect   41
Sample Processing   42
    Preparation of Platelet-Poor Plasma   42
    Effects of Temperature and Time   42
    Factors Affecting Specimen Quality   44
Suggested Reading   45

CHAPTER 5. Clinical History in Hemostasis   **47**
Introduction   47
Clinical History in Bleeding Patients   47
    Determine Whether the Patient Has Excess Bleeding   47
        Extent of Bleeding   47
        Duration of Bleeding Abnormality   49
        Type of Bleeding   49
        Bleeding Due to Vascular Disorders   50
        Posttraumatic and Spontaneous Bleeding   50
        Hemarthrosis and Hematomas   50
        Epistaxis   50
        Surgical Procedures   50
        Gingival/Dental Bleeding History   51
        Menstrual History   51
        Obstetric History   51
    Bleeding Due to Organ System Disorders   51
        Medication and Food Habits   52
        Family History   53
        Previous Coagulation Test Results   54
        Pediatric Considerations   54
Clinical History in Patient with Thrombosis   54
    Age of Onset   55

Ethnicity   55
Location and Type of Thrombosis (Venous, Arterial, or Embolic)   57
    Recurrences   57
Acquired Thrombophilic Risk Factors   57
    Medical History   57
    Malignancy   58
    Obstetrical History and Hormone Therapy   58
    Medication History   59
    Surgery and Trauma   59
    Elevated Levels of Homocysteine   59
Summary   59
Suggested Reading   59

CHAPTER 6. Coagulation Testing   **63**
Introduction   63
Laboratory Measurement of Clot-Based Coagulation Tests   63
    Coagulation Analyzers   63
        Principles of Performance   63
        How to Select and Validate a Coagulation Analyzer   64
    Prothrombin Time   64
        Reagent Considerations   65
        Reporting of Prothrombin Time   66
        Calculation of International Normalized Ratio   66
        Clinical Utility of Prothrombin Time   70
        Point-of-Care Monitors for Oral Anticoagulant Therapy   70
    Activated Partial Thromboplastin Time   70
        Principle   70
        Reagents   71
        Method   71
        Reference Range   71
        Clinical Utility of Activated Partial Thromboplastin Time   71
        Steps to Take in the Evaluation of a Prolonged Activated Partial Thromboplastin Time   72
        Evaluation of Unfractionated Heparin Sensitivity   72
    Thrombin Time   73
        Method   73
        Reagents   73
        Reference Ranges   73
        Clinical Utility of Thrombin Time   73
    Fibrinogen Assay   74
        Clinical Utility of Fibrinogen Assays   75
    Assay of Specific Factors   75
        Principle   75
        Factor II (Prothrombin)   76
        Factors V, VII, and X   76
        Factors VIII, IX, XI, and XII, and High-Molecular-Weight Kininogen and Prekallikrein   76
        False-Positives and False-Negatives in Factor Assays   77
        Factor XIII   77
        Von Willebrand Factor   78
    Mixing Studies   78
        Principle   78
        Technical Considerations   78
        Procedure   80
        Criteria for Interpretation of Mixing Studies   80

Specific Inhibitor Studies 80
    Bethesda Assay 80
    Nijmegen Modification 81
    Reporting of Results 81
Establishing Reference Ranges and Validating Coagulation Laboratory Testing 82
    Preanalytic and Postanalytic Considerations 82
        Reference Range 82
        Accuracy 82
        Precision 83
        Linearity 83
        Interfering Conditions 83
Quality Control 85
    Frequency of QC Testing 86
Results Reporting 87
    Turnaround Time 87
    Reporting of Results 87
Quality Assurance 88
    External Proficiency Testing 88
Summary 89
Suggested Reading 89

CHAPTER 7. Platelet Testing 93
Patient History 93
    Clinical History for Patients with a Bleeding Diathesis 93
    Clinical History for Patients with Thrombosis 93
Platelet Count, Indices, and Morphology 94
    Sample Collection and Processing Considerations 94
    Test Performance and Interpretation 95
Bone Marrow Analysis 96
    Sample Collection and Processing Considerations 96
    Bone Marrow Interpretation 97
Platelet Function Testing 97
    Bleeding Time 98
    PFA-100 98
        Specimen Considerations 98
        Test Performance and Interpretation 98
Platelet Aggregation 101
    Specimen Considerations 101
    Test Performance and Interpretation 101
    Plateletworks 103
    VerifyNow 104
Release Assays 105
    Lumiaggregation 105
    Release Markers 106
    P-selectin 106
Adhesion Assays 106
    Platelet Adhesion Chambers 106
    Impact 106
Platelet Mechanical Assays 106
    Hemostasis Analysis System 106
Whole Blood Viscoelastometry 107
Flow Cytometry 107
    Platelet Turnover (Platelet Reticulocyte Analysis) 109
Electron Microscopy 109

Platelet Genetic Testing 109
ADAMTS-13 Assays 110
Summary 110
Suggested Reading 110

CHAPTER 8. Fibrinolysis Testing **113**
Clinical History Suggesting the Need for Fibrinolytic Testing 113
Sample Collection and Processing for Fibrinolytic Testing 113
D-Dimer and Fibrin Split Product Assays 115
Plasminogen Assays 117
Alpha 2-Antiplasmin Assays 118
Tissue Plasminogen Activator Assays 118
Plasminogen Activator Inhibitor-1 Assays 120
Global Tests of Fibrinolysis 121
Lipoprotein (a) 122
Research Assays 122
Suggested Reading 123

CHAPTER 9. Thrombophilia Testing **125**
Introduction 125
Protein C 125
    Protein C Functional Assays 125
        Clotting Method 125
        Spectrophotometric Method 126
    Protein C Antigen Assays 127
Protein S 128
    Protein S Functional Assays 129
        Protein S Activity Assays 129
        Protein S Antigenic Assays 129
Antithrombin 130
    Antithrombin Assays 131
        Antithrombin Functional Assays 131
        Antithrombin Antigen Assay 131
Activated Protein C Resistance 131
    Activated Protein C Resistance Assays 131
Factor VII Activity 132
Fibrinogen Abnormalities 133
Lupus Anticoagulant 133
Factor V Leiden and Prothrombin G20210A Mutations 133
Plasma Homocysteine Concentration 134
Global Tests of Hypercoagulability 135
    Endogenous Thrombin Potential 135
    Undiluted Activated Protein C Resistance 136
    Activated Partial Thromboplastin Time 136
    Activation and Degradation Peptides 136
Conclusion 137
Suggested Reading 137

Section 3. Algorithmic Approach to Bleeding Disorders **141**

CHAPTER 10. Abnormal Activated Partial Thromboplastin Time **143**
Introduction 143
Clinical Etiologies 143
    Medications 143
    Coagulation Factor Deficiency Associated with Significant Hemorrhage 144

Von Willebrand Disease 144
Factor VIII Deficiency 145
Factor IX Deficiency 145
Factor XI Deficiency 145
Other Inherited and Acquired Deficiencies 146
Coagulation Factor Deficiencies Not Associated with Bleeding 146
Lupus Anticoagulant 146
Specific Coagulation Factor Inhibitors 146
Alloantibodies in Hemophilic Patients 146
Autoantibodies to Factor VIII and Other Factors 147
Spurious Causes of an Elevated aPTT 147
Laboratory Testing 148
Accelerated aPTT 150
Summary 151
Suggested Reading 151
Case Study 143

CHAPTER 11. Prolonged Prothrombin Time   153
Introduction 153
Differential Diagnosis of a Corrected Mixing Study 153
Dilutional Coagulopathy 154
Disseminated Intravascular Coagulation 155
Drugs 155
Warfarin 155
Factor Deficiency 155
Factor II Deficiency 155
Factor V Deficiency 156
Factor V and Factor VIII Deficiency, Combined 157
Factor VII Deficiency 157
Factor X Deficiency 157
Fibrinogen Deficiency 158
Vitamin K-Dependent Coagulation Factor Deficiency, Hereditary 158
Liver Disease 159
Superwarfarin Poisoning 159
Vitamin K Deficiency 159
Newborns 159
Adults 160
Differential Diagnosis of a Noncorrected Mixing Study 160
Drugs 160
Direct Thrombin Inhibitors 160
Heparin 160
Factor Inhibitors 161
Factor II Inhibitor 161
Factor V Inhibitor 161
Factor VII Inhibitor 162
Factor X Inhibitor 162
Lupus Anticoagulant 162
Algorithm 162
Suggested Reading 164
Case Studies: Prolonged PT with Normal aPTT; Prolonged PT with Prolonged aPTT   154

CHAPTER 12. Normal Prothrombin Time and Activated Partial Thromboplastin Time   169
Introduction 169
Clinical Evaluation 169

False-Negative (Normal) PT and aPTT   169
   Patient-Specific Causes   170
   Preanalytic Causes   170
   Analytic Causes   170
   Postanalytic Causes   171
Additional Testing   171
   Primary Hemostasis Defects   171
   Secondary Hemostasis Defects   171
      Factor Assays   171
      Factor XIII Deficiency   171
   Fibrinolysis Defects   172
   Vascular Disorders   172
Algorithm   173
Suggested Reading   173
Case Study   *170*

CHAPTER 13. Fibrinolytic Bleeding Disorders   **175**
Clinical Evaluation of Fibrinolytic Bleeding Disorders   175
Plasminogen Activator Inhibitor-1 Deficiency   175
Alpha-2 Antiplasmin Deficiency   177
Acquired Causes of Fibrinolytic Bleeding   177
Disseminated Intravascular Coagulation   178
Suggested Reading   182
Case Study: Postoperative Bleeding and PAI-1 Deficiency   *175*
Case Study: Tissue Plasminogen Activator Release During Renal Transplantation *177*
Case Study: DIC Associated with Gastric Carcinoma   *178*

CHAPTER 14. Platelet Disorders   **185**
Introduction   185
Platelet Dysfunction with Normal Platelet Count (Qualitative Platelet Disorders) 185
   Glycoprotein Disorders   189
      Glanzmann Thrombasthenia (GPIIb/IIIa Deficiency) 189
      Glycoprotein Ib/IX/V (Bernard-Soulier Syndrome) 190
      Platelet-Type von Willebrand Disease   190
      Glycoprotein IV   191
      Collagen Receptor Disorders (GPIa/IIa and GPVI) 191
      Adenosine Diphosphate Receptor Abnormalities 191
   Platelet Release Defects   191
      Storage Pool Disorders (Alpha and Dense Granule Disorders) 191
      Signal Transduction Disorders   194
      Disorders of Platelet Procoagulant Activity (Scott Syndrome) 195
Platelet Disorders with Thrombocytosis   195
   Reactive Thrombocytosis   195
   Myeloproliferative Disorders Associated with Thrombocytosis 196
Platelet Disorders with Thrombocytopenia   196
   Thrombocytopenia with Increased Platelet Size   197
      Macrothrombocytopenias with Neutrophilic Inclusions (MYH9 Disorders) 197
      Bernard-Soulier Syndrome   199
      The Gray Platelet Syndrome (Alpha Granule Storage Pool Disorder) 200
   Thrombocytopenia with Decreased Platelet Size   201
      Wiskott-Aldrich Syndrome and X-Linked Thrombocytopenia 201
   Thrombocytopenia with Normal Platelet Size   201
   Peripheral Platelet Destruction   202
      Immune Destructive Thrombocytopenias   202

Nonimmune Destructive Thrombocytopenias 204
Decreased Platelet Production 207
    Hereditary Thrombocytopenias with Decreased Platelet Production 207
    Acquired Thrombocytopenias with Decreased Platelet Production 208
    Platelet Dysfunction Associated with Other Illnesses 208
Platelet Disorders Associated with Thrombosis or Platelet Activation 208
    Platelet Activation 209
    Genetic Polymorphisms Associated with Thrombosis and Cardiovascular Disease 209
    Cardiovascular Devices 210
Summary 210
Suggested Reading 211
Case Study *185*

CHAPTER 15. Abnormal Thrombin Time **217**
Introduction 217
Clinical Indications 217
Differential Diagnosis 218
    Amyloidosis 218
    Autoantibodies 219
    Drugs 219
        Bovine Thrombin, Topical 219
        Direct Thrombin Inhibitors 219
        Heparin 219
        Plasminogen Activators 220
        Volume Expanders 220
    Fibrin Degradation Products 220
    Fibrinogen Abnormalities 220
        Dysfibrinogenemia 220
        Hypofibrinogenemia 222
        Afibrinogenemia 222
        Hyperfibrinogenemia 222
    Radiocontrast Agents 222
Algorithm 222
Suggested Reading 222
Case Study *217*

CHAPTER 16. Von Willebrand Disease **225**
Introduction 225
Von Willebrand Factor 225
Clinical Features 225
    Classification 226
    Inheritance 227
Laboratory Testing 227
    Laboratory Findings in Type 1 von Willebrand Disease 230
    Laboratory Findings in Type 2 von Willebrand Disease 230
    Laboratory Findings in Type 3 von Willebrand Disease 231
The Molecular Basis of von Willebrand Disease 231
Acquired von Willebrand Disease 232
Treatment 233
Suggested Reading 234

Section 4. Algorithmic Approach to Thrombophilic Disorders **237**

CHAPTER 17. Arterial and Venous Thrombosis in Adults **239**
Introduction 239

General Considerations 240
    Consent and Counseling 240
    Assay Calibration, Reference Range, and Assay Reliability 240
    Combined Deficiency 241
Testing for Thrombophilic Risk in Patients with Venous or Arterial Thrombosis 241
    Thrombophilic Risk Factor Test Selection in Provoked and Spontaneous
      Venous Thromboembolism 241
        What Tests Should Be Performed? 242
        When Should Thrombophilia Testing Be Performed? 244
    Recommendations for Testing of Individual Analytes 244
      Protein C, Protein S, and Antithrombin Deficiency 244
      Factor V Leiden and Prothrombin G20210A Mutations 245
      Hyperhomocysteinemia and High Concentrations of Factor VIII 246
    Acquired Thrombophilic Risk Factors 246
      Antiphospholipid Antibodies 246
      Heparin-Induced Thrombocytopenia 246
    Thrombophilic Risk Factor Test Selection in Arterial Thrombosis 247
      Antiphospholipid Antibodies (Lupus Anticoagulant and Anticardiolipin Antibodies) 248
      C-Reactive Protein 248
      Lipoprotein (a) 248
      Homocysteine 249
      Prothrombin G20210A Mutation 250
      Factor V Leiden (Activated Protein C Resistance) 250
      Deficiencies of Protein C, Protein S, and Antithrombin 250
      Coagulation Factors 251
      Heparin-Induced Thrombocytopenia 251
    Thrombophilic Risk Factor Test Selection in Neurovascular Thrombosis 251
      Antiphospholipid Antibodies (Lupus Anticoagulant and Anticardiolipin Antibodies) 251
      Lipoprotein (a) 251
      Homocysteine 252
      Prothrombin G20210A Mutation 253
      Factor V Leiden (Activated Protein C Resistance) 253
      Deficiencies of Protein C, Protein S, and Antithrombin 253
      Heparin-Induced Thrombocytopenia 254
Suggested Reading 254
Case Studies 242

CHAPTER 18. Unique Issues of Thrombophilia in Women and Children **257**
Introduction 257
Pregnancy and Hormonal Therapy 257
    Oral Contraceptives and Risk of Venous Thromboembolism 257
    Hormone Replacement Therapy and Venous Thromboembolism 258
    Pregnancy-Related Venous Thromboembolism 259
    Thrombophilia and Recurrent Fetal Loss 259
    Thrombophilia and Late Gestational Vascular Complications 260
    Treatment of Pregnant Women with Thrombophilias 261
Thrombophilia in Neonates, Infants, and Children 262
    Description of the Problem 262
    Testing for Thrombophilia in Children 263
Suggested Reading 265
Case Study 258
Case Study 262

CHAPTER 19. Laboratory Diagnosis of Inherited Thrombophilia 267
Introduction 267
Algorithmic Approach to Laboratory Diagnosis of Thrombophilic Risk Factors 267
    Protein C 268
    Protein S 270
    Antithrombin 271
    Factor VIII 272
    Activated Protein C Resistance and Factor V Leiden 273
    Prothrombin G20210A Mutation 274
    Homocysteine 274
Thrombophilia Testing, Duration of Anticoagulation Therapy, and Screening First-Degree Relatives 274
Suggested Reading 277
Case Study 268

CHAPTER 20. Fibrinolytic Thrombotic Disorders 279
Introduction 279
Thrombosis and Defects of Fibrinolysis 279
    Plasminogen Deficiency 279
    Tissue Plasminogen Activator Deficiency 279
    Plasminogen Activator Inhibitor Excess 280
    Alpha 2-Antiplasmin Excess 280
    Euglobulin Clot Lysis Time Assay 280
    Thrombin-Activatable Fibrinolysis Inhibitor 280
    Dysfibrinogenemia 281
    Factor XII (Hageman Factor) Deficiency 281
    Lipoprotein (a) Excess 281
    Disseminated Intravascular Coagulation 281
    Antifibrinolytic Medications 282
The Role of D-dimer in Excluding Venous Thromboembolism 282
    D-Dimer Testing 283
        Variations in the Types of D-Dimer Assays 283
        Selecting the Appropriate Cutoff and Units for Excluding Venous Thromboembolism 284
Summary 284
Suggested Reading 284
Case Study 282

CHAPTER 21. Heparin-Induced Thrombocytopenia 287
Introduction 287
Clinical Presentation 287
Pathophysiology 289
Clinical Laboratory Diagnosis and Algorithm 289
Treatment 292
Suggested Reading 293
Case Study 287

CHAPTER 22. Antiphospholipid Antibodies 295
Introduction 295
Laboratory Testing for Antiphospholipid Antibodies 295
    Lupus Anticoagulant Testing Considerations and International Society on Thrombosis
      and Haemostasis (ISTH) Criteria 295
        Preanalytical Variables 297
        Testing for a Lupus Anticoagulant 297
        Exclusion of Other Abnormalities (Factor Inhibitors) 300
        Indeterminate Results: When Samples Do Not Meet All International Society
          on Thrombosis and Haemostasis Criteria 301

Anticardiolipin Antibody Testing and Other Antiphospholipid Antibody Immunoassays 301
Clinical Features and Diagnosis 301
Monitoring Anticoagulation in Patients with Lupus Anticoagulants 303
Suggested Reading 303
Case Study 296

Section 5. Antiplatelet and Anticoagulant Drugs 305

CHAPTER 23. Antithrombotic Agents 307
Introduction 307
Target Therapeutic Range and Monitoring Considerations 308
Common Clinical Settings Where Antithrombotic Agents Are Used 309
    Acute Venous Thromboembolism 309
    Prophylaxis for Venous Thromboembolism 309
    Prophylaxis in Patients with Atrial Fibrillation 309
    Acute Coronary Syndromes 309
    Peripheral Vascular and Cerebrovascular Disease 310
    Extracorporeal Circulation 310
    Implantation of Mechanical Heart Valves 310
Effects on Tests of Hemostasis 310
Summary 310
Suggested Reading 311

CHAPTER 24. Anticoagulant Agents 313
Introduction 313
Unfractionated Heparin 313
    Laboratory Monitoring of Unfractionated Heparin 315
        Acute Venous Thromboembolism 316
        Acute Coronary Syndromes 319
        Renal Dialysis, Percutaneous Coronary Intervention, and Cardiopulmonary Bypass 320
Low-Molecular-Weight Heparins 320
    Laboratory Monitoring of Low-Molecular-Weight Heparin 320
Fondaparinux 321
    Laboratory Monitoring of Fondaparinux 322
Hirudins (Peptide Direct Thrombin Inhibitors) 322
    Laboratory Monitoring of Hirudin Analogues 324
Argatroban 325
    Laboratory Monitoring of Argatroban 326
Oral Anticoagulants 327
    Laboratory Monitoring of Vitamin K Antagonists 328
Suggested Reading 329

CHAPTER 25. Antiplatelet Agents 333
Introduction 333
Parenteral Agents 333
    Glycoprotein IIb/IIIa Inhibitors 333
        Mechanism of Action 333
        Effect on Platelet Function Tests 336
        Clinical Uses 336
        Role of Monitoring 336
        Special Issues 337
Oral Agents 337
    Aspirin 337
        Mechanism of Action 337
        Clinical Uses 337

Laboratory Tests Used to Assess Effect 338
Laboratory Monitoring 339
Thienopyridines 340
Mechanism of Action 340
Clinical Uses 341
Laboratory Monitoring 342
Clopidogrel Resistance 342
Dipyridamole 342
Mechanism of Action 342
Clinical Uses 343
Monitoring Issues 343
Iloprost 344
Cilostazol 344
Mechanism of Action 344
Clinical Uses 344
Monitoring Issues 344
New Antiplatelet Drugs in Development 344
Antiplatelet Effect of Other Drugs and Herbal Remedies 345
Summary 346
Suggested Reading 346

Section 6. Appendices and Index 353

Appendix A. Abbreviations 355
Appendix B. Example Reference Ranges 359
Index 361

List of Algorithms

Detection of Interfering Drug Effects Prior to Performing aPTT Mixing Studies 79
Platelet Functional Analyzer-100 (PFA-100) Testing 99
Thrombophilia Risk Factor Screening 126
Evaluation of a Prolonged aPTT 148
Evaluation of Slight Prolongation of the aPTT 149
Investigation of a Prolonged PT with or without a Prolonged aPTT 163
Diagnosis of a Bleeding Etiology with a Normal PT and aPTT 172
Use of D-Dimer Assay to Assess DIC 180
Distinguishing Platelet Abnormalities with a Normal or Increased Platelet Count 187
Platelet Aggregation Studies to Distinguish Between Acquired Platelet Disorders 187
Distinction of Platelet Disorders with Thrombocytopenia with Increased or Decreased Platelet Size 197
Diagnosis of Thrombocytopenic Disorders Associated with Increased Platelet Destruction
    or Decreased Production 202
Evaluation of an Abnormal Thrombin Time 221
Inherited Thrombophilia Testing 244
Inherited Thrombophilia Risk Factor Confirmatory Testing 267
Protein C Testing 269
Protein S Testing 270
Antithrombin Testing 272
Activated Protein C Resistance/Factor V Leiden Testing 273
Homocysteine Testing 275
Diagnosis of Heparin-Induced Thrombocytopenia, Typical Onset 290
Diagnosis of Heparin-Induced Thrombocytopenia, Rapid Onset 291
Diagnosis of the Presence of a Lupus Anticoagulant 298

Preface

Bleeding and thrombotic disorders are common causes of morbidity and are associated with many primary disease processes, such as malignancy and cardiovascular disease. Antithrombotic drugs are among the most commonly prescribed pharmaceuticals, but many require monitoring to assure therapeutic effect and to prevent bleeding or thrombotic complications. Most hemostasis textbooks focus on the pathophysiology of bleeding and thrombotic disorders and their treatment, but often little emphasis is placed on the laboratory diagnosis of these disorders.

Laboratory testing for both bleeding and thrombotic disorders often requires the performance and interpretation of many assays due to the complexity of the hemostatic system, which involves numerous coagulation proteins, platelets, von Willebrand factor, the fibrinolytic system, and the vasculature. Furthermore, the appropriate collection, processing, and storage of blood samples for hemostasis testing offers more challenges than any other area in the laboratory. *An Algorithmic Approach to Hemostasis Testing* is a new textbook that offers a practical and comprehensive guide for pathologists and laboratories engaged in hemostasis testing, and will serve as a test interpretation resource for pathologists and clinicians alike.

This textbook grew out of many prior publications by members of the Coagulation Resource Committee of the College of American Pathologists that were directed toward best practices for hemostasis testing and use of algorithms to direct testing and patient diagnosis. Originally conceived as a compilation of these publications, the Committee instead produced a cohesive and completely re-written text. It includes sections on hemostasis physiology, the basics of hemostasis laboratory testing and interpretation, an algorithmic approach to bleeding disorders and venous thrombophilic disorders, and monitoring anticoagulant drugs, together with illustrative patient case studies. Sample collection and processing as well as patient history and diagnostic criteria also are covered.

I am deeply grateful to the College of American Pathologists for their support of this project and for the extensive effort and expert contributions by my coauthors, section editors, and Coagulation Resource Committee members, past and present. I would also like to thank my husband, Dr. Roger Marchant, for his patience and understanding during this project, which consumed many weekends of "family time." I hope that this text will be a useful guide to hemostasis testing and will become a desired resource for hemostasis laboratories.

Kandice Kottke-Marchant, MD, PhD

Section I

Hemostasis Physiology

Jerry B. Lefkowitz, MD, Editor

Chapter 1. Coagulation Physiology

Chapter 2. Platelet Structure and Function

Chapter 3. Fibrinolytic System Physiology

Coagulation Pathway and Physiology

Jerry B. Lefkowitz, MD

Introduction

Our understanding of blood clotting is intimately tied to the history of civilization. With the advent of writing 5000 years ago, it could be argued that the first symbols used for blood, bleeding, or clotting represented the first published coagulation pathway. The ancient peoples of the world always held blood in utmost mystical esteem. Through the ages, this esteem has been transmitted to modern times in the many expressions that use "blood," such as "blood is thicker than water," "blood of our fathers," and others.

Mysticism aside, the study of blood clotting and the development of laboratory tests for blood clotting abnormalities are historically inseparable. The workhorse tests of the modern coagulation laboratory, the prothrombin time (PT) and the activated partial thromboplastin time (aPTT), are the basis for the published extrinsic and intrinsic coagulation pathways, even though it is now known that these pathways do not accurately reflect the function of blood clotting in a living organism. In this chapter, and ultimately this textbook, the many authors hope to present a clear explanation of coagulation testing and its important place in the medical armamentarium for diagnosing and treating disease.

Constituents of the Hemostatic System

With the evolution of vertebrates and their pressurized circulatory system, there had to arise some method to seal the system if injured—hence the hemostatic system. Interestingly, there is nothing quite comparable to the vertebrate hemostatic system in invertebrate species. In all vertebrates studied, the basic constituents of the hemostatic system appear to be conserved.

Figure 1-1 illustrates the three major constituents of the hemostatic pathways and how they are interrelated. Each element of the hemosta-

tic system occupies a site at the vertex of an equilateral triangle. This representation implies that each system constituent interacts with and influences all other constituents. In the normal resting state, these interactions conspire to maintain the fluidity of the blood to ensure survival of the organism. Normally, only at the site of an injury will the fluidity of the blood be altered and a blood clot form.

Endothelium

Figure 1-2 shows some of the basic properties of the endothelium. The endothelium normally promotes blood fluidity, unless there is an injury. With damage, the normal response is to promote coagulation at the wound site while containing the coagulation response and not allowing it to propagate beyond this site.

Until recently, the dogma of blood clotting suggested that the single, major procoagulant function of the endothelium is to make and express tissue factor with injury. Endothelial cells do not normally make tissue factor but may synthesize it following cytokine stimulation or acquire the material from activated monocytes in the circulation. Tissue factor is a glycosylated intrinsic membrane protein that is expressed on the surface of

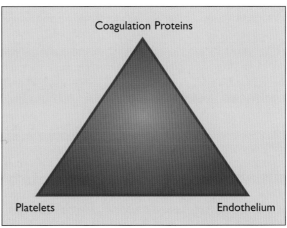

Figure 1-1. Basic representation of the elements of hemostasis.

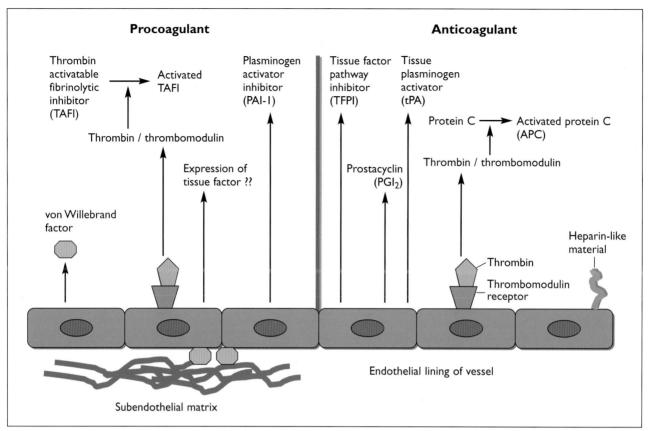

Figure 1-2. A stylized view of endothelial functions related to procoagulation and anticoagulation. The subendothelial matrix, represented by the purple interlocking lines, is a complex of many materials. The most important constituents of the subendothelial matrix related to coagulation function are collagen and von Willebrand factor.

adventitial vascular wall cells and is exposed to flowing blood during vascular injury or endothelial denudation. Tissue factor, when bound to factor VIIa, is the major activator of the extrinsic pathway of coagulation. Classically, tissue factor is not present in the plasma but only presented on cell surfaces at a wound site. Because tissue factor is "extrinsic" to the circulation, the pathway was thusly named.

Endothelium is the major synthetic and storage site for von Willebrand factor (vWF). Von Willebrand factor is secreted from the endothelial cell both into the plasma and also abluminally into the subendothelial matrix. It is a large multimeric protein that acts as the intercellular glue binding platelets to one another and also to the subendothelial matrix at an injury site. In addition, the second major function of vWF is to act as a carrier protein for factor VIII (antihemophilic factor). Von Willebrand factor is synthesized in the endothelial cell as a dimeric protein, with a molecular weight of the monomer approximately 250,000 daltons. The propeptide of vWF, the N-terminal portion of the protein that is cleaved off prior to secretion or storage, is important in directing the multimerization of vWF. Von Willebrand factor multimers can range in size up to 20×10^6 daltons. Von Willebrand factor binds to the platelet glycoprotein Ib/IX/V receptor and mediates platelet adhesion to the vascular wall under shear, which is discussed in more detail in chapter 2.

The remaining endothelial procoagulant functions discussed in this brief review and listed in Figure 1-2 are all parts of the fibrinolytic system, discussed in more detail in chapter 3. The fibrinolytic system is responsible for proteolysis and solubilization of the formed clot constituents to allow its removal. Plasminogen activator inhibitor acts to block the ability of tissue plasminogen activator to turn on plasmin, the primary enzyme of fibrinolysis. The thrombin activatable fibrinolytic inhibitor (TAFI), similar to the thrombin regulatory enzyme protein C, is cleaved to its activated form by the thrombin/thrombomodulin complex. The endothelial cell surface receptor thrombomodulin is a 450,000-dalton protein whose only known ligand is thrombin. Thrombin, once bound to thrombomodulin, loses some proteolytic capabilities but gains the ability to either activate TAFI or protein C. Relevant to this discussion is that the activated form of TAFI catalyzes the removal of lysine residues from the fibrin clot, making it less recognizable as a substrate for plasmin and hence more persistent.

The endothelium and ultimately the entire vessel are normally geared to maintain blood fluidity in the absence of injury. To that end, the endothelium possesses a variety of functions whose ultimate goal is to promote blood fluidity. Although Figure 1-2 portrays a typical endothelium, evidence is accumulating suggesting that endothelium from different vessels and from various parts of the vascular tree is not homogeneous.

The anticoagulant functions of the endothelium inhibit all aspects of the coagulation pathways. Tissue factor pathway inhibitor (TFPI) is a 33,000-dalton low-concentration inhibitor of the extrinsic pathway that is secreted into the plasma by the endothelium. This material rapidly inhibits the tissue factor/factor VIIa complex. Prostacyclin (PGI$_2$), a prostaglandin produced by the endothelial-specific cyclooxygenase enzyme system, is a potent inhibitor of platelet aggregation. Tissue plasminogen activator (tPA) is the main enzymatic activator of the potent fibrinolytic enzyme plasmin. The fibrinolytic system is discussed in more detail in chapter 3.

Once the enzyme thrombin is bound to thrombomodulin, it also gains the ability to activate protein C to activated protein C (APC). APC, along with its cofactor protein S, phospholipid surface, and calcium ions, acts to down-regulate thrombin generation by proteolyzing the two protein cofactors in the coagulation pathways, factor Va and factor VIIIa. Finally, the last anticoagulant property of endothelium listed in Figure 1-2 is the presence of the heparin-like material on the surface. Heparan sulfate is a glycosaminoglycan that is attached to the luminal surface of the endothelium by a protein backbone. The cell-surface heparan sulfate acts as a cofactor for one of the main direct inhibitors of many of the coagulation enzymes, antithrombin. Antithrombin is a 58,000-dalton serpin (serine protease inhibitor) that has a five-stranded central β-sheet (the A-sheet), together with a heparin-binding D-helix and a mobile reactive site loop containing an Arginine393-Serine394 bond that resembles the substrate for thrombin and other serine proteases such as factor Xa, factor IXa, and factor XIa. Once thrombin cleaves the bond, the protease is inhibited by a covalent link to the antithrombin. In its native state, antithrom-

bin inactivates the proteinases inefficiently, due to conformational inaccessibility of the arginine-serine bond. Inhibition is accelerated approximately 1000-fold by the binding of heparin to arginine residues in the D-helix of antithrombin, with a resultant conformational change and exposure of the reactive center.

Platelets

Platelets are discoid anucleate subcellular fragments that can vary in size up to 3 μm in diameter. They arise from the megakaryocyte in the bone marrow and circulate in blood at a platelet count that ranges from 200,000 to 400,000/μL. The platelet has a complex ultrastructure that includes many different surface receptors, several types of storage granules, and a network of actin and myosin filaments. At the site of an injury, the platelets contact extracellular matrix components, causing a series of metabolic changes that result in the formation of a platelet plug. These metabolic changes are usually termed aggregation. Ultimately, this platelet plug is stabilized by the formation of a fibrin clot. Platelet structure and function are described in chapter 2.

Coagulation Proteins

Table 1-1 lists the proteins involved in the formation of the fibrin clot. Factors II, VII, IX, and X (as well as proteins C, S, and Z) are the zymogen forms of vitamin K-dependent serine proteases. Vitamin K is a necessary cofactor for a post-translational modification that adds a carboxyl group to the 10 to 12 glutamic acid residues in the amino terminal portion of each of these proteins. The vitamin K-dependent proteins utilize these clusters of γ-carboxyl glutamic acid (gla) residues to adhere to phospholipid surfaces and assemble multimolecular coagulation complexes. Without this important post-translational modification, the assembly of cell-based coagulation complexes is impaired, leading to ineffective fibrin formation. Another reason to understand this biochemical fact is that the most commonly used oral anticoagulant, warfarin, exerts its anticoagulant effect by inhibiting this modification and ultimately producing dysfunctional vitamin K-dependent factors. These dysfunctional factors affect coagulation tests, and the warfarin anticoagulant effect is therefore monitored by the clot-based PT assay.

Since the identification of all the factors listed in Table 1-1, accumulating epidemiologic evidence has called into question whether some of these factors truly participate in the formation of a fibrin

Table 1-1. Coagulation Factors

Name	Description	Function
Fibrinogen (Factor I)	Molecular Weight (MW) = 340,000 daltons (Da); glycoprotein	Adhesive protein that forms the fibrin clot
Prothrombin (Factor II)	MW = 72,000 Da; vitamin K-dependent serine protease	Activated form is main enzyme of coagulation
Tissue factor (Factor III)	MW = 37,000 Da; also known as thromboplastin	Lipoprotein initiator of extrinsic pathway
Calcium ions (Factor IV)	Necessity of Ca++ ions for coagulation reactions described in 19th century	Metal cation necessary for coagulation reactions
Factor V (Labile factor)	MW = 330,000 Da	Cofactor for activation of prothrombin to thrombin
Factor VII (Proconvertin)	MW = 50,000 Da; vitamin K-dependent serine protease	With tissue factor, initiates extrinsic pathway
Factor VIII (Antihemophilic factor)	MW = 330,000 Da	Cofactor for intrinsic activation of factor X
Factor IX (Christmas factor)	MW = 55,000 Da; vitamin K-dependent serine protease	Activated form is enzyme for intrinsic activation of factor X
Factor X (Stuart-Prower factor)	MW = 58,900 Da; vitamin K-dependent serine protease	Activated form is enzyme for final common pathway activation of prothrombin
Factor XI (Plasma thromboplastin antecedent)	MW = 160,000 Da; serine protease	Activated form is intrinsic activator of factor IX
Factor XII (Hageman factor)	MW = 80,000 Da; serine protease	Factor that nominally starts aPTT-based intrinsic pathway
Factor XIII (Fibrin stabilizing factor)	MW = 320,000 Da	Transamidase that cross-links fibrin clot
High-molecular-weight kininogen (Fitzgerald, Flaujeac, or William factor)	MW = 110,000 Da; circulates in a complex with factor XI	Cofactor
Prekallikrein (Fletcher factor)	MW = 85,000 Da; serine protease	Activated form that participates at beginning of aPTT-based intrinsic pathway

clot in vivo. Although discussed more fully in the following sections, evidence suggests that factor XII and prekallikrein may not normally participate in clotting in vivo but are important in the in vitro laboratory clot assays.

The Extrinsic Pathway and the PT

The first description of the extrinsic pathway was reported by Dr. Paul Morawitz in 1905. Dr. Morawitz produced a hemostasis model incorporating all of the scientific information of his day. Figure 1-3 illustrates a version of this model.

In 1935, Dr. Armand Quick published his method for the prothrombin time—with minor variations, the same laboratory test that is still in use today. Dr. Quick, using the classic four-component extrinsic pathway model of Dr. Morawitz,

essentially made "thrombokinase" with calcium ions. This "thrombokinase" was prepared from a saline extract of rabbit brain with the addition of calcium. The more modern nomenclature for this material is thromboplastin. The basis for Dr. Quick's assay was that adding calcium ions with

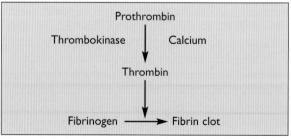

Figure 1-3. A representation of the original extrinsic pathway proposed in 1905.

an excess of thromboplastin to anticoagulated plasma was a direct measure of the prothrombin amount in the plasma—hence the name of the assay, prothrombin time. Only in the 1950s and early 1960s, with the discovery of additional coagulation factors, did the true nature of the extrinsic pathway become known. This is discussed in more detail below in the section, "The PT and aPTT Pathways."

The Intrinsic Pathway and the aPTT

Dr. Quick, in his first publication, observed that his new PT assay was not sensitive to the hemophilic defect. Patients with the symptoms of hemophilia did not usually have an abnormal PT. Evidence had been accumulating that the four-component extrinsic pathway model of blood clotting was not complete. The plasma had the potential to clot without the addition of an extrinsic material. The thromboplastin, thrombokinase, or what we now call tissue factor was not always needed to make blood clot, especially in vitro. Therefore, it appeared that plasma had within it or intrinsic to it all the factors necessary to cause blood clotting.

In 1953, Drs. Langdell, Wagner, and Brinkhous published a paper detailing a clot-based assay that was sensitive to the defect in hemophilic plasma. Instead of using a complete tissue extract, such as the prothrombin time thromboplastin reagent, their assay used only a partial extract. Hence these researchers called their assay a partial thromboplastin time (PTT). In this group's initial assay, the activator necessary to cause clotting was added separately from the PTT reagent and calcium ions. Other workers modified the assay, adding an activator to the PTT reagent, producing the modern activated thromboplastin time assay.

The PT and aPTT Pathways

The PT and aPTT assays were developed based on theories and specific testing needs, without complete knowledge of all the proteins involved in coagulation. In the period from 1935 and the inception of the PT until the early 1970s, all of the procoagulant proteins involved in forming a fibrin clot were identified. Many of these factors were identified because patients were found with deficiency states. Some of these patients had congenital bleeding disease, while others presented with an abnormal prolongation in the PT and/or the aPTT. It became clear that a fresh model of coagulation other than the classic extrinsic pathway was needed.

In the early 1960s, a new synthesis of all hemostasis knowledge was put together, and the PT or extrinsic and aPTT or intrinsic coagulation pathways were published. This provided a framework of how many of the proteins listed in Table 1-1 interact to form a blood clot. This pathway is illustrated in Figure 1-4. Although there have been some modifications since the original papers, these are the pathways with which most workers in hemostasis are familiar. From the initial publication, accumulating epidemiologic evidence suggests this formulation of the intrinsic and extrinsic hemostasis pathways might not be a correct representation of blood clotting in vivo. For example, patients deficient in factor XII, prekallikrein, or high-molecular-weight kininogen do not have a bleeding or thrombotic phenotype. All of these proteins are present at the start of the intrinsic pathway, and deficiencies of each factor can cause a significantly prolonged aPTT assay. Logically, it would make sense that a deficiency of a factor at the start of a pathway would cause bleeding pathology. However, all evidence suggests this is not so. The intrinsic and extrinsic pathways as they have existed since their inception are based on in vitro testing. The in vivo function appears to be different. **It is important to be familiar with the older pathway model because the PT and the aPTT are still useful as diagnostic tests.**

Newer Coagulation Model

Figure 1-5 illustrates a newer model of coagulation. In this figure, thrombin is depicted as the center of the coagulation universe; all aspects of hemostasis feed into the regulation and control of thrombin generation, which in turn forms the definitive clot at the site of an injury. Of note, the figure lacks several proteins normally considered part of the classic intrinsic coagulation pathway: factor XII and prekallikrein. These proteins are not currently thought to be important for in vivo coagulation activation. Although high-molecular-weight kininogen deficiency may not be associated with a bleeding diathesis, it is still part of the new coagulation pathway as it circulates in plasma bound to factor XI. Also important in this newer concept is that the majority of the steps in the coagulation cascade take place by the formation of multimolecular coagulation protein complexes on phospholipid cell surfaces.

This new coagulation model has extrinsic and intrinsic pathway limbs, but the in vivo process of hemostasis is thought only to be initiated by cell-based tissue factor expressed at an injury site.

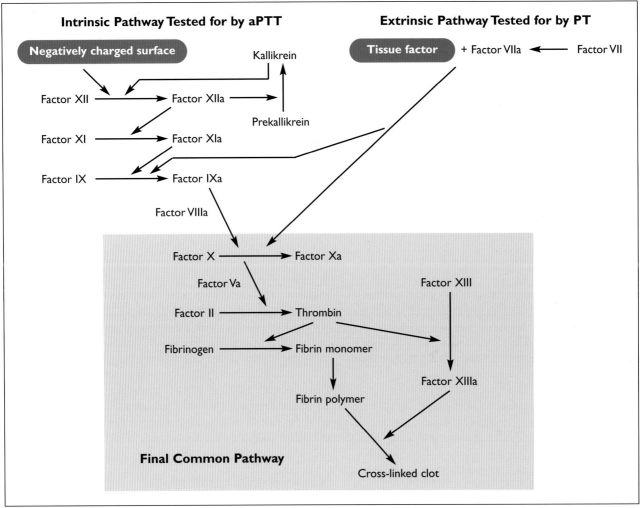

Figure 1-4. A model of the classic extrinsic and intrinsic coagulation pathways. For the sake of clarity, calcium ions and phospholipids, two important cofactors for most coagulation reactions, have been omitted from the figure. Ca⁺⁺ and phos-pholipids are necessary for most reactions except the activation of factor XII and the activation of prekallikrein. The cofactor for the activation of prekallikrein, high-molecular-weight kininogen, has also been omitted from the figure for clarity.

Tissue factor binds to factor VII or VIIa in a 1:1 complex. Limited proteolysis leads to a tissue factor/factor VIIa complex that can activate factor X or factor IX to activated serine proteases through cleavage of an activation peptide. Once the pathway commences, the tissue factor/factor VIIa activation of factor X is rapidly shut down by an inhibitor produced by endothelial cells, tissue factor pathway inhibitor (TFPI). The newly activated factor IXa then binds to its cofactor, factor VIIIa, on a phospholipid surface to form the tenase complex that results in the activation of factor X to factor Xa.

The activation of factor X to Xa starts the final, common pathway for thrombin activation. Factor Xa combines with the cofactor, factor Va, together with calcium on phospholipid surfaces to form the prothrombinase complex. This complex then effects the conversion of prothrombin to thrombin by cleavage of an activation peptide, prothrombin F1.2.

The generation of a small amount of thrombin initiated by extrinsic means appears to be enough to start the coagulation mechanism and, if conditions are right, the expansion of thrombin generation through an intrinsic mechanism. The intrinsic limbs of the pathway include activation of factor XI to factor XIa by thrombin, with the ultimate generation of more thrombin using factor IXa and factor VIIIa to activate factor X. This pathway model also explains the mechanism of how two of the important cofactors, factor V and factor VIII, are activated by thrombin. It was known that both factor V and factor VIII had to be partially proteolyzed or activated to participate in the formation of a blood clot, but the mechanism was never heretofore part of the coagulation model.

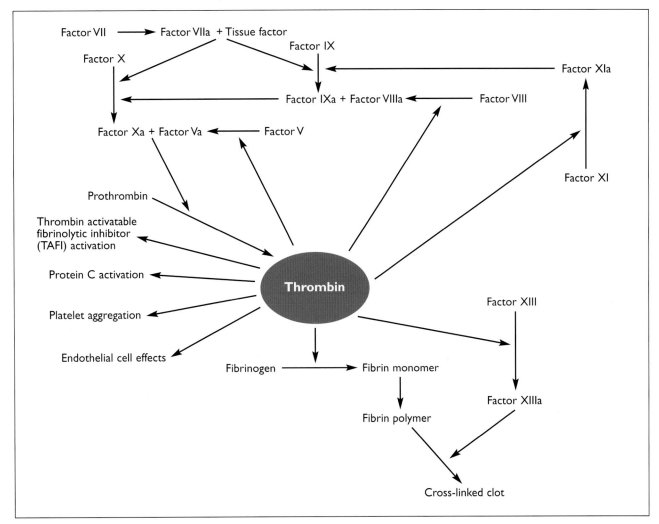

Figure 1-5. A newer model of the coagulation pathway. For the sake of clarity, Ca⁺⁺ and phospholipids have been omitted from the figure. These two cofactors are necessary for all of the reactions listed in the figure that result in the activation of prothrombin to thrombin. The pathway is initiated by an extrinsic mechanism that generates small amounts of factor Xa, which in turn activate small amounts of thrombin. The tissue factor/factor VIIa proteolysis of factor X is quickly inhibit-ed by tissue factor pathway inhibitor (TFPI). The small amounts of thrombin generated from the initial activation feedback to create activated cofactors, factors Va and VIIIa, which in turn help to generate more thrombin. Tissue factor/factor VIIa is also capable of indirectly activating factor X through the activation of factor IX to factor IXa. Finally, as more thrombin is created, it activates factor XI to factor XIa, thereby enhancing the ability to ultimately make more thrombin.

Activation of factors V and VIII by thrombin results in a further burst of coagulation activity through increased activity of the tenase and prothrombinase complexes.

Fibrinogen is the ultimate substrate protein of the coagulation cascade and forms the principal structural protein of the fibrin clot. Fibrinogen, produced in the liver, is a dimer composed of three pairs of protein chains, Aα, Bβ, and γ, that are disulfide-linked at their N-terminal ends. Fibrinogen, as viewed by molecular imaging techniques, is composed of three globular domains, a central E domain flanked by two identical D domains (Figure 1-6). Thrombin cleaves small peptides, termed fibrinopeptides A and B, from the Aα and Bβ chains, respectively, to form a fibrin monomer. These monomers assemble into protofibrils in a half-staggered, side-to-side fashion that is stabilized by noncovalent interactions between fibrin molecules. The protofibrils laterally associate into thicker fibrin fibers and form the fibrin clot. This clot, however, is not stable and ultimately will come apart if not covalently cross-linked. Thrombin activates factor XIII to the transglutaminase enzyme factor XIIIa. Factor XIIIa, acting upon the glutamic acid and lysine side chains

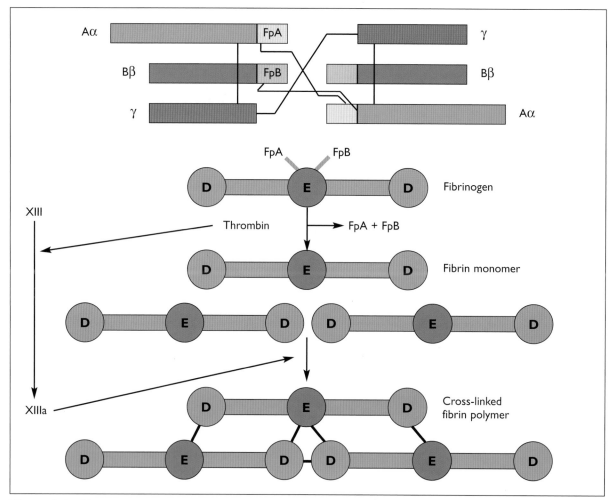

Figure 1-6. Fibrinogen is an abundant plasma protein that is a dimer of the Aα, Bβ, and γ chains connected by disulfide bonds. The fibrinogen dimer is composed of two flanking D globular domains with a central E domain. Fibrinogen forms the main structure of the fibrin clot, after cleavage of fibrinopeptides A (FpA) and B (FpB) by thrombin. The fibrin monomer assembles in a half-staggered overlap with adjoining fibrin monomers and is then covalently cross-linked into a fibrin polymer by the transamidase factor XIIIa.

in the fibrin amino acid sequence, creates covalent bonds between fibrin monomer γ chains, creating a stable clot. In addition, factor XIIIa can covalently cross-link a variety of other materials into the forming fibrin clot, including plasminogen and antiplasmin. This property of factor XIIIa is important for the penultimate purpose of the clot: wound healing and tissue repair.

Finally, Figure 1-5 alludes to the fact that there are many properties of thrombin other than the formation of the fibrin clot. Thrombin has direct effects on the other constituents of the coagulation triad: platelets and endothelial cells. Additionally, thrombin participates in its own downregulation. In the next section, some of these coagulation regulatory processes will be mentioned.

Regulatory Mechanisms

When a clot is formed, there has to be some mechanism to limit the clot to the site of an injury and ultimately to remove the clot when that injury has healed. The clot removal system, or fibrinolysis pathway, consists of the zymogen plasminogen, a variety of activators, and several inhibitors. Primary among these activators is tissue plasminogen activator (tPA), a product of endothelial cells. The fibrinolytic pathway is discussed in detail in chapter 3.

The most important coagulation proteins that are involved in regulation of thrombin generation are summarized in Table 1-2. Tissue factor pathway inhibitor has been mentioned in both the "Endothelium" and "Newer Coagulation Model"

Table 1-2. Coagulation Factor Inhibitors

Name	Description	Function
Tissue factor pathway inhibitor (TFPI)	Molecular Weight (MW) = 33,000 daltons (Da)	Inhibits the tissue factor/factor VIIa complex
Protein C	MW = 62,000 Da; vitamin K-dependent serine protease	Activated form cleaves coagulation cofactors Va and VIIIa
Protein S	MW = 75,000 Da; vitamin K-dependent protein	Cofactor for protein C
Antithrombin	MW = 58,000 Da	Serpin that directly inhibits several of the serine proteases; requires heparin as a cofactor

sections. Antithrombin also has been discussed in the "Endothelium" section of this chapter.

Once thrombin is formed, it can, as Figure 1-5 shows, be directed along an assortment of paths, which can result in a variety of important coagulation functions. If thrombin binds to the endothelial cell receptor thrombomodulin, its ability becomes limited to two paths. One of those paths is the protein C pathway. The thrombin/thrombomodulin receptor can activate protein C to its active form. Once activated, protein C and its cofactor, protein S, along with a phospholipid surface and calcium ions, can cleave factor Va and factor VIIIa to inactive forms. The presence of the cofactors factor Va and factor VIIIa is necessary for coagulation to function. Limiting the amounts of factor Va or factor VIIIa at the site of a growing clot essentially shuts down the ability to activate more thrombin and increase the amount of blood clot.

Operation of the Hemostasis and Thrombosis Pathway

The coagulation pathway is a complex interaction of many elements—the endothelium, coagulation factors, and platelets—with the ultimate goals of stemming the loss of blood at the site of an injury and laying the groundwork for injury repair and healing. Figure 1-7 illustrates a flow-chart diagram of how all the elements of the hemostatic pathways function together.

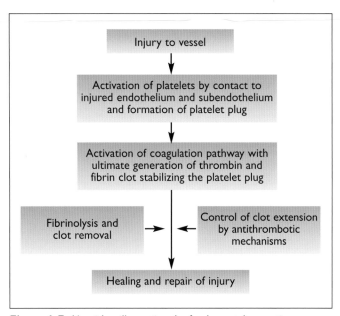

Figure 1-7. Algorithm illustrating the fundamental operation of the coagulation pathway.

Suggested Reading

General References

Colman RW, Clowes AW, George JN, Goldhaber SZ, Marder VJ, eds. Overview of hemostasis. In: *Hemostasis and Thrombosis: Basic Principles and Clinical Practice.* 5th ed. Philadelphia, Pa: JB Lippincott Co; 2006: 3-16.

Goodnight SH Jr, Hathaway WE, eds. Mechanisms of hemostasis and thrombosis. In: *Disorders of Hemostasis and Thrombosis.* 2nd ed. New York, NY: McGraw-Hill; 2001: 3-19.

Endothelium

Aird WC. Spatial and temporal dynamics of the endothelium. *J Thromb Haemost.* 2005:1392-1406. Epub 2005 May 9.

Carrell RW, Perry DJ. The unhinged antithrombins. *Brit J Haematol.* 1996;93:253-257.

Eilertsen KE, Osterud B. Tissue factor: (patho)physiology and cellular biology. *Blood Coagul Fibrinolysis.* 2004;15:521-538.

Jin L, Abrahams JP, Skinner R, et al. The anticoagulant activation of antithrombin by heparin. *Proc Natl Acad Sci USA.* 1997;94;14683-14688.

Michiels C. Endothelial cell functions. *J Cell Physiol.* 2003;196(3):430-443.

Price GC, Thompson SA, Kam PC. Tissue factor and tissue factor pathway inhibitor. *Anaesthesia.* 2004; 59(5):483-492.

Steffel J, Luscher TF, Tanner FC. Tissue factor in cardiovascular diseases: molecular mechanisms and clinical implications. *Circulation.* 2006;113(5):722-731.

Van de Wouwer M, Conway EM. Novel functions of thrombomodulin in inflammation. *Crit Care Med.* 2004;32(suppl 5):S254-S261.

Verhamme P, Hoylaerts MF. The pivotal role of the endothelium in haemostasis and thrombosis. *Acta Clin Belg.* 2006;61(5):213-219.

Coagulation Proteins

Butenas S, Orfeo T, Brummel-Ziedins KE, Mann KG. Tissue factor in thrombosis and hemorrhage. *Surgery.* 2007;142(suppl 4):S2-14.

Davie, EW, Kulman, JD. An overview of the structure and function of thrombin. *Semin Thromb Haemost.* 2006;32(suppl 1):3-15.

Duga S, Asselta R, Tenchini ML. Coagulation factor V. *Int J Biochem Cell Biol.* 2004;36(8):1393-1399.

Laurens N, Koolwijk P, de Maat MP. Fibrin structure and wound healing. *J Thromb Haemost.* 2006;4(5):932-939.

Lorand L. Factor XIII and the clotting of fibrinogen: from basic research to medicine. *J Thromb Haemost.* 2005;3(7):1337-1348.

Mosesson MW. Fibrinogen and fibrin structure and functions. *J Thromb Haemost.* 2005;3(8):1894-1904.

Stafford DW. The vitamin K cycle. *J Thromb Haemost.* 2005;3:1873-1878.

Coagulation History (The Extrinsic Pathway and the PT)

Langdell RD, Wagner RH, Brinkhous KM. Effect of antihemophilic factor on one-stage clotting tests: a presumptive test for hemophilia and a simple one-stage antihemophilic factor assay procedure. *J Lab Clin Med.* 1953;41:637-647.

Morawitz P. Die chemie der blutgerinnung. *Ergebn Physiol.* 1905;4:307-422.

Owen CA Jr. *A History of Blood Coagulation.* Nichols WL, Bowie EJW, eds. Rochester, Minn: Mayo Foundation for Medical Education and Research; 2001.

Quick AJ, Stanley-Brown M, Bancroft FW. A study of the coagulation defect in hemophilia and in jaundice. *Am J Med Sci.* 1935;190:501.

Wintrobe MM, ed. *Blood, Pure and Eloquent: A Story of Discovery, of People, and of Ideas.* New York, NY: McGraw-Hill; 1980: 601-657.

Newer Coagulation Model

Hoffman M. Remodeling the blood coagulation cascade. *J Thromb Thrombolysis.* 2003;16(1-2):17-20.

Hoffman MM, Monroe DM. Rethinking the coagulation cascade. *Curr Hematol Rep.* 2005;4(5):391-396.

Mann KG. Thrombin formation. *Chest.* 2003;124(suppl 3):4S-10S.

Roberts HR, Hoffman M, Monroe DM. A cell-based model of thrombin generation. *Semin Thromb Hemost.* 2006;32(suppl 1):32-38.

Stassen JM, Arnout J, Deckmyn H. The hemostatic system. *Curr Med Chem.* 2004;11(17):2245-2260.

Regulatory Mechanisms

Dahlback B. Blood coagulation and its regulation by anticoagulant pathways: genetic pathogenesis of bleeding and thrombotic diseases. *J Intern Med.* 2005; 257(3):209-223.

Dahlback B, Villoutreix BO. The anticoagulant protein C pathway. *FEBS Lett.* 2005;579(15):3310-6. Epub 2005 Mar 13.

Lwaleed BA, Bass PS. Tissue factor pathway inhibitor: structure, biology and involvement in disease. *J Pathol.* 2006;208(3):327-339.

Rigby AC, Grant MA. Protein S: a conduit between anticoagulation and inflammation. *Crit Care Med.* 2004; 32(suppl 5):S336-S341.

Platelet Structure and Function

Kandice Kottke-Marchant, MD, PhD

Platelet Production

Megakaryocyte Development and Maturation

Platelets are small (2-μm diameter), non-nucleated blood cells that play a vital role in hemostasis and are produced in the bone marrow from megakaryocytes (Figure 2-1). Megakaryocytes are polyploid bone marrow cells that develop by repeated rounds of endomitosis, increasing cellular DNA content without proliferation. Mature megakaryocytes are the largest cell in the bone marrow, with abundant granular cytoplasm and large multilobated nuclei (Figure 2-2). The polyploid DNA content may reach 32 to 64 times a normal diploid cell.

Megakaryocytes are descended from pluripotent hematopoietic progenitors through a bipotential erythroid/megakaryocytic cell. Megakaryoblasts are small cells, present in low number in the hematopoietic bone marrow, that have a high nuclear-to-cytoplasmic ratio, with dark blue, agranular cytoplasm, often with cytoplasmic blebs. They express CD34, CD41, the thrombopoi-

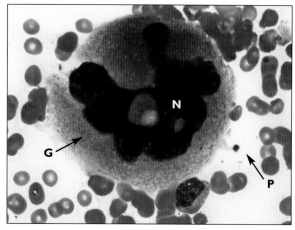

Figure 2-2. Megakaryocytes in the bone marrow aspirate are extremely large cells that can be up to 100 to 200 μm in diameter. The nucleus (N) is typically large and lobulated, reflecting the polyploidy nature of the cells. The cytoplasm of mature megakaryocytes is filled with abundant granules (G). Platelets are produced by budding of granule-filled sacs from cytoplasmic protrusions called protoplatelets (P). (Bone marrow aspirate, original magnification X100, Wright stain.)

etin receptor (c-Mpl), and the alpha chemokine receptor CXCR4. After this stage, the megakaryoblasts undergo committed maturation to megakaryocytes, with expression of the β3 integrin CD61 and increased levels of $\alpha_{IIb}\beta_3$ (glycoprotein IIb/IIIa [GPIIb/IIIa]). This phase is stimulated by thrombopoietin and other cytokines and chemokines, such as interleukin (IL)-3, IL-6, IL-11, and c-Kit ligand. Early maturation is dependent upon transcription factors GATA-1 and GATA-2 together with the cofactor FOG-1. This maturation is accompanied by endomitosis, a shortened mitosis caused by a block in anaphase without cytoplasmic division, resulting in an increase in DNA content. When the megakaryocytic precursors are diploid or tetraploid, they begin to express the von Willebrand factor receptor, glycoprotein Ib/IX/V (GPIb/IX/V). Megakaryocytes expressing GPIb/IX/V become less responsive to thrombopoietin and undergo fewer endomitotic cycles

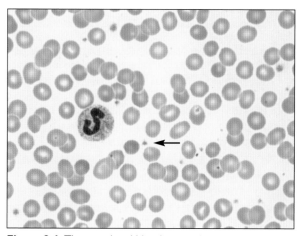

Figure 2-1. This peripheral blood smear demonstrates normal platelet number and morphology. Platelets (arrow) are small non-nucleate cells that have a pale cytoplasm with purple granules by Wright stain. (Original magnification X100, Wright stain.)

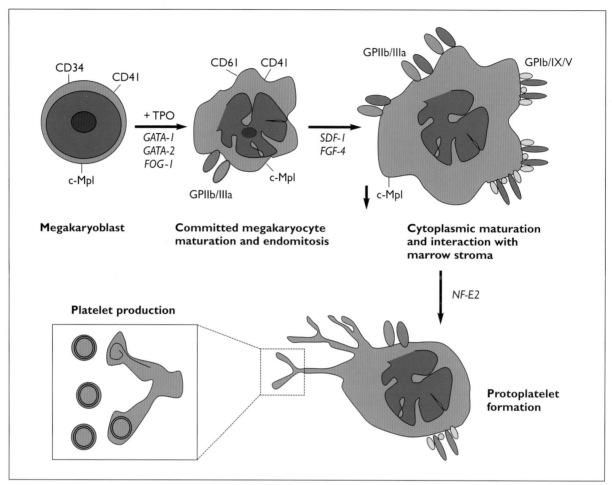

Figure 2-3. Platelet maturation process. Megakaryoblasts express CD34, CD41, and the thrombopoietin (TPO) receptor (c-Mpl). After this stage, the megakaryoblasts undergo committed maturation to megakaryocytes, with expression of the β_3 integrin CD61 and increased levels of $\alpha_{IIb}\beta_3$ (GPIIb/IIIa), dependent upon transcription factors GATA-1 and GATA-2 together with the cofactor FOG-1. This maturation is accompanied by endomitosis, resulting in an increase in DNA content accompanied by expression of the von Willebrand factor receptor, GPIb/IX/V. The cytoplasm enlarges (up to 100 to 150 μm) and becomes filled with platelet-specific organelles, proteins, and membrane systems. During a terminal phase of megakaryocyte differentiation, platelets are produced from cytoplasmic projections known as protoplatelets that form along one pole of the megakaryocyte cytoplasm under the guidance of transcription factor NF-E2. Boluses of platelet granules and membrane systems travel from the megakaryocyte cytoplasm down the tubules, yielding bulbous projections at tubule ends and along their length, giving a beaded appearance to the developing tubules. Longitudinal microtubule bundles composed of tubulin $\alpha\beta$-dimers form along the tubules and can move down the tubule. The microtubule bundles reorganize in the protoplatelet tip to form a microtubule coil that envelopes and defines a newly matured protoplatelet.

than more immature megakaryocytes that are GPIb/IX/V negative (Figure 2-3).

During and after DNA synthesis, cytoplasmic maturation of the megakaryocytes ensues. The cytoplasm enlarges (up to 100 to 150 μm) and becomes filled with platelet-specific organelles, proteins, and membrane systems. Two platelet-specific granules, the alpha and dense granules, form at this time. The alpha granules store many protein constituents involved in platelet adhesion and coagulation function. The dense granules store adenine nucleotides, calcium, and serotonin, which function to recruit additional platelets to a developing thrombus. During maturation, much of the cytoplasm becomes filled with the demarcation membrane system (DMS), a characteristic feature of megakaryocytes, consisting of a network of membrane channels composed of flattened cisternae and tubules. It was previously believed that fragmentation of the DMS was responsible for platelet formation, but it is now known that the DMS functions as a membrane repository for protoplatelet function. In order for one megakaryocyte to transform into thousands of tiny membrane-bound platelets, a large store of excess membranes is necessary to support the large increase in membrane surface area.

Continued megakaryocyte differentiation requires interaction with bone marrow stromal cells. Megakaryocytes migrate near to stromal endothelial cells, where platelet release into the circulation is thought to occur. This migration is likely guided by stromal-derived factor (SDF)-1 chemokine and fibroblast growth factor (FGF)-4. Megakaryocytes also affect other bone marrow cell lineages, with stimulation of bone formation by osteoprotegerin and the receptor-activated NF-κB (RANK) ligand. Megakaryocytes can also stimulate stromal fibrosis, with expression of transforming growth factor, beta 1 (TGFB1), but the mechanism that links abnormal proliferation of megakaryocytes with clinical development of myelofibrosis is not known.

Platelet Production from Megakaryocytes

During a terminal phase of megakaryocyte differentiation, platelets are produced from cytoplasmic projections known as protoplatelets, which form along one pole of the megakaryocyte cytoplasm under the guidance of transcription factor NF-E2. Wide pseudopods branch and elongate to yield thin tubules that further branch repeatedly to give the megakaryocyte an appearance resembling an octopus. The branching is regulated by actin filaments and myosin II polymerization, which may explain the macrothrombocytopenia observed in genetic disorders of the myosin heavy chain gene.

Boluses of platelet granules and membrane systems travel from the megakaryocyte cytoplasm down the tubules, yielding bulbous projections at tubule ends and along their length, giving a beaded appearance to the developing tubules. Longitudinal microtubule bundles composed of tubulin αβ-dimers form along the tubules and can move down the tubule. The β1-tubulin gene shows megakaryocyte-specific expression and is under the regulation of NF-E2. The microtubule bundles reorganize in the protoplatelet tip to form a microtubule coil that envelopes and defines a newly matured protoplatelet. Indeed, knockout mice deficient in β1-tubulin fail to form protoplatelets in vitro, and the platelets formed are spherical, not discoid. At this time, the protoplatelet bleb fills with platelet granules and organelles traveling down the tubules from the megakaryocyte cytoplasm (Figure 2-3).

The precise mechanism of release of the newly formed platelets into the circulation is still being elucidated. While circulating platelets appear discoid, it is not clear whether they assume this shape immediately upon release from the protoplatelet tubule or whether they are released with a long cytoplasmic tail or even have a dumbbell shape.

Platelet Structure

Surface Glycoproteins and Plasma Membrane

Upon release from the bone marrow, platelets circulate in a quiescent state but are activated rapidly upon blood vessel injury and play a crucial role in the primary hemostatic response. In their unactivated state, platelets are roughly discoid in shape, with numerous intrinsic glycoproteins attached to the outer surface of the plasma membrane, which serve as receptors for adhesive ligands ranging from fibrinogen (GPIIb/IIIa), collagen (GPIa/IIa and GPVI), and thrombospondin (GPIV) to von Willebrand factor (GPIb/IX/V), fibronectin (GPIc/IIa), and vitronectin ($\alpha_v\beta_3$).

Two alternate nomenclatures are used to describe receptors of the integrin family. The older nomenclature gave each protein on the platelet surface, including integrins, a glycoprotein or "GP" identifier, while the integrin nomenclature gives each integrin, including platelet integrins, an integrin identifier. For example, the fibrinogen receptor is alternatively identified as GPIIb/IIIa and $\alpha_{IIb}\beta_3$. For simplicity, we will refer to all the platelet receptors by the GP nomenclature, with reference to the integrin nomenclature when they are described in detail. Platelets also have numerous membrane receptors involved in signal transduction and biochemical activation, such as adenosine diphosphate (ADP) receptors ($P2Y_1$, $P2Y_{12}$, and $P2X_1$), thrombin receptors (PAR-1 and PAR-4), thromboxane receptors, and adrenergic receptors.

Von Willebrand Receptor (GPIb/IX/V)

Von Willebrand factor (vWF) is secreted by endothelial cells into the extracellular matrix, where it becomes a ligand that facilitates platelet adhesion upon vascular injury. The receptor on the platelet surface responsible for the initial adhesion of platelets to the extracellular matrix through vWF is GPIb/IX/V, which is a nonintegrin receptor that is a member of the leucine-rich repeat

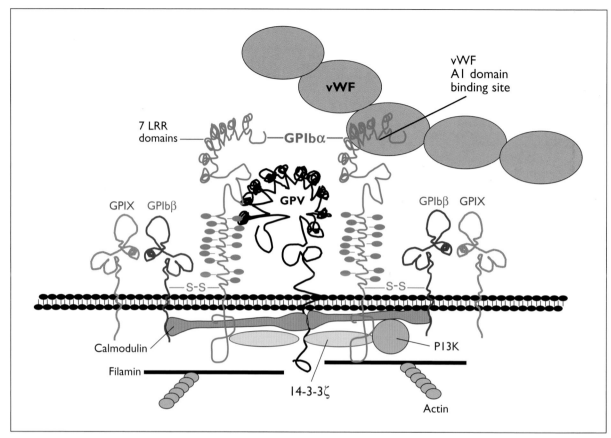

Figure 2-4. The receptor on the platelet surface responsible for the initial adhesion of platelets to the extracellular matrix through von Willebrand factor (vWF) is glycoprotein (GP) Ib/IX/V, which is a nonintegrin receptor that is a member of the leucine-rich repeat (LRR) protein family. The GPIb/IX/V receptor is a transmembrane glycoprotein composed of four subunits, a central GPV, flanked on either side by one subunit of GPIbα, GPIbβ, and GPIX.

The GPIbα N-terminal globular region, containing seven LRR domains, serves as the binding site for the A1-domain of vWF. The cytoplasmic tail of GPIbα has binding sites for two cytoskeletal proteins, filamin and 14-3-3ζ, as well as the signaling protein phosphoinositide 3-kinase (PI3K). In resting platelets, GPIbβ and GPV bind calmodulin through their cytoplasmic tails, but the calmodulin dissociates from the complex upon activation.

(LRR) protein family. The GPIb/IX/V receptor is a transmembrane glycoprotein composed of four subunits, a central GPV, flanked on either side by one subunit of GPIbα, GPIbβ, and GPIX (Figure 2-4). The GPIbα N-terminal globular region, containing seven LRR domains, serves as the binding site for the A1-domain of vWF. The cytoplasmic tail of GPIbα has binding sites for two cytoskeletal proteins, filamin and 14-3-3ζ as well as the signaling protein phosphoinositide 3-kinase (PI3K). A proteolytic fragment of GPIbα that circulates in plasma is known as glycocalicin and is observed in some thrombocytopenic disorders. In addition to a role as the vWF receptor, GPIb is thought to function as one of the thrombin receptors on the platelet surface, and thrombin is also known to cleave GPV. In resting platelets, GPIbβ and GPV bind calmodulin through their cytoplasmic tails,

but the calmodulin dissociates from the complex upon activation.

Binding of vWF to GPIb/IX/V is thought to be initiated by a conformational change in the von Willebrand molecule that can be initiated by fluid shear stress or vWF adhesion to extracellular matrix proteins, or stimulated by ristocetin and botrocetin. The vWF conformational change is thought to expose a GPIb binding epitope in the vWF A1 domain that is responsible for initiating early platelet adhesion to subendothelium.

Collagen Receptors (GPVI and GPIa/IIa)

Platelets interact with collagen during platelet adhesion to the vascular wall, but collagen also serves as a strong platelet agonist. It is now known that there are two platelet receptors for collagen, GPVI and the integrin GPIa/IIa ($\alpha_2\beta_1$). The GPIa/IIa integrin serves as a primary receptor for

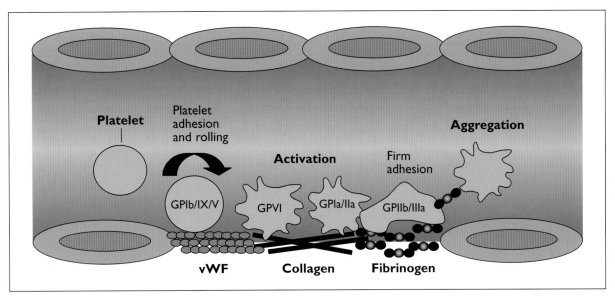

Figure 2-5. In the process of platelet activation, unactivated circulating platelets initially encounter an area of vascular injury and adhere rapidly to exposed von Willebrand factor (vWF) through the glycoprotein (GP) Ib/IX/V membrane receptor. Collagen in the extracellular matrix is also engaged through receptors GPVI and GPIa/IIa, leading to platelet shape change and activation. Cell signaling results in conformational change in the fibrinogen receptor GPIIb/IIIa, with binding to fibrinogen and fibrin leading to platelet aggregation and thrombus formation.

platelet adhesion to collagen in the extracellular matrix, while GPVI, together with the Fc receptor γ (FcRγ), is thought to function as the primary agonist receptor by which collagen initiates platelet aggregation and granule release. The FcRγ is noncovalently associated with GPVI, forming a signaling complex. The FcRγ contains an immunoreceptor type activation motif (ITAM), which is phosphorylated by a Src family kinase following platelet adhesion to collagen. Recent studies have suggested that the initial step in platelet interaction with collagen is due to binding to GPVI/FcRγ leading to platelet activation, followed by activation of the GPIa/IIa receptor that leads to stable platelet adhesion (Figure 2-5). There are several known polymorphisms of the GPIa/IIa receptor that are associated with altered integrin density on the platelet surface that correspond to changes in collagen-induced platelet aggregation.

Adenosine Diphosphate Receptors

Platelet aggregation by adenosine diphosphate (ADP) involves two receptors, the $P2Y_1$, coupled to phospholipase Cβ via G_q, and the $P2Y_{12}$, a purinoreceptor coupled to G_{i2} and associated with adenylate cyclase inhibition. The $P2Y_1$ receptor is a classic G-coupled protein receptor with seven transmembrane regions. It is the high-affinity receptor responsible for the initial platelet shape change, and it coupling to G_q leads to phospholipase Cβ activation, mobilization of cytoplasmic

Ca^{++} and mediating shape change and an initial wave of aggregation, but it is not able to support a full platelet aggregation response. The $P2Y_1$ receptor is present in only 150 copies per platelet and is widely distributed in many tissues, including heart, blood vessels, and smooth muscle cells. The $P2Y_{12}$ receptor is a low-affinity receptor that results in activation of the PI3K pathway and is involved in completion of platelet aggregation by ADP, potentiation of platelet secretion, and thrombus stabilization. It also leads to an inhibition of adenylate cyclase, with decreased cytoplasmic cyclic adenosine monophosphate (cAMP) levels, which further accelerates platelet activation (Figure 2-6). The $P2Y_{12}$ receptor has a more narrow tissue distribution than $P2Y_1$, being found almost solely in platelets. The P2X1 receptor is an adenosine triphosphate (ATP)-gated cation channel that participates in raising cytoplasmic calcium levels during platelet activation and may also bind to collagen under shear.

Thrombin Receptors (PAR-1 and PAR-4)

Thrombin is a proteolytic enzyme with wide-ranging hemostatic function. It is responsible for converting fibrinogen to fibrin, activating factor XIII, and activating protein C. Thrombin is also known to activate platelets, but specific receptor-mediated activation was under question until a family of protease-activated receptors (PARs) was described. There are four members of the PAR

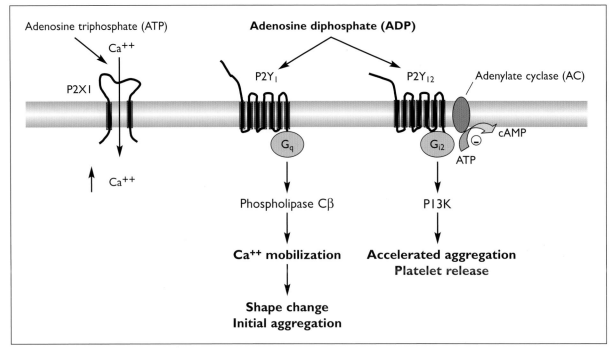

Figure 2-6. Adenosine diphosphate (ADP)-induced platelet aggregation. P2Y₁ is involved in Ca⁺⁺ mobilization and shape change through the G protein G_q and stimulation of phospholipase Cβ. The P2Y₁₂, coupled to adenylate cyclase (AC) and G_i2, enhances P2Y₁. P2Y₁₂ leads to activation of phospho-inositide 3-kinase (PI3K) with an acceleration of platelet activation and granule release. Engagement of ADP by P2Y₁₂ results in downregulation of AC, with decreased levels of cyclic adenosine monophosphate (cAMP), further stimulating platelet activation. P2X1 is associated with Ca⁺⁺ influx.

family, but PAR-1 and PAR-4 are found on platelets. The thrombin receptors are unique among platelet receptors in that thrombin activates the receptor by proteolytic cleavage of a peptide fragment from the exodomain, with the new N-terminal end of the tethered fragment (peptide SFLLRN) binding to the receptor and triggering intracellular signaling (Figure 2-7). This small SFLLRN peptide, called thrombin receptor-activating peptide (TRAP), can also initiate platelet activation through binding to the PARs because it mimics the tethered ligand sequence. Recent work has suggested that thrombin binding to the GPIbα subunit of the vWF receptor promotes thrombin-dependent activation of PAR-1. The PARs are coupled to G proteins of the $G_{12/13}$, G_q, and $G_{i/z}$ families and can initiate a cascade of signaling responses. Because PARs are activated by proteolytic cleavage, their activity cannot easily be shut down. The PAR activation response is thought to be downregulated by internalization of the receptors with degradation in the lysosomes.

Thromboxane Receptor and Prostaglandin Receptors

Platelets have receptors for various prostanoids that regulate platelet function. Thromboxane A₂ (TxA₂), the metabolic product of the platelet cyclo-oxygenase pathway, is released from platelets and interacts with TxA₂ receptors (TP) on adjacent platelets to stimulate platelet activation through G_q and phospholipase C with subsequent calcium mobilization, but also through Rho activation. In contrast, platelets have receptors for prostaglandin I2 (IP receptor), prostaglandin E2 (EP3 receptor), and prostaglandin D2 (DP receptor) that lead to decreased platelet activation through stimulation of adenylate cyclase, resulting in elevation of cAMP levels in the platelet cytosol.

Fibrinogen Receptor (GPIIb/IIIa)

The platelet receptor for fibrinogen, GPIIb/IIIa, is a member of the integrin family ($\alpha_{IIb}\beta_3$) that is the most abundant platelet surface protein, being present in over 80,000 copies per platelet. The receptor is a heterodimer of integrin α_{IIb} and β_3 chains. The α_{IIb} subunit participates solely in the fibrinogen receptor, while the β_3 subunit can combine with different alpha chains and form other receptors, such as the vitronectin receptor ($\alpha_v\beta_3$). Expression of both α_{IIb} and β_3 chains is necessary to form the complete fibrinogen receptor, whose expression is restricted to megakaryocytes and

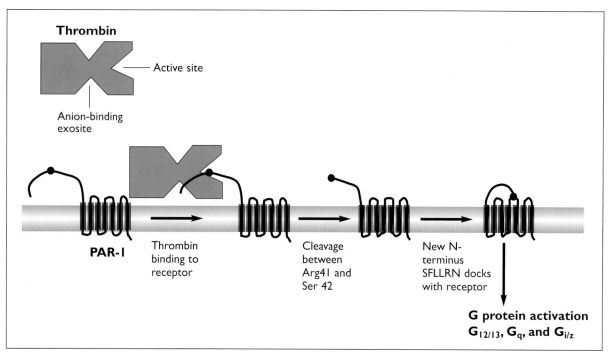

Figure 2-7. Thrombin activates platelets through a family of protease-activated receptors (PARs). There are four members of the PAR family, but PAR-1 and PAR-4 are found on platelets. The thrombin receptors are unique among platelet receptors in that thrombin activates the receptor by proteolytic cleav-

age of a peptide fragment from the exodomain, with the new N-terminal end of the tethered fragment (peptide SFLLRN) binding to the receptor and triggering intracellular signaling. The PARs are coupled to G proteins of the $G_{12/13}$, G_q, and $G_{i/z}$ families and can initiate a cascade of signaling responses.

platelets. The extracellular domain of the α_{IIb} chain has a beta propeller domain, with seven molecular "propeller blades" that coordinate binding to the ligand. The I-domain of the β_3 chain contains a cation-binding site; this explains the Ca^{++} and Mg^{++} binding requirement of the integrin. Both α_{IIb} and β_3 chains transverse the platelet plasma membrane, with the cytoplasmic tails interacting both with the platelet cytoskeleton and with each other.

In resting platelets, the GPIIb/IIIa receptor maintains an unactivated conformation that can not bind to its ligand, fibrinogen. The favored structural basis is that the ligand-binding head group is bent down facing the plasma membrane. Upon platelet activation by an agonist or binding of von Willebrand factor to GPIb/IX/V, the receptor undergoes a conformational change through molecular signaling mechanisms that are still being elucidated. A molecular movement of the cytoplasmic tails, perhaps initiated by talin binding to the β_3 chain, is thought to transmit to a straightening of the globular head group and opening of the ligand-binding pocket. Alternatively, the conformational change may be regulated by receptor clustering and interaction of transmembrane domains (Figure 2-8). The confor-

mational change in the ligand binding pocket enables binding to fibrinogen through an arginine-glycine-aspartic acid (RGD) tripeptide from the fibrinogen alpha chain and a terminal dodecapeptide from the fibrinogen gamma chain. Since fibrinogen is a dimeric molecule that contains two alpha chains, each of which has two RGD sequences, two beta chains, and two gamma chains, the bivalent or multivalent binding of one fibrinogen molecule to two or more platelets is thought to facilitate platelet-platelet attachment, or platelet aggregation.

Platelet Cytoskeleton

Beneath the plasma membrane, platelets have an important cyotskeletal system and a complex system of membranes and granules (Figure 2-9). Interactions between the surface glycoproteins and the cytoskeletal system play a role in platelet activation, signal transduction, and platelet aggregation.

The cytoskeleton is composed of a circumferential band of microtubles and microfilaments that defines the platelet's discoid shape, along with other proteins, such as actin, spectrin, adducing, and filamin A. The cytoskeleton has two major forms: a membrane-associated skeleton and a

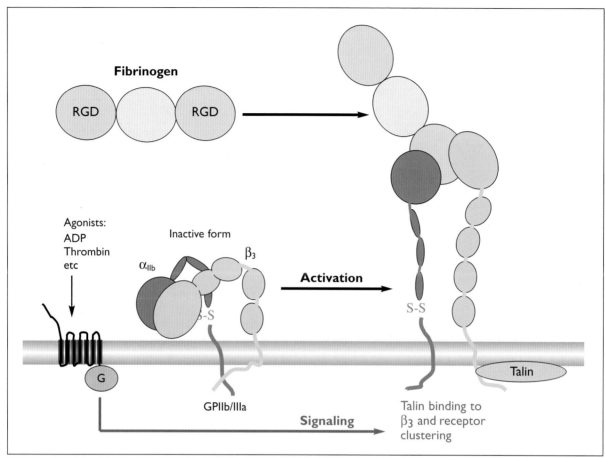

Figure 2-8. The platelet receptor for fibrinogen GPIIb/IIIa is a member of the integrin family ($\alpha_{IIb}\beta_3$). The receptor is a heterodimer of integrin α_{IIb} and β_3 chains. In resting platelets, the receptor maintains an unactivated conformation that can not bind to its ligand, fibrinogen. The favored structural basis is that the ligand-binding head group is bent down facing the plasma membrane. Upon platelet activation by an agonist or binding of von Willebrand factor to GPIb/IX/V, the receptor undergoes a conformational change. A molecular movement of the cytoplasmic tails, perhaps initiated by talin binding to the β_3 chain, is thought to transmit to a straightening of the globular head group and opening of the ligand-binding pocket. The conformational change in the ligand-binding pocket enables binding to fibrinogen through an arginine-glycine-aspartic acid (RGD) tripeptide from the fibrinogen alpha chain and a terminal dodecapeptide from the fibrinogen gamma chain.

structural cytoplasmic scaffold. The predominant component of the cytoskeleton is actin, which exists as a linear polymer as well as a storage form that can be polymerized upon platelet activation. The cytoskeleton also consists of a meshwork of spectrin polymers, with adducin creating an intermolecular link between the spectrin meshwork and actin filaments. The membrane-associated cytoskeleton is located immediately beneath the plasma membrane, and the cytoplasmic tails of many surface glycoproteins interact with one or more cytoskeletal elements. In the structural cytoplasmic scaffold, the linear actin polymers are thought to be cross-linked by proteins such as spectrin, filamin, and α-actinin. There is an interaction between the cytoskeletal protein filamin and the cytoplasmic tail of the GPIbα subunit,

which anchors the vWF receptor to the cytoskeleton. The cytoskeleton is responsible for maintaining the discoid shape of resting platelets and is involved in development of filipods and lamellae during platelet adhesion and activation.

A microtubule band is present just under the cytoskeleton. The microtubule is a relatively rigid hollow rod that is approximately 100 μm in length and is wound circumferentially around the platelet many times to form a microtubule loop along the edge of the platelet disc. The microtubule is composed of polymerized β-tubulin proteins that are present in a linear aggregate of 13 stacks of subunits, each arranged in protofilaments. During platelet activation, there is reorganization of the microtubules, with constriction of the microtubule band into the center of the cell and formation of some smaller microtubule fragments.

Platelet Membrane Systems

Unlike most other cells, platelets possess an invaginating open-canalicular membrane system (OCS) that reaches from the outer plasma membrane to the depths of the platelet cytosol and gives the platelet a sponge-like structure (Figure 2-9). The open canalicular system forms pit-like openings in the otherwise flat platelet membrane, when observed by scanning electron microscopy. The OCS has three important functions: it provides a passageway to facilitate the rapid release of platelet granule contents; it is thought to be a storage repository for plasma membrane necessary for platelet filipod formation and cell spreading; and it serves as a storehouse for excess copies of some surface receptors.

There is also a dense tubular system that stores metabolic enzymes and calcium. It is the remnant of the endoplasmic reticulum of megakaryocytes and has functions analogous to the sarcoplasmic reticulum of muscle cells. The calcium stored by the dense tubular system is released into the cytosol during platelet activation and is an important mediator of platelet function. In addition to calcium, the dense tubular system stores ATPases and is the major site for prostanoid synthesis and adenylate cyclase activity.

Platelet Granules

The platelet cytosol contains typical cellular organelles, such as mitochondria, lysosomes, and glycogen, but also contains platelet-specific granules, called alpha and dense granules, which play a major role in amplification of platelet activation and interaction with coagulation (Figure 2-9).

Alpha Granules

Alpha granules are abundant, protein-containing granules that are specific to platelets and are approximately 200 to 500 µm in diameter. They are delimited by a single membrane and contain two morphologically distinct compartments: a dark nucleoid and an electron-lucent matrix. The dark nucleoid region contains the platelet-specific proteins platelet factor 4 and β-thromboglobulin. The electron-lucent area is thought to contain high-molecular-weight von Willebrand factor, multimerin, and factor V. The zone between these two regions contains other proteins, such as fibrinogen, coagulation factors, adhesive glycoproteins, protease inhibitors, and growth factors (Table 2-1).

Table 2-1. Content of Platelet Granules

Dense Granules	Alpha Granules
Adenine nucleotides Adenosine diphosphate (ADP) Adenosine triphosphate (ATP)	**Proteoglycans** Platelet factor 4, β-thromboglobulin, serglycin, histidine-rich glycoprotein
	Adhesive glycoproteins Fibronectin, thrombospondin, vitronectin, von Willebrand factor
Amines Serotonin Histamine	**Coagulation factors** Fibrinogen, factors V, VII, XI, XIII, kininogens, protein S, plasminogen
	Cellular mitogens Platelet derived growth factor (PDGF), transforming growth factor β (TGFβ), endothelial cell growth factor (ECGF), epidermal growth factor (EGF), vascular endothelial cell growth factor (VEGF), vascular permeability factor (VPF), insulin-like growth factor (IGF), interleukin-b (IL-b), growth arrest specific 6 (Gas-6)
Cations Calcium Magnesium	**Protease inhibitors** α_2-macroglobulin, α_2-antitrypsin, α_2-antiplasmin, plasminogen activator inhibitor-1 (PAI-1), tissue factor pathway inhibitor (TFPI), C1 inhibitor
	Miscellaneous Immunoglobulins, albumin, GPIa/multimerin, amyloid precursor protein (APP)

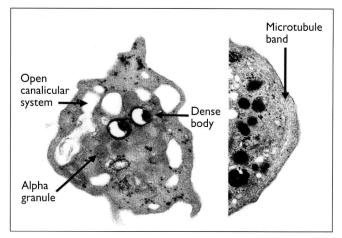

Figure 2-9. By electron microscopy, platelets are non-nucleated cells with a circumferential band of microtubules beneath the plasma membrane. The platelet cytoplasm is filled with alpha and dense granules, together with glycogen, mitochondria, and lysosomes. There is also an invaginating system of membrane channels, the open canalicular system, which serves to facilitate the release of granule components. (Original magnification 40,500X.)

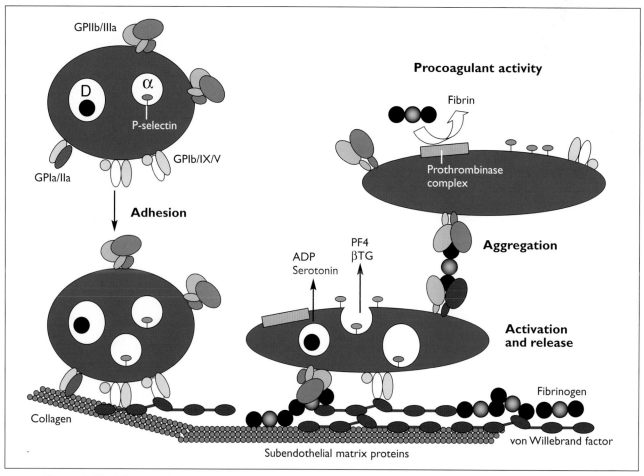

Figure 2-10. Schematic diagram of the role of platelets in hemostasis. Platelet activation is stimulated by vascular injury through exposure of platelets to extracellular matrix proteins and adsorbed plasma proteins, including von Willebrand factor and fibrinogen. Platelets adhere to von Willebrand factor through the surface glycoprotein (GP) Ib/IX/V and to fibrinogen through GPIIb/IIIa. Platelet adhesion stimulates intracellular signaling, leading to degranulation of alpha granules (platelet factor 4 [PF4] and β-thromboglobulin [βTG]), and phospholipid reorganization, with formation of coagulation complexes and fibrin formation. Adenosine diphosphate (ADP) release from dense granules and activation of the GPIIb/IIIa receptor also occur, leading to aggregation of platelets to the adherent layer.

Only some of the alpha granule proteins are synthesized by the megakaryocyte; others are internalized by endocytosis from the circulating plasma pool. The internal table of the alpha granule membrane is the site of some granule-specific intrinsic glycoproteins, such as P-selectin, osteonectin, and GMP33. Upon platelet activation, the contents of the alpha granules are released into the OCS for rapid extracellular release.

Dense Granules

Dense granules are small platelet granules (150 to 300 μm) that are electron dense and are present in low numbers, typically less than 10 per platelet. They are membrane-limited granules with submembrane clear space and a characteristic dense central core that can be observed by transmission electron microscopy (Figure 2-9). Due to the low number of these granules in each platelet, not every electron microscopic platelet cross-section will display a dense granule. Dense granules contain calcium, magnesium, serotonin, histamine, and a nonmetabolic pool of adenine nucleotides ADP and ATP. During platelet activation, the dense granule contents are released directly through fusion with the plasma membrane.

Platelet Activation

Platelets promote hemostasis by four interconnected mechanisms: (1) *adhering* to sites of vascular injury; (2) undergoing an *activation process*, leading to the release of compounds from their granules; (3) *aggregating* together to form a hemostatic platelet plug; and (4) providing a *procoagulant* surface for activated coagulation protein complexes on their phospholipid membranes (Figure 2-10). The following discussion gives an overview of these four processes.

Platelet Adhesion

Role of GPIb/IX/V and von Willebrand Factor

Platelets circulate in the vasculature as nonactivated discoid cells, largely due to production of platelet inhibiting factors, such as ecto-ADPase, nitrous oxide, and prostaglandin I_2, by the endothelial lining. Platelet function is also inhibited by high cytoplasmic levels of cyclic adenosine monophosphate (cAMP), which is produced by the enzyme cAMP phophodiesterase. Upon encountering a site of vascular injury or endothelial denudation, adhesion of the unactivated platelets to the vessel wall is the initial step in platelet activation. The subendothelium is exposed when the endothelial layer is disrupted; it is composed of extracellular matrix proteins, such as collagen, fibronectin, vitronectin, von Willebrand factor, thrombospondin, and laminin, many of which bind to receptors expressed on the platelet surface. Due to the large number of extracellular matrix proteins and a high density of platelet surface receptors, platelet adhesion to areas of vascular injury is extremely rapid.

For example, von Willebrand factor (vWF), a large multimeric protein secreted into the extracellular matrix from endothelial cells, facilitates platelet rolling and adhesion by binding to the platelet surface glycoprotein Ib/IX/V, especially at high shear rates (Figure 2-4). The binding is thought to be mediated between the A1 domain of vWF and the GPIbα subunit of the platelet receptor, and only occurs when the binding domain of the vWF A1 domain is exposed. This is essentially a "molecular switch" that is turned on when vWF is exposed to shear stress, interacts with extracellular matrix proteins such as collagen, or is induced by the pharmacologic effect of ristocetin or botrecetin.

The signaling induced by vWF binding to GPIb/IX/V is a subject of active study, but evidence suggests that signaling is mediated through Src family kinases that lead to phospholipase Cγ2 activation, production of 1,4,5-inositol 3 phosphate (1,4,5-IP3) and 1,2-diacylglycerol, with subsequent release of Ca^{++} from dense tubular stores (Figure 2-11). The increased levels of cytoplasmic calcium play a role in activation of the GPIIb/IIIa receptor, as outlined below (see "Platelet Aggregation"). Recent evidence suggests that gly-

cosphingolipid-enriched membrane microdomains, or "lipid rafts," with decreased membrane fluidity that limits membrane diffusion may play an important role in localizing these signaling complexes.

Role of Collagen Receptors

The collagen receptors detailed above, GPVI/FcRγ and GPIa/IIa, participate in the adhesion of platelets to the extracellular matrix. Under shear, the initial platelet adhesion to collagen is thought to be mediated through the GPIa/IIa receptor, but further platelet activation and collagen-induced aggregation requires GPVI/FcRγ. The binding of collagen to GPVI/FcRγ induces clustering of the receptors with phosphorylation of the γ-chain by the Src family tyrosine kinases and subsequent activation of phospholipase Cγ2, analogous to GPIb/IX/V signaling outlined previously (Figure 2-11). The platelet adhesion to collagen and vWF is closely related both spatially and functionally, because collagen binding by vWF in the extracellular matrix is thought to result in the conformational change in vWF that allows binding to the platelet membrane. The initial stages of platelet adhesion to the vessel wall entail relatively weak binding to vWF and stronger binding to collagen, resulting in platelet rolling and tethering. Firm adhesion of platelets is via vascular wall-associated RGD peptides supplied by vWF, fibrin, or fibrinogen through interaction with the activated platelet surface GPIIb/IIIa.

Cytoskeletal Changes

Upon activation, the increased cytoplasmic calcium leads to gelsolin activation, resulting in cytoskeletal actin remodeling, whereby the platelet cytoskeleton is disassembled and the membrane-associated actin network becomes disrupted. This is thought to be due, in part, to dissociation of adducin from the actin and spectrin filaments. Actin is repolymerized from the actin filaments that bind to the plasma membranes through the action of gelsolin and cofilin. During this stage, the platelets protrude filopodia, which are thin membrane-bound processes with an actin core. After a longer adhesion time, the platelets form large networks of actin filaments and protrude broad lamellae, and eventually the entire membrane spreads, with the adherent platelets assuming a flat, pancake-like conformation. During the latter stages of the platelet adhesion phase, the GPIb/IX/V is translocated into the OCS, while there are increased in surface expres-

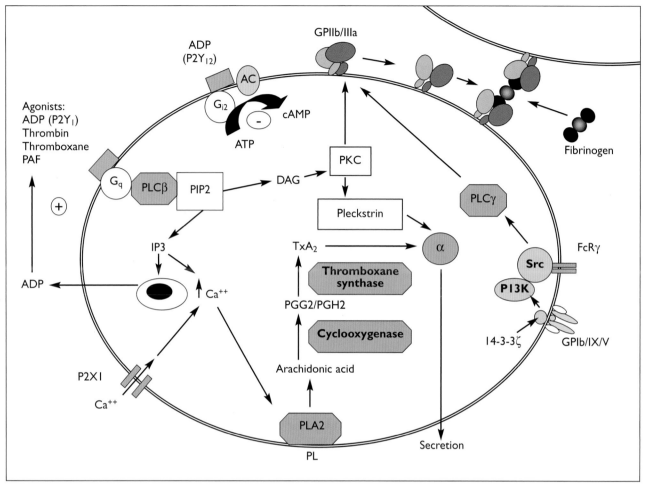

Figure 2-11. Schematic diagram of platelet activation. Platelet adhesion to von Willebrand factor through glycoprotein (GP) Ib/IX/V leads to activation of phosphoinositide-3 kinase (PI3K) and binding to Src, which is facilitated by the Fc receptor γ chain (FcRγ). This leads to downstream activation of phospholipase Cγ (PLCγ). Stimulation of platelet receptors by agonists, such as adenosine diphosphate (ADP), thrombin, thromboxane or platelet activating factor (PAF), signal through GTP binding proteins (Gq or Gi), leading to activation of phospholipase C (PLCβ). PLCβ converts phosphatidylinositol bisphosphate (PIP2) to diacylglycerol (DAG) and inositoltrisphosphate (IP3). This leads to activation of protein kinase C (PKC) and phosphorylation of pleckstrin. The activity of PKC and PLCγ leads to molecular reorganization of GPIIb/IIIa, with activation of the receptor, allowing binding to fibrinogen and subsequent platelet aggregation. IP3 production leads to release of calcium from the dense tubular system and calcium flux into the cytoplasm, facilitated by the P2X1 receptor. The increased cytoplasmic calcium activates phospholipase A2 (PLA2) to liberate arachidonic acid from the platelet membrane. Arachidonic acid is acted upon by cyclooxygenase and thromboxane synthase to produce thromboxane A_2 (TxA2). TxA2, together with phosphorylated pleckstrin and increased cytoplasmic calcium, leads to degranulation of alpha and dense granules. The process is amplified by release of TxA2 and ADP from the activated platelets, because they are potent platelet agonists that can initiate the activation of nearby platelets.

sion of PAR-1 and PAR-4 as well as GPIIb/IIIa and GPIV. This change facilitates the transition of the platelet from an adhesive to an aggregating phenotype.

Platelet Activation and Granule Release

Activation of the platelet signaling mechanism starts with the initial stages of platelet adhesion, as outlined above. Further platelet activation is stimulated by continued binding to collagen, by production of thrombin during coagulation, by release of ADP from the platelet dense granules, and by platelet production of thromboxane A_2.

Phospholipase Pathway

After adhering to the subendothelium, platelets are stimulated by thrombin and collagen via binding to distinct receptors. Binding of thrombin to its receptors, PAR-1 and PAR-4, leads to activation of

G proteins from the $G_{12/13}$, G_q, and $G_{i/z}$ families, which initiates a cascade of signaling responses. This cascade includes activation of phospholipase Cβ, with cleavage of polyphosphoinositides to diacylglycerol and inositol-1,4,5-trisphosphate, resulting in activation of protein kinase C and increased cytosolic calcium due to mobilization of calcium from intracellular depots and by entry of calcium from the plasma. This leads to phosphorylation of pleckstrin and molecular reorganization of GPIIb/IIIa, with activation of the receptor, allowing binding to fibrinogen and subsequent platelet aggregation (Figure 2-11). Activation of the thrombin receptors leads to activation of Rho-based pathways that result in shape change and activation of PI3K. Engagement of the ADP receptors $P2Y_1$ and $P2Y_{12}$ after release of ADP from the dense granules also leads to stimulation of signaling pathways. The $P2Y_1$ receptor is thought to signal through the G protein Gq, with activation of phospholipase Cβ and calcium release, while the $P2Y_{12}$ receptor is coupled through G_{i2} to activation of PI3K. There is a marked level of redundancy between various pathways of platelet activation, as can be seen from the preceding discussion; this helps to explain why similar platelet functional responses are observed with platelet stimulation by a variety of agonists.

Cyclooxygenase Pathway

The increased cytoplasmic calcium activates phospholipase A2 (PLA2) to liberate arachidonic acid from the platelet membrane. Arachidonic acid is acted upon by cyclooxygenase (COX-1) to form prostaglandin G2 and H2, with further action by thromboxane synthase needed to produce thromboxane A_2 (Figure 2-11). Both TxA_2 and the other prostaglandin intermediates are very evanescent species, with half-lives of seconds. TxA_2 is released from the platelet through the dense tubular system and binds to the TxA_2 receptor (TP) on adjacent platelets, where it stimulates platelet activation and aggregation. The production of TxA_2 occurs in parallel to other mechanisms of platelet activation and is thought to be important in the amplification of the platelet activation process. The cyclooxygenase pathway is a noted target in the pharmacologic regulation of platelet function, as the drug aspirin irreversibly inhibits cyclooxygenase through acetylation.

Platelet Granule Release

The increased cytoplasmic calcium resulting from the binding of agonists to their platelet membrane receptors initiates a secretory *release* reaction, whereby products from the alpha granules (platelet factor 4, β-thromboglobulin, thrombospondin, platelet derived growth factor, fibrinogen, and vWF, among others) and dense granules (ADP, ATP, serotonin, calcium) are released into the surrounding milieu. The granule membranes contain many integral glycoproteins on their inner leaflet, such as P-selectin (CD62p) in the alpha granule and LAMP (lysosome associated membrane protein [CD63]) in the lysosome, which become expressed on the outer platelet membrane after the release reaction.

Platelet granule release is an exocytosis process similar to that observed in neuronal neurotransmitter release. Granule release uses the molecular machinery of the SNARE (**s**oluble **N**-ethyl-maleimide-sensitive factor **a**ttachment protein **r**eceptor) complex, with cooperation of the t-SNAREs, v-SNAREs, and soluble components. The alpha granules degranulate directly into the OCS channels, while the dense granules merge with the plasma membrane during degranulation. During activation, platelets are known to shed microparticles that are tiny (100-nm) membrane-bound particles rich in surface glycoproteins. The microparticles are thought to further amplify the thrombotic process, as they express the activated form of GPIIb/IIIa and have an activated phospholipid surface to support formation of coagulation complexes and generation of thrombin.

Platelet Aggregation

The release of ADP from the dense granules, together with calcium mobilization, leads to a conformational change of the fibrinogen receptor, the GPIIb/IIIa receptor (Figure 2-8). A molecular movement of the cytoplasmic tails, perhaps initiated by talin binding to the $β_3$ chain, is thought to transmit to a straightening of the globular head group and opening of the ligand-binding pocket. Alternatively, the conformational change may be regulated by receptor clustering and interaction of transmembrane domains. This conformational change of the fibrinogen receptor initiates the process of *aggregation*, whereby a glycoprotein IIb/IIIa receptor on one platelet is bound in a homotypic fashion to the same receptor on adjacent platelets via a central fibrinogen molecular bridge (Figure 2-8). Since fibrinogen is a dimeric molecule that contains two alpha chains, each of which has two RGD sequences, two beta chains, and two gamma chains, the bivalent or multivalent binding of one fibrinogen molecule to two or more platelets is thought to facilitate platelet-platelet attachment, or platelet aggregation.

Besides ADP, other agonists, such as epinephrine, thrombin, collagen, and platelet activating factor, can initiate platelet aggregation by interaction with membrane receptors. This platelet release reaction and aggregation leads to the recruitment of many other platelets to the vessel wall, with the formation of a hemostatic platelet plug.

Platelet Procoagulant Activity

Activated platelets play a vital procoagulant role that serves as a link between platelet function and coagulation activation. Platelet membrane phospholipids undergo a rearrangement during activation, with a transfer of phosphatidyl serine from the inner table to the outer table of the platelet membrane, providing a binding site for phospholipid-dependent coagulation complexes that activate both factor X and prothrombin. Additionally, the glycoprotein Ib/IX/V complex is thought to participate in activation of factor XI and XII. The platelet alpha granules release many procoagulant factors during platelet activation, including fibrinogen, von Willebrand factor, and factor V.

Suggested Reading

Platelet Production

Hartwig J, Italiano JR. The birth of the platelet. *J Thromb Haemost.* 2003;1:1580-1586.

Italiano JE Jr, Lecine P, Shivdasani R, Hartwig JH. Blood platelets are assembled principally at the ends of proplatelet processes produced by differentiated megakaryocytes. *J Cell Biol.* 1999;147:1299-1312.

Kaushansky K. Historical review: megakaryopoiesis and thrombopoiesis. *Blood.* 2008;111:981-986.

Levin J. Thrombopoietin: clinically realized? *N Engl J Med.* 1997;336:434-436.

Schulze H, Shivdasani RA. Molecular mechanisms of megakaryocyte differentiation. *Semin Thromb Hemost.* 2004;30:389-398.

Schulze H, Shivdasani RA. Mechanisms of thrombopoiesis. *J Thromb Haemost.* 2005;3:1717-1724.

Platelet Structure

Hartwig JH. The platelet: form and function. *Semin Hematol.* 2005;43(suppl 1):S94-S100.

Italiano JE Jr, Hartwig JH. Megakaryocyte and platelet structure. In: Colman RW, Marder VJ, Clowes AW, George JN, Goldhaber SZ, eds. *Hemostasis and Thrombosis: Basic Principles and Clinical Practice.* 5th ed. Philadelphia, Pa: JB Lippincott Co; 2006: 1872-1880.

Plow EF, Abrams CS. The molecular basis for platelet function. In: Hoffman R, Benz E, Shattil S, Furie B, Cohen H, eds. *Hematology: Basic Principles and Practice.* 4th ed. London: Churchill Livingstone; 2004: 1881-1897.

Platelet Surface Glycoproteins

Bennett JS. Structure and function of the platelet integrin $\alpha_{IIb}\beta_3$. *J Clin Invest.* 2005;115:3363-3369.

Bergmeier W, Chauhan AK, Wagner DD. Glycoprotein Ibalpha and von Willebrand factor in primary platelet adhesion and thrombus formation: lessons from mutant mice. *Thromb Haemost.* 2008;99:264-270.

Clemetson KJ, Clemetson JM. Platelet collagen receptors. *Thromb Haemost.* 2001;86:189-197.

Coughlin SR, Camerer E, Hamilton JR. Protease-activated receptors in hemostasis, thrombosis and vascular biology. In: Colman RW, Marder VJ, Clowes AW, George JN, Goldhaber SZ, eds. *Hemostasis and Thrombosis: Basic Principles and Clinical Practice.* 5th ed. Philadelphia, Pa: JB Lippincott Co; 2006: 555-566.

Farndale RW, Siljander PR, Onley DJ, et al. Collagen-platelet interactions: recognition and signaling. *Biochem Soc Symp.* 2003;70:81-94.

Fullard JF. The role of the platelet glycoprotein IIb/IIIa in thrombosis and haemostasis. *Curr Pharm Des.* 2004;10(14):1567-76

Hechler B, Cattaneo M, Gachet C. The P2 receptors in platelet function. *Semin Thromb Hemost.* 2005;31:150-161.

Herbert JM, Savi P. P2Y12, a new platelet ADP receptor, target of clopidogrel. *Semin Vasc Med.* 2003;3:113-122.

Hollopeter G, Jantzen H-M, Vincent D, et al. Identification of the platelet ADP receptor targeted by antithrombotic drugs. *Nature.* 2001;409:202-207.

Lundblad RL, White GC II. The interaction of thrombin with blood platelets. *Platelets.* 2005;16:373-385.

Nieswandt B, Watson SP. Platelet-collagen interaction: is GP VI the central receptor? *Blood.* 2003;102:449-461.

Oury C, Toth-Zsamboki E, Vermylen J, Hoylaerts MF. The platelet ATP and ADP receptors. *Curr Pharm Des.* 2006;12:859-875.

Peerschke EIB. Platelet membrane glycoproteins: functional characterization and clinical applications. *Am J Clin Pathol.* 1992;98:455-463.

Suehiro K, Smith JW, Plow EF. The ligand recognition specificity of beta3 integrins. *J Biol Chem.* 1996; 271:10365-10371.

Platelet Cytoskeleton

Cramer EM, Fontenay MA. Platelets: Structure related to function. In: Colman RW, Marder VJ, Clowes AW, George JN, Goldhaber SZ, eds. *Hemostasis and Thrombosis: Basic Principles and Clinical Practice.* 5th ed. Philadelphia, Pa: JB Lippincott Co; 2006: 463-481.

Fox JE. The platelet cytoskeleton. *Thromb Haemost.* 1993;70:884-893.

Hartwig JH, DeSisto M. The cytoskeleton of the resting human blood platelet: structure of the membrane skeleton and its attachment to actin filaments. *J Cell Biol.* 1991;112:407-425.

Platelet Granules

Furie B, Furie BC, Flaumenhaft R. A journey with platelet P-selectin: the molecular basis of granule secretion, signaling and cell adhesion. *Thromb Haemost.* 2001;86:214-221.

Harrison P, Cramer EM. Platelet alpha-granules. *Blood Rev.* 1993;7:52-62.

Polasek J. Procoagulant potential of platelet α granules. *Platelets.* 2004;15:403-407.

Rendu F, Brohard-Bohn B. The platelet release reaction: granules' constituents, secretion and functions. *Platelets.* 2001;12:261-273.

Platelet Adhesion

Bodin S, Giuriato S, Ragab J, et al. Producion of phosphatidylinositol 3,4,5-trisphosphate and phosphatidic acid in platelet rats: evidence for a critical role of cholesterol-enriched domains in human platelet activation. *Biochemistry.* 2001;40:15290-15299.

Ozaki Y, Asazuma N, Suzuki-Inoue K, Berndt MC. Platelet GPIb-IX-V-dependent signaling. *J Thromb Haemost.* 2005;3:1745-1751.

Savage B, Saldivar E, Ruggeri ZM. Initiation of platelet adhesion by arrest onto fibrinogen or translocation on von Willebrand factor. *Cell.* 1996;84:289-297.

Simons K, Ikonen E. Functional rafts in cell membranes. *Nature.* 1997;387:569-572.

Varga-Szabo D, Pleines I, Nieswandt B. Cell adhesion mechanisms in platelets. *Arterioscler Thromb Vasc Biol.* 2008;28:403-412.

Platelet Activation and Release

Abrams CS. Intracellular signaling in platelets. *Curr Opin Hematol.* 2005;12:401-405.

Martorell L, Martinez-Gonzalez J, Rodriguez C, Gentile M, Calvayrac O, Badimon L. Thrombin and protease-activated receptors (PARs) in atherothrombosis. *Thromb Haemost.* 2008;99:305-315.

Reed GL, Fitzgerald ML, Polgar J. Molecular mechanisms of platelet exocytosis: insights into the "secrete" life of thrombocytes. *Blood.* 2000;96;3334-3342.

Reed GL. Platelet secretory mechanisms. *Semin Thromb Hemost.* 2004;30:441-450

Smyth EM, Funk CD. Platelet eicosanoids. In: Colman RW, Marder VJ, Clowes AW, George JN, Goldhaber SZ, eds. *Hemostasis and Thrombosis: Basic Principles and Clinical Practice.* 5th ed. Philadelphia, Pa: JB Lippincott Co; 2006: 583-589

Platelet Aggregation

Tracy PR. Role of platelets and leukocytes in coagulation. In: Colman RW, Hirsh J, Marder VJ, Clowes AW, George JN, eds. *Hemostasis and Thrombosis: Basic Principles and Clinical Practice.* 4th ed. Philadelphia, Pa: JB Lippincott Co; 2001: 575-596.

CHAPTER 3

Fibrinolytic System Physiology

Wayne L. Chandler, MD

History

As far back as Hippocrates, it was noted that clotted blood from a living person would spontaneously dissolve, and that once dissolved, the blood could not be made to clot again. This phenomenon was rediscovered 300 years ago by Malpighi, who reported that after death the blood clotted then subsequently liquefied. Since then, it has been determined that the lytic factors in blood are primarily aimed at fibrin, are an intrinsic part of blood itself, and that lysis of fibrin is regulated by a series of activators and inhibitors.

Proteins and Function

Fibrin is a temporary structure designed to aid hemostasis and wound healing, after which it is removed. Fibrin can be removed both through cellular processes involving macrophages and fibroblasts, and by the fibrinolytic system, a series of proteins that ultimately produce the enzyme plasmin, which cleaves fibrin into soluble fragments. The fibrinolytic system degrades fibrin in two settings: intravascular fibrinolysis regulates the formation and removal of fibrin at sites of vascular injury, while extravascular fibrinolysis is key to tissue remodeling and cell migration. Intravascular fibrinolysis has been studied most extensively due to the ease of measuring fibrinolytic factors in blood and the association between pathologic thrombosis, cardiovascular disease, and the development of thrombolytic therapy for the treatment of acute thrombosis.

The primary role of the fibrinolytic system is to regulate the rate and extent of fibrin deposition during vascular repair, prevent the formation of occlusive thrombi, and, along with leukocytes, remove residual fibrin during healing. The fibrinolytic system is probably most important in regulating the initial formation of small thrombi. Formation of a large thrombus indicates that fibrinolytic regulation has been overcome either due to

excessive clotting stimulus or decreased fibrinolytic activity. Intravascular fibrinolysis (Figure 3-1) is initiated by tissue plasminogen activator (tPA), which is released by vascular endothelial cells into the plasma. Plasminogen and tPA bind to specific lysine sites on the fibrin molecule and to receptors on the endothelial cell surface. tPA cleaves plasminogen, converting it into the active enzyme plasmin, which in turn degrades fibrin to dissolve the clot.

Under normal circumstances, only a small fraction of the fibrin being formed is removed by the fibrinolytic system. Fibrin is primarily removed during the healing process by leukocytes and fibroblasts as they grow into the clot, healing the wound. Plasmin cuts fibrin in many different places, leading to a wide range in the size of degradation fragments, from large fragments consisting of multiple partially degraded, still polymerized fibrin molecules, to fragments of one or two fibrin molecules like D-dimers. When plasmin activity is excessive, it may degrade fibrinogen as well as fibrin, adding to the mix of fragments in blood. Some assays detect both fibrinogen and fibrin degradation fragments.

D-dimer is a fragment consisting of the factor XIIIa cross-linked ends of two fibrin molecules. For D-dimer to appear, three processes must occur: (1) formation of polymerized fibrin, (2) cross-linking of the fibrin by factor XIIIa, and (3) degradation of cross-linked fibrin by plasmin. The concentration of D-dimer in the blood is an indication of the amount of cross-linked fibrin in the vascular system. D-dimers can be released from thrombi attached to the vascular wall or from soluble fibrin circulating in the blood.

Many different types of plasminogen activators have been described, including tPA (released from vascular endothelium), urokinase-plasminogen activator (uPA, released from a variety of cell types), streptokinase and staphylokinase (bacterial plasminogen activators), bat salivary plasminogen activator, and contact-phase plasminogen acti-

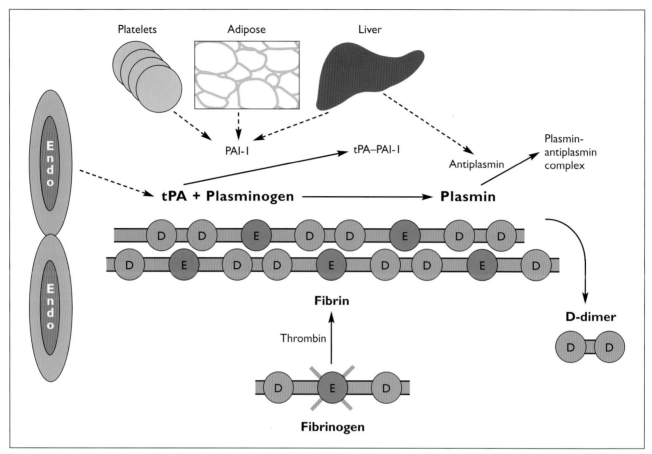

Figure 3-1. Diagram of the fibrinolytic system. Tissue plasminogen activator (tPA) is released by vascular endothelium (Endo). Fibrinogen, plasminogen, and antiplasmin (AP) are secreted by the liver. Plasminogen activator inhibitor-1 (PAI-1) is secreted by the liver and adipose tissue and released from platelets and activated endothelial cells. Fibrinogen is converted to fibrin by thrombin. tPA and plasminogen bind to polymerized fibrin, followed by conversion of plasminogen to plasmin by tPA. Plasmin in turn lyses fibrin, producing a variety of degradation fragments including D-dimers. tPA is inhibited by PAI-1, forming tPA–PAI-1 complex, which is inactive. Plasmin is inhibited by antiplasmin forming plasmin-antiplasmin complex, which is inactive.

vation (a hybrid system less well characterized). tPA is a 65,000-dalton (molecular weight) serine protease produced primarily by vascular endothelial cells. tPA is released in an active single-chain form, which can be converted to a two-chain form through proteolytic cleavage by plasmin and other enzymes. Both the one-chain and two-chain forms of tPA are active but slow at converting plasminogen into plasmin in the absence of fibrin. When bound to fibrin, both the one-chain and two-chain forms of tPA activate plasminogen approximately 1000-fold faster than free tPA in the circulation. Thus, endogenous tPA is essentially inactive unless bound to fibrin. uPA is a 54,000-dalton serine protease secreted as an inactive single-chain zymogen (scuPA). scuPA is cleaved by plasmin, kallikrein, and other proteases, forming the active two-chain form of the enzyme (tcuPA or urokinase), which activates plasminogen in solution at a rate similar to the tPA on the fibrin surface. As there is essentially no circulating active plasmin in normal blood (due to rapid inhibition by α_2-antiplasmin), most tPA and uPA circulates in a single-chain form. While both one- and two-chain tPA can be inhibited by plasminogen activator inhibitor-1 (PAI-1), only the active two-chain form of uPA (urokinase) is inhibited by PAI-1, forming uPA–PAI-1 complexes. uPA is the primary extravascular plasminogen activator secreted by a variety of cell types associated with cell migration and tissue remodeling.

Activation of the contact or kallikrein-kinin system results in formation of kallikrein from prekallikrein, activated factor XII from the zymogen form of factor XII, and bradykinin from high-molecular-weight kininogen. Activated contact system factors have been reported to play a variety of roles in fibrinolysis, including kallikrein

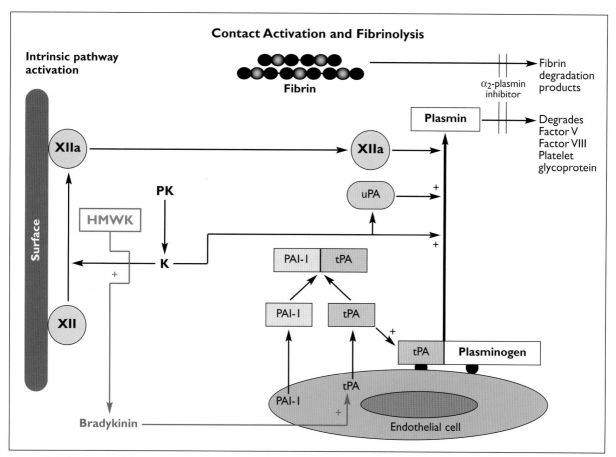

Figure 3-2. Activation of the contact or kallikrein-kinin system results in formation of kallikrein (K) from prekallikrein (PK), activated factor XII from the zymogen form of factor XII, and bradykinin from high-molecular-weight kininogen (HMWK). Activated contact system factors have been reported to play a variety of roles in fibrinolysis, including kallikrein activation of scuPA, bradykinin stimulation of tPA release from endothelial cells, and factor XIIa activation of plasminogen. Abbreviations: PAI-1, plasminogen activator inhibitor type 1; tPA, tissue plasminogen activator; uPA, urokinase plasminogen activator.

activation of scuPA, bradykinin stimulation of tPA release from endothelial cells, and factor XIIa activation of plasminogen (Figure 3-2). Homozygous deficiency of contact system proteins, including factor XII, prekallikrein, and high-molecular-weight kininogen, are associated with abnormal in vitro coagulation screening tests but not with a clinical bleeding state. Case studies and small series have suggested that complete deficiency of contact system proteins might be an increased risk factor for thrombosis, but data is limited. While uPA and contact system activation of fibrinolysis have been studied in the blood, there is no clear clinical association with bleeding or thrombosis, and they will not be considered further.

As with most reactive enzyme systems, fibrinolysis is regulated in part by a series of inhibitors, including PAI-1, which binds to and inhibits tPA, forming tPA–PAI-1 complexes; α_2-antiplasmin (α_2-plasmin inhibitor), which binds to and inhibits

plasmin, forming plasmin-antiplasmin (PAP) complexes; and thrombin activatable fibrinolysis inhibitor (TAFI), an enzyme that modifies the fibrin surface, slowing plasminogen activation.

Most in vivo inhibitors of the fibrinolytic system are members of the serine protease inhibitor (SERPIN) family. PAI-1 is the most important inhibitor of tPA in plasma. PAI-1 is a 52,000-dalton single-chain glycoprotein released by the liver, adipose tissue, megakaryocytes, and possibly vascular endothelium during an acute phase response. PAI-1 binds to and inhibits tPA and uPA, forming inactive tPA–PAI-1 and uPA–PAI-1 complexes. PAI-1 exists in three forms in blood: (1) active PAI-1 able to bind and inhibit tPA, (2) inactive or latent PAI-1, and (3) PAI-1 complexed with tPA or uPA. Active PAI-1 is unstable at 37°C, undergoing a slow conformational change over several hours into the inactive or latent form of PAI-1. Platelets contain substantial amounts of

PAI-1, which is released during platelet activation. The majority of the PAI-1 in platelets is in the inactive or latent form, but enough active PAI-1 is present to slow fibrinolysis near an actively forming platelet clot. Plasma and platelet PAI-1 binds to fibrin, and binds to and is stabilized by vitronectin on the extracellular matrix. Free active PAI-1, vitronectin bound PAI-1, and fibrin bound PAI-1 are all rapid inhibitors of tPA.

Plasminogen activator inhibitor-2 (PAI-2) is a rapid inhibitor of two-chain uPA and a slow inhibitor of two-chain tPA, but does not inhibit single-chain tPA to a significant extent under physiologic conditions. PAI-2 is secreted by the placenta. PAI-2 is not measurable in normal plasma but reaches high levels in plasma during pregnancy. Its role in regulating intravascular fibrinolysis is unclear; it may play a role in regulating fibrin removal in the placenta. Other slow inhibitors of tPA include C1-inhibitor, α_2-antiplasmin, and α_2-macroglobulin. None of these other inhibitors of tPA are fast enough to replace PAI-1 in vivo. A deficiency of PAI-1 leads to a moderate bleeding syndrome due to uncontrolled plasminogen activation.

Plasminogen is a 92,000-dalton single-chain glycoprotein secreted by the liver. Plasminogen is proteolytically cleaved by tPA or uPA to the active two-chain molecule plasmin. Plasmin is very active at degrading fibrin but is capable of cleaving many other proteins, including hemostatic factors V and VIII. The principle inhibitor of plasmin is α_2-antiplasmin, a 70,000-dalton single-chain glycoprotein secreted by the liver. α_2-antiplasmin binds and is cross-linked to fibrin in an active form. When excess plasmin generation occurs, as during thrombolytic therapy or disseminated intravascular coagulation, plasmin is also inhibited to a small extent by α_2-macroglobulin, antithrombin, and C1-inhibitor. As discussed for PAI-1, none of these other inhibitors of plasmin are fast enough to replace α_2-antiplasmin in vivo. A deficiency of α_2-antiplasmin leads to a moderate to severe bleeding syndrome due to uncontrolled fibrinolysis.

Both tPA and plasminogen bind to lysine residues in fibrin. As plasmin begins cleaving the C-terminal alpha-chain of fibrin, it exposes additional lysine binding sites, which increases the activity and binding of tPA, plasminogen, and plasmin. A new protein that regulates fibrinolysis in a novel fashion has recently been discovered. Thrombin activatable fibrinolysis inhibitor (TAFI) is a carboxypeptidase that removes newly exposed lysine binding sites in fibrin, slowing fibrinolysis. TAFI is a 55,000-dalton protein secreted by the liver. While some studies suggest that reduced levels of TAFI may be associated with an increased risk of venous thrombosis, further research is needed before TAFI measurements will become a routine part of thrombophilia workups.

Regulation of Fibrinolysis

Tissue plasminogen activator is released from vascular endothelial cells. The majority of vascular endothelium is in the capillary beds. Thus, most of the tPA circulating in blood is added as the blood passes through capillary beds. This results in higher levels of tPA activity in veins versus arteries. If a tourniquet is placed on the arm for 5 to 10 minutes, tPA released by the capillary beds in the arm will be trapped, resulting in a rapid increase in active tPA in the venous blood. PAI-1 is primarily secreted by liver and adipose tissue. Active PAI-1 is lower in venous blood and can be eliminated by venous occlusion due to trapping of secreted tPA, which reacts with PAI-1 trapped in venous circulation. Under normal circumstances, little if any PAI-1 is secreted by endothelial cells in vivo. It is possible that during an acute phase response associated with inflammation or infection, endothelial cells may secrete PAI-1 to some extent. Large vessels, including arteries and veins, also have an endothelial lining, but the ratio of blood volume and flow to endothelial surface area is so low that large vessel endothelium do not appreciably affect circulating levels of tPA or PAI-1. Release of tPA from large vessel endothelium may have a local effect if a thrombus forms on the vessel surface.

Three processes regulate the plasma levels of fibrinolytic factors: (1) the release of fibrinolytic proteins into the plasma, (2) inhibition of active proteins, and (3) clearance of proteins from the plasma. Under normal circumstances, tPA appears to be released at a relatively constant rate during the day. tPA release can be increased in two ways. First, endothelial cells have a storage pool of tPA, which can be released in seconds when the endothelial cells are stimulated, typically by vasoactive substances like epinephrine, bradykinin, or vasopressin. Stimulated release can rapidly increase active tPA, but the storage pool is limited in extent and can be depleted if the stimulus lasts for too long. tPA production in the endothelial cell can also be increased, usually associated with an acute phase inflammatory response. This requires new protein production;

thus, there is a lag phase of several hours between the stimulus and increasing tPA levels.

PAI-1 release is highly variable in humans. PAI-1 levels show a circadian variation, with peak PAI-1 activity in the morning and nadir PAI-1 activity in the afternoon or evening. The difference between morning and evening is on average about 2-fold but can be up to 10-fold in some individuals. PAI-1 is also an acute phase reactant like tPA, but while tPA levels typically increase 2- to 3-fold during an inflammatory response, PAI-1 levels can increase 10- to 100-fold.

PAI-1 binds to and inactivates tPA, forming tPA–PAI-1 complexes. Higher concentrations of active PAI-1 in blood are associated with lower levels of active tPA and slower fibrinolysis. Both active tPA and tPA–PAI-1 complex are cleared from the blood by the liver. tPA clearance is rapid. Active tPA has a clearance half-life of 2 to 4 minutes, while tPA–PAI-1 complex has a half-life of 4 to 6 minutes—slightly longer. tPA is cleared from the blood in proportion to liver blood flow. Approximately 50% of the tPA is cleared from the blood as it passes through the liver. tPA clearance is delayed in patients with reduced liver blood flow, as occurs in cirrhosis. The ultimate version of delayed tPA clearance is during the anhepatic phase of liver transplantation, when tPA clearance is essentially eliminated, and plasma tPA levels transiently rise 10-fold or more until the new liver is connected and begins clearing tPA from the blood.

To summarize, the level of active tPA in blood is a function of the rate of tPA release, the rate of PAI-1 release and thus tPA inhibition, and the rate of liver clearance of tPA. In healthy subjects with normal PAI-1 levels, approximately 20% to 50% of total tPA is active. Increased release of tPA increases the level of active tPA in blood and accelerates initiation of fibrinolysis. An example of this is increased tPA release during cardiopulmonary bypass. Increased PAI-1 secretion leads to more rapid inhibition of tPA, reducing active tPA levels in blood while increasing tPA–PAI-1 levels. Increased PAI-1 levels slow fibrinolysis. Patients with elevated PAI-1 levels associated with the metabolic syndrome or acute phase increases in PAI-1 secretion may have only a few percent of total tPA circulating in an active form. Deficiency of PAI-1 results in the majority of the tPA circulating in an active form, accelerating fibrinolysis. Overall, fibrinolysis is accelerated by increased levels of active tPA in the blood and slowed by reduced levels of active tPA, typically due to increased secretion of PAI-1.

In vivo and in vitro studies suggest that lipoproteins and the fibrinolytic system are interrelated. An important factor affecting PAI-1 release is the metabolic syndrome (insulin resistance). Patients with the metabolic syndrome have increased levels of fasting insulin and glucose, elevated triglycerides, and elevated PAI-1. Levels of PAI-1 in the blood are positively correlated with intra-abdominal fat, waist/hip ratios, and weight. Similar findings are seen in patients with type 2 diabetes. Weight loss or treatment with metformin reduces insulin, glucose, triglyceride, and PAI-1 levels. In vitro studies suggest that endothelial release of PAI-1 may be increased and tPA release decreased by high concentrations of apolipoprotein-B-containing lipoproteins, such as low density lipoprotein (LDL) and very low density lipoprotein.

A further connection between lipoproteins and fibrinolysis was the discovery of lipoprotein (a), an LDL-like particle containing an additional protein moiety called apolipoprotein (a) that is homologous to plasminogen. The level of lipoprotein (a) is an independent risk factor for atherosclerosis. Apolipoprotein (a) consists of a variable number of tandem repeat domains similar to plasminogen kringle-4, a single region similar to kringle-5, and an inactive pseudo-protease segment. In vitro studies show that lipoprotein (a) binds to fibrin and co-localizes with fibrin in atheromatous tissue. Further in vitro studies suggest that lipoprotein (a) and apolipoprotein (a) compete with plasminogen and tPA binding to endothelial cells and inhibit fibrin-dependent plasminogen activation. The degree to which lipoprotein (a) inhibits fibrinolysis in vivo is still controversial.

The renin-angiotensin system also plays a role in regulating fibrinolysis. Angiotensin II stimulates increased production of PAI-1. Angiotensin converting enzyme breaks down bradykinin, potentially slowing tPA release. Treatment with angiotensin converting enzyme inhibitors is associated with a reduction in PAI-1 levels.

Fibrinolysis at the Site of a Thrombus

Blood flowing through an undamaged vessel contains ~20 to 40 pmol/L active tPA when PAI-1 activity is low, or as little as 1 to 2 pmol/L of active tPA when PAI-1 activity is increased. In healthy individuals, soluble fibrin levels are very low, and little if any plasminogen activation is occurring. Any plasmin that is formed in the circulating blood is rapidly inhibited by α_2-antiplasmin.

When a vessel is injured, procoagulant factors are released and exposed. Platelets bind to exposed collagen, activate and release granular contents, expose anionic phospholipids, and activate and aggregate other platelets. PAI-1 is released from platelet granules. Tissue factor is exposed during vascular injury, activating the coagulation system, resulting in thrombin generation. Thrombin, in turn, proteolyzes fibrinogen to fibrin, forming a meshwork that stabilizes the growing platelet thrombus. tPA and plasminogen bind to the fibrin surface, where tPA converts plasminogen into plasmin, which begins lysing the fibrin. PAI-1 and α_2-antiplasmin bind to fibrin as well, slowing the activation of plasminogen and increasing the inhibition of plasmin. tPA and plasmin bound to fibrin are inhibited more slowly than the free forms of these enzymes. Under normal conditions, the procoagulant stimulus is greater than anticoagulant and fibrinolytic forces, resulting in formation of a hemostatic clot. As the wound is filled, procoagulants like collagen and tissue factor are covered, slowing the rate of new activation and leading to a steady-state clot size controlled by balanced clot formation versus lysis. Healing then ensues with migration of endothelium, fibroblasts, and macrophages. The clot is covered over, replaced with collagen and other proteins, and the fibrin and platelet scaffold removed as the wound heals.

If active tPA levels are too high, as occurs during cardiopulmonary bypass, too much plasmin is formed, leading to excessive loss of fibrin, incomplete hemostasis, and increased bleeding. If active tPA levels are too low, typically due to elevated PAI-1 activity, fibrinolysis is slowed, and the forming thrombus may grow too large, occluding the vessel, leading to ischemia and infarction.

Suggested Reading

Fibrinolytic Proteins and Function

Booth NA, Bachmann F. Plasminogen-plasmin system. In: Colman RW, Marder VJ, Clowes AW, George JN, Goldhaber SZ, eds. *Hemostasis and Thrombosis: Basic Principles and Clinical Practice.* 5th ed. Philadelphia, Pa: JB Lippincott Williams & Wilkins; 2006: 335-364.

Cale JM, Lawrence DA. Structure-function relationships of plasminogen activator inhibitor-1 and its potential as a therapeutic agent. *Curr Drug Targets.* 2007; 8(9):971-981.

Regulation of Fibrinolysis

Andreotti F, Davies GJ, Hackett DR, et al. Major circadian fluctuations in fibrinolytic factors and possible relevance to time of onset of myocardial infarction, sudden cardiac death and stroke. *Am J Cardiol.* 1988;62:635-637.

Chandler WL, Velan T. Secretion of tissue plasminogen activator and plasminogen activator inhibitor 1 during cardiopulmonary bypass. *Thromb Res.* 2004; 112:185-192.

Cheema SP, Webster NR, Dunn F, Bellamy MC. Mediators of fibrinolysis in orthotopic liver transplantation. *Clin Transplant.* 1996;10:24-27.

Huber K, Kirchheimer C, Korninger C, Binder BR. Hepatic synthesis and clearance of components of the fibrinolytic system in healthy volunteers and in patients with different stages of liver cirrhosis. *Thromb Res.* 1991;62:491-500.

Jovin IS, Müller-Berghaus G. Interrelationships between the fibrinolytic system and lipoproteins in the pathogenesis of coronary atherosclerosis. *Atherosclerosis.* 2004;174:225-233.

Kitchens CS. The contact system. *Arch Pathol Lab Med.* 2002;126:1382-1386.

Marder VJ, Francis CW. Physiologic regulation of fibrinolysis. In: Colman RW, Marder VJ, Clowes AW, George JN, Goldhaber SZ, eds. *Hemostasis and Thrombosis: Basic Principles and Clinical Practice.* 5th ed. Philadelphia, Pa: JB Lippincott Williams & Wilkins; 2006: 419-436.

Mavri A, Alessi MC, Juhan-Vague I. Hypofibrinolysis in the insulin resistance syndrome: implication in cardiovascular diseases. *J Intern Med.* 2004;255:448-456.

Medcalf RL. Fibrinolysis, inflammation, and regulation of the plasminogen activating system. *J Thromb Haemost.* 2007;5(suppl 1):132-142.

Rau JC, Beaulieu LM, Huntington JA, Church FC. Serpins in thrombosis, hemostasis and fibrinolysis. *J Thromb Haemost.* 2007;5(suppl 1):102-115.

Skurk T, Hauner H. Obesity and impaired fibrinolysis: role of adipose production of plasminogen activator inhibitor-1. *Int J Obesity.* 2004;28:1357-1364.

Section 2

Basics of Laboratory Testing in Hemostasis

Kandice Kottke-Marchant, MD, PhD, Editor

Chapter 4. Sample Collection and Processing

Chapter 5. Clinical History in Hemostasis

Chapter 6. Coagulation Testing

Chapter 7. Platelet Testing

Chapter 8. Fibrinolysis Testing

Chapter 9. Thrombophilia Testing

Sample Collection and Processing

Jayashree Krishnan, MD

Introduction

Quality laboratory results begin with proper collection and handling of blood samples. For most coagulation tests and factor assays, platelet-poor plasma anticoagulated with sodium citrate and prepared by removal of cellular elements is necessary. Sodium citrate is a mild calcium chelator that complexes with free Ca^{++}, thus acting to inhibiting calcium-dependent coagulation steps. In order to produce accurate results, specimen integrity is crucial, so appropriate collection, handling, and processing of hemostasis samples are very important. This chapter will discuss proper techniques for sample collection, processing, storage, and handling of specimens for coagulation testing. Sample collection and handling for platelet testing will be covered in chapter 7, and sample issues for fibrinolytic testing will be covered in chapter 8.

Sample Collection Technique for Peripheral Venipuncture

The ideal specimen for coagulation testing is a properly collected, atraumatic venous sample obtained by direct venipuncture. Great care should be taken in collecting coagulation samples to avoid hemolysis, excessive tissue injury with thromboplastin release, improper mixing with in vitro clotting, or carryover from other anticoagulant tubes. The venipuncture technique recommended by the Clinical and Laboratory Standards Institute (CLSI, formerly NCCLS) involves the following steps:

1. Verify that the computer-printed labels match the requisition.

2. Check the patient identification band against labels and the test requisition. Ask the patient for their full name or verify the patient's identity from another reliable source established by hospital protocol. Do not draw any specimen without properly identifying the patient.

3. Address the patient and inform the patient of the procedure to be performed.

4. Position the patient properly. Body posture is known to influence the concentration of blood constituents. Changing from the supine to upright position can reduce the plasma volume approximately 21%. Therefore, the supine position is the recommended posture. A rest of up to 30 minutes is recommended for fibrinolysis tests.

5. Assemble equipment and supplies, including collection tubes, tourniquet, syringes, etc.

6. Follow standard hospital safety precautions.

7. Ask the patient to make a fist so that veins are more palpable.

8. Select a suitable vein for puncture. Veins of the ante-cubital fossa (median cubital, cephalic, and basilic vein, in that order of priority) are preferred. A small diameter vein can collapse, which may lead to blood clotting and/or underfilling of the tube.

9. The choice of the venipuncture site is important. Blood should preferably be drawn from an arm without indwelling intravenous infusion lines or arterio-venous fistula in hemodialysis patients. Contamination of the samples by the infusion solution, especially those containing heparin, is the most common preanalytic error in hospitalized patients.

10. Cleanse the venipuncture site with 70% isopropyl alcohol solution or 1% iodine-saturated swab stick. Allow the area to dry.

11. Apply a tourniquet several inches above the planned puncture site. The purpose of using a tourniquet is to help locate the vein. A tourniquet should be used to locate the vein unless it would interfere with the test results. Tapping the vein is no longer recommended.

12. Anchor the vein properly.

13. Perform the venipuncture by entering the skin with the needle at approximately a 15-degree

angle to the arm, with the bevel of the needle up. Insert the needle smoothly and fairly rapidly to minimize discomfort to the patient.

14. If using a syringe, pull back on the barrel with a slow even tension as the blood flows into the syringe. Hemolysis or collapsing of the vein by too quick a pull should be avoided. Repeated venipunctures into a single vein or traumatic venipunctures should be avoided, if possible.

15. After the required blood has been drawn, have the patient relax their fist. The patient should avoid opening and closing their hand in a pumping motion to prevent bleeding from the venipuncture site. Place a sterile cotton ball or gauze lightly over the site. Withdraw the needle, then apply pressure to the site, and apply an adhesive/bandage strip to stop bleeding.

16. The sample should mixed by gentle inversion. Shaking, mixing aggressively, or inverting more than five times should be avoided.

Needle Selection for Venipuncture

The needle size should be adapted to the diameter of the vein. A needle with a diameter greater than 1 mm (lower than 19-gauge needle) may be traumatic for the vein, leading to release of vascular fragments or thromboplastin, which can modify hemostasis in the tube. A needle with a diameter under 0.7 mm (higher than 22-gauge needle) may prolong blood collection due to a high pressure gradient in the needle and may thereby cause hemolysis and platelet activation.

Tourniquet Effect

Good venipuncture technique should avoid prolonged use of the tourniquet. If the tourniquet is applied for more than 1 minute, the levels of factor VIII, von Willebrand factor, and tissue plasminogen activator (tPA) are increased, and fibrinolysis can become activated. In addition, there is a time-dependent hemoconcentration reaching its peak approximately 5 to 6 minutes after applying a tourniquet, which will increase the levels of all blood proteins. When placing the tourniquet, it should not be too tight or it could block systolic blood flow, leading to ischemia of the extremity. As soon as blood begins to flow into the collection tube, the tourniquet should be released.

Collection Sets

Butterfly collection sets are used commonly in young patients or in patients with a difficult venous access. These systems typically consist of a needle connected to a length of tubing that then connects to an adapter for the collection tube. This system can reduce the blood flow and thereby increase the risk of platelet activation. In addition, the void volume can reach 500 to 700 µL, which may modify the anticoagulant to blood ratio. This may be more of a concern with low-volume samples. Therefore, collection sets should be chosen with a void volume of less than 10% of the draw volume.

Specimen Labeling

Collected blood specimens should be labeled clearly with the following information at the patient's bedside, as pertinent to the individual institution:

❑ Patient's first and last name, sex (optional), date of birth.
❑ Medical record number or laboratory reference number.
❑ Collection date and time.
❑ Type of specimen submitted.
❑ Test(s) to be performed.
❑ Name of person collecting the specimen (optional).
❑ If a test requires special handling during transport, indicate the correct temperature at which the specimen is to be maintained during transport.

The following information should be included with the test request form or in the computerized patient test order:

❑ A unique identification number.
❑ An accessioning number.
❑ Patient's first and last name.
❑ Patient's date of birth.
❑ Assay(s) to be performed.
❑ Date and time of collection.
❑ Specimen type (whole blood, plasma, etc).
❑ Relevant clinical and laboratory information.
❑ The identity of the phlebotomist.
❑ The ordering physician's name.
❑ The department or location where the specimen was collected.
❑ Billing information (if applicable).
❑ Other information, as needed.

Blood Collection from Indwelling Lines and Catheters

Collection of blood for coagulation testing through intravenous lines that have been previously flushed with heparin should be avoided, if possible. If the blood must be drawn through an indwelling catheter, possible heparin contamination and specimen dilution should be considered. When the samples are obtained from indwelling arterial or venous catheters, the heparinized fluid infusion should be stopped. The solution used to maintain patency of the vein, usually heparin, must be cleared. It is recommended that the line be flushed with 5 mL of saline, and a minimum of 5 mL or 6 times the line volume (dead space volume of the catheter) should be drawn off and discarded or used for other tests before the coagulation tube is filled. The appropriate discard volume should be established by each laboratory.

CLSI guidelines allow drawing blood from above the intravenous site with great caution but recommend following the established guidelines and standards of the respective institution. If a coagulation specimen is collected from a line, it is important to note this fact in the laboratory information system or on the requisition slip accompanying the specimen. Studies have shown that there can be a good correlation between blood from a catheter and routine venipuncture for prothrombin time (PT) and fibrinogen levels. However, the results for activated partial thromboplastin time (aPTT) and thrombin time (TT) may be discordant between the two samples, especially if heparin is present in the catheter and the heparin solution is not appropriately removed.

Blood Collection Tubes and Anticoagulant Considerations

Blood Collection Tubes

Blood collection tubes are made from various materials and are available in a wide variety of sizes. Evacuated blood collection tubes are tubes with a rubber cap that are sealed with a vacuum that is set to draw a fixed amount of blood. These are commonly called "vacutainers" and are recommended for collection of coagulation specimens, rather than collecting in a syringe. Commercially available coagulation blood collection tubes are sterile, and the sterility is validated by the manufacturer for a certain period of time. Tubes for collecting coagulation specimens contain sodium citrate.

The tube components should minimize activation or adsorbtion of clotting factors. The inside of commercially available glass tubes is typically siliconized. Some forms of silicone, such as copolymer L-720, can activate platelets. Although siliconized glass tubes are standard tubes, evacuated plastic tubes are gaining popularity due to their light weight, decreased fragility, and cost. It is important to recognize that there are many different plastic materials with different chemical properties that may have different effects on clotting times, heparin adsorption, interference with fibrinogen assays, etc. Plastic tubes also require stringent storage temperature conditions. Their evacuation may be incomplete, especially after prolonged storage.

The volume evacuated should represent at least 80% of the total tube volume in order to reduce platelet activation caused by decrease in pH observed in partially evacuated tubes. This is more of a concern with glass tubes compared to plastic collection tubes. The tube closure should be designed to decrease the risk of aerosol formation and blood spillage while opening. The stopper materials should be free of any substance that could interfere with the anticoagulant and the clotting factors. Typical citrated (light blue top) vacutainer tubes contain 0.5 mL of anticoagulant, and the vacuum is precisely set to draw 4.5 mL of blood, although tubes with lower volume draws are widely available. It is critical that the tubes are adequately filled so that the ratio of blood to sodium citrate is 9:1. Underfilled tubes will yield falsely increased clotting times due to excess of anticoagulant. Anything that disturbs this ratio of 9:1 can alter the clotting time results significantly. Two circumstances in which this can occur are (1) when the vacuum is not used (eg, filling the tube from another syringe manually) or (2) when the patient has an abnormal plasma volume due to high hematocrit or severe anemia.

Many different types and sizes of citrate tubes are commercially available, but the ratio of 9:1 for blood to anticoagulant should be maintained. If pediatric patients are drawn, a pediatric size tube may be used. These typically draw 2.7 mL blood to 0.3 mL of citrate. The minimum fill volumes are as follows: 4.0 mL for the 4.5-mL tube; 2.5 mL for the 2.7-mL tube; and 1.6 mL for the 1.8-mL tube.

Some blood collection tubes are designed with only partial filling (reduced volume of anticoagulant and partial evacuation). The ratio of blood volume to surface area may have an impact on the blood flow in the tube and consequently may have an effect on platelet activation. This can cause significant differences in clotting times, especially for the aPTT, due to increased platelet activation with release of platelet factor 4 and neutralization of heparin.

Blood Tube Draw Order

The order in which blood collection tubes are filled during phlebotomy can influence blood coagulation testing results. This may be due to carryover from other tubes, such as those filled with ethylenediaminetetraacetic acid (EDTA) or heparin. The order of blood draw when multiple tubes (sequential order) are drawn as recommended by the current CLSI guidelines is as follows:

1. Blood culture tubes
2. Sodium citrate tube (light blue top)
3. Serum tube with or without a clot activator or gel separator
4. Heparin tube
5. EDTA tube
6. Glycolic inhibitor

This order of draw is sufficient to prevent carryover without the use of a clearing tube and regardless of whether one uses glass or plastic tubes. However, institutional protocols may vary and should be followed.

Some authors advocate the use of a discard tube drawn before the tube for coagulation studies when blood is drawn for hemostasis testing only (random draw). However, studies have shown that PT and aPTT results are not adversely affected if tested on the first tube collected. Recent CLSI guidelines do not indicate that a discard tube is necessary when drawing coagulation specimens. However, the use of a clearing or discard tube is the safest and best way to avoid specimen contamination.

Anticoagulant

The recommended anticoagulant for hemostasis testing is tri-sodium citrate due to its mild calcium-chelating properties. Of the two commercially available forms of citrate, 3.2% buffered sodium citrate (109 mmol/L of the dehydrate form of tri-sodium citrate $Na_3C_6H_5O_7.2H_2O$) is the recommended anticoagulant for coagulation testing. It can be mixed with other additives, if necessary. In 1997, NCCLS (now CLSI) recommended collecting blood in sodium citrate with a concentration of 3.2%. The previously-used 3.8% citrate concentration currently is not recommended because the

higher concentration of citrate may result in falsely lengthened clotting times with calcium-dependent coagulation tests (PT and aPTT) in slightly underfilled samples and in samples with high hematocrits. The international sensitivity index (ISI) of some of the thromboplastins varies when tested with 3.2% versus 3.8% citrate. The calibration of the reference thromboplastin is performed with plasma from blood drawn in a 3.2% concentration of sodium citrate, so use of 3.2% citrate is recommended for consistency.

Citrate concentrations may not have a significant effect on clotting tests from normal individuals, but a higher concentration of citrate can prolong the coagulation time of patients with anticoagulant therapy in a reagent-dependent manner. The difference in INR (international normalized ratio) can be as high as 0.2 to 0.3 INR units. If the laboratory does not adhere to recommendations for use of 3.2% buffered sodium citrate, it must have data on file to demonstrate that the alternative citrate concentration produces accurate and precise coagulation results. Heparinized tubes are not appropriate due to the inhibitory effect of heparin on multiple coagulation proteins.

Anticoagulant tubes with additives or other anticoagulants are available for research and specialty coagulation testing. These include tubes with a citrate-theophylline-adenosine-dipyridamole (CTAD) mixture, which may be used for monitoring heparin therapy either by aPTT or anti-Xa assays and is also used for flow cytometric analysis of platelets. It reduces ex vivo platelet activation and heparin neutralization by platelet factor 4. However, CTAD inhibits platelet function, is very sensitive to light, and requires stringent storage requirements. Another tube available is the SCAT-1 tube (Hematologic Technologies Inc, Essex Junction, Vermont), which contains the fibrinolytic inhibitor aprotinin and the thrombin inhibitor D-Phe-Pro-Arg chloromethylketone (PPACK) along with EDTA. These tubes may be useful for collection of markers of coagulation activation, such as fibrinopeptide A or prothrombin F1.2. Acidified citrate tubes may be needed for accurate measurement of plasma tissue plasminogen activator (tPA), as described in more detail in chapter 8.

Volume Effect

The recommended proportion of blood to the sodium citrate anticoagulant volume is 9:1.

Inadequate filling of the collection device will decrease this ratio and may lead to inaccurate results for calcium-dependent clotting tests, such as the PT and aPTT.

Tubes should be filled to capacity, or at least 90%. With assays using a "moderately sensitive" thromboplastin (ISI of 2.06), normal individuals will have accurate results in tubes filled to 65% or more. In patients receiving anticoagulants, results from tubes filled to 80% or less are not accurate. A more sensitive thromboplastin (ISI 1.01) requires that the tubes be filled to 90% capacity to obtain an accurate PT. In an underfilled tube, the citrate concentration is increased and can lengthen the aPTT by about 3% for a 10% underfilling, and about 10% for a 20% underfilling. The PT test can be prolonged by about 5% for a 20% underfilling, and by about 15% for a 30% underfilling.

Overfilling of tubes can occur if the tube was filled either after opening it or with a syringe previously used for blood drawing. Overfilling of the collection tubes will increase the citrate to anticoagulant ratio and may lead to clotted samples or inaccurate results.

Citrate is usually buffered with citric acid to a pH of 5.1 to 5.3 in order to maintain the pH of the plasma sample between 7.3 and 7.45. If the pH is outside this range, spurious prolongation of the PT may occur. In unbuffered citrate, the plasma pH will be outside the range since the pH of trisodium citrate is slightly alkaline and the plasma does not have the buffering capacity of red cells removed by centrifugation.

Hematocrit Effect

A hematocrit value greater than 55% may lead to spurious coagulation results. The citrate anticoagulant distributes only in the plasma and not into the blood cells. For this reason, plasma citrate concentration will be increased if the patient's hematocrit is greater than 55%, leading to prolonged PT and aPTT results, as well as erroneous results for other calcium-dependent clotting tests, such as protein C and protein S testing. Each laboratory should have a documented procedure for detection and special handling of specimens with high hematocrits. If possible, a new phlebotomy should be performed using a reduced volume of sodium citrate, adjusted for the elevated hematocrit. Conversely, there are no current data to support a recommendation for adjusting the citrate concentration in the presence of severe anemia (hematocrit <20%).

Table 4-1. Quick Reference for Citrate Volume Adjustment for High Hematocrit

Patient's Hematocrit	Volume of Citrate for 5 mL Total Volume (blood + citrate)	Volume of Citrate to Remove from 5-mL Vacuum Tube (starting citrate volume 0.5 mL)
55%	0.42 mL	0.08 mL
60%	0.37 mL	0.13 mL
65%	0.33 mL	0.17 mL
70%	0.29 mL	0.21 mL
75%	0.24 mL	0.26 mL
80%	0.19 mL	0.31 mL

If the patient has a known hematocrit above 55%, the amount of anticoagulant in the collection tube should be adjusted before drawing the blood according to CLSI guidelines. To calculate the amount of citrate required in the collection tube or syringe, use this formula:

$$C = (1.85 \times 10^{-3})(100 - HCT)(V\ blood)$$

where:

C is the volume of citrate in the tube (mL).

HCT is the hematocrit of the patient (%).

V is the volume of blood added (mL); for a 5.0-mL tube, the volume of blood drawn is actually 4.5 mL.

1.85×10^{-3} is a constant (taking into account the citrate volume, blood volume, and citrate concentration).

Example: If a patient has a hematocrit of 60%, and the total anticoagulated blood volume is 5 mL, the adjusted citrate volume is calculated as follows: $C = (1.85 \times 10^{-3})(100 - 60)(5.0\ mL) = 0.37$ mL of sodium citrate anticoagulant.

A chart of calculated citrate volumes for a standard 5.0-mL blue top tube with varying high hematocrit values is shown in Table 4-1.

Sample Processing

Platelets can be activated or disrupted by freeze-thawing, cooling, hemolysis, slow or difficult blood draws, movement through pneumatic tube systems, and other mechanisms. Activated or disrupted platelets can neutralize lupus inhibitors and heparin in plasma as well as cause other interference in coagulation assays. Because of this, it is routinely recommended that coagulation samples should be free of hemolysis or fibrin strands and centrifuged to remove platelets to a count of less than 10,000/μL.

Most coagulation tests are performed on platelet-poor plasma (PPP), which is prepared by centrifugation to remove the cellular elements. Specimens should be centrifuged at high speeds, typically 1500g for 15 minutes. It is important to obtain "platelet-poor" plasma because platelet membranes form a procoagulant surface that can accelerate coagulation and spuriously shorten clotting times. The platelet count of prepared plasma should be verified periodically to ensure that it is less than 10,000/μL.

The CLSI guidelines indicate that aPTT, PT/INR, and TT performed on fresh plasma samples are not affected by platelet counts of up to 200,000/μL. While routine aPTT, PT/INR, and TT results on fresh, warm, unactivated platelet-rich plasma may not be significantly different than in platelet-poor plasma, this may not be the case for samples being evaluated for heparin therapy or lupus inhibitors. In both of these cases, platelet-poor plasma is required. Additionally, it is difficult to predict which samples will have to be frozen for later testing. Thus, the recommended approach is to centrifuge all samples to remove platelets.

Samples for coagulation testing can be tested using direct tube sampling, and some automated instruments use a cap-piercing instrument. For manual testing, the plasma should be removed from the pellet and transferred to a clean tube to avoid plasma contamination by platelets during the testing process.

Preparation of Platelet-Poor Plasma

1. Centrifuge the specimen for at least 15 minutes at 1500g at room temperature.
2. Remove an aliquot of platelet-poor plasma from the red cells by using a plastic pipette.
3. Place the plasma into properly labeled plastic vials (polypropylene is preferred over polycarbonate or polystyrene).
4. Perform testing as soon as possible.

Special precautions are suggested when processing platelet-poor plasma for lupus anticoagulant testing. The most important step in diagnosis of lupus anticoagulant is to take special precaution in obtaining platelet-poor plasma. Since lupus anticoagulants are antiphospholipid antibodies, the presence of residual phospholipid from platelets or other cells can neutralize lupus anticoagulants and result in false-negative results. Therefore, the more platelet free the sample is, the greater is the sensitivity of the test results. In order to be considered platelet-poor plasma, a residual platelet count of less than 10,000/μL is recommended. To achieve this, the plasma can be filtered through a 0.2-micron filter to remove platelets. Alternatively, a double-spin technique can be used, as follows:

1. Centrifuge the specimen at 1500g for 15 minutes.
2. Transfer the plasma to a clean plastic tube by using plastic pipette. It is important to remove the aliquot of plasma by keeping the pipette tip away from the buffy coat layer.
3. Centrifuge the plasma portion again at 1500g for 15 minutes.
4. Transfer the plasma into another clean plastic tube with another clean plastic pipette, staying clear of the bottom of the tube where the platelet pellet will be located.

Effects of Temperature and Time

Coagulation tests should be promptly performed on fresh plasma or platelet-poor plasma that has been frozen until testing can be performed. For fresh plasmas, most testing should be performed within 4 hours of phlebotomy. Prothrombin times can be performed within 24 hours provided the collection tubes have remained unopened. After blood collection, there is progressive degradation of the labile coagulation factors V and VIII, leading to increasing prolongation of the aPTT and PT. The allowable time interval between specimen collection and sample testing depends on the temperature encountered during transport and storage of the specimen.

Optimally, for specimens that are uncentrifuged or centrifuged with plasma remaining on top of the cells in an unopened tube kept at room temperature, the following guidelines should be observed. These are further detailed in Table 4-2.

❑ PT: 24 hours.
❑ aPTT: 4 hours.
❑ Specimens containing unfractionated heparin should be centrifuged within 1 hour, and plasma should be tested within 4 hours.

Table 4-2. Recommended Storage Conditions and Intervals for Coagulation Testing

Assay	Stored as Whole Blood			Processed and Plasma Aliquoted			
	Room temperature	Refrigerated	Frozen	Room temperature	Refrigerated	Frozen -20°C	Frozen -70°C or colder
PT	Up to 24 hr	Unacceptable	Unacceptable	Up to 24 hr	Unacceptable	2 wk	12 mo
aPTT	Up to 4 hr	Unacceptable	Unacceptable	4 hr	4 hr	2 wk	12 mo
aPTT for unfractionated heparin analysis	1 hr	Unacceptable	Unacceptable	4 hr	4 hr	2 wk	Unknown
aPTT for von Willebrand factor and factor VIII analysis	4 hr	Unacceptable	Unacceptable	4 hr	4 hr	2 wk	6 mo
Other	4 hr	Unacceptable	Unacceptable	4 hr	4 hr	Depends on analyte	

Previously frozen samples must be thoroughly mixed before testing. Adapted from CLSI document H21-A5, *Collection,* *Transport, and Processing of Blood Specimens for Testing Plasma-Based Coagulation Assays and Molecular Hemostasis Assays.*

❏ Other coagulation assays, such as thrombin time, protein C, factor V, and factor VIII: 4 hours. For samples to be tested for factor VIII (FVIII) and/or von Willebrand factor (vWF), storage at room temperature is recommended, as storage at 2 to 4°C may lead to precipitation of these large proteins.

❏ Following venipuncture, platelets continue to activate in vitro. Therefore platelet aggregation studies should be completed within 4 hours from the time of phlebotomy. (See chapter 7 for more complete information regarding processing samples for platelet function testing.)

If a laboratory has established an allowable time interval different than that detailed above, data verifying that coagulation testing is valid for the time interval should be generated and saved.

If the plasma sample cannot be tested within 4 hours of phlebotomy, it should be frozen and stored for future testing. Both PT and aPTT results are reliable when the platelet-poor plasma is stored at -20°C for up to 2 weeks for short-term storage, or -70°C for up to 6 months. Prior to freezing, plasma should be platelet poor, because freeze-thawing will rupture platelets and affect the aPTT results, especially in samples containing heparin or those with a lupus anticoagulant. A frost-free freezer should not be used because of warming cycles that induce repeated freeze/thaw cycles.

Samples previously frozen should be thawed quickly in a 37°C waterbath prior to testing. Typically, the samples should be placed in the waterbath for no more than 5 minutes and then removed and kept at room temperature. Defrosting frozen plasma at cold temperatures is essentially how cryoprecipitate is prepared. If the sample is defrosted at cold temperatures, the high-molecular-weight proteins, such as factor VIII, von Willebrand factor, and fibrinogen, will precipitate and the clotting times may be prolonged due to decreased levels of these proteins. Thus, thawing at 37°C rather than room temperature or 4°C is important to yield a plasma sample with all coagulation proteins in suspension.

Coagulation specimens should be checked for clots of any size before reporting results. Clotted specimens will have extremely low levels of fibrinogen and variably decreased levels of other coagulation proteins, so that results of the PT, aPTT, fibrinogen, and other coagulation assays will be inaccurate or unobtainable. Checking for clots may be done with applicator sticks or by visual inspection of centrifuged plasma for small clots. This may also be performed by analysis of results by delta checks or waveform analysis. When a clot is not detected during PT and aPTT testing, and where the fibrinogen level is <25 mg/dL, the sample should be suspected of being serum. This may be important when the coagulation specimens are received as centrifuged

Table 4-3. Common Collection and Handling Problems that Affect Coagulation Test Results

Short draw	Prolonged PT and PTT	Anticoagulant to blood ratio exceeds 1:9
Failure to mix specimen gently	Prolonged PT and PTT	Even small blood clots will affect the assay
Excessive mixing	Shortened PT and PTT	Hemolysis and platelet activation affect the assay
Hemolysis caused by slow collection	Shortened PT and PTT	Hemolysis and platelet activation affect the assay
Improper storage; wrong temperature	Prolonged PT and PTT	—
Undue chilling or placing the sample on ice	Shortened PT and PTT	Activation of factor VII on chilling
Prolonged tourniquet application	False elevation of factor VIII and von Willebrand factor (vWF)	Due to venous stasis, which elevates the concentration of large molecules
Inadequate centrifugation	Inaccurate factor assay results; false-negative lupus anticoagulant results	Plasma is not platelet poor; platelets release phospholipids, which neutralize lupus anticoagulants

frozen "plasma." Centrifuged plasma and serum cannot be distinguished by visual inspection. There should be a mechanism in place to identify these specimens appropriately and/or reject the sample as a probable serum sample.

It is often necessary to ship coagulation samples to another laboratory for testing. Unless a rapid courier system is in place that can ensure the testing be completed within four hours, the samples should be prepared and frozen, as outlined above. Additional considerations for shipping plasma samples are as follows:

❑ Plasma should be transferred into a plastic tube using a plastic pipette. Do not use glass tubes or glass pipettes as glass can activate the clotting process.

❑ If freezing of the sample is necessary, freeze the sample immediately. Quick freezing of the samples is ideal. This can be achieved by freezing with liquid nitrogen, freezing in a mixture of dry ice and methanol, or freezing in a -70°C freezer.

❑ Each assay requested should be submitted in a separate plastic vial and marked clearly as "Citrate Plasma." In order to facilitate additional testing if necessary, it is advisable to submit an additional vial of plasma.

❑ Transport the frozen tubes in a container with dry ice. The samples should remain frozen at all times during the shipping process.

❑ It is recommended to freeze the specimen in plastic by laying the tube in the freezer at a 45-degree angle to avoid tube breakage caused by expansion during freezing.

❑ Samples that thaw during shipping are not acceptable for analysis.

❑ The specimen should be free of any significant lipemia or hemolysis.

Sample collection and processing on ice is required for homocysteine levels. The sample is immediately put on crushed ice and then centrifuged as soon as possible to prevent an increase in homocysteine concentration due to continued production of homocysteine in the sample by the ongoing metabolism in blood cells, mostly in the red cells.

Factors Affecting Specimen Quality

Coagulation tests are highly susceptible to error introduced by suboptimal specimen quality. This is due to the fact that the act of obtaining a blood sample initiates the hemostatic response. The complexity of biochemical and cellular reactions even in routine assays such as PT and aPTT, the lability of coagulation proteins, calcium dependency of many of the reactions, and the excitable state of platelets are all factors that make the coagulation assays dependent on excellent sample quality. The advances in the instrumentation and reagent systems in coagulation testing have considerably reduced the analytical variability. However, the test results continue to depend more and more on preanalytical factors, such as proper sample collection, storage, and processing. Hemolysis caused by improper mixing or collection techniques, microclots, improper storage, interfering substances, etc, can have a significant impact on test results (Table 4-3).

An essential component of quality assurance in any laboratory is the establishment of criteria for rejection of specimens of poor or suboptimal quality. Rejection criteria also give guidance to those who submit the blood specimens to improve collection and transportation of the samples. Sample acceptability for one type of assay may not be appropriate for another. Therefore, all samples must be handled as if further testing was to be performed so that the specimen is not compromised. The list shown in Table 4-4 is a general guideline for specimen rejection. In addition, any other institutional requirements should also be included.

In summary, the preanalytical phase of coagulation testing is prone to errors that could compromise the outcome of coagulation tests and proper patient care. Preanalytical variables include how the phlebotomy is performed, anticoagulant concentration and proper ratio of blood to anticoagulant in the specimen tube, hematocrit, specimen handling, storage, and centrifugation to obtain

platelet-poor plasma. Eliminating or minimizing the preanalytical variables is an important initial step in achieving a good outcome of the overall coagulation testing.

Suggested Reading

General Reviews and Guidelines

Collection, Transport and Processing of Blood Specimens for Testing Plasma-Based Coagulation Assays and Molecular Hemostasis Assays. Approved Guideline. 5th ed. Wayne, Pa: CLSI; 2008: Document H21-A5.

Kiechle FL, ed. *So You're Going to Collect a Blood Specimen: An Introduction to Phlebotomy.* 12th ed. Northfield, Ill: College of American Pathologists; 2007.

Lippi G, Franchini M, Montagnana M, Salvagno GL, Poli G, Guidi GC. Quality and reliability of routine coagulation testing: can we trust that sample? *Blood Coagul Fibrinolysis.* 2006;17(7):513-519.

Procedures for the Collection of Diagnostic Blood Specimens by Venipuncture. Approved Standard. 6th ed. Wayne, Pa: CLSI; 2007: Document H3-A6.

Sample Collection Technique for Peripheral Venipuncture

Adcock DM, Kressin DC, Marlar RA. Are discard tubes necessary in coagulation studies? *Lab Med.* 1997;28:530-533.

Bamberg R, Cottle J, Williams J. Effect of drawing a discard tube on PT and APTT results in healthy adults. *Clin Lab Sci.* 2003;16:16-19.

Brigden ML, Graydon C, McLeod B, Lesperance M. Prothrombin time determination: the lack of need for a discard tube and 24-hour stability. *Am J Clin Pathol.* 1997;108(4):422-426.

Felding P, Tryding N, Hyltoft Petersen P, Holder M. Effects of posture on concentrations of blood constituents in healthy adults: practical application of blood specimens collection procedure recommended by the Scandinavian Committee on Reference values. *Scand J Clin Lab Invest.* 1980;40:615-621.

Gottfried EL, Adachi MM. Prothrombin time and activated partial thromboplastin time can be performed on the first tube. *Am J Clin Pathol.* 1997;107:681-683.

Iverson LH. Pre-analytical variation in the measurement of sensitive markers of coagulation and fibrinolysis: the influence of venipuncture and mixing of blood. *Haemostasis.* 1997;27:119-124.

McGlasson DL, More L, Best HA, Norris WL, Doe RH, Ray H. Drawing specimens for coagulation testing: is a second tube necessary? *Clin Lab Sci.* 1999;12(3):137-139.

Table 4-4. Criteria for Rejection of Coagulation Testing Specimens

- ❑ Specimen not labeled properly.
- ❑ Specimen missing patient identification.
- ❑ Sample spilled or the tube broken.
- ❑ Illegible test request form.
- ❑ Hemolyzed or lipemic sample. Lipemic samples, samples from patients receiving intravenous fat emulsions and those with low fibrinogen concentrations can give inaccurate results.
- ❑ A clot in the sample.
- ❑ Inadequately filled tube.
- ❑ Specimens received more than 4 hours after collection (for testing other than PT only).
- ❑ Hematocrit over 55% and not adjusted for a high hematocrit.
- ❑ Frozen whole blood sample.
- ❑ Insufficient volume (quantity not sufficient [QNS]).

McPhedran P, Clyne LP, Ortoli NA, Gagnon PG, Sanders FJ. Prolongation of the activated partial thromboplastin time associated with poor venipuncture technique. *Am J Clin Pathol.* 1974;62:16-20.

Yawn BP, Loge C, Dale J. Prothrombin time: one tube or two. *Am J Clin Pathol.* 1996; 1055:794-797.

Blood Collection From Indwelling Lines and Catheters

Baer D. Guidelines for phlebotomy in patients receiving IV fluids in both arms. *MLO Med Lab Obs.* "Tips from the clinical experts." November 1997.

Konopad E, Grace M, Johnson R, Noseworthy T, Shustack A. Comparison of PT and aPTT values drawn by venipuncture and arterial line using three discard volumes. *Am J Crit Care.* 1992;1(3):94-101.

Laxson CJ, Titler MG. Drawing coagulation studies from arterial lines: an integrative literature review. *Am J Crit Care.* 1994;3(1):16-22.

Watson KR, O'Kell RT, Joyce JT. Data regarding blood drawing sites in patients receiving intravenous fluid. *Am J Clin Pathol.* 1983;79:119-121.

Zlotowski SJ, Kupas DF, Wood GC. Comparison of laboratory values obtained by means of routine venipuncture versus peripheral intravenous catheter after a normal saline solution bolus. *Ann Emerg Med.* 2001;38:497-504.

Blood Collection Tubes and Anticoagulant Considerations

Adcock DM, Kressin DC, Marlar RA. Effect of 3.2% vs 3.8% sodium citrate concentration on routine coagulation testing. *Am J Clin Pathol.* 1997;107(1):105-110.

Adcock DM, Kressin DC, Marlar RA. Minimum specimen volume requirements for routine coagulation testing: dependence on citrate concentration. *Am J Clin Pathol.* 1998;109:595-599.

Biron-Adreani C, Mallol C, Seguret F, et al. Plastic versus siliconized glass tubes: evaluation in current laboratory practice. *Thromb Haemost.* 2000;83:800-801.

Contant G, Gouault-Heilmann M, Martinoli JL. Heparin inactivation during blood storage: its prevention by blood collection in citric acid, theophylline, adenosine, dipyridamole-C.T.A.D mixture. *Thromb Res.* 1983;31:365-374.

Hunt BJ, Parratt R, Cable M, et al. Activation of coagulation and platelets is affected by hydrophobicity of artificial surfaces. *Blood Coagul Fibrinolysis.* 1997; 8:223-231.

Kratz A, Stanganelli N, Van Cott EM. A comparison of glass and plastic blood collection tubes for routine and specialized coagulation assays: a comprehensive study. *Arch Pathol Lab Med.* 2006;130(1):39-44.

Marlar RA, Potts RM, Marlar AA. Effect on routine and special coagulation testing values of citrate anticoagulant adjustment in patients with high hematocrit values. *Am J Clin Pathol.* 2006;126:400-405.

Peterson P, Gottfried EL. The effects of inaccurate blood sample volume on prothrombin time (PT) and activated partial thromboplastin time (aPTT). *Thromb Haemost.* 1982;47(2):101-103.

Polack B, Barro C, Mossuz P, Pernod G. Inadequate quality of a blood collection tube containing an anticoagulant platelet inhibitor mixture. *Thromb Haemost.* 1997;77:1035-1036.

Rahr HB, Sorensen JV, Danielsen D. Markers of coagulation and fibrinolysis in blood drawn into citrate with and without D-Phe-Pro-Arg-Chloromethylketone (PPACK). *Thromb Res.* 1994;73(5):279-284.

Reneke J, Etzell J, Leslie S, Ng VL, Gottfried EL. Prolonged prothrombin time and activated partial thromboplastin time due to underfilled specimen tubes with 109 mmol/L (3.2%) citrate. *Am J Clin Pathol.* 1998;109(6):754-757.

Siegel JE, Bernard DW, Swami VK, Sazama K. Monitoring heparin therapy: APTT results from partial- vs full-draw tubes. *Am J Clin Pathol.* 1998; 110(2):184-187.

Sample Processing

Barnes PW, Eby CS, Lukoszyk M. Residual platelet counts in plasma prepared for routine coagulation testing with Beckman Coulter power processor. *Lab Hematol.* 2002;8:205-209.

Carroll WE, Wollitzer Ao, Harris L, et al. The significance of platelet counts in coagulation studies. *J Med.* 2001;32:83-96.

Lippi G, Salvagno GL, Montagnan M, Manzato F, Guidi GC. Influence of the centrifuge time of primary plasma tubes on routine coagulation testing. *Blood Coagul Fibrinolysis.* 2007;18(5):525-528.

Plumhoff EA, Thompson CK, Fisher PK, Bowie EJW, Nichols WL. Effects of specimen storage and handling on coagulation testing [abstract]. *Thromb Haemost.* 1993; 69:866.

Woodhams B, Girardot O, Blanco M-J, Colosse G, Gourmeln Y. Stability of coagulation proteins in frozen plasma. *Blood Coagul Fibrinolysis.* 2001; 12:229-236.

Factors Affecting Specimen Quality

Lawrence JB. Preanalytical variables in the coagulation laboratory. *Lab Med.* 2003;34:49-57.

Narayanan S. Preanalytical aspects of coagulation testing. *Haematologica.* 1995;80(suppl 2):1-6.

Siegel JE, Swami VK, Glenn P, Peterson P. Effect (or lack of it) of severe anemia on PT and APTT results. *Am J Clin Pathol.* 1998;110(1):106-110.

Clinical History in Hemostasis

Jayashree Krishnan, MD

Introduction

Clinically useful interpretation of coagulation tests can be derived best in the context of appropriate history. The information gathered by clinical history and physical examination ultimately guides the appropriate testing and determines how potential hemostatic complications can be prevented or managed. One may choose to formulate a questionnaire that would require either simple "yes" or "no" answers, or may require a more detailed answer format to get adequate information. An example questionnaire is shown in Figure 5-1. Simple questionnaires can be used as a guide for uniformity of history taking. However, to be most valuable, a history should not be a simple list of facts as spontaneously described by the patient or his/her relatives. It should be the result of a careful medical interview conducted by a physician posing critical questions to the patient while appreciating the perception of the symptoms by the patient.

The basic information should include whether the patient has a problem of hemorrhage or thrombosis, history of medications, age and gender of the patient, family history of bleeding or thrombosis, results of previous coagulation tests, and information about any coexisting diseases. The clinical history should be obtained, with the goal of determining whether the patient may have a defect in primary hemostasis, secondary hemostasis, or both. The questions to be asked of patients will vary depending upon whether the patient has a history of bleeding or thrombosis.

Clinical History in Bleeding Patients

The occurrence, frequency, severity, and other inherent characteristics of bleeding episodes in a patient should be completely investigated. The presence and absence of symptoms should be recorded. It is, however, impossible to report in detail every episode of bleeding, as some may be trivial. Therefore, in order to accurately record the overall picture, symptoms that are clinically significant and episodes that are of trivial importance should be carefully separated and recorded.

Determine Whether the Patient Has Excess Bleeding

An important first step is to determine whether or not the patient truly has a history of excessive bleeding. Surrogate markers of bleeding, such as declining hemoglobin and hematocrit, hypotension, hemodilution, and hemolysis, may be helpful but are all subject to some degree of misinterpretation and may be the result of other etiologies. Certain elements in the history, such as extent, type, pattern, and duration of bleeding, are extremely useful in determining whether the bleeding is caused by an underlying hemostatic disorder or by a local anatomic defect.

Some bleeding symptoms are common in the general population, and if the symptoms are mild, patients may not pay significant attention to them. Eliciting a bleeding history thus may be difficult in mild bleeding disorders. Mild bleeding disorders may not be apparent until the individual is exposed to significant challenges such as surgery.

Extent of Bleeding

Assessment of severity and site of bleeding, determination of duration of bleeding, and evaluation of the clinical presentation of the patient are the next important steps. Quantification of excessive bleeding is difficult and is often assessed through inexact methods. One has to determine whether bleeding that has occurred is inappropriate for a normal individual or to that particular clinical situation.

Bleeding that is sufficiently severe to require blood transfusion merits special attention. A positive history of bleeding during and after surgical or dental procedures, a return to the operating room, blood transfusion, or a need for packing or suture

CLINICAL HISTORY QUESTIONNAIRE

DATE: NAME:

DATE OF BIRTH: SEX:

MARITAL STATUS: Single / Married / Divorced   HEIGHT: WEIGHT:

ETHNICITY: White / African American / Asian / American Indian / Other

OCCUPATION:

REASON FOR THIS DOCTOR'S VISIT:

This questionnaire is designed to assist your doctor in evaluating your blood clotting mechanism.

1.	Do you think you have a bleeding disorder?	Yes	No
2.	If so, how long have you had this problem?		
3.	Have you been previously diagnosed as having a bleeding disorder?	Yes	No
4.	If so, please state when and where it was diagnosed.		
5.	Do you bruise easily?	Yes	No
6.	Have you bled or bruised spontaneously without trauma or with minimal trauma?	Yes	No
7.	Have you ever had nose bleeds?	Yes	No
	If yes, how often?		
	Do you remember having nose bleeds only in certain months of the year?	Yes	No
	If yes, when?		
8.	Did you see a doctor for your nose bleeds?	Yes	No
9.	Can you explain how the nose bleeds stopped?	Yes	No
10.	Have you had dental extractions?	Yes	No
	If yes, when?		
11.	Did you notice unusual or prolonged bleeding during or after a dental procedure?	Yes	No
12.	Have you ever had bleeding from your mouth, stomach, or bowels?	Yes	No
	If yes, please circle one or more of the following:		
	Vomiting blood / Red blood in stools / Black tarry stools		
13.	Have you ever noticed blood in your urine?	Yes	No
14.	Have you ever bled into a joint?	Yes	No
15.	Have you ever bled into a muscle or abdomen?	Yes	No
16.	Have you ever had red skin spots (petechiae)?	Yes	No
	If yes, please explain the site(s) of petechiae and duration.		
17.	Have you ever had a stroke or bleeding in the head?	Yes	No
18.	Have you ever had surgery?	Yes	No
	If yes, please list the type(s) and date(s).		
19.	Did you have abnormal bleeding during surgery?	Yes	No
20.	Have you ever had a blood transfusion?	Yes	No
	(Please answer yes if you received whole blood, packed red cells, plasma, platelets, cryoprecipitate, or coagulation factor concentrate.)		
	If yes, list the number of units and how often.		
21.	Did the doctor tell you that fibrin glue was applied during surgery?	Yes	No
22.	Are you taking any blood thinners?	Yes	No
	If yes, list the medication(s) and dose(s).		
23.	Do you take any medication not prescribed by the doctor?	Yes	No
	If yes, list the medication(s) and dose(s).		
24.	Do you take cold medication, aspirin, sinus medication, or sedatives?	Yes	No
	If yes, list the medication(s).		
25.	Do you take any special diet products, muscle builders, or soy products?	Yes	No
	If yes, list product(s).		

26.	Do you drink alcohol?	Yes	No
	If yes, approximately how much and how often.		
27.	Question for men only: Have you been circumcised?	Yes	No
	If yes, did anyone tell you that you bled a lot when circumcised?	Yes	No
28.	Questions for women only:		
	a. Duration of periods:		
	b. Do you have heavy menstrual periods?	Yes	No
	c. Number of pregnancies: miscarriages:		
	d. Any history abnormal bleeding during delivery?	Yes	No
	e. Are you taking any estrogen hormone or birth control pills?	Yes	No
	f. Number of children and their ages:		
	g. Have your children had any miscarriages?	Yes	No
29.	Do you have any history of cancer/chemotherapy?	Yes	No
	If yes, please explain.		
30.	Have you ever had a bone marrow procedure?	Yes	No
	If yes, please explain why you had the procedure.		
31.	Does anyone in your family have a bleeding disorder?	Yes	No
	If yes, indicate the disease and the family member's relationship to you.		
	Hemophilia / von Willebrand disease / Platelet disorders / Other (specify)		

Figure 5-1. Sample clinical history questionnaire for diagnosis of bleeding disorders. This questionnaire has been constructed with many Yes or No questions for ease of completion. Devising a questionnaire with more open-ended questions is also acceptable.

of dental sockets would increase the likelihood of an underlying bleeding disorder. A history of excessive bleeding after dental extractions, circumcision, tonsillectomy, other previous surgical operations, and childbirth should always be sought, as should a history of unexplained anemia, gastrointestinal bleeding without the demonstration of a cause, and previous blood transfusion.

Duration of Bleeding Abnormality

Some bleeding disorders are present at birth and are caused by rare inherited disorders; others are acquired during certain illnesses, such as vitamin K deficiency, with hepatic disease, or during treatment with anticoagulant drugs. Patients with inherited disorders usually present with symptoms in infancy or childhood and frequently experience postoperative or posttraumatic hemorrhage. Patients with acquired disorders may present as adults and have negative family history, and often have evidence of an associated underlying disorder. Some patients referred for an evaluation of mild bleeding problems may have an underlying acquired condition that may not have been recognized as the cause of bleeding. These acquired causes may be diverse and may or may not be obvious at presentation. They include autoimmune thrombocytopenia, acquired bone marrow disorders, hemorrhagic side effects of

antiplatelet or anticoagulant drugs, acquired inhibitors of factor VIII or von Willebrand factor, thyroid disease, Cushing syndrome, liver disease, and renal disease. Assessment of bleeding duration and age at presentation provides only a general guideline and does not always distinguish between a mild inherited disorder and a recent acquired disorder.

Type of Bleeding

A thorough bleeding history should determine the types and sites of bleeding. The most common sites where bleeding is observed are the skin and mucus membranes. The pattern of bleeding may be helpful in identifying the underlying cause. Spontaneous bleeding without an identifiable cause or trauma occurring from the mucous membranes is commonly associated with severe thrombocytopenia or other platelet disorders. Spontaneous bleeds into the joints or muscle are more characteristic of certain severe coagulation factor deficiencies. If the bleeding is multifocal, an acquired etiology, such as disseminated intravascular coagulation (DIC), should be suspected.

Collections of blood in the skin are called purpura. Small pinpoint hemorrhages due to leakage of red cells through capillaries are called petechiae. Petechiae are characteristic of platelet disorders/severe thrombocytopenia. Larger subcutaneous collections of blood due to leakage of blood from small vessels are called ecchymoses, which are common bruises. Deeper and palpable bruises form hematomas. Ecchymoses and hematomas are common in patients with platelet defects, but also they can result from minor trauma. Any spontaneous bruise or hematoma larger than 3 cm or considered disproportionate to the type of trauma is of importance. Large firm bruises with nodular centers are commonly seen in congenital factor deficiencies. A hemarthrosis results from bleeding into a joint and is characteristically observed with severe factor deficiencies.

Bleeding Due to Vascular Disorders

Clinical bleeding may be due to vascular disorders. Vascular disorders may be hereditary or acquired, and are characterized by abnormalities in the vascular blood interface, which impair platelet adhesion. Examples of hereditary disorders include hereditary hemorrhagic telangiectasia, Ehlers-Danlos syndrome, Marfan syndrome, osteogenesis imperfecta, pseudoxanthoma elasticum, and homocystinuria. Dilated capillaries or telangiectasia may cause bleeding without any hemostatic defect. The loss of connective tissue support for capillaries and other small blood vessels can result in increased fragility of the superficial blood vessels in the elderly and cause bleeding from the superficial blood vessels (senile purpura). Acquired vascular disorders result from insults to the vascular blood interface. Examples of these include malignant paraprotein disorders, amyloidosis, lupus, scleroderma, rheumatoid arthritis, and Cushing syndrome.

Posttraumatic and Spontaneous Bleeding

A self-reported history of excessive bleeding after trauma may be difficult to evaluate since most individuals tend to believe they suffered excessive bleeding during trauma. The bleeding manifestations of patients with vascular disorders, thrombocytopenia, or functional platelet disorders are manifested by bleeding that begins at the time of injury and continues for hours, but once controlled does not recur. Bleeding that is truly spontaneous suggests a bleeding disorder, although it is still possible that spontaneous bleeding may have an occult cause, such as self-inflicted injuries or factitious purpura.

Hemarthrosis and Hematomas

Bleeding into body cavities, joints, retroperitoneum, etc, is a common manifestation of plasma coagulation defects. Repeated bleeding into the joints may cause significant inflammation and erosion of osteochondrous tissue and result in joint deformities. Joint deformities are particularly common in patients with deficiencies of factors VIII and IX, the two sex-linked coagulation disorders (hemophilias). Hemarthroses are much less common in patients with other plasma coagulation defects, the reason for which is not known. Retroperitoneal hematomas can cause nerve compression and necrosis of soft tissue, thus mimicking a tumor.

Epistaxis

Any nosebleeds that could not be managed by the patient or family, especially those occurring in the postpubertal age or requiring medical attention, are important to record. Very frequent epistaxis (at least once a week) should be recorded and should raise the suspicion of a bleeding disorder. Epistaxis is rare in infancy and infrequent after puberty. Epistaxis may occur much more frequently in the late fall and winter months, often due to the drying of the nasal mucous membranes from climate changes or from dry indoor air due to central heating. Repeated nose bleeds lasting more than 10 minutes despite compression suggest a local cause or an underlying bleeding disorder.

Thrombocytopenia is the most common cause of nosebleeds. The most common inherited bleeding disorder associated with nosebleeds is von Willebrand disease, although other inherited bleeding disorders that may lead to nosebleeds include hemophilia, Glanzmann thrombasthenia, hepatic disease, severe vitamin K deficiency or malabsorption, vitamin C deficiency, and drugs such as aspirin and nonsteroidal anti-inflammatory drugs (NSAIDs), warfarin, and heparin.

Surgical Procedures

The initial assessment of postoperative bleeding should distinguish between incomplete surgical ligation and the presence of an underlying hemostatic defect. It is appropriate to obtain a description of all wounds and venipuncture sites and estimated blood loss.

The timing of the hemorrhage in relation to the procedure, ie, immediate versus delayed, can provide important information. Intraoperative and immediate postoperative bleeding at the surgical site is often due to defects in primary hemostasis, such as platelet dysfunction. Spontaneous or continuous oozing of the surgical field could be related to a defect in primary hemostasis (platelets, vessel wall, von Willebrand factor). On the other hand, delayed postoperative bleeding at the surgical site is more likely due to coagulation factor deficiencies, qualitative or quantitative disorders of fibrinogen, fibrinolytic abnormalities, vascular abnormalities, or defects due to collagen structure. Bleeding from the umbilical cord at birth or bleeding after circumcision strongly suggests an inherited disorder, while tolerance of tonsillectomy, abdominal procedures, and dental extractions of adult teeth make it far less likely that the patient suffers from an inherited bleeding disorder. Although bleeding is usual from any surgical procedure, some mild bleeding disorders, such as mild von Willebrand disease, factor XI deficiency, and mild Ehlers-Danlos syndrome, may be diagnosed later in life for the first time due to mucosal bleeding or bleeding after a routine surgical procedure.

Poor wound healing is suggestive of factor XIII deficiency, fibrinogen deficiency, and other collagen disorders. With fibrinolytic disorders or with deficiency of factor XIII, a clot will develop but will remain unstable. The clot will eventually break down and cause recurrent bleeding. This process will result in prolonged bleeding after trauma, delayed wound healing, delayed bleeding after surgery, spontaneous abortions in women, infertility in men, soft tissue bleeds, menorrhagia, and persistent bleeding after circumcision.

Bovine topical thrombin is used as a surgical hemostatic agent. Some patients with prior surgical exposure to bovine topical thrombin may develop bovine thrombin antibodies and also factor V inhibitors. Therefore, a clinical history of previous surgery and the use of bovine topical thrombin should be sought.

Gingival/Dental Bleeding History

Any spontaneous gum bleeding for a minute or longer, frankly bloody sputum, or persistent or profuse bleeding after brushing the teeth are of significance, assuming that it is not secondary to tooth/periodontal disease. Any bleeding lasting more than 5 minutes due to bites to the lip, cheek, or tongue is of significance. Any bleeding (after tooth extraction) that lasts after leaving the dentist's office, or a prolonged bleeding at the dentist's office causing a delay in the procedure, should be recorded. Questions about excessive bleeding, and its timing, with surgery, dental extractions, and trauma are important. Most often patients can recall the details of bleeding during procedures such as dental work, which are done without general anesthesia.

Menstrual History

Menorrhagia is a sensitive indicator of a bleeding disorder but is associated with a high false-positive rate. The perception of menstrual loss is variable and also influenced by family norms. Nonetheless, menstrual histories often provide useful information in diagnosing bleeding disorders, especially in women with persistent menorrhagia. Different bleeding scales are used to assess the extent of menstrual blood loss, such as the frequency of pad/tampon change, duration of bleeding (longer than 7 days), or duration of heavy flow lasting more than 3 days.

Persistent menorrhagia sufficient to cause iron-deficiency anemia may indicate a bleeding disorder if no structural uterine abnormality is present. The clinical history of severe iron-deficiency anemia in a young woman, the use of packed red cell transfusions for an anemia of undetermined etiology, the need for dilatation and curettage procedure for persistent bleeding in a young woman, or the need for oral contraceptives or hysterectomy to treat menorrhagia should raise the suspicion of an underlying bleeding disorder. Oral contraceptive therapy may lead to increased levels for factor VIII and von Willebrand factor, so diagnosis of von Willebrand disease may be difficult in women on oral contraceptive therapy.

Obstetric History

Menorrhagia commencing after pregnancies and child birth and uninfluenced by hormone supplements is unlikely to be due to an inherited hemostatic disorder. Multiparous women should be carefully questioned about each pregnancy.

Multiple spontaneous miscarriages and infertility can be associated with maternal coagulopathies, such as dysfibrinogenemia and antiphospholipid antibody syndromes. Patients with mild to moderate von Willebrand disease do not usually experience easy bruising or other bleeding manifestations during pregnancy and childbirth, or when they are taking oral contraceptives, because of the increased synthesis of von Willebrand factor and factor VIII as an acute phase reactant. Some women may develop postpartum bleeding due to acquired autoantibodies directed against factor VIII.

Bleeding Due to Organ System Disorders

A number of organ system disorders can give rise to acquired hemostatic defects, so a complete medical history, inquiring about the presence of general medical disorders, may help to elucidate the cause of bleeding. Most proteins of the coagulation cascade and their regulators and inhibitors are synthesized in the liver. Acquired abnormalities can be due to impaired synthesis, increased consumption, or rarely the formation of autoantibodies against coagulation proteins. Bleeding associated with organ system disorders can be due to single coagulation factor disorders or due to an abnormality of multiple coagulation proteins.

Deficiency of single or multiple coagulation factors can occur in various disease states or due to therapy. Isolated factor II deficiency can occur in association with lupus anticoagulants. Isolated factor V deficiency can occur in association with myeloproliferative disorders or myelodysplastic syndromes, probably reflecting increased absorption by blood and bone marrow cells. Acquired von Willebrand disease can occur in association with a variety of disorders: monoclonal gammopathies, lymphoproliferative disorders, myeloproliferative disorders, myelodysplastic syndromes, Wilms tumor, autoimmune disorders, hypothyroidism, and severe aortic stenosis. Isolated factor X deficiency can occur in systemic amyloidosis, especially when there is perivascular amyloid deposition in liver and spleen, which may absorb factor X. Amyloid-induced gastrointestinal malabsorption can cause vitamin K deficiency and bleeding. Gaucher disease can result in factor IX deficiency.

Acquired causes of multiple factor deficiencies include leukemia therapy with L-asparaginase, T-cell lymphoproliferative disorders, liver disease, vitamin K deficiency, and DIC. Multifactorial acquired bleeding disorders are most commonly due to chronic liver disease, which can result in a variety of hemostatic abnormalities ranging from deficiency of vitamin K-dependent coagulation factors, factor deficiencies, or low-grade DIC to thrombocytopenia. Decreased hepatic synthesis of coagulation factors may affect all factors except factor VIII, which is not synthesized in liver. Factor VII has the shortest half-life and therefore is often the earliest and most severely decreased factor. Factors II, XI and XII have the longest half-lives and are therefore the last to be affected. End-stage hepatic disease can lead to defects in both primary and secondary hemostasis.

Malabsorption of vitamin K from the gut or dietary deficiency can cause a coagulation disorder similar to that caused by ingestion of warfarin.

When a vitamin K deficiency is suspected, a thorough history of diet, gastrointestinal disorders, and herbal medicine should be taken. Green leafy vegetables contain high amounts of vitamin K; a diet lacking in these foods might suggest vitamin K deficiency.

Patients with renal failure/uremia may have a significant qualitative platelet dysfunction. Patients with myeloproliferative disorders may have significant abnormalities of platelet function. Patients with lymphoproliferative disorders may develop autoantibodies to coagulation factors or platelets. Paraproteins in patients with plasma cell dyscrasia may interfere with platelet function or fibrin formation. Circulating heparin-like anticoagulants of endogenous origin can be seen occasionally in patients with certain malignancies such as multiple myeloma, dysproteinemias, or urothelial malignancies. These patients can have severe life-threatening bleeding. Patients with lupus anticoagulant may develop autoantibodies to prothrombin, causing a decrease in factor II.

Disseminated intravascular coagulation affects the coagulation system at multiple levels. DIC occurs with activation of both coagulation and fibrinolytic systems, resulting in thrombosis and hemorrhage. This may occur secondary to trauma, placental abruption, sepsis, and a variety of other disorders.

Medication and Food Habits

Many drugs, foods, and dietary supplements can affect hemostatic function through interference with coagulation proteins or platelets. A thorough drug history should be part of any hemostasis evaluation. Specific history of food supplements, vitamin supplements, and food habits should be carefully obtained. For example, aspirin and other nonsteroidal anti-inflammatory drugs cause platelet dysfunction. This occurs due to inhibition of synthesis of thromboxane A_2, which is a potent stimulator of platelet activation. Other antiplatelet agents, such as cilostazol, clopidogrel, dipyridamole, ticlopidine, and monoclonal antibody inhibitors that are directed against the platelet glycoprotein IIb/IIIa complex, are also of importance. Cytotoxic drugs cause thrombocytopenia. Some herbal and nonconventional supplements (soy products, cranberry juice, ginseng, etc) can interfere with normal coagulation processes.

Anticoagulant therapy either with oral vitamin K antagonists (warfarin) or heparin is a common cause of acquired bleeding abnormalities. A clini-

Table 5-1. Medications and Herbal Products Associated with Hemostatic Disorders

Product	Effect
Aspirin, aspirin-containing products, nonsteroidal anti-inflammatory drugs (NSAIDs)	Platelet dysfunction by inhibition of thromboxane A_2 synthesis with bleeding
Warfarin	Decreased vitamin K-dependent factors (II, VII, IX, X), protein C, protein S
Direct thrombin inhibitors (lepirudin, bivalirudin, argatroban)	Inhibits thrombin; can interfere with clot-based and lupus anticoagulant tests
Antibiotics: cephalosporins, Ciprofloxacin, Erythromycin, sulfonamides, tetracyclins, Neomycin, Chloramphenicol	Decreased vitamin K gastrointestinal absorption; decreased vitamin K-dependent factors; bleeding
Clopidogrel, ticlopidine	Platelet dysfunction; rarely thrombotic thrombocytopenic purpura (TTP)
Heparin	Can interfere with clot-based tests; increased bleeding; paradoxical thrombosis (heparin-induced thrombocytopenia [HIT])
Low-molecular-weight heparin, fondaparinux	Increased bleeding
Cytotoxic drugs	Thrombocytopenia
Sweet clover	Natural coumarin or coumarin derivative
Brodifacoum (rat poison)	Long-acting "super warfarin"
Geum japonicum	May inhibit coagulation serine proteases; increased bleeding
Green tea, coenzyme Q10, soy milk, ginkgo biloba	Coumadin antagonist
Gold, quinine, quinidine, cocaine	Increased platelet destruction
Danshen, garlic, turmeric, ginger, feverfew, clove, ginseng, ginkgo biloba, capsaicin	May decrease platelet aggregation

cal history of anticoagulant therapy should be obtained prior to coagulation testing. Particular attention should be paid to indwelling lines and catheters, which may be flushed with heparin and spuriously affect clotting results. Occasionally, therapeutic anticoagulants can be used surreptitiously. It may be difficult to obtain clinical history in such instances as well as in anticoagulant malingerers. Rodenticides are potent inhibitors of vitamin K epoxide reductase; they have a very long biologic half-life and can be easily obtained. Poisoning with these "super warfarins" (such as brodifacoum, the active compound in DeCon®) characteristically produces severe, persisting, or recurring coagulation abnormalities. A clinical history should be carefully obtained regarding the usage and type of over-the-counter medication and exposure to pesticides. Drugs and food supplements that can affect hemostasis are listed in Table 5-1.

Family History

Family history is important in the assessment of inherited bleeding disorders and to identify others at risk for similar problems. A family history of bleeding that can not be linked to surgery or trauma suggests a systemic disorder. An inherited hemostatic defect should be documented by interviewing the patient and family members, and by examination of hospital records. Information about the type and amount of bleeding of not only the patient, but the other family members, is critical. Some questions to ask are: Have any members of your family experienced severe bleeding? If so, did any of them require blood transfusions or packed red cells? How was the bleeding treated or managed? Is there a racial or ethnic heritage (eg, factor XI-Eastern European Jewish background)? While obtaining the history of the family members, the privacy of individual family members should be kept in mind.

The pattern of inheritance provides a clue to the nature of the inherited disorder. A sex-linked recessive pattern of inheritance typically affects the male members of the family, and the females are usually carriers. So the maternal uncle, grandfather, and cousins of the male patient may be affected. This type of X-linked inheritance pattern is seen in hemophilia A or B, or a platelet disorder such as Wiskott-Aldrich syndrome. With autosomal recessive disorders, the parents of the affected person are heterozygous and typically have a 50% plasma concentration of the relevant clotting factor. The parents are usually unaffected, but siblings may be affected. This genetic pattern is usually seen in deficiency of factors II, V, VII, X, XI, and XIII, prekallikrein, and high-molecular-weight kininogen.

An autosomal dominant pattern of inheritance is seen in von Willebrand disease, some of the qualitative platelet defects, dysfibrinogenemia, and hereditary hemorrhagic telangiectasia. These disorders are present in 50% of the family members and do not usually skip a generation. However, this may not always be true because of variable penetrance and expression. For example, the different forms of von Willebrand disease can vary in their presentation, and plasma factor VIII levels can vary among the family members. The genetic pattern of inheritance can be obscured in certain disorders. Carriers of hemophilia A and B can have low levels of factor VIII or factor IX and may bleed. This is because of the random inactivation of one X chromosome in some females (lyonization).

Although a positive family history is of great value in evaluating bleeding disorders, a negative family history does not exclude the possibility of an inherited disorder. One may not be able to obtain relevant family history because of various reasons, such as lack of communication between family members, adoption, etc, and the family history may be negative in mild bleeding disorders or in cases of spontaneous mutation, which account for up to 20% of the patients with hemophilia A.

Previous Coagulation Test Results

The results from any previously performed coagulation tests should be obtained along with the reason why the tests were done. Prior test results should be obtained for the family members, if possible. Careful review of laboratory tests to evaluate for worsening anemia, thrombocytopenia, or alterations in prothrombin time (PT), activated partial thromboplastin time (aPTT), and thrombin time (TT) are very helpful in assessing the presence or absence of a hemostatic disorder.

Pediatric Considerations

Bleeding in neonates may present with oozing from the umbilicus or stump, cephalhematoma, bleeding from peripheral venipuncture or procedure sites, bleeding into scalp, petechiae, bleeding after circumcision, intracranial hemorrhage, bleeding from mucus membranes, and unexplained anemia and hypotension. For neonatal patients with bleeding, particular points that should be included in the history are maternal diseases, such as idiopathic thrombocytopenic purpura (ITP), preeclampsia, neonatal alloimmune thrombocytopenia, and diabetes; maternal exposure to drugs, such as aspirin, anticonvulsants, rifampicin, and isoniazid; family history of bleeding disorders; and previous affected siblings. Causes of bleeding in a "well neonate" include immune thrombocytopenia (alloimmune or autoimmune, maternal ITP), vitamin K deficiency, inherited coagulation factor deficiencies, and bleeding from anatomic lesions, such as hemangioma and arteriovenous malformation. Causes of bleeding in a "sick" neonate include DIC (usually associated with sepsis, asphyxia, etc) and liver failure. Unfortunately, the possibility of spurious coagulopathy (Munchausen syndrome by proxy, abuse) should be kept in mind.

Clinical History in Patients with Thrombosis

Thrombophilia is a hereditary or acquired disorder predisposing to either arterial or venous thrombosis. Arterial thrombosis results in myocardial infarction, stroke, and peripheral vascular disease. Venous thrombosis results in deep vein thrombosis and pulmonary embolism, and is due to a combination of stasis and hypercoagulability. Risk factors for venous thrombosis include immobility, surgery/trauma, pregnancy/puerperium, long-distance travel, and use of oral contraceptives. Other acquired conditions associated with hypercoagulable states include trauma, postoperative state, advancing age, estrogen therapy, antiphospholipid syndrome, malignancy, nephrotic syndrome, heparin-induced thrombocytopenia, thrombotic thrombocytopenic purpura (TTP), myeloproliferative disorders, paroxysmal nocturnal hemoglobinuria, hyperlipidemia, diabetes mellitus, hyperviscosity syndromes, and congestive heart failure. Inherited causes of venous thrombosis include antithrombin (AT) deficiency, protein C deficiency, protein S deficiency, factor V Leiden mutation, and the prothrombin G20210A mutation. These are covered in detail in chapters 17 through 19.

In approaching a patient with a suspected thrombophilia, it is useful clinically to distinguish patients with acquired versus hereditary hypercoagulable states. In many patients, more than one risk factor may be responsible. A sample questionnaire for use in collecting clinical history from patients with thrombosis is presented in Figure 5-2.

Age of Onset

Clinical features of familial thrombophilia include family history of venous thromboembolism, first episode at an early age, recurrence of venous thromboembolism, unusual sites of thrombosis (cerebral, mesenteric), spontaneous venous thrombosis without environmental or acquired risk factors, heparin resistance, coumarin necrosis, venous thrombosis in pregnancy/puerperium, and recurrent superficial thrombophlebitis. Some types of inherited thrombophilia, such as deficiencies of antithrombin, protein C, or protein S, are associated with venous thrombosis in people less than 50 years of age, while others, such as factor V Leiden or the prothrombin mutation, can predispose to a first thrombosis in all age groups. Venous thrombosis is infrequent before puberty in patients with inherited thrombophilia. The cumulative lifetime incidence of thrombosis among carriers of the most common familial thrombophilia (factor V Leiden mutation) is only about 10%. Therefore, most venous thromboembolism patients with a familial thrombophilia do not have a family history of thrombosis.

Thrombosis during infancy or childhood is significant and can be associated with severe deficiencies of natural anticoagulants. For example, homozygous protein C deficiency presents in newborns with extensive life-threatening thrombotic complications. The initial features include purpura fulminans, severe DIC, central nervous system thrombosis, blindness, and venous thrombosis. Other risk factors of thrombosis during infancy and childhood are listed in Table 5-2.

A heterozygous deficiency of the natural anticoagulants rarely produces thrombosis in newborns and children except in the presence of another pathologic event, such as trauma, prolonged immobilization, use of oral contraceptives, local infection, or cancer and its treatment. Some chil-

Table 5-2. Factors Predisposing to Prothrombotic States in Children

- ❑ Indwelling vascular catheters
- ❑ Infection
- ❑ Trauma
- ❑ Surgery
- ❑ Dehydration
- ❑ Obesity
- ❑ Immobilization
- ❑ Vascular malformation
- ❑ Malignancy
- ❑ Chemotherapy with L-asparaginase
- ❑ Cardiac disease
- ❑ Systemic lupus erythematosus
- ❑ Crohn's disease
- ❑ Rheumatoid arthritis
- ❑ Ulcerative colitis
- ❑ Polycythemia
- ❑ Sickle cell anemia and other hemoglobinopathies
- ❑ Renal disease
- ❑ Diabetes mellitus
- ❑ Myelomeningocele
- ❑ Appendicitis
- ❑ Inflammatory disorders
- ❑ Prosthetic cardiac valves

dren with protein S deficiency can present with strokes. Secondary thromboembolic complications occur in children due to abnormalities of vessel wall (thrombosis due to intravascular catheters), disturbances of blood flow (an increased viscosity due to polycythemia seen with cyanotic congenital heart disorders), and alterations in the coagulability of blood (secondary to shock or infection).

Ethnicity

The prevalence of hereditary thrombophilia (factor V Leiden or prothrombin G20210A mutations) among patients with thrombophilia differs among ethnic groups. In the Western European Caucasian population, the carrier frequency of factor V Leiden is as high as 5%. However, in African, American Indian, and Asian populations, the carrier frequency of these mutations is low.

THROMBOSIS HISTORY QUESTIONNAIRE

DATE: NAME:

DATE OF BIRTH: SEX:

MARITAL STATUS: Single / Married / Divorced   HEIGHT: WEIGHT:

ETHNICITY: White / African American / Asian / American Indian / Other

OCCUPATION:

REASON FOR THIS DOCTOR'S VISIT:

This questionnaire is designed to assist your doctor in evaluating your blood clotting mechanism.

1.   Do you think you have ever had deep vein thrombosis?   Yes   No
     If yes, at what age?

2.   Have you ever been admitted to the hospital for pulmonary embolism?   Yes   No
     If yes, at what age?

3.   Do you remember that a physician ever diagnosed you as having:
     i.   Thrombosis or thrombophlebitis   Yes   No
          If yes, where: veins of the legs or veins of other sites?
     ii. Pulmonary embolism   Yes   No
     iii. Postphlebitic syndrome   Yes   No

4.   Are you currently receiving or have you ever received blood thinners?   Yes   No
     If yes, list the name(s) and dose(s).

5.   Do you remember whether your parents or any other family members were ever
     diagnosed as having a clotting disorder or thrombosis?   Yes   No
     If yes, list the family member(s) and age(s) at which they were diagnosed.

6.   Do you remember whether your parents or any other family members were ever
     admitted to the hospital for pulmonary embolism?   Yes   No
     If yes, list the family member(s) and age(s) at which they were diagnosed.

7.   Have you or anyone in your family been diagnosed as having lupus, rheumatoid arthritis,
     liver disease, or kidney disease?   Yes   No
     If yes, describe briefly.

8.   Do you have high blood pressure or diabetes?   Yes   No

9.   Are you on oral contraceptives, hormonal therapy, estrogens, or corticosteroids?
     If yes, list type(s) and length(s) of time you have been on the therapy.   Yes   No

10. Did you travel recently?   Yes   No
     If yes, describe briefly.

11. Do you smoke or have you ever smoked?   Yes   No
     If yes, how long and how much?
     If you quit, when did you quit?

12. Do you drink alcohol?   Yes   No
     If yes, approximately how much and how often.

13. Has anyone told you that you have a hemoglobin disorder such as sickle cell disease?   Yes   No

14. Have you had any surgery (major or minor)?   Yes   No
     If yes, please list the type(s) of surgery and date(s).

15. Have you ever had a stroke or heart attack?   Yes   No
     If yes, describe briefly.

16. Number of children and their ages.

17. Question for women only: Have you had any miscarriages, abortions, or stillbirths?   Yes   No
     If yes, describe briefly.

17. Do you have any history of cancer?   Yes   No
     If yes, describe briefly.

18. Do you have any history of chemotherapy or radiotherapy?   Yes   No
     If yes, describe briefly..

Figure 5-2. Sample clinical history questionnaire for diagnosis of thrombotic disorders. Note that questions attempt to distinguish between arterial and venous thrombosis and investigate family history of thrombophilia.

Location and Type of Thrombosis (Venous, Arterial, or Embolic)

Arterial thrombosis is usually linked to a dysfunction of vascular wall or platelets. Primary, large vessel, arterial occlusive disease is usually associated with atheromatous thrombi; dissection; large vessel arteritis, such as Takayasu disease and Buerger disease; migraine; and drug-induced thrombosis. Primary small vessel occlusive diseases are usually associated with arteritis, eclampsia, drug-induced thrombosis, antiphospholipid antibody, etc. Arterial thrombosis in infancy is a feature of homozygous AT deficiency, and cerebral arterial thrombosis has been associated with protein C and protein S deficiency. Embolic disorders are either cardiogenic in origin, due to atheromatous disease, due to deep vein thrombosis with pulmonary embolization, or paradoxical (deep vein thrombosis with a patent foramen ovale leading to cerebral emboli).

Venous thrombosis can limit blood flow through the vein, causing swelling and pain. Most commonly, venous thrombosis occurs in deep veins in the legs, thighs, or pelvis. Venous thrombosis can form anywhere in the venous system, but presentation in the upper extremities, liver, and mesentery is uncommon. Deep vein thrombosis (DVT) and pulmonary embolism (PE) are the most common manifestations of venous thrombosis. Venous thrombosis is typically associated with deficiencies of the natural anticoagulants, factor V Leiden, or the prothrombin G20210A mutation, but may also be seen with the lupus anticoagulant, elevated levels of factor VIII, and hyperhomocysteinemia.

Superficial thrombophlebitis can occur in inherited thrombophilias, particularly in protein C and protein S deficiency, but less commonly in AT deficiency. Thrombotic events in unusual locations, such as portal, cerebral, and retinal veins, are usually attributed to acquired risk factors, such as myeloproliferative disorders, oral contraceptives, or recent surgical procedures.

Recurrences

Studies have shown that the presence of dual hereditary defects significantly increases the likelihood of recurrent thrombosis. In idiopathic venous thrombosis, the recurrence rate may be 7% to 10%. Previous venous thrombosis in patients is an important risk factor for venous thrombosis.

Acquired Thrombophilic Risk Factors

There are many associations between acquired risk factors and venous thrombosis, including obesity, immobilization, smoking, malignancy, obstetric history, and medications.

The association between obesity and venous thrombosis is controversial. Some suggest that for overweight patients, the incidence of venous thrombosis is twice as much as for average and underweight patients. In contrast, others, using leg scanning, have found no increase in venous thrombosis among obese patients.

Over the past decades, a substantial amount of basic and clinical research has documented an association between chronic cigarette smoking and injury to the endothelium and promotion of vascular thrombosis. Clinical and epidemiological studies have documented that cigarette smokers have an increased risk for venous thromboembolism, but it is unclear whether the risk of thromboembolic disease differs for smokers according to gender.

Clinically important venous thromboembolism after air travel is rare. Most cases of travel-related thrombosis affected people at risk because of previous venous thromboembolism or other predisposing factors. However, a history of long-haul air travel should be sought in a complete clinical history.

Medical History

Bleeding and thrombosis can occur in patients with any myeloproliferative disorder. Thrombosis complicates polycythemia vera and essential thrombocythemia but is less common in chronic myelogeneous leukemia or chronic idiopathic myelofibrosis. Myeloproliferative syndromes have been associated with cerebral or mesenteric venous thrombosis, or thrombosis of the splenic vein, portal vein, and hepatic vein. A variety of arterial thrombotic events, such as strokes, myocardial infarction, neurologic symptoms, and ischemia of fingers and toes, can also occur. Transient cerebral ischemia and digital ischemia occur in essential thrombocythemia.

Paroxysmal nocturnal hemoglobinuria (PNH) is a disorder associated with hemolysis, bone marrow failure, and venous thrombosis. Thrombosis in PNH occurs in visceral vessels more often than deep vein thrombosis or pulmonary embolism. Painful splenomegaly due to visceral thrombosis can occur. PNH should be suspected and the diagnosis pursued in thrombophilic patients with a negative family history for thromboembolism, but evidence of pancytopenia and iron deficiency.

Rheumatologic disorders, including systemic lupus erythematosus, are associated with an increased risk of both arterial and venous thrombosis. Venous thrombosis is usually manifested as deep vein thrombosis of the lower extremities, with or without pulmonary emboli. In addition, unusual sites of thrombosis, such as hepatic veins, inferior vena cava, mesenteric veins, renal veins, cerebral venous sinuses, and axillary veins, can occur. Arterial thrombosis is usually manifested as stroke, transient ischemic attacks, or myocardial infarction. Unusual presentations, such as gangrene of the extremities and digits, multi-infarct dementia, and bowel infarction, can also occur. Livedo reticularis, cerebrovascular disease, and transient thromboembolic events can also occur in patients with systemic lupus erythematosus.

Information regarding neurologic disorders, including stroke, is helpful to elicit in obtaining a history in a patient with thrombophilia. Thromboembolism occurs in approximately 12% of patients with neurological disorders associated with immobility and paralysis. Cerebral vein thrombosis is due to thrombosis of superficial and deep cerebral veins. The symptoms are headache, papilledema, fever, seizures, and altered mental status. Thrombotic stroke may be associated with hematological causes, including hemoglobinopathies, sickle cell disease, and myeloproliferative disorders. The role of hypercoagulable risk factors, such as factor V Leiden mutation, protein C deficiency, protein S deficiency, antithrombin deficiency, and G20210A prothrombin mutation, in stroke is controversial; however, in patients who do not have any other obvious risk factors, a hypercoagulable state should be considered.

Malignancy

A history of malignancy should be sought in patients with thrombosis. An association between cancer and venous thrombosis has been well described. Cancer is also associated with a rapidly progressive thrombotic process of both superficial and deep veins (thrombophlebitis migrans), and there are reports of failed anticoagulant therapy (particularly oral anticoagulants) in patients with cancer and venous thromboembolism. Patients with pancreatic cancer, ovarian cancer, lung cancer, and mucin-producing gastrointestinal carcinoma can develop unusual forms of thrombosis, including migratory superficial thrombophlebitis,

nonbacterial thrombotic endocarditis, and thrombosis in unusual sites, such as renal veins, inferior vena cava, and portal and hepatic veins. There are also reports of an association between venous thrombosis and subsequent development of cancer. Cancer screening is appropriate in patients who present with idiopathic venous thrombosis or venous thrombosis in unusual sites.

Obstetrical History and Hormone Therapy

There is a known increased risk of thrombosis during normal pregnancy; however, there are significant associations between pregnancy complications, thrombosis, and underlying thrombophilic risk factors, particularly spontaneous abortions, preeclampsia, placental insufficiency, intrauterine growth retardation, and peripartum thrombosis. Recent evidence points to maternal/fetal thrombotic/thrombophilic disorders as a cause of both neonatal and antenatal disorders of the central nervous system, including neonatal stroke, central venous thrombosis, and porencephaly; they may also play a role in terminal limb malformations and late fetal loss, as well as other pregnancy complications. Therefore, a thorough clinical obstetric history should include particular questions to elicit this information.

Placental insufficiency is frequently cited as the cause and is known to be associated with maternal thrombophilic conditions predisposing to thromboembolism. The antiphospholipid antibody syndrome is well known to be associated with spontaneous abortions and peripartum thrombosis. The factor V Leiden mutation has been associated with third-trimester as well as first- and second-trimester fetal losses. Women with methylenetetrahydrofolate reductase (MTHFR) C677T variant have also been shown to be at an increased risk of fetal loss as well as various obstetrical complications, such as preeclampsia, abruptio placentae, and intrauterine growth retardation.

Women taking hormonal treatment either for contraception or as replacement therapy have a 2- to 4-fold increased risk of venous thromboembolism compared with control women. This risk is dramatically increased by the carrier status of a thrombophilic abnormality. Therefore, a detailed history of the use of oral contraceptives, hormone replacement therapy, and estrogen usage should be obtained. History regarding osteoporosis and treatment with selective estrogen receptor modulators such as raloxofene should be obtained, since such treatment predisposes to thrombosis. The risk of thromboembolism with raloxofene treatment is considered similar to that with estrogen-containing hormonal therapy.

Medication History

Many drug therapies, including phenothiazines, chemotherapy, heparin, oral contraceptives, hormones, tamoxifen, and thalidomide treatment for multiple myeloma, can increase the risk of thrombosis. A complete history should include a detailed history of medications, including those that are administered for existing diseases or previous disease states. Approximately one-third of individuals who sustain the rare complication of warfarin-induced skin necrosis have underlying hereditary protein C deficiency. This syndrome can also be seen in individuals with protein S deficiency and the factor V Leiden mutation. Heparin therapy and development of heparin-induced thrombocytopenia is unfortunately associated with a risk of both arterial and venous thrombosis. Increasing use of thalidomide as initial therapy for multiple myeloma has been associated with an increased incidence of deep vein thrombosis and other thrombotic events. Cocaine is associated with a large but transient increase in risk of acute myocardial infarction in patients who otherwise would not be regarded as at risk.

Surgery and Trauma

A history of prior surgical procedures is important to elucidate. Hip and knee surgery and major abdominal surgery are more frequently associated with a risk of thromboembolism. Tissue injury from trauma (and surgery) increases the frequency of venous thromboembolism. Coagulation serine proteases are activated and platelets are triggered by the exposure of tissue factor to flowing blood. This, coupled with prolonged immobilization postoperatively, can further elevate venous thrombosis risk.

Elevated Levels of Homocysteine

Mild to moderate elevation of plasma homocysteine is a risk factor for occlusive arterial disease (myocardial infarction, stroke, carotid artery stenosis, and peripheral arterial disease) and venous thrombosis. Elevated homocysteine levels can be found with congenital enzyme abnormalities and can be acquired in patients with chronic renal failure and in patients on methotrexate, anticonvulsants, and other such drugs that interfere with folate metabolism.

Summary

A thorough clinical and medication history for patients with either bleeding or thrombotic disorders is important in the evaluation of underlying causative etiologies. Often, clues from the clinical or family history can be very helpful in directing the laboratory workup and in the judicious selection of testing to be performed.

Suggested Reading

General References

Girolami A, Luzzatto G, Varvarikis C, Pellati D, Sartori R, Girolami B. Main clinical manifestations of a bleeding diathesis: an often disregarded aspect of medical and surgical history taking. *Haemophilia*. 2005;11:193-202.

Khair K, Liesner R. Bruising and bleeding in infants and children: a practical approach. *Br J Haematol*. 2006;133:221-331.

Lillicrap D, Nair SC, Srivastava A, Rodeghiero F, Pabinger I, Federici AB. Laboratory issues in bleeding disorders. *Haemophilia*. 2006;12(suppl 3):68-75.

Lippi G, Franchini M, Guidi GC. Diagnostic approach to inherited bleeding disorders. *Clin Chem Lab Med*. 2007;45:2-12.

Wahlberg T, Blomback M, Hall P, Axelsson G. Applications of indicators, predictors and diagnostic indices in coagulation disorders, I: evaluation of a self-administered questionnaire with binary questions. *Methods Inf Med*. 1980;9:194-200.

Clinical History for Bleeding Disorders

Furie B, Voo L, McAdam K, Furie BC. Mechanism of factor X deficiency in systemic amyloidosis. *N Engl J Med*. 1981;304:827-830.

Gitter MJ, Jaeger TM, Patterson TM, et al. Bleeding and thrombosis during anticoagulant therapy: a population based study in Rochester, Minnesota. *Mayo Clin Proc*. 1995;70:725-733.

Kadir RA, Economides DL, Sabin CA, Owens D, Lee CA. Frequency of inherited bleeding disorders in women with menorrhagia. *Lancet*. 1998;351:485-489.

Kadir RA, Economides DL, Sabin CA, Pollard D, Lee CA. Assessment of menstrual blood loss and gynaecological problems in patients with inherited bleeding disorders. *Haemophilia*. 1999;5:40-48.

Kouides PA, Conard J, Peyvandi F, Lukes A, Kadir R. Hemostasis and menstruation: appropriate investigation for underlying disorders of hemostasis in women with excessive menstrual bleeding. *Fertil Steril*. 2005;84:1345-1351.

Mammen EF. Coagulation abnormalities in liver disease. *Hematol Oncol Clin North Am.* 1992;6:1247-1257.

Mannucci PM, Duga S, Peyvandi F. Recessively inherited coagulation disorders. *Blood.* 2004:104:1243-1252.

Michiels JJ. Acquired hemophilia A in women postpartum: clinical manifestations, diagnosis and treatment. *Clin Appl Thromb Hemost.* 2000;6:82-86.

Pollmann H, Richter H, Ringkamp H, Jurgens H. When are children diagnosed as having severe haemophilia and when do they start to bleed? A 10-year single-centre PUP study. *Eur J Pedatr.* 1999;158(suppl 3): S166-170.

Rodriguez-Merchan EC. Common orthopaedic problems in haemophilia. *Haemophelia.* 1995;5(suppl 1):5-9.

Sagripanti A, Barsotti G. Bleeding and thrombosis in chronic uremia. *Nephron.* 1997;75(2):125-139.

Sagripanti A, Cozza V, Baicchi U, Camici M, Barsotti G. Increased thrombin generation in patients with chronic renal failure. *Int J Clin Lab Res.* 1997;27(1):72-75.

Samuels N. Herbal remedies and anticoagulant therapy. *Thromb Haemost.* 2005;93:3-7.

Sramek A, Eikenboom JC, Briet E, Vandenbroucke JP, Rosendaal FR. Usefulness of patient interview in bleeding disorders. *Arch Intern Med.* 1995;155:1409-1415.

Violi F, Basili S, Ferro D, et al. Association between high values of D-dimer and tissue plasminogen activator activity and first gastrointestinal bleeding in cirrhotic patients. CALC Group. *Thromb Haemost.* 1996;76(2):177-183.

History for Thrombotic Disorders

Abdollahi M, Cushman M, Rosendaal FR. Obesity: risk of venous thrombosis and the interaction with coagulation factor levels and oral contraceptive use. *Thromb Haemost.* 2003;89:493-498.

Anderson FA, Spencer FA. Risk factors for venous thromboembolism. *Circulation.* 2003;107(suppl 1):9-16.

Auletta MJ, Headington JT. Purpura fulminans: a cutaneous manifestation of severe protein C deficiency. *Arch Dermatol.* 1988;124:1387-1391.

Bajzar L, Chan AK, Massicotte MP, Mitchell LG. Thrombosis in children with malignancy. *Curr Opin Pediatr.* 2006;18(1):1-9.

Cruickshank JM, Gorlin R, Jennett B. Air travel and thrombotic episodes: the economy class syndrome. *Lancet.* 1988;2:497-499.

Cushman M, Kuller LH, Prentice R, et al. Estrogen plus progestin and risk of venous thrombosis. *JAMA.* 2004;292:1573-1580.

D'Angelo A, Vignano-D'Angelo S, Esmon CT, et al. Acquired deficiencies of protein S: protein S activity during oral anticoagulation, in liver disease and in disseminated intravascular coagulation. *J Clin Invest.* 1988;81:1445-1454.

Dietrich JE, Hertweck SP, Perlman SE. Efficacy of family history in determining thrombophilia risk. *J Pediatr Adolesc Gynecol.* 2007;20:221-224.

Griffin JH, Evatt B, Zimmerman TS, et al. Deficiency of protein C in congenital thrombotic disease. *J Clin Invest.* 1981;68:1370-1373.

Gunther G, Junker R, Strater R, et al. Symptomatic ischemic stroke in full-term neonates: role of acquired and genetic prothrombotic risk factors. *Stroke.* 2000;31(10):2437-2441.

Haan J, Caekebeke JF, van der Meer FJ, Wintzen AR. Cerebral venous thrombosis as presenting sign in myeloproliferative disorders. *J Neurol Neurosurg Psychiatry.* 1988;51:1219-1220.

Karpatkin M, Mannuccio Mannucci P, Bhogal M, Vigano S, Nardi M. Low protein C in the neonatal period. *Br J Haematol.* 1986;62:137-142.

Lane DA, Mannucci PM, Bauer KA, et al. Inherited thrombophilia: part 1. *Thromb Haemost.* 1996;76: 651-662.

Lane DA, Mannucci PM, Bauer KA, et al. Inherited thrombophilia: part 2. *Thromb Haemost.* 1996;76: 824-834.

Male C, Julian JA, Massicotte P, Gent M, Mitchell L. PROTEKT Study group. Significant association with location of central venous line placement and risk of venous thrombosis in children. *Thromb Haemost.* 2005;94(3):516-521.

Manco-Johnson MJ, Marlar RA, Jacobson LJ, et al. Severe protein C deficiency in newborn infants. *J Pediatr.* 1988;113:359-363.

Mannucci PM, Tripodi A, Bertina RM: Protein S deficiency associated with "juvenile" arterial and venous thrombosis. *Thromb Haemost.* 1986;55:440.

Marongiu F, Cauli C, Mariotti S. Thyroid, hemostasis and thrombosis. *J Endocrinol Invest.* 2004;27:1065-1071.

Mitchell L, Piovella F, Ofosu F, Andrew M. Alpha-2 macroglobulin may provide protection from thromboembolic events in antithrombin-III-deficient children. *Blood.* 1991;78:2299.

Nieuwenhuis HK, Albada J, Banga JD, et al. Identification of risk factors for bleeding during treatment of acute venous thromboembolism with heparin or low molecular weight heparin. *Blood.* 1991;78:2337-2343.

O'Donnell J, Tuddenham EGD, Manning R, et al. High prevalence of elevated factor VIII levels in patients referred for thrombophilia screening: role of increased synthesis and the relationship of acute phase reaction. *Thromb Haemost.* 1997;77(5):825-828.

Prandoni P, Lensing AWA, Buller HR, et al. Deep-vein thrombosis and the incidence of subsequent symptomatic cancer. *N Engl J Med.* 1992;327:1128-1133.

Sagripanti A, Ferriti A, Nicolini A, Carpi A. Thrombotic and hemorrhagic complications in chronic myeloproliferative disorders. *Biomed Pharmacother.* 1996; 50(8);376-382.

Vandenbroucke JP, Koster T, Briet E, Reitsma PH, Bertina RM, Rosendaal FR. Increased risk of venous thrombosis in oral-contraceptive users who are carriers of factor V Leiden mutation. *Lancet.* 1994; 344:1453-1457.

Vigano-D'Angelo S, D'Angelo A, Kaufman CE, et al. Protein S deficiency occurs in nephritic syndrome. *Ann Intern Med.* 1987;107:42-47.

von Kaulla E, von Kaulla KN. Antithrombin III and diseases. *Am J Clin Pathol.* 1967;48:69-80.

Webber J, Kline RA, Lucas CE. Aortic thrombosis associated with cocaine use: report of two cases. *Ann Vasc Surg.* 1999;13:302-304.

Zangari M, Barlogie B, Anaissie E, et al. Deep vein thrombosis in patients with multiple myeloma treated with thalidomide and chemotherapy: effects of prophylactic and therapeutic anticoagulation. *Br J Haematol.* 2004;126:715-721.

Coagulation Testing

Jayashree Krishnan, MD

Introduction

Hemostasis testing is performed in the laboratory during the evaluation of the cause of bleeding to determine the etiology of a previously abnormal coagulation test, evaluate the risk factors for a hypercoagulable state, monitor patients on anticoagulant therapy, and assess platelet function. The number and complexity of hemostasis tests is increasing due to an increased knowledge of the pathophysiology of hemostasis and the development of new testing methodologies. The interpretation of these test results is also increasingly complex, especially due to the development of newer drugs and their interference with hemostasis tests. For the most part, testing in the coagulation laboratory can be divided into two broad categories: screening assays that assess the function of a range of analytes and tests that identify/quantify specific analytes.

In evaluating a patient with a coagulopathy, the most common screening tests performed are the prothrombin time (PT), activated partial thromboplastin time (aPTT or PTT), fibrinogen assay, the thrombin time (TT), and platelet count. The prothrombin time measures the extrinsic clotting system and the common pathway, which is the confluence of intrinsic and extrinsic pathways at the point of conversion of prothrombin to thrombin. The activated partial thromboplastin time measures the intrinsic and common pathways. The thrombin time measures conversion of fibrinogen to fibrin polymer. Specific analyte tests include assays for fibrinogen and clotting factors. These can be used to identify the etiology of a factor deficiency. Specific assays to diagnose von Willebrand disease are covered in chapter 16. Further steps in hemostasis testing include mixing studies and specific assays, such as the Bethesda assay, to identify and quantify coagulation inhibitors. Assays for testing platelet function, fibrinolysis, and thrombophilia are covered in chapters 7, 8, and 9, respectively.

Laboratory Measurement of Clot-Based Coagulation Tests

Good-quality clot-based coagulation testing requires proper sample collection and processing and the use of citrate-anticoagulated platelet-poor plasma, as described in chapter 4. Since the coagulation process requires both calcium and phospoholipids, the removal of platelets and chelation of calcium during sample processing put the plasma into a "suspended animation" that prevents coagulation until the addition of the appropriate reagents in the various coagulation assays. In addition to sample processing adequacy, it is also important to understand the mechanism and capabilities of the coagulation analyzers used to perform the testing and how to select a suitable instrument to meet the needs of the laboratory.

Coagulation Analyzers

Principles of Performance

The clotting end-point for clot-based assays, such as the PT, is reached at the formation of the fibrin clot, which can be detected optically by a change in turbidity, or mechanically, due to the conversion of a liquid plasma sample to a gel. Clot detection can be determined manually or by semiautomated or automated methods. With manual methods, all reagents, samples, and clot formation timing are controlled by the operator. With semiautomated methods, reagents and samples are added manually. The clot formation is detected automatically. With fully automated methods, reagent pipetting, timing, and clot detection are automated completely.

Manual clot detection usually is performed by a "tilt tube" method, where the operator pipettes the plasma and reagents into a clear test tube, starts a timer, and tilts the tube back and forth until a clot is formed, at which point the timer is stopped. This is the original method for clot-based coagulation testing and is still considered the "gold standard" for instrument comparison. Mechanical endpoint detection (eg, Fibrometer) detects a change in electrical conductivity with one stationary probe and one moving probe. Fibrin formed acts as conductor between the two probes, which stops the instrument timer. Other mechanical clot-detection instruments use a steel ball for clot detection. The steel ball is magnetically moved back and forth in the plasma. When the clot forms, the viscosity stops the ball motion and the timer is tripped. For photo-optical endpoint detection, a change in optical density of the test sample is detected. Fibrin formation causes plasma to become opaque or turbid, and the amount of light detected decreases. Both mechanical and optical instruments give accurate results for clot-based coagulation tests. The optical instruments suffer from more interference by lipemia, icterus, and hemolysis, but these interferences can be detected by proper specimen handling and technologist education. Factors to be considered in the choice of an instrument are cost, reliability, service, ease of use, maintenance requirements, volume of tests, reagent availability, cost of disposables, throughput, and training support.

How to Select and Validate a Coagulation Analyzer

Coagulation analyzers used in clinical laboratories have become increasingly complex. In addition to the common clot-based screening tests (prothrombin time, activated partial thromboplastin time, thrombin clotting time, and Clauss fibrinogen assay), most coagulation analyzers also perform chromogenic and immunoturbidometric assays for assays such as antithrombin and von Willebrand factor antigen. Some analyzers may also have the capability to perform chemiluminescent assays.

The Clinical and Laboratory Standards Institute (CLSI, formerly NCCLS) has developed an updated guideline for the evaluation of coagulometers, *Protocol for the Evaluation, Validation, and Implementation of Coagulometers* (H57-P, 2008). Earlier evaluation protocols compared instrument performance to the manual clotting technique. Although the tilt-tube method remains the international reference method for PT/INR (prothrombin time/international normalized ratio), it has now largely disappeared from routine laboratories. Contemporary reagents frequently are developed exclusively for use with a given instrument and consequently are unsuitable for instruments employing other end-point detection methods.

With this trend towards tailored reagent/instrument systems, instrument evaluation is best achieved by comparison against reference reagents and reference instruments, selected for their suitability to the instrument and reagents under evaluation.

The evaluation of a coagulation analyzer typically progresses through several stages. The process starts with literature review and market research, during which several desirable instrument platforms are selected, and proceeds to a preliminary preacquisition evaluation of one or more instruments. This is followed by a more detailed postacquisition validation of a single platform, culminating in implementation of the testing system for clinical use. Upon the decision to acquire new instruments, the laboratory should determine their testing needs in terms of testing volume, menu and turnaround time expectations, then match those needs with hemostasis instrumentation, reagent, test menu, and system parameters that are available.

The range of assays evaluated will depend upon the volume and the range of assays performed by the laboratory concerned. A basic performance assessment should always examine the precision and comparability of the coagulation screening tests: one-stage prothrombin time (PT) assays; activated partial thromboplastin time (aPTT) test; thrombin time (TT) and/or the Clauss fibrinogen assay.

For a larger evaluation, a broader range of assays, including the testing of normal and abnormal plasma, is required. This may include assays used to screen for a bleeding tendency or thrombophilia, and assays used to monitor anticoagulation. It is generally impractical to evaluate every one-stage factor assay and chromogenic substrate assay available on an analyzer. The CLSI guidelines recommend assessing a single one-stage PT-based assay (eg, factor V), a single one-stage aPTT-based assay (eg, factor VIII), a direct chromogenic assay (eg, protein C), and an indirect chromogenic assay (eg, antithrombin), so it is possible to obtain a good indication of the instrument's capabilities. Ultimately, the range of tests evaluated will depend upon the range of work performed by the individual laboratory and the types of samples tested.

Prothrombin Time

The prothrombin time (PT) test was first described by Armand Quick in 1935. The PT is a global test of the extrinsic and common coagulation path-

Table 6-1. Purposes for Performing Prothrombin Time (PT) Testing

- ❑ Evaluation of bleeding
- ❑ Detection of factor deficiencies
- ❑ Assessment of liver disease
- ❑ Monitoring of oral anticoagulation (warfarin) therapy

ways, and is sensitive to decreased levels of factors VII, X, V, II, and fibrinogen. It is commonly used to screen for inherited and acquired coagulation disorders and to monitor warfarin therapy, as shown in Table 6-1. The PT is typically performed on a platelet-poor plasma sample that is anticoagulated with 3.2% sodium citrate to chelate calcium. The test is started by the addition of thromboplastin (tissue factor plus phospholipid) with the addition of $CaCl_2$. The time to formation of a fibrin clot is then measured by either an optical or mechanical method (Figure 6-1). The PT results are expressed in seconds, with typical values of 10 to 13 seconds. This is faster than the usual 25 to 40 seconds for the activated partial thromboplastin time; the rapidity of clotting is the result of tissue factor complexing with factor VII and direct activation of factor X, bypassing the relatively slower intrinsic pathway activation of clotting. The normal reference range should be established by each laboratory using the methods described below.

For PT/INR testing, samples may be tested up to 24 hours from the time of collection, provided the sample is uncentrifuged or centrifuged with plasma remaining on top of the cells. The tube must remain capped and stored at 2 to 4°C or at room temperature. If testing cannot be completed within 24 hours, the plasma should be removed from the cells and frozen at -70°C.

Reagent Considerations

The PT varies in its sensitivity to individual factor deficiencies. It is most sensitive to factor VII deficiency and least sensitive to factor II (prothrombin) and fibrinogen deficiency. Levels of sensitivity vary among different thromboplastin reagents. Different thromboplastins may vary in the elevation of the PT in response to warfarin and in the assessment of liver disease severity.

The wide use of the PT has resulted in the availability of numerous thromboplastin reagents and coagulation instruments. These reagents all con-

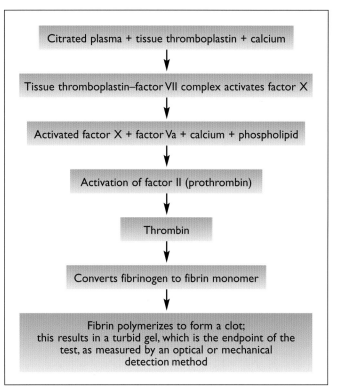

Figure 6-1. The prothrombin time (PT) measures the extrinsic and common pathways of coagulation, as shown. Platelet-poor plasma samples are anticoagulated with sodium citrate, which chelates calcium; the plasma is also missing tissue factor and phospholipid. Thus, the PT is performed by adding back these factors and measuring the time to the formation of a fibrin clot, which is detected by transformation of a clear liquid plasma to a turbid gel.

tain thromboplastin (tissue factor plus phospholipid) and calcium chloride. They vary in their phospholipid source and concentration and type of tissue factor. Both lyophilized and liquid products are available. Many thromboplastins are extracted from biological sources (ie, rabbit brain or human placenta), while others are derived from a recombinant human tissue factor source. The sensitivity of the reagents to factor deficiencies varies, as reflected by the International Sensitivity Index (ISI). The ISI is a measure of the sensitivity of a reagent and is a measure of a reagent's responsiveness to depressed functional levels of vitamin K-dependent coagulation factors compared to the primary World Health Organization (WHO) International Reference Preparation (IRP). Each thromboplastin/instrument combination typically is assigned an International Sensitivity Index by comparison of PT results of the reagent compared to the WHO IRP using plasma sets from normal and oral-anticoagulated individuals, as shown in Figure 6-2. Reagents with an ISI near 1.0 are simi-

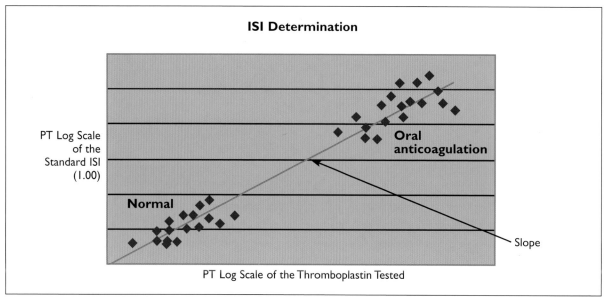

Figure 6-2. The ISI is calculated for an individual reagent by comparison to the World Health Organization International Reference Preparation (WHO IRP) using sets of plasma from normal individuals and individuals receiving oral anticoagulants.

lar to the WHO IRP and are considered sensitive reagents. ISIs can range up to 3.0. Reagents with higher ISIs are considered less sensitive.

Low heparin sensitivity is desirable in a PT reagent. Some commercial PT reagents include a heparin-neutralizing substance, such as polybrene, in the reagent to neutralize heparin's effect on the PT. The heparin neutralization is typically only effective up to a heparin concentration of 1.0 U/mL. In reagents that do not contain a heparin-neutralizing agent, the PT will be progressively prolonged at higher heparin concentrations.

Minimal sensitivity to lupus anticoagulants is also desirable in a PT reagent, since patients with lupus anticoagulants are often treated with oral anticoagulants and monitored using the PT. The lupus insensitivity of PT reagents is typically due to the high phospholipid concentration of the PT reagent. However, some strong lupus anticoagulants prolong the PT. When a lupus anticoagulant prolongs the PT result, an alternate assay, such as the chromogenic factor X assay, should be used to monitor oral anticoagulant therapy.

Reporting of Prothrombin Time

The PT result is reported in seconds and/or international normalized ratio (INR). The INR was introduced in 1983 as a means of standardizing the PT test and improving management of patients receiving warfarin-based oral anticoagulant therapy. It is well understood that the various reagents used for PT testing can yield widely differing PT clotting time results in patients receiving oral anti-

coagulants. The INR system primarily minimizes this variation and improves the quality of laboratory monitoring of oral anticoagulant therapy. By using the ISI value assigned by manufacturers to each thromboplastin and instrument combination, the PT result in seconds can be normalized to a ratio that reflects a result that would have been obtained if the WHO IRP had been used. The INR allows the comparability of PT results among different laboratories or among different reagent lots when PT is used for monitoring oral anticoagulation therapy.

Calculation of International Normalized Ratio

The INR calculation is given by the formula:

$$INR = \left[\frac{PTpatient}{PTnormal} \right]^{ISI}$$

where:

PTpatient is the patient's PT in seconds.

PTnormal is the geometric mean PT determined from a group of normal samples.

ISI is an experimentally determined value indicating the sensitivity of the thromboplastin to warfarin-induced factor deficiencies of the vitamin K-dependent factors (factors II, VII, and X). This is typically supplied by the manufacturer and may be specific for various instrument/reagent combinations.

The geometric mean normal population value may change when the specimen collection process, instrument, reagent lot, or reagent changes. The geometric and arithmetic means may not be identical, and the geometric mean is the preferred value. When the distribution of values is distributed normally, the geometric mean, the arithmetic mean, the median and mode of the population being studied are identical. These values diverge from each other, however, as the population distribution becomes more skewed. The geometric mean is a more appropriate estimate of the average value than the arithmetic mean when the population of interest is log-normally distributed, because the geometric mean takes skewing into account.

The geometric mean (GM) is the antilog of the arithmetic mean of the logarithms of the individual normal PT values of interest. The GM should be calculated from at least 20 normal plasmas with each change of reagent lot, reagent type, or instrument. The calculation of the geometric mean is indicated below. This calculation is available in many spreadsheet programs.

$$GM = antilog[log(X1)+log(X2)+log(X3)+....log(Xn/n]$$

Causes of variation in INR calculation include incorrect determination of the mean normal PT; local effect of the reagent/instrument combinations; incorrect ISI value used; incorrect choice of IRP for reagent calibration, which would result in an inaccurate ISI; and inaccuracy and imprecision in the calibration of the commercial reagent against the appropriate IRP, causing an inaccurate ISI value. It is important to remember that the INR calculations can be performed by the coagulation analyzer, the laboratory information system, or the hospital information system; accurate and current values for the ISI and GM should be used in patient INR calculations.

International Sensitivity Index. The ISI represents the slope of an orthogonal regression line, comparing prothrombin time values from 20 fresh plasma samples obtained from normal subjects and 60 fresh plasma samples from stabilized patients on oral anticoagulation therapy, using both an international reference thromboplastin and the thromboplastin in question. The slope and ISI must be determined using orthogonal regression analysis based on a logarithmic-logarithmic plot (see Figure 6-2 for ISI calibration). ISI values are instrument and reagent specific. Therefore, coagulation instruments can have significant effects on the ISI of thromboplastins. A generic ISI is an ISI determined for a thromboplastin that is not instrument specific. Generic ISI is used when the ISI for a specific thromboplastin/instrument

combination is not available. The ISI difference between different optical and mechanical instruments can be significant. A human thromboplastin can have an ISI of 1.3 for one brand of optical instrument but can have an ISI of 1.1 for another instrument. The use of thromboplastin/instrument-specific ISI can therefore give a more accurate INR than a generic ISI. It is important that changes in ISI occurring with changes in reagent lot, reagent type, or instrument are programmed correctly in the analyzer or laboratory information system.

Validation and Calibration of the International Normalized Ratio. Despite the widespread use of the INR system, interlaboratory variations persist. Factors influencing INR include derivation of geometric mean normal PT (MNPT), method of clot detection, the difference in ISI value of test thromboplastin and international reference preparations. Verification of the PT/INR system is done by using a minimum of three certified plasmas with known INRs in the therapeutic range of 1.5 to 4.5. CLSI recommends this verification process for laboratories using thromboplastin/instrument-specific ISIs and suggests that it is mandatory for laboratories using a generic ISI. INRs determined from the verification process should compare within 15% of the true INRs for each certified plasma; if not, corrective action in the form of local ISI determination should be undertaken. Verification should be performed with any change in PT reagent, PT reagent lot number, or instrument, if the instrument has undergone a major repair, or with any aberrant PT result from external proficiency testing. Use of certified plasmas also allows direct calculation of the INR from a line of best fit calculated from a plot of assigned INR versus local PT. Although this method of INR calculation is used internationally, certified plasmas are not yet available in the United States. Some College of American Pathologists (CAP) checklist questions regarding performance and reporting of the PT/INR are shown in Table 6-2.

Reference Ranges for Prothrombin Time. The PT reference range varies significantly among different reagent and instrument combinations and may also vary from lot to lot of the same reagent. Each laboratory should establish an in-house reference range each time the reagent lot or one component of the testing system changes. The approximate lower limit of the normal reference range is 10 to 11 seconds; the approximate upper limit of normal is 12 to 14 seconds. Newborns usually

Table 6-2. Checklist/Actions for International Normalized Ratio

In order to maximize patient safety, the CAP Coagulation Resource Committee suggests the following checklist/actions to assist laboratories with prevention of problems with the INR calculation:

❏ Has an appropriate reference range for your current reagent/instrument been identified?

❏ Is the geometric mean of this reference range being used in calculation?

❏ Is the correct ISI being used in the calculation? The ISI varies not only with the lot of reagent but also with the instrument being used.

> If in doubt, the reagent manufacturer should be contacted.

> When performing the calculation, the values for the patient PT and the PT reference range mean should include one decimal place (eg, 12.1) and ISI should be included using two decimal places (eg, 1.05)—in each case, three significant digits.

❏ How is the calculation being performed? Manually? By the instrument? By the LIS?

> The calculation should only be performed by one (not more than one) of these methods. Having more than one method in place increases the maintenance efforts and increases the risk of a miscalculation.

❏ If a manual method of calculation is used, how are errors minimized?

❏ Is the calculation tested periodically to assure than the correct INR is being produced? In particular, this calculation should be tested immediately after changing any of the components of the formula (ISI or geometric mean of the reference range). Actual patient reports should periodically be reviewed to assure that the appropriately calculated result is being received by patient care personnel.

❏ Are specimens collected into a 3.2% citrate, not 3.8%? Published reports have documented that INR results differ between the two citrate concentrations.

have prolonged PT compared to adults. The PT is slightly longer at birth and gradually shortens to adult normal range by 6 to 12 months of age (Table 6-3).

Plasma from at least 40 normal individuals, preferably 50% males and 50% females, over a wide age range, should be tested for development of the normal reference range. The normal individuals should not have a history of liver disease or be taking oral anticoagulant therapy. It is helpful to perform both PT and aPTT testing on each specimen. Samples with an elevated aPTT should not be used in the calculation of a PT reference range. The reference range is typically calculated as the arithmetic mean +/- 2 standard deviations (SD). A 2-SD range reflects normalcy for 95% of the population, whereas a 3-SD range includes 99.7%.

Comparison Between the New and Old Lot Numbers of PT Reagent. The laboratory should draw samples from patients with PT/INR results within the therapeutic range and from abnormal patients in order to perform comparison studies. This is done to verify the consistency of the new reagent. If the PT/INR results are found to be running higher or lower with the new reagent, the appropriate physicians and pharmacy should be informed of the change in reagent and the changes in PT/INR results. A manual check of the INR calculation should be done with each new lot number

and periodically thereafter. It may also be helpful to perform parallel factor sensitivity studies between lots of reagents to determine the sensitivity of a reagent to a decrease in fibrinogen or factors II, V, VII, or X. This is typically performed by making a series of dilutions of normal plasma and factor-deficient plasma with factor levels ranging from 0% to 100% and performing the PT value. Typically, the PT value prolongs above the normal reference range when the factor level is between 40% to 50%.

Quality Control. There are commercially available lyophilized and frozen control plasmas available for performing quality control (QC). At least two levels of controls should be performed every 8 hours during PT testing. Controls should be prepared according to the manufacturer's instructions. Control ranges must be set for each new lot number of control and reagent. Preliminary operating ranges are established for each level of control material based on the mean of at least 20 between day determinations +/- 2 SDs. These ranges must be observed for trends and adjusted accordingly. The mean and SD are calculated from the set.

Critical Value. An INR > 5 is the most commonly used critical value, but the value varies depending on the reagent-instrument combination and individual laboratory guidelines.

Table 6-3. Example Reference Ranges for Pediatric and Adult Coagulation Assays

PT (s) AGE	LOW	HI
0-1 d	8.4	14.8
2-5 d	7.9	14.2
6-30 d	7.8	13.3
1-3 mo	8.0	13.2
4-11 mo	8.9	12.9
1 y - adult	9.0	13.0

INR AGE	LOW	HI
0-1 d	0.7	1.4
2-5 d	0.7	1.3
6-30 d	0.7	1.2
1-3 mo	0.7	1.2
4-11 mo	0.8	1.1
1 y - adult	0.8	1.2

PTT (s) AGE	LOW	HI
0-1 d	28.9	44.2
2-5 d	23.5	48.6
6-30 d	23.7	44.8
1-3 mo	22.3	40.7
4-11 mo	26.0	34.8
1 y - adult	24.6	32.8

TT (s) AGE	LOW	HI
0-1 d	15.4	19.3
2-5 d	14.0	19.9
6-30 d	16.0	19.9
1-3 mo	16.9	20.3
4-11 mo	16.3	21.3
1 y - adult	16.2	20.7

Fibrinogen (mg/dL) AGE	LOW	HI
0-1 d	214	399
2-5 d	208	462
6-30 d	208	378
1-3 mo	137	379
4-11 mo	147	387
1 y - adult	200	400

Factor II (%) AGE	LOW	HI
0-1 d	26	67
2-5 d	33	88
6-30 d	34	97
1-3 mo	45	100
4-11 mo	60	110
1 y - adult	71	138

Factor V (%) AGE	LOW	HI
0-1 d	29	108
2-5 d	36	145
6-30 d	50	134
1-3 mo	38	132
4-11 mo	44	127
1 y - adult	50	150

Factor VII (%) AGE	LOW	HI
0-1 d	20	109
2-5 d	26	150
6-30 d	31	145
1-3 mo	29	150
4-11 mo	35	134
1-5 y	41	122
6-10 y	38	126
11-16 y	43	121
Adult	50	150

Factor VIII (%) AGE	LOW	HI
0-1 d	22	161
2-5 d	22	139
6-30 d	25	142
1-3 mo	33	113
4-11 mo	37	98
1 y - adult	49	134

vWF AG (%) AGE	LOW	HI
0-1 d	36	315
2-5 d	50	279
6-30 d	19	270
1-3 mo	57	226
4-11 mo	32	216
1-5 y	60	132
6-10 y	44	158
11-16 y	46	168
Adult	50	173

Factor IX (%) AGE	LOW	HI
0-1 d	17	81
2-5 d	17	81
6-30 d	24	73
1-3 mo	24	101
4-11 mo	42	121
1-5 y	55	93
6 y - adult	65	145

Factor X (%) AGE	LOW	HI
0-1 d	12	67
2-5 d	19	77
6-30 d	32	85
1-3 mo	36	105
4-11 mo	39	115
1-5 y	51	113
6-10 y	48	99
11-16 y	44	114
Adult	62	148

Factor XI (%) AGE	LOW	HI
0-1 d	7	78
2-5 d	17	103
6-30 d	20	94
1-3 mo	30	115
4-11 mo	28	159
1-5 y	41	178
6-10 y	38	142
11-16 y	37	115
Adult	50	150

Factor XII (%) AGE	LOW	HI
0-1 d	12	97
2-5 d	10	86
6-30 d	16	84
1-3 mo	24	113
4-11 mo	37	120
1 y - adult	50	170

Prekallekrein (%) AGE	LOW	HI
0-1 d	4	64
2-5 d	16	70
6-30 d	18	84
1-3 mo	33	97
4-11 mo	45	107
1 y - adult	50	150

Note: These are example ranges only. Each laboratory should establish its own reference ranges. Adapted from Andrew M et al, 1987.

Clinical Utility of Prothrombin Time

The PT has many clinical uses, as listed below. The reader is also referred to a further discussion in chapter 11.

The PT is an important screening test to detect the deficiency of one or more of the clotting factors of the extrinsic pathway of the coagulation cascade. This deficiency may be hereditary or due to acquired causes, such as liver disease, vitamin K deficiency, or a specific factor inhibitor. A prolonged PT indicates deficiencies of plasma factors VII, X, V, II, or fibrinogen, as well as inhibitors to these factors. The test is more sensitive to factor VII deficiency than to decreases in either prothrombin or fibrinogen. The PT is significantly prolonged when the factor VII, V, or X is less than approximately 40%, but it is not significantly prolonged until the fibrinogen level is less than about 100 mg/dL or until the prothrombin concentration is less than 30%. Levels of sensitivity to detect these defects vary among different thromboplastin reagents.

The PT is useful in monitoring oral anticoagulant therapy and thereby to maintain a patient in a safe clinical range due to its sensitivity to decreases in factors II, VII, and X typically observed with warfarin.

A number of prescription medications, over-the-counter medications, herbal products, hormone replacement therapy, oral contraceptives, nonsteroidal anti-inflammatory drugs, vitamin K, anticoagulant therapy, and antibiotics can alter the INR.

Therapeutic Range. One single therapeutic range for coumarins will not be optimal for all indications. A moderate intensity INR of 2.0 to 3.0 is effective for most indications, including treatment and prevention of venous thromboembolism, hip surgery, major gynecologic surgery, atrial fibrillation, and prosthetic heart valves, and for treatment and prevention of recurrent thromboembolic events in antiphospholipid syndrome and in patients with heart valve disorders. These guidelines are developed by the American College of Chest Physicians and are updated periodically. There are certain conditions where clinicians may decide to use a lower or higher therapeutic range, eg, INR 1.5 to 1.9 for secondary prevention of idiopathic venous thromboembolism and an INR 2.5 to 3.5 for older-generation mechanical heart valves. The INR is calibrated for standardization of prolonged PT results for monitoring oral anticoagulation therapy and not for prolonged PT results due to other causes such as liver disease, hereditary factor deficiencies etc.

Point-of-Care Monitors for Oral Anticoagulant Therapy

Over recent years, near patient testing by portable monitors for PT/INR has gained popularity. The advantages of these point-of-care monitors are ease of use, decreased test duration, faster turn-around time for monitoring dosages, and increased convenience for patients. Many models for point-of-care testing are available in the United States. However, accuracy and precision have to be carefully determined for these devices. These devices can not be used interchangeably, and individual device performance is not the same for all similar devices.

Activated Partial Thromboplastin Time

The partial thromboplastin time (PTT) was first devised by Langdell and colleagues in 1953. By activating the plasma to a maximum level, Proctor and Rappaport developed the activated partial thromboplastin time (aPTT or APTT) modification in 1961. It is a kinetic test that derives its name from the use of a partial thromboplastin, such as cephalin or phosphatide, which activates the clotting mechanism. The aPTT is a global test of the intrinsic and common coagulation pathways. It is commonly used to screen for inherited and acquired coagulation disorders, as well as to monitor unfractionated heparin therapy.

Principle

The aPTT is performed by recalcifying plasma in the presence of a standardized amount of platelet-like phosphatides and an activator of the contact factors of the intrinsic pathway of the coagulation cascade. The aPTT can detect deficiencies of all clotting factors except factors VII and XIII; is very sensitive to deficiencies of the contact activation factors XII, prekallekrein, and high-molecular-weight kininogen; and has lower sensitivity to the fibrinogen level and common pathway factors. The aPTT result is usually normal if all coagulation factors are present in a concentration greater than 30%, but there is considerable variability between aPTT reagents. The aPTT is also prolonged in the presence of heparin, direct thrombin inhibitors, or circulating anticoagulants such as lupus-like anticoagulant or specific factor inhibitors.

Reagents

Several commercial aPTT reagents are available. They vary in the source of phospholipids and in the type and concentration of the activator used. Both liquid and lyophilized reagents are available. Reagents should be chosen by careful evaluation of the individual laboratory and the instrumentation system. Commercial reagents utilize various activators, such as Celite, kaolin, micronized silica, and ellagic acid. A common property of these agents is that they are negatively charged and are of high molecular mass. Exposure of blood to negatively charged surfaces leads to rapid binding of all contact factors to the surfaces. The predominant mechanism for the initiation of contact activation during the aPTT test is considered to be the activation of factor XII and prekallikrein in the presence of high-molecular-weight kininogen.

aPTT reagents also vary in the sources and concentrations of phospholipids, which may be synthetic or biologic in origin. Most aPTT reagents include calcium chloride as a set; if not included, calcium chloride, either 0.025 M or 0.020 M, is also available commercially. Due to the variability in aPTT reagent composition, the different reagents vary in their sensitivities to heparin and factor deficiencies. The aPTT values also vary in sensitivity to the lupus anticoagulant, depending on the type and amount of phospholipid in the reagent. In general, lupus-insensitive reagents typically have high phospholipid concentrations.

Method

One part of aPTT reagent and one part of patient's plasma are incubated together at 37°C for a specified period of time, which is usually about 3 to 5 minutes (Figure 6-3). This incubation step is necessary for the activator to begin to activate the intrinsic pathway, which occurs in the absence of calcium. Following incubation with the activator, calcium chloride is added, and the clotting time is measured in seconds.

Reference Range

The aPTT normal reference range will vary depending on the instrument, anticoagulant, tube type, reagent type, and reagent lot. Calculation of an in-house normal reference range that is specific to the current reagent lot and testing system is essential for aPTT testing. However, as a guideline, aPTT values in normal individuals range from approximately 25 to 40 seconds (this varies depending upon the reagent and instrumentation). Newborns have prolonged aPTT values in

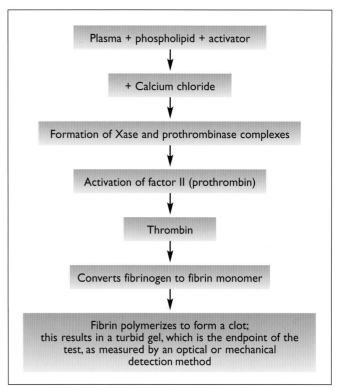

Figure 6-3. The activated partial thromboplastin time (aPTT) measures the intrinsic and common pathways of coagulation. It is performed using citrate anticoagulated platelet-poor plasma, with addition of an activator and phospholipids, followed by calcium chloride.

comparison to adults. The aPTT can range up to 45 seconds at birth and decreases into the adult normal range by about 6 to 12 months (Table 6-3). A critical value for aPTT testing should be established. This can be somewhere in the range of >100 to 150 seconds, but this may vary depending on the reagent and instrumentation and the individual laboratory policy.

Clinical Utility of Activated Partial Thromboplastin Time

The aPTT can be used to monitor heparin and direct thrombin inhibitor anticoagulant therapy. The aPTT is also used as a screening procedure to determine von Willebrand disease and coagulation factor deficiencies in the intrinsic and common pathway, namely fibrinogen and factors II, V, VIII, IX, X, XI, XII, prekallikrein, and high-molecular-weight kininogen.

The aPTT is very useful for detection of lupus anticoagulant and other coagulation inhibitors. In some institutions, aPTT is used to assess the overall competence of a patient's ability to clot in preoperative screening.

False-positive and false-negative aPTT results can occur from many preanalytic variables. Prolongation of the aPTT can be seen when blood is drawn from intravenous catheters that have been flushed with heparin or from "underfilled" tubes. Polycythemic samples with an elevated hematocrit lead to an elevated aPTT as well as an elevated PT due to excess plasma citrate concentration.

Steps to Take in the Evaluation of a Prolonged Activated Partial Thromboplastin Time

The following steps should be followed when a prolonged aPTT is discovered:

1. Repeat the assay result on the plasma sample to ensure reproducibility.

2. Next, determine whether the patient is receiving heparin or a direct thrombin inhibitor (DTI), or whether the sample could have been collected from an intravenous line containing heparin.

3. If the aPTT result is still prolonged on repeat testing and the patient is not receiving heparin or a DTI, further testing should be performed, including mixing studies and factor assays, as described in chapter 10, to evaluate the intrinsic pathway factors and exclude coagulation inhibitors.

Evaluation of Unfractionated Heparin Sensitivity

It is well known that the response of the aPTT to heparin varies greatly due to the variability of the reagents, instruments, patient response to heparin, and heparin itself. The in vitro heparin sensitivity differs between aPTT reagents and is influenced by the clot detection method used. Each laboratory should determine the sensitivity of the aPTT reagent to heparin and establish a new heparin therapeutic range each time there is a change in the method. Establishing lot-specific heparin therapeutic ranges is important; otherwise, patients may end up receiving too much or too little heparin, with the complication of bleeding or thrombosis.

The following methods may be used to determine the heparin therapeutic range:

Validation of Heparin Sensitivity of the aPTT Using an Assay of Heparin. This is the recommended method for determining the heparin-sensitivity of an aPTT reagent system. This can be done by simultaneously determining the aPTT (seconds) and heparin concentration (U/mL) by an anti-Xa assay using fresh samples collected from patients receiving unfractionated heparin for treatment of thromboembolism. A dose-response curve can be calculated from the data using regression analysis, and the aPTT range corresponding to a heparin concentration of 0.3 to 0.7 U/mL (by a factor Xa inhibition assay) can be derived. The use of fresh and not previously frozen samples to perform the aPTT is recommended, as residual platelets in the plasma may release platelet factor 4 during freezing and neutralize the heparin, thus decreasing the apparent aPTT.

Laboratory validation of the therapeutic range by the performance of simultaneous aPTTs and heparin assays may be overwhelming, if not technically infeasible, in small hospitals. Some large reference laboratories now offer a service to perform the anti-Xa assay and calculate heparin therapeutic ranges for aPTT when in-house testing is not feasible.

Validation of Heparin Sensitivity of the aPTT Using Ex Vivo Heparin Specimens: Comparison with an Existing Validated aPTT Reagent. The goal of this method is for the laboratory to select a reagent that has the same (or nearly the same) heparin responsiveness as the one currently in use. By doing so, the clinician behavior need not change. It is important to control for and prevent drift with multiple changes over time. This can be done by:

Accumulating patient specimens prior to the time that the new reagent would be evaluated in the laboratory. Specimens should be carefully centrifuged to remove all platelets and frozen at -70°C, in aliquots for future aPTT reagent comparisons. This should not be difficult, even for a small laboratory.

Selecting a new aPTT reagent to test. Laboratories can obtain reagents from suppliers, with responsiveness to heparin that would be predicted to be similar to the reagent that is currently being used in the laboratory. This depends upon the manufacturer knowing the heparin sensitivities of their aPTT reagents in order to supply a reagent with similar sensitivity to heparin when requested.

Comparison testing. Once a potential replacement reagent has been selected for testing, split specimens can be used to perform the aPTT using each of the two reagents on the instrument used in production. The comparison data are plotted with the old reagent on the X-axis and the new reagent on the Y-axis. Visual or regression analysis can be used to judge the acceptability of the comparison data and identify discrepant and outlier results. The data for each aPTT reagent are summed, and the mean and standard deviation determined. The difference between the means of the new and old aPTT reagents are then recorded for future reference.

A cumulative summation of differences. Each time that there is a change in reagents or instrument, the above method should be performed. In addition to recording the difference in the mean, the laboratory should prepare a cumulative summation of the differences that have occurred in the past. In doing so, the cumulative shift in the reagent performance in the presence of heparin can be determined. A difference between reagent means or a cumulative change of more than 7 seconds is reason for concern and necessitates a suitable action: finding a different reagent with an acceptable level of variation; informing the clinicians using heparin of the change in the therapeutic range, recommending that they change their thresholds; or reverifying the aPTT, using comparison with heparin concentration.

It is important to keep in mind that reagents of the same name from the same manufacturer, but with different lot numbers, usually have different sensitivities to heparin. These differences can be substantial.

Validation of Heparin Sensitivity of the aPTT Using Instrument Recalibration. Some coagulation analyzers can be calibrated to make the response of a new reagent match the response of the prior reagent. Using this method, the variability of the reagents to heparin can be compensated by adjusting the instrument itself.

Thrombin Time

The thrombin time (TT) test, which was first developed by Jim and Goldfein in 1957, assesses fibrinogen activity in plasma. The TT test measures the time for clot formation when thrombin is added to citrated plasma. Thrombin catalyzes the conversion of fibrinogen to fibrin, by cleaving fibrinopeptides A and B. Polymerization of fibrin to form a clot occurs. By adding exogenous thrombin, the phospholipid-dependent intrinsic, extrinsic, and common pathways are bypassed. A normal result requires effective release of fibrinopeptides from fibrinogen and unimpeded polymerization of fibrin monomers. The TT test is sensitive to the amount and functionality of fibrinogen in the plasma but is also sensitive to unfractionated heparin, direct thrombin inhibitors, fibrin(ogen) split products, and paraproteins.

Method

1. An aliquot of plasma is warmed to 37°C.
2. A volume of thrombin reagent is added.
3. The sample is observed for clot formation, which is the endpoint of the test.

Reagents

Several TT reagents are commercially available. They vary in source (bovine or human), thrombin concentration, and heparin sensitivity. Manufacturer's instructions should be followed for reagent preparation and handling. If laboratories choose to prepare their own thrombin reagent, typically it consists of thrombin diluted with buffered saline to a concentration of 1 to 5 U/mL. Lower concentrations of thrombin are more sensitive to heparin, dysfibrinogenemias, and other abnormalities than are higher concentrations.

Reference Ranges

The TT normal reference range varies according to the concentration of the thrombin reagent used in the assay. With a lower concentration of thrombin, it is typically 16 to 25 seconds, but may be as low as 8 to 12 seconds with a higher concentration of thrombin. Laboratories should establish their own normal reference range for their thrombin time assay methodology. The source of the thrombin for the assay may be either human thrombin or bovine thrombin. Reagents containing bovine thrombin may give elevated thrombin times in patients who develop bovine thrombin inhibitors after exposure to topical hemostatic agents.

Clinical Utility of Thrombin Time

A prolonged TT is seen in:

❑ Hypofibrinogenemia (usually 90 mg/dL or less).
❑ An abnormal fibrinogen (dysfibrinogenemia).
❑ The presence of fibrinogen or fibrin split products, which inhibits the conversion of fibrinogen to fibrin.
❑ The presence of heparin-like anticoagulants.
❑ The presence of inhibitors to thrombin, such as direct thrombin inhibitor drugs.
❑ The presence of high levels of immunoglobulin paraprotein, which can interfere with fibrin formation.

The thrombin time may also be useful to screen samples with a prolonged aPTT result for heparin contamination.

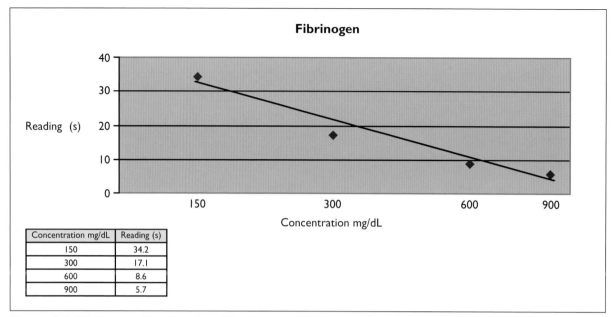

Figure 6-4. Fibrinogen calibration curve. The fibrinogen activity of the sample is derived from a standard curve relating the clotting time to plasma standards of known fibrinogen activity. In the Clauss fibrinogen assay, the concentration of fibrinogen is inversely proportional to the clotting time.

Fibrinogen Assay

Fibrinogen levels can be measured by isolation methods, such as heat precipitation, salt precipitation, or clotting with thrombin, and then weighing the clot or measuring its protein content. It can also be measured by immunological methods, optical density techniques, or a modified thrombin clotting time.

The most commonly performed fibrinogen assay is a modified thrombin time test, known as the Clauss fibrinogen assay. It is used to measure the qualitative amount of fibrinogen in patient's plasma. The Clauss method measures the rate of clot formation after adding a high concentration of thrombin to citrated plasma. The fibrinogen activity of the sample is derived from a standard curve relating the clotting time to plasma standards of known fibrinogen activity (Figure 6-4). In the Clauss fibrinogen assay, the concentration of fibrinogen is inversely proportional to the clotting time. The reference range should be established by each laboratory for their particular assay and reagent lot, but the typical fibrinogen reference range is 170 to 410 mg/dL. In the newborn, the fibrinogen level may be slightly lower due to the presence of fetal fibrinogen, which is not assayed by the functional assay. Fibrinogen levels reach adult levels by 6 months to 1 year of age.

Therapeutic heparin levels and direct thrombin inhibitor drugs typically do not affect the test results because the thrombin is present in excess. However, high levels of DTI drugs may falsely suppress the clottable fibrinogen levels. Low functional fibrinogen levels may be seen in disseminated intravascular coagulation (DIC), during massive bleeding, in liver disease, or in the rare congenital afibrinogenemia, hypofibrinogenemia, or dysfibrinogenemia.

Other functional fibrinogen assays include a turbidimetric method, where a lower amount of thrombin is added to undiluted patient plasma, and the change in turbidity is measured by a spectrophotometer. In a PT-based fibrinogen assay, tissue factor with phospholipids is added to undiluted patient plasma to generate endogenous thrombin, and the turbidity/light scatter is measured. The measured optical change is proportional to the amount of fibrinogen.

In the immunologic (antigenic) fibrinogen assay method, patient plasma is introduced into agar containing antibody to fibrinogen, and zones of precipitation are measured by radial immunodiffusion. The diameter of the zone is directly proportional to the concentration of fibrinogen. Other immunologic fibrinogen assays include enzyme-linked immunosorbent assays (ELISA), sulfite precipitation, and total thrombin-clottable fibrinogen. The immunologic fibrinogen assays are not functional assays, and therefore a normal amount of immunologic fibrinogen does not assure normal functional fibrinogen level. An immunologic fibrinogen assay typically is used to assess the quantitative abnormality of fibrinogen in an individual with a decreased fibrinogen functional assay. Individuals with low functional fibrinogen and normal immunological fibrinogen may have a dysfunctional fibrinogen, or dysfibrinogenemia. Individuals with a low functional and immunological fibrinogen likely have decreased fibrinogen production or increased destruction. The disparity between functional and antigen assays may be less pronounced by PT-based assays.

The fibrinogen activity/antigen ratio may be used as a confirmatory test for dysfibrinogenemia. The test is performed by determining the fibrinogen functional activity and fibrinogen antigen levels on the same plasma sample, and then calculating the ratio. In a healthy normal population, the ratio typically is 0.80 to 1.7. A ratio that is below the reference range is usually indicative of dysfibrinogenemia, assuming that the thrombin time and reptilase time are prolonged.

Clinical Utility of Fibrinogen Assays

Fibrinogen assays are useful in:
❑ The diagnosis of dysfibrinogenemia and hypofibrinogenemia.
❑ The diagnosis of DIC, liver failure, and primary fibrinolysis.
❑ Guiding transfusion therapy with cryoprecipitate.

Assay of Specific Factors

Coagulation factors are necessary for the normal function of the extrinsic and intrinsic coagulation systems. A deficiency of any of these factors will limit the function of the coagulation cascade. Specific coagulation factor assays are performed to diagnose factor deficiencies and the degree of deficiency of the individual factors. It allows the clinician to make a more definitive diagnosis of a bleeding tendency. Specific assays can be performed for factor VIII complex, factor IX, and factors II, V, VII, X, XI, and XII. Automated assays based on clotting, chromogenic and immunological methods are available, but the clot-based assays are most commonly performed.

Principle

Factor assays are performed by determining the extent to which the patient's plasma corrects the clotting time of plasma known to be deficient only in that particular clotting factor. The performance of a one-stage factor assay lies on the ability of a reference, control, or an unknown plasma to cor-

rect the PT or aPTT of factor-deficient substrate plasma as compared to that of patient plasma. The analysis is performed by making serial dilutions, such as 1:10, 1:20, 1:40, and 1:80, of a reference plasma (with known assay values for the factor that is being tested) in buffer. Each dilution is mixed with an equal volume of "substrate" plasma that is known to contain normal levels of all factors, except the one that is being assayed, in which it will be deficient (less than 1%). A PT test (for factors II, V, VII, and X activities) or aPTT test (for factors VIII, IX, XI, and XII activities) is then performed on the serial dilutions. The results are then plotted on a log/log graph as a polynomial curve with percent factor on the abscissa and the time in seconds on the ordinate. The connected points from the above dilutions form a straight line. To analyze a factor level in an unknown sample, dilutions are prepared (1:10, 1:20, 1:40, 1:80) and mixed with substrate plasma. The percent activity is read from the abscissa of the graph by finding where the time obtained on the unknown intercepts the standard curve. Clotting times of patients are thus compared to the reference line (Figure 6-5).

The limits of linearity differ for each factor and correspond to the value of the percent activity of the highest and lowest dilutions of the calibrator. If the limits of linearity for any factor are exceeded with the 1:10 dilution, the sample in the instrument is automatically rediluted 1:2 (if the activity is lower than 14) or 1:50 (if the activity is over 120). If a dependent test result (dilution other than 1:10) falls outside the limits of linearity, the percent activity is extrapolated or Vmin/Vmax is given. Usually at least two dilutions' run should fall within the linear portion of the curve. For an elevated factor activity, the 1:80 dilution clotting time should fall on the curve for an activity up to ~800%. The 1:160 should include a high activity up to ~1600%. When possible, reference material used to construct reference line should be referenced to the WHO international standard.

The CAP requires that assay calibration be performed minimally at 6-month intervals and whenever major maintenance is performed on the instrument or there is a change in the lot number of reagents. The multiple dilutions for each assay, when calculated from reference line and multiplied by the appropriate dilution factor, should agree within approximately 20% CV, depending on the observed factor level. Occasionally, the factor levels show an increasing value with increasing dilution, which is called a "dilutional effect."

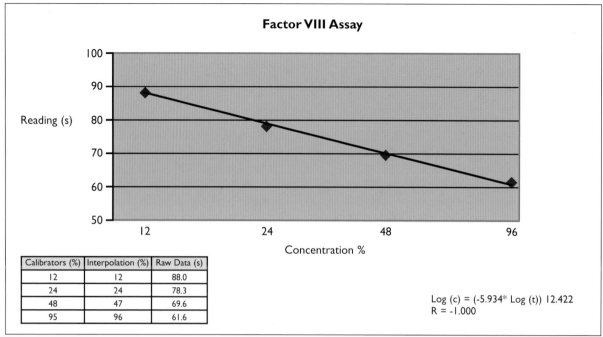

Factor VIII Assay

Calibrators (%)	Interpolation (%)	Raw Data (s)
12	12	88.0
24	24	78.3
48	47	69.6
95	96	61.6

Log (c) = (-5.934* Log (t)) 12.422
R = -1.000

Figure 6-5. Calculation of factor VIII. To analyze a factor level in an unknown sample, dilutions are prepared (1:10, 1:20, 1:40, 1:80) and mixed with substrate plasma. The percent activity is read from the abscissa of the graph by finding where the time obtained on the unknown intercepts the standard curve. Clotting times of patients are thus compared to the reference line. For example, in this figure, a clotting time of 70 seconds would yield a factor VIII level of approximately 48%.

This pattern suggests the presence of a coagulation inhibitor that is diluted out by subsequent buffer dilutions, yielding an apparent increase in factor level.

Factor levels are expressed as percent of normal plasma factor concentrations. Normal plasma typically contains 100% of each of the factors (1.0 unit/mL). The reference range for factors is typically 50% to 150%, but each laboratory should calculate normal reference ranges for each factor using the specific instrument/reagent system in use. At birth, factor levels may be below adult levels (Table 6-3). They reach adult levels at approximately 6 months of age or may remain slightly lower during childhood.

Factor II (Prothrombin)

Prothrombin is a vitamin K-dependent protein. Since it participates in the common pathway, prothrombin activity is most commonly measured by using a PT-based assay or Taipan venom time.

Factors V, VII, and X

Assay for specific activity of factors V, VII, and X is done by measuring the prothrombin time of plasma that lacks any of these factors with diluted patient plasma. Chromogenic factor X assays are useful in monitoring warfarin in the presence of lupus anticoagulant, hirudin, or argatroban, since the chromogenic assay for factor X does not have interference from a lupus anticoagulant or direct thrombin inhibitors. The chromogenic assay principle utilizes an activator of factor X and a peptide substrate for factor Xa that is coupled to a chromogenic marker. Cleavage of the peptide by factor Xa releases the marker and yields a colored product, the intensity of which is directly proportional to factor X levels. Each laboratory should determine its own chromogenic factor X therapeutic range.

Factors VIII, IX, XI, and XII, and High-Molecular-Weight Kininogen and Prekallikrein

Dilutions of the patient's plasma are mixed with substrate factor-deficient plasma in various dilutions, and an aPTT test is done. The degree of correction of the clotting time is compared with that of the normal plasma.

The factor VIII activity assay is important in the diagnosis and therapy of factor VIII deficiency (hemophilia A), acquired factor VIII inhibitors, and von Willebrand disease. This assay is also used to quantify factor VIII concentrations in drug products that are used for treating hemophilia A. The factor IX assay is used to diagnose and monitor treatment for factor IX deficiency (hemophilia

B). Assay for high-molecular-weight kininogen is done only in specialized centers due to the limited availability of deficient plasma.

In addition to the one-stage assay, factor VIII assay levels can also be performed by a two-stage or chromogenic assay. In a two-stage assay, factor VIII dilutions are added to a mixture of reagents, which results in the formation of prothrombinase, which is proportional to the amount of factor VIII added. The chromogenic assay is similar to the two-stage factor VIII assay, with the factor Xa produced assessed by a specific chromogenic substrate instead of a clot-based endpoint. The chromogenic assay shows the best interassay imprecision for factor levels near 150 IU/dL. The best correlation is usually obtained between one-stage clot-based and chromogenic factor VIII assays. The lowest correlation is between the one-stage clot-based and antigenic factor assays. The presence of lupus inhibitors or heparin can cause major interference in the one-stage clot-based assay but not chromogenic or antigenic factor VIII assay. Chromogenic factor activity assays show good precision and good correlation with other assays and no interference from heparins, lepirudin, or lupus inhibitors.

Monitoring Factor Replacement Therapy. Replacement of factor VIII or factor IX to achieve adequate plasma levels to prevent acute bleeding is the basis of treatment in the management of hemophilia. Both factor VIII and factor IX replacement products are either plasma derived or genetically engineered (recombinant). Albumin-free formulations of recombinant "full-length" and B-domain deleted factor VIII concentrates, such as ReFacto® (Genetics Institute) have become available. With the newer recombinant antihemophilic factors, the chromogenic assay typically yields results higher than the results obtained with the more commonly used one-stage clotting assay. Expected factor VIII increments are not achieved when monitored with a typical one-stage clotting time, but the accuracy of the one-stage assay can be improved by employing a concentrate-specific reference preparation for the standard curve. Thus, when monitoring factor VIII response for therapy with newer factor VIII products, using either a chromogenic factor VIII assay or product-specific reference standard material when preparing the assay standard curve is recommended. The discrepancies most likely reflect differences in phospholipid content between both assay systems. When switching between products, it is important to individually titrate each patient's dose in order to ensure an adequate therapeutic response.

False-Positives and False-Negatives in Factor Assays

Variations in instrument or reagents and dilution errors can cause variations in the results obtained for percent factor activity. The presence of the lupus anticoagulant or therapy with heparin or direct thrombin inhibitors may cause falsely decreased results. Such an interference can be overcome by making serial dilutions of the factor to be assayed and observing a plateau or dilutional effect, where the apparent factor level increases with increased dilution, due to progressively lower inhibitor levels.

Factor XIII

Factor XIII is a transglutaminase enzyme that serves as the fibrin-stabilizing factor. It forms cross-links with fibrin in the clot. Factor XIII deficiency usually causes poor wound healing, traumatic bleeding, and postsurgical bleeding. Factor XIII activity cannot be assessed by either the PT or aPTT assays, as the clotting endpoint detected in those assays occurs prior to clot stabilization or cross-linking. Thus, deficiency of factor XIII typically is accompanied by a normal PT and aPTT result. For screening for factor XIII activity, a qualitative screening assay is performed, where the plasma is recalcified and allowed to clot. The clot is suspended in 5M urea or 2% acetic acid, which disrupts the noncovalent hydrogen bonds but does not degrade any factor-XIII-dependent covalent cross-links. If factor XIII is decreased, then there are no covalent bonds in the fibrin clot. Therefore, in factor XIII deficiency (less than 5% level), the clot will dissolve, while normal cross-linked clots do not dissolve in 5M urea or acetic acid.

Quantitative factor XIII assays are not routinely available but can be performed by a variety of techniques, including a dansylcadaverine assay, microtiter assay utilizing a bound substrate, and monoclonal antibody-enzyme conjugate.

The indications for factor XIII testing include delayed bleeding with normal PT, aPTT, and platelet count. The qualitative factor XIII test can not be used to identify heterozygous factor XIII deficiency or homozygous patients with mild deficiency. The test may also give false results if the concentration of urea and calcium chloride are not optimal.

Von Willebrand Factor

Von Willebrand factor (vWF) is a large, multimeric carrier protein for factor VIII that is involved in platelet adhesion to sites of vascular injury. Deficiency of vWF results in a bleeding tendency. Testing for von Willebrand disease involves assay of factor VIII, vWF antigenic activity, and multimer structure, and is performed for the diagnosis, classification, and therapy of von Willebrand disease. This testing is discussed in detail in chapter 16.

Mixing Studies

Mixing studies are performed to detect inhibitors of clotting and to determine whether the prolongation of the aPTT or PT is due to deficiency in factor levels or due to a circulating inhibitor, often referred to as a circulating anticoagulant. Circulating inhibitors can be specific factor inhibitors, nonspecific inhibitors such as the lupus anticoagulant, or even circulating inhibitor drugs such as heparin or direct thrombin inhibitors. Specific factor inhibitors are mostly immunoglobulins of the IgG class. They typically develop in patients with severe factor deficiencies, but the underlying condition for the inhibitor may not always be obvious. Some clinical conditions where a circulating anticoagulant can develop as an autoantibody include malignancy, lymphoma, autoimmune disorders, monoclonal gammopathies, myeloma, drug reactions, old age, and the postpartum period. In other instances, the underlying condition may not be known, and the circulating anticoagulant can develop suddenly without a preceding illness. Specific factor inhibitors are associated with clinical bleeding, which can be severe or, in milder cases, can present with ecchymoses or easy bruising. Alternatively, the lupus anticoagulant is a type of antiphospholipid antibody that also acts as an inhibitor in plasma but is associated with a risk of thrombosis, as described in chapter 22. Inhibitors of coagulation can be a very challenging problem to evaluate in any coagulation laboratory.

Before a mixing study is started for evaluation of a prolonged PT or aPTT, some common preanalytical variables must be excluded. The plasma specimen must be evaluated to exclude any fibrin strands, which could preactivate the sample. A high hematocrit can result in falsely prolonged PT and aPTT due to excess citrate anticoagulant. Heparin contamination may be detected by screening samples with a thrombin time or anti-Xa

inhibition assay. A heparin effect should be excluded by a removal/neutralization procedure, such as protamine neutralization, or heparinase. Ideally, the aPTT should be reassessed after heparin neutralization to ensure that the aPTT has corrected into the normal range. If the aPTT is still prolonged and no residual heparin activity is detected, then further studies to exclude an inhibitor are warranted. If the thrombin time is elevated with a negative anti-Xa assay, this suggests sample contamination by a direct thrombin inhibitor drug. These drugs cannot be neutralized or removed by any of the heparin removal techniques, and further assessment of a circulating anticoagulant is not possible. An algorithm is presented in Figure 6-6.

The rationale for performing mixing studies is to differentiate between a factor deficiency and an inhibitor. The laboratory can incorporate mixing studies into an appropriate algorithm that will lead to identification of a specific factor deficiency or determine the type of inhibitor present. Such an algorithm and testing process is discussed in chapter 10.

Principle

Inhibitors are typically present in excess, such that if present, they will inhibit normal and patient plasma. Approximately 50% of any factor is sufficient to yield a normal clotting time result. In practice, 30% to 40% of the normal level of a factor is sufficient to yield a normal PT or aPTT in a 1:1 mixing study. If there is correction to within the normal range after mixing, a factor deficiency is indicated. If there is no correction, the presence of inhibitor is indicated. This scheme is shown in Figure 6-7. A 1:1 aPTT mixing study that shows correction does not exclude the presence of an inhibitor. A 4:1 mixing study may help to detect a mild inhibitor. However, if the inhibitor is time dependent, such as factor VIII inhibitors and about 15% of lupus anticoagulant inhibitors, timed incubation mixing study is necessary.

In order to perform a mixing study, either or both PT and aPTT results must exceed the upper limit of a laboratory's defined reference range. Depending on which test result exceeds the upper limit, one or both mixing studies are performed.

Technical Considerations

❑ Normal pooled plasma should be a pool from a minimum of 20 donors (preferably 30 donors) with ~100% factor levels, and fresh frozen, cell free. Either an in-house prepared pooled nor-

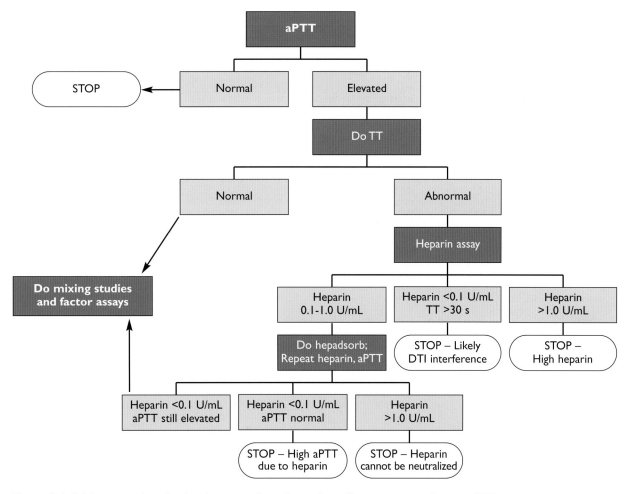

Figure 6-6. A laboratory algorithm for detection of interfering drug effects prior to performing aPTT mixing studies.

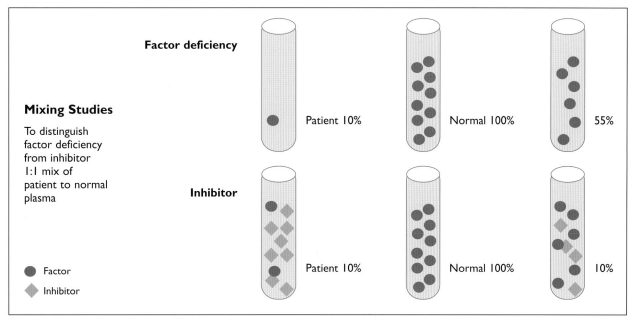

Figure 6-7. A schematic diagram for performing and interpreting a plasma mixing study. With factor deficient plasma, addition of normal plasma brings the factor level to over 50%, which would normalize the PT and aPTT. With plasma with excess inhibitor, all the available factor would be bound by the excess inhibitor and the aPTT will remain prolonged.

mal plasma or commercial normal pooled plasma (NPP) can be used. It is recommended that the laboratory ensure that the aPTT and PT of the plasma pool used falls within the laboratory's normal reference ranges.

❑ The use of lyophilized products is not recommended.

❑ The use of single normal patient plasmas is not recommended, since a normal PT and aPTT do not guarantee normal levels of all clotting factors.

❑ Factor sensitivity of the chosen PT or aPTT reagent should be ~30% to 40% for all factors. A mixing study is only as good as the PT and aPTT on which it is based. Therefore, it is critical for the laboratory to know at what factor levels their reagents will cause prolongation of the PT and aPTT and how sensitive their reagents are to the lupus anticoagulant. Sensitivity to lupus anticoagulant is determined by concentration and type of phospholipid in the aPTT reagent.

Procedure

In the classical 1:1 mixing study, the clotting time of the patient and normal plasma alone is determined. Then, equal amounts of patient plasma and normal plasma are mixed, and the clotting time of the mixture is measured. Immediately after preparation of a 1:1 mix, the mixture should be tested in the test system that was originally prolonged. A 1:1 mix usually discriminates between factor deficiencies and inhibitors if initial clotting times are prolonged, ie, aPTT greater than 10 seconds and PT greater than 3 seconds above the upper limit of their respective reference ranges.

Another type of mixing study is the 4:1 mix, where 4 parts of patient plasma is mixed with 1 part of NPP. This mixing study is usually used for aPTT and only if the aPTT is minimally prolonged (3 to 10 seconds above the upper limit of normal range).

Mixing studies can be performed in two time frames: immediately upon mixing and following incubation at 37°C. Incubation is used to tease out time- and temperature-dependent inhibitors, especially factor VIII inhibitors. These mixtures can be tested immediately and after incubation for 1 to 2 hours at 37°C. If one incubates a 50:50 mix, one should also run a mixture of patient and NPP separately incubated and then mixed after 2 hours at 37°C to serve as a control for factor V and factor VIII lability.

Criteria for Interpretation of Mixing Studies

The most common criterion to assess correction (or lack thereof) is to use an established reference range in the laboratory or to gauge correction against NPP that is tested in conjunction with the mixing study. Different targets are used to determine correction, and there is no general consensus on how to interpret the results of mixing studies. Commonly used targets include correction to within the reference range, correction to within 5 seconds of NPP, 10% of the NPP value, or lack of correction at greater than 15% of the NPP value.

One may also use the Rosner index or Chang percentage to determine correction. The Rosner index is based on a 1:1 mix of patient to NPP. The Chang percentage is based on 4:1 mix of patient to NPP, which may increase the sensitivity.

$$\text{Rosner index} = \frac{\text{Clotting time of 1:1 mix} - \text{Clotting time of NPP}}{\text{Clotting time of patient}} \times 100$$

$$\text{Chang percentage} = \frac{\text{Initial clotting time} - \text{Clotting time of 4:1 mix}}{\text{Initial clotting time} - \text{Clotting time of NPP}} \times 100$$

With the Rosner index, an index >15 suggests the presence of an inhibitor. With the Chang percentage, a result of <50% suggests the presence of an inhibitor.

Although the principle of the mixing study may seem simple, in practice, the interpretation of mixing studies can be more complex than the actual performance of the test. There is no uniform agreement as to what criteria should be used for correction. Therefore, each laboratory has to determine its own guidelines as to when to perform the test, whether to perform a mixing study for a prolonged PT, what the criteria would be for correction, and how to report it (see Table 6-4). A consultative interpretation method may be very useful in this regard.

Specific Inhibitor Studies

Specific coagulation inhibitors can develop in patients with severe deficiency of a factor after treatment with plasma products but also can develop idiopathically and in association with some malignancies and drug therapies.

Bethesda Assay

If a patient sample with a long aPTT has been identified as containing an inhibitor, after positive mixing studies and demonstration of low factor levels, it is important to measure the intensity of the inhibitor. The standard assay for this is the Bethesda assay in the USA. The Bethesda assay is based on the ability of the patient's plasma containing the factor VIII (or other factor) inhibitor to inactivate factor VIII present in normal pooled plasma (NPP). Dilutions of the patient's plasma are incubated with NPP for 2 hours at 37°C, and then the residual factor VIII is measured. Although the Bethesda assay is commonly performed for detection of an inhibitor to factor VIII, it can also be used to quantify inhibitors to factor V and factors IX, X, XI, and XII. The test is performed as follows:

1. Serial dilutions of a patient's plasma are mixed with normal plasma containing a known amount of factor VIII (or other factor).

2. The mixture is incubated for 2 hours at 37°C.

3. Residual factor VIII activity is determined by a factor VIII assay and compared to a normal control.

The control uses equal amounts of NPP and imidazole buffer and is processed similarly. Calculation of Bethesda titer is obtained off a log-log graph, where the y axis equals the percentage residual factor VIII:C after incubation, and the x axis represents the dilution of patient's plasma. In some patients, the calculations may be approximate and may underestimate the titer. One Bethesda unit (BU) is defined as the quantity of inhibitor that neutralizes 50% of the factor VIII in normal plasma in 2 hours at 37°C.

Although it is a standardized test, the results of the assay depend on the specific aPTT reagent, contact activating agents used, and the nature of the phospholipids. Although the precision within the laboratory is excellent for this assay, there is significant variation in instruments used, sources of phospholipids, factor deficient substances, and activating agents. Heparin contamination and the presence of a lupus anticoagulant may interfere with the Bethesda assay. With the use of heparinase enzyme, this interference can be minimized significantly.

Nijmegen Modification

In 1995, a modification of the Bethesda assay which improved the specificity and reliability of the system especially for low-titer inhibitors was reported. When the inhibitor titer is low, a modified method is used. The Nijmegen variant of the Bethesda assay involves two modifications. First,

Table 6-4. Guidelines for Interpretation of Mixing Study Results

Complete correction
- ❑ Suggestive of factor deficiencies
- ❑ Correction on immediate mix does not rule out the presence of time-dependent inhibitor

Partial correction or no correction
- ❑ Suggestive of an inhibitor
- ❑ Time dependent inhibitors (factor VIII and some lupus anticoagulants) give prolonged clotting time results on incubated mix compared to control
- ❑ Immediate acting inhibitors (lupus anticoagulants, factor IX, von Willebrand factor, heparin, direct thrombin inhibitors)

Borderline correction and no time dependency
- ❑ Suspicious for lupus anticoagulants
- ❑ Partial correction could indicate factor deficiency

the pooled buffered plasma is stabilized with imidazole buffer. Next, the control sample is mixed with deficient plasma instead of buffer. Both the buffering effect of imidazole buffer and the protein concentration of the factor-deficient plasma maintain the pH of the reaction mixture for 2 hours. Incubation is at 37°C. Assay for residual factor VIII activity is done by either a clotting or chromogenic method. The use of buffered normal plasma stabilizes factor VIII in the plasma during the incubation, which reduces the possibility of getting false-positive results of low titer and reduces the coefficient of variation compared to the classical Bethesda method. If a high titer is suspected (or if the titer is not known), the sample should be diluted with factor-deficient plasma.

Reporting of Results

The results of the Bethesda assay are reported in Bethesda units (BU), where 1 BU is defined as that amount of inhibitor that results in 50% residual factor VIII activity. A low inhibitor titer is 0.5 to 5 BU, and a high titer is over 5 BU. A low inhibitor titer by Nijmegen modification is <0.4. Low responders include patients with a low inhibitor titer that does not rise on re-exposure to factor VIII. High responders are patients with an inhibitor titer that rises sharply on re-exposure. Spontaneous (acquired) factor VIII inhibitors in nonhemophiliacs often demonstrate a complex non-linear relationship, and the Bethesda assay may underestimate the potency of the inhibitor.

Factor VIII inhibitor assays show a high inter-laboratory variation. There are a number of variables that influence test outcome: pH stability of incubation mixtures, type of control sample, type of deficient plasma used in the assay, liquid handling, incubation time, factor VIII content of normal pooled plasma, type of factor VIII assay, and type of reagent.

Establishing Reference Ranges and Validating Coagulation Laboratory Testing

Good-quality hemostasis laboratory testing requires proper sample collection and processing, as discussed in chapter 4. The tests should also be performed accurately and the results reported correctly. In order to validate an assay in the hemostasis laboratory, several parameters need to be determined for each assay, including reference range, accuracy, precision, linearity, and interfering substances. Periodic quality control and calibration is necessary to verify optimal ongoing performance of the test.

Preanalytic and Postanalytic Considerations

Reference Range

The reference range, or reference interval, is the range of laboratory test results expected for testing in normal individuals. Calculation of the reference range has been discussed above for individual assays, but it is typically established by collecting specimens from at least 40 healthy individuals, performing the test, and calculating the mean and standard deviation of the test results. The reference range usually consists of a central 95% of population of interest, calculated as the mean value +/-2 SDs of the mean. Individuals to be excluded from reference range testing would be those receiving anticoagulant medication or oral contraceptives/hormone replacement therapy, individuals with known bleeding or thrombotic disorders, and individuals with illnesses such as liver disease, malignancy, renal disease, or acute infections. With regard to coagulation screening tests, due to the variability between different types of PT and aPTT reagents, there may be significant changes in the reference interval after a change of the type or lot of reagent. For this reason, the laboratory should establish or verify the reference

intervals with each lot or change in reagent or change in instrument/reagent combinations.

Periodic re-evaluation of reference ranges can be used to validate test procedure stability within a laboratory. This can also be useful in verifying agreement between laboratories that use similar procedures. Reference range determination must be clinically appropriate for the population served by the laboratory. This is especially important if testing is performed on infants and children, as the reference range for many analytes is different in the pediatric population compared to adults, as shown in Table 6-3. Establishing an in-house pediatric reference range may be problematic due to sample acquisition. Pediatric reference ranges have been published along with adult reference ranges. These can be adapted to pediatric reference range values for an individual laboratory by comparison of the laboratory's adult reference range with the published adult reference range and pediatric reference range values. An example ratio formula is shown below.

$$\text{Lab pediatric range} = \frac{\text{Published pediatric range}}{\text{Published adult range}} \times \text{Lab adult range}$$

Accuracy

Accuracy is a measure of agreement between the estimate of a value and a true value. Any departure from accuracy is expressed as systematic error or bias. Accuracy of an analytical process is expressed as the difference between the average result obtainable by a method under specified conditions and the result accepted as true or standard. It is expressed in the same units as the result or as a percentage of the standard result. The lower the difference, the greater the accuracy or lower the inaccuracy. The difference includes contributions not only from process inaccuracy, but also from process imprecision. In practice, accuracy testing for a new reagent, lot, or instrument should include testing a group of samples spanning the analytical range by both the old and new methods and plotting the results with respect to each other. The data can be analyzed by a linear regression analysis, but the correlation coefficient, R, is not the only measure of accuracy. For analytes with identical units, attention should be paid to the slope (ideally this should be close to 1.0) and the intercept, which should be close to zero. The data can also be analyzed by use of a bias plot, whereby the results with the old method are plotted against the difference between the new and old methods (Figure 6-8). For an accurate method,

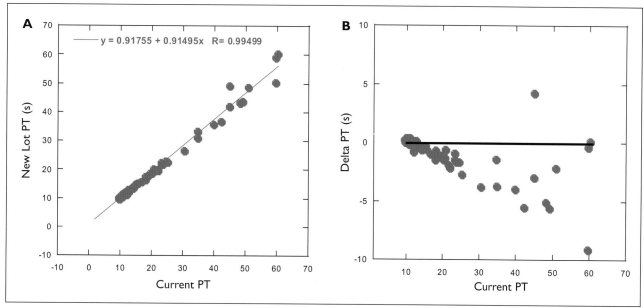

Figure 6-8. A graphical representation of accuracy for a laboratory test can be displayed by graphing the current results with the comparison method (A). Attention should be paid to the slope (which should be close to 1.0) and the intercept (which should ideally be zero). In the example shown for the prothrombin time and two lots of reagents, the slope is 0.91 and the intercept is 0.92 seconds.

A bias plot (B) is helpful to detect accuracy trends with increasing values. For a bias plot, the current method is plotted on the X axis, while the Y axis is the difference of the comparison method minus the current method. In the example shown, a bias plot derived from the data in A, the comparison method yields slightly shorter prothrombin time (PT) results at high PT values.

there should be no trends or bias between the methods for increasing analyte values.

Precision

Precision is the closeness of agreement between independent test results obtained under stipulated conditions. A test is said to be precise when repeated analyses on the same sample give similar results. When the test is precise, random variation is small. Precision is usually expressed quantitatively in terms of imprecision, such as the coefficient of variation (CV) of the results in a set of replicate measurements. The CV is a measure of dispersion of a probability distribution. It is defined as the ratio of the standard deviation to the mean:

$$CV = Standard\ deviation\ /\ Mean$$

The CV is a dimensionless number. For distributions of positive-valued random variables, it allows comparison of the variation of populations that have significantly different mean values. It is often reported as a percentage (%) by multiplying the above calculation by 100. For coagulation assays, determination of intra-run and between-run precision is useful to determine assay performance. Intra-run precision is predominantly a measure of assay/instrument performance, while

between-run precision also investigates technologist variability and instrument quality control.

Linearity

Linearity is the ability, within a given range, to provide results that are directly proportional to the concentration of the analyte in the test sample. Linearity refers to overall system response and thereby the final analytical answer. Most coagulation screening assays, such as the PT and aPTT, do not give a linear response over a range of analyte concentrations, so establishment of linearity is not generally required for these tests. However, linearity is important to establish for quantitative analyte assays, such as fibrinogen. To establish linearity for an assay, a sample with a known high concentration is serially diluted and the concentrations measured on the diluted samples. A graph of the dilution versus the concentration should give a linear response.

Interfering Conditions

With optical instruments used for clot detection, test results may be affected by interfering conditions such as lipemia, hemolysis, and icteric samples.

Lipemia refers to an abnormally high concentration of lipid in the blood. Large lipid particles, especially chylomicrons and lipoproteins, can interfere with laboratory tests, including coagulation testing. These particles increase the turbidity, thus interfering with light detection or scatter systems of the instruments. In optical clot-detection methods used for PT and aPTT determination, lipemia results in artificial prolongation of clotting times. The degree of lipemic interference varies with different instruments. In optical detection methods, if a sample is lipemic and the baseline optical density (OD) is too high, the instrument will not report the result. If a baseline OD is within an acceptable tolerance limit, the result will be reported by the instrument. Some institutions report the result if the result can be duplicated within 5%, even if the sample is lipemic. Turbidimetric and nephelometric methods, which measure scattered light, may also be affected by lipemia and other interfering substances. The degree of interference by lipemia may depend on the degree of dilution. In assays that utilize a higher dilution, the interference may be less. Mechanical or electromechanical methods of clot detection are not affected by lipemia. According to CLSI guidelines on processing blood specimens for coagulation testing, clotting of lipemic samples should ideally be measured by mechanical or electromechanical methods to minimize interference by lipemia.

Extraction of lipids from blood specimens with n-hexane before coagulation testing has been described, but it is not a common practice in most laboratories. Ultracentrifugation is suggested by some for clearing lipids. However, no published data is available that compares the cleared samples to uncleared samples with regard to coagulation testing.

In circumstances when a mechanical or electromechanical detection system is not available, the following guidelines may be useful:

❑ Visualize a specimen that is flagged lipemic to see whether it is lipemic.

❑ Determine if the optical properties are within instrument specifications.

❑ If the specimen successfully passes the review, the result can be reported with a comment about lipemia.

❑ If the specimen fails the review process, an aliquot of the sample may be ultracentrifuged to separate the lipid. The cleared plasma sample can be retested by the optical instrument, and the result may be reported with a comment about lipemia.

❑ Another option would be to re-collect a sample from the patient after the patient has been fasting for 12 hours. However, this may not be practical in some instances.

Hemolysis, causing leakage of hemoglobin and other internal components from the erythrocyte membrane into the surrounding fluid, is usually defined as extracellular hemoglobin concentrations greater than 0.3 g/L. Hemolysis confers a detectable pink-to-red color to serum or plasma and is clearly visible in samples containing as little as 0.5% hemolysate. Hemolysis is caused by biochemical, immunologic, physical, or chemical mechanisms. While in-vitro blood cell lysis can be prevented by proper sample collection techniques and storage, in-vivo hemolysis, which can arise from various disease processes, does not depend on the blood collection technique or procedure and is unavoidable.

Hemolysis is the most frequent cause of specimen rejection, as indicated by a CAP Q-Probes study on chemistry specimen acceptance (Jones BA et al, 1997). Release of hemoglobin and additional intracellular components from erythrocytes, white blood cells, and platelets into the surrounding fluid may cause false elevations of substances in plasma, may cause dilution effects, may increase the optical absorbance, or may change the blank values producing spectrophotometric interference. Hemolyzed samples are not suitable for coagulation assays because of the release of hemoglobin, intracellular components, and thromboplastic substances from damaged blood cells. Although slightly hemolyzed samples may still be analyzable, a moderate blood cell lysis as low as 0.9% can influence the reliability of coagulation testing. Increases in PT and dimerized plasmin fragment D and decreases in aPTT and fibrinogen can occur in hemolyzed samples. Because the interference in coagulation assays can have a wide interindividual bias, lysis correction is not usually recommended, and the most appropriate corrective measure would be sample recollection. Prevention of hemolysis can be achieved through standardization of the preanalytical phase, ie, use of standardized needles, closed tubes, and calibrated centrifuges.

Icteric samples. Bilirubin has a high absorbance between 340-nm and 500-nm wavelengths. Therefore, the presence of bilirubin in a sample can be a limiting factor in spectrophotometric procedures using this wavelength range, because of the constantly high background absorbance

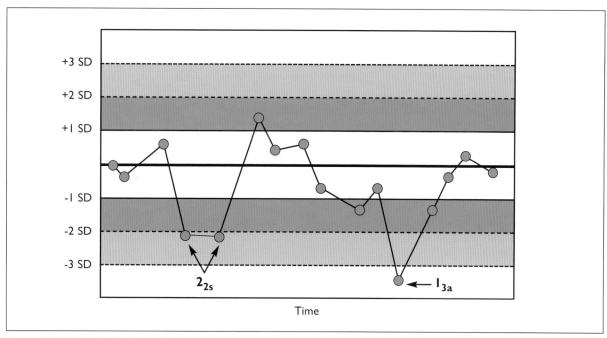

Figure 6-9. Example of a Levey-Jennings chart for laboratory quality control, displaying quality control (QC) values over time. Two Westgard rules for adequacy of QC were violated in this example: 1_{3a}, because one QC result is more than 3 standard deviations (SD) from the target mean, and 2_{2s}, because two control results were more than 2 SD from the target mean.

caused by bilirubin. In turbidometric coagulation assays, a bilirubin concentration exceeding 25 μmol/L can lead to clinically relevant changes of the measured values of antithrombin. In higher concentrations, the interference can be much more significant.

Quality Control

Quality control (QC) is an organized system for continuously monitoring the analytical processes to ensure true and precise data. It is designed to detect random errors and trends in test processes. This involves preanalytical variables (collection and processing of the specimens), the analytical process, and reporting of the results (postanalytical component). Quality control can be long term or short term. In routine practice, QC is done through the use of Levey-Jennings charts or multivariate rules programmed into the instruments. Levey-Jennings charts display QC data over time. One axis represents time or sequential run numbers. The other axis is the value of each quality control result; this axis is typically centered around the target mean, with upper and lower control limits marked, and allows laboratories to visualize QC results over weeks or months.

The goal of quality control is to provide assurance that the assay system is working to specifica-

tions and to determine significant assay errors without repeating the tests unnecessarily. A significant error is a wrong test result that would impact the diagnosis and treatment of the patient. In order to detect clinically significant errors, it is important to determine specific QC rules for an assay based on its total allowable error (TEa) and its specific performance. TEa (allowable error) is either based on the Clinical Laboratory Improvement Amendments of 1988 (CLIA '88) or CAP requirements for proficiency testing, or determined from individual and group biological variance. CLIA defines criteria for acceptable performance for prothrombin time (TEa=15%), activated partial thromboplastin time (TEa=15%), and fibrinogen (TEa=20%). CAP defines a TEa criterion of 20% for the acceptable performance of INR.

A few simple guidelines to follow in QC of an assay are:

1. Validate an assay to determine imprecision (% coefficient of variation [CV]) and accuracy (bias).
2. Establish the mean and standard deviation using the CV for the assay.
3. Determine TEa for the analyte from CLIA regulations or biological variation tables.
4. Using the Operational Process Specification (OPspecs) charts (can be obtained from the Westgard web site, www.westgard.com), calcu-

late Westgard rules for the assay based on the precision and accuracy of each analyte.

Westgard rules (Multiple rules) are a set of rules to help increase the error detection rate without increasing the false-rejection rate. A graphical display of an example of a monthly QC chart is shown in Figure 6-9. The Westgard rules were based on the fact that the use of simple upper and lower control limits lacked the power to identify analytical problems. To attain an acceptable rate of error detection, control limits have to be narrow and control testing frequency has to be high. The symbolic description of Westgard multirule set is as follows:

1_{3s}: Any control result more than 3 SD from the target mean.

2_{2s}: The last two control results (or two results from the same run, even on different control materials) are more than 2 SD from the target mean.

R_{4s}: The difference between successive control results is 4 SD.

4_{1s}: The last four consecutive control results exceed x + 1s or are below x.

10_x: The last ten consecutive control results are on the same side of the target mean.

Westgard multirule sets can be modified by adding or subtracting rules, changing the action limits, or changing the number of control results tested by the rules. Each modification alters the error detection rate and the false-rejection rate.

Laboratory personnel with appropriate experience should inspect the quality control results daily to evaluate for trends, shifts, and outliers. Individual patient results should be reviewed for unusual or unlikely patterns. Such unusual variations may indicate clerical errors or instrument or system errors or malfunction. Maintenance of all instruments should be carried out in accordance with manufacturers' directions. All actions taken must be appropriately documented. Manufacturers' instructions for reagents and equipment should be followed. There should also be a periodic review of quality control data to look for long-term changes in the analytical systems and comparison of those of the peer group.

In general, collection of quality control data should be simple and low cost. Data should be collected in the routine course of testing by the same personnel who would produce results during routine patient care setting, and saved in a form of documentation acceptable to the licensing or accrediting agency. Data should be summarized in a form that is easy to review and to identify unac-

ceptable results. Each laboratory should keep accurate and complete records of the lot numbers of reagents, reference materials, and blood collection devices. All reagents and solutions should be marked with expiration dates, dates received, and reconstitution dates, where applicable.

If there are trends, shifts, or out-of-limit values of quality control results, or if there are spurious patient results, the entire coagulation testing system should be reviewed to determine if there is a problem with phlebotomy, the technologist performing the testing, the reagent and/or instrument, or if an unexpected or unknown component is involved. It is imperative when troubleshooting to investigate whether the problem involves one test or a group of tests, is due to sample integrity or interfering substances, or involves one particular type of assay, such as a clot-based or chromogenic assay.

Delta checks can be used to identify a patient's result that has deviated significantly from his/her previous results. Delta-checks are considered as part of the usual QC in most laboratories. They can help to assess whether the significant change in the results is due to a change in the patient's status or treatment effect or whether they are due to an instrument or procedural malfunction in the laboratory.

Frequency of QC Testing

For all nonmanual coagulation testing systems and photo-optical or electromagnetic coagulation testing systems, the laboratory should include at least two different levels of control material during each 8 hours of patient testing and each time there is a change in reagents. When documenting control results, the laboratory should identify shifts. The manufacturer's recommendations regarding controls and calibrators should be followed. The two levels of control material should encompass both the abnormal and normal range of reportable patient values. Manual coagulation systems should be checked by each testing person with two different levels of control material in duplicate during each 8 hours of patient testing and each time there is a change of reagents. Controls once thawed or reconstituted should not be refrozen or reused unless otherwise stated by the manufacturer. If the test values for the control samples are not within established limits, appropriate action must be taken, such as testing new controls and method troubleshooting. Acceptable results must be obtained before patient samples are tested and reported.

Results Reporting

Determinations are usually performed in duplicate for manual coagulation testing, and the mean of the two values reported. With the much improved automated and semiautomated coagulation instruments, single testing is acceptable.

Turnaround Time

Each laboratory should define an expected turnaround time for each test performed. The turnaround time is the interval between the time a test is ordered by the physician or a designated person to the time the results reach the patient's records. In reality, the turnaround time mostly reflects the interval between sample collection/sample receipt in the laboratory and the time the result is reported. Periodic assessment of actual turnaround times should be performed, especially for testing ordered in a "stat" fashion or by the emergency department.

Reporting of Results

The final report of laboratory results is the culmination of the entire testing process. Depending on the laboratory information system and the test ordered, reports may include preliminary and final reports. The final report should be in a format determined by the individual laboratory, which can be easily understood by the medical staff. The final report should include all pertinent information, namely, patient's name; unique identifier; location, date, and time of sample collection; person ordering the test; name of the test ordered; date and time of receipt of the specimen in the laboratory; result of the test, with reference range/ranges; interpretation of results, where necessary; appropriate comments, such as adequacy; name and location of the laboratory performing the test; and all required signatures. There must be a process to verify accurate transcription and transmittal of the results to the physician and the medical records. It is important for the laboratory to work closely with the information system's staff to ensure that the hemostasis assays are reported in an acceptable and easily understandable format. When the hemostasis testing instrument transmits data electronically to a laboratory or hospital information system through an interface, it is crucial for the laboratory to verify that the results are transmitted accurately and that the results are identical in the laboratory instrument, laboratory information system, and hospital information system(s).

Patient care staff should be notified of abnormal laboratory values that could affect patient care or treatment. For example, patients being treated with heparin whose aPTT values are supratherapeutic are at risk for hemorrhage, and the patient care staff and/or physician should be notified of this value in a timely fashion so the heparin dose can be adjusted. The laboratory staff should have explicit instructions about when and how to notify appropriate personnel when results are in the laboratory's "alert" or "panic/critical" values or "courtesy call" values. These values should be predetermined by each laboratory, and there should be a process in place to confirm patient identification before verbal reports are given. It is imperative to make sure that the verbal reports are heard clearly by the receiving personnel and the identity of the recipient documented. Such verbal reports should be well documented, and the final report should include any preliminary reports issued, either verbal or written. Verbal reports of abnormal values should not be left on voice mail or telephone answering machines.

In coagulation testing, especially when an algorithmic approach is used to evaluate a bleeding patient, one may choose to develop an interpretive reporting system. Consultative coagulation testing is "goal-oriented," and the report can give an interpretation of results, indicating cause and significance of the abnormalities. This system can give flexibility in reporting as well as consistency in reporting, especially if coded comments are used, thus enabling the ordering physician to understand the test results better in comparison to receiving mere numbers as results. The advantages of consultative coagulation testing and reporting include efficiency, streamlining of testing, cost effectiveness, decreased sample volume, rapid results, exclusion of interfering drugs, and the provision of an interpretation of results, including an estimation of bleeding risk and recommendation for therapy. However, consultative or interpretive coagulation testing comments should be rendered by pathologists or laboratory physicians with an expertise in hemostasis testing and should take factors other than test results into consideration, such as patient history and medication or transfusion therapy.

There should be an established mechanism for correcting any erroneous test results, notifying the appropriate personnel, and issuing a corrected report. There should be clear instructions for personnel about how to reconcile any disparities between preliminary or verbal reports and final test results.

Quality Assurance

Quality assurance (QA) involves all aspects of the coagulation testing process and appropriate response to problems one may encounter during the analytical as well as preanalytical and postanalytical phases. This includes the test order, completion of requisition forms, specimen collection, transport and processing, accessioning, instrument function and operator proficiency, transmission of results, and interpreting results. In contrast, quality control (QC) is designed to detect random error, shifts, and trends of the testing process in the laboratory. In other words, QC assesses only the analytical phase of the testing process, while QA assesses the entire testing system. Recommendations for QA processes in hemostasis testing are shown in Table 6-5.

To evaluate the quality of test performance, the clinical laboratory should use two different kinds of quality assurance: external quality assessment and internal quality assessment. Internal quality assessment is designed to establish the stable performance of an assay system on any single day as well as from day to day. External quality assessment, on the other hand, provides laboratories with a tool to evaluate the performance of the individual laboratory in relation to other laboratories. Each system has its own factors, which may or may not vary between different analytical runs. Internal quality assurance can only control differences in batches of reagents and calibrators but can not control the assay principle. External quality assurance methods can help the laboratory to measure their test results by the assay principle used in their laboratory as compared to those used by other laboratories. External quality assurance methods can also control variable factors and help to evaluate the quality of test performance over time. However, the two methods are complimentary; one system can not replace the other.

External Proficiency Testing

Each laboratory should participate in an external quality assessment/proficiency testing program that will provide an external means to verify the test method and results for each analyte tested. The program should include a means to determine the reliability of patient test results. External quality assurance is an important component of total quality assurance to any coagulation laboratory.

Some of the most prominent external quality assurance programs include the CAP proficiency testing program (www.cap.org), North American Specialized Coagulation Laboratory Association (NASCOLA/ECAT) proficiency testing program (www.nascola.org), United Kingdom National External Quality Assessment Service (UKNEQAS) in the UK, ECAT in Europe, and Royal College of Pathologists of Australasia (RCPA) in Australia. The mission of such programs is to provide external quality assessment for tests used to diagnose bleeding or prothrombotic disorders, thereby promoting high standards of performance and practice for coagulation laboratories. They help to meet the requirement of external quality assurance by providing a proficiency testing program that evaluates a broad range of hemostasis methods and analytes. Through such programs, laboratories improve the quality of their practice, improve patient care and safety, can characterize test accuracy and precision across multiple methods and test centers, can correlate specific method variables with accuracy and precision, and can identify interfering substances for certain tests and quantify their effects for multiple methods.

Laboratories in the US are required by CLIA to participate in proficiency testing programs in order to maintain accreditation with professional organizations and to retain their licenses. These external quality assessment programs provide three to four surveys per year. One or two plasma samples per analyte are usually provided to participating laboratories. Participants are asked to treat the sample, using their own laboratory methods, similar to their clinical laboratory samples. Results are usually reported in standard forms. Standard statistical analysis of the results are performed after each survey, and the results are provided to the participants. The results usually include standard deviation, mean and median, and comparison of the laboratory results using same or similar instrumentation or reagents. This usually enables the participating laboratory to determine their test performance and the derivation of the laboratory result from the consensus value. Usually, consensus values are used in coagulation surveys for comparison because of the lack of a comprehensive reference system in hemostasis testing.

Each proficiency testing program (governmental or independent accreditation agency) has requirements that may vary as to the scope, schedule, and evaluation process. When laboratories participate in external quality assessment pro-

Table 6-5 Recommendations for Quality Assurance Evaluations in Hemostasis Testing

❑ There should be a system to periodically measure the actual platelet concentration of the usual "platelet-poor" plasma.

❑ If the laboratory has more than one instrument for performing the same coagulation assays, the instruments/reagents should be checked against each other at least twice a year for correlation of patient results.

❑ Guidelines and criteria should be set for use of alternate procedures (such as manual method) when the readable range of the instrument is exceeded.

❑ If the electromechanical system has reusable probes to detect a clot, there must be documented guidelines for cleaning probes. Reusable probes can develop significant protein build-up over a short period. Such protein build-ups interfere with sensing clot formation. Routine cleaning of the probes must be performed to ensure accurate coagulation tests.

❑ Manual clotting assays must be performed in duplicate to verify the results. The duplicates must agree within 10% of the shorted clotting time. If duplicates do not agree within these criteria, the test must be repeated. Three of the four times should then agree within 10% of the shortest clotting time.

❑ Coagulation assays are performed usually at 37°C. The temperature of the water baths or incubator should be verified with a certified thermometer or equivalent technique.

❑ The laboratory should periodically verify that samples for coagulation testing are reported within 4 hours of phlebotomy.

grams, the database created yields statistical information which is useful in intralaboratory and interlaboratory imprecision, laboratory bias relative to the peer group, and relationship of analytical and statistical parameters of imprecision and relative bias.

For analytes that do not have external proficiency testing available, laboratories should find an alternate method of verifying the reliability of test performance. Split sample testing may be used, with results compared between different laboratories. Alternatively, the results of testing can be compared to other known patient diagnostic parameters, such as a decreased factor VIII level in a patient with known hemophilia A.

Summary

Hemostasis testing is one of the most complex areas of the medical diagnostic laboratory. The coagulation protein system is quite complex in its activation and regulation, and requires very specific procedures for test collection, sample preparation, test performance, and quality assurance. Furthermore, many factors, ranging from anticoagulant therapy to liver disease and infections, can influence the results of hemostasis tests. This chapter has focused on the performance of common hemostasis assays but has also tried to communicate an appreciation of other preanalytical and analytical factors to be aware of when ordering and interpreting hemostasis testing.

Suggested Reading

Laboratory Techniques: General Reviews

Bockenstedt PA. Laboratory methods in hemostasis. In: Loscalzo J, Schafer AI, eds. *Thrombosis and Hemorrhage.* 3rd ed. Philadelphia, Pa: Lippincott Williams and Wilkins; 2003: Ch 21, 363-378.

Bowie EJW, Thompson JH Jr, Didisheim P, Owen CA. *Laboratory Manual of Hemostasis.* New York, NY: WB Saunders Company; 1971.

Jones BA, Calam RR, Howanitz PJ. Chemistry specimen acceptability: a College of American Pathologists Q-Probes study of 453 laboratories. *Arch Pathol Lab Med.* 1997;121:19-26.

Koepke JA. Coagulation testing systems. In: Koepke JA. *Practical Laboratory Hematology.* New York, NY: Churchill Livingston; 1991: 329-345.

Kottke-Marchant K. Laboratory diagnosis of hemorrhagic and thrombotic disorders. *Hematol Oncol Clin North Am.* 1994;8:809-853.

Jasperson J, Betina RM, Haverkate F, eds. *Laboratory Techniques in Thrombosis. A Manual.* 2nd ed. Boston, Mass: Luwer Academic Publishers; 1999.

Lillicrap D, Nair SC, Srivastava A, Rodeghiero F, Pabinger I, Federici, AB. Laboratory issues in bleeding disorders. *Haemophilia*. 2006;12(suppl 3):68-75.

Miale JB. *Laboratory Medicine. Hematology*. 5th ed. St. Louis, Mo: CV Mosby Company; 1977.

Van Cott EM, Laposata M. Coagulation, fibrinolysis and hypercoagulation. In: Henry JB, ed. *Clinical Diagnosis and Management by Laboratory Methods*. 20th ed. New York, NY: WB Saunders; 2001: 642-659.

Laboratory Measurement of Clot-Based Coagulation Tests

Coagulation Analyzers

Aller R, Sheridan B. Coagulation analyzers: service above all. *CAP Today*. 1999;13(1):39-40, 44-48.

Alter RD. Survey of instruments. Coagulation analyzers. *CAP Today*. 2002;16(1):28-40.

Gardiner C, Kitchen S, Dauer RJ, Kottke-Marchant K, Adcock DM. Recommendations for evaluation of coagulation analyzers. *Lab Hematol*. 2006;12(1):32-38.

Kottke-Marchant K, Alter RD, Chandler WL. *CAP Today*. 2003;17(1):18-22, 26-28, 30-33.

Protocol for the Evaluation, Validation, and Implementation of Coagulometers. Approved Guideline. Wayne, Pa: CLSI; 2008. Document H57-P.

Walenga JM, Fareed J. Automation and quality control in the coagulation laboratory. *Clin Lab Med*. 1994;14(4):709- 728.

Prothrombin Time and Activated Partial Thromboplastin Time

Adcock DM, Johnston M. Evaluation of frozen plasma calibrants for enhanced standardization of the international normalized ratio (INR): a multi-center study. *Thromb Haemost*. 2002;87(1):74-79.

Ansell J, Hirsh J, Poller L, Bussey H, Jacobson A, Hylek E. The pharmacology and management of the vitamin K antagonists: the Seventh ACCP Conference on Antithrombotic and Thrombolytic Therapy. *Chest*. 2004;126(suppl 3):204S-233S.

Denis-Magdelaine A, Flahault A, Verdy E. Sensitivity of sixteen APTT reagents for the presence of lupus anticoagulants. *Haemostasis*. 1995;25(3):98-105.

Eikelboom JW, Hirsh J. Monitoring unfractionated heparin with the aPTT: time for a fresh look. *Thromb Haemost*. 2006;96(5):547-552.

Kitchen S, Cartwright I, Woods TAL, Jenning I, Preston FE. Lipid composition of seven APTT reagents in relation to heparin sensitivity. *Brit J Haematol*. 1999;106:801-808.

Kitchen S, Jennings I, Woods TA, Preston FE. Wide variability in the sensitivity of APTT reagents for monitoring of heparin dosage. *J Clin Pathol*. 1996;49(1): 10-14.

One-Stage Prothrombin Time (PT) Test and Activated Partial Thromboplastin Time (APTT) Test. Approved Guideline. Wayne, Pa: NCCLS (now CLSI); 1996. Document H47-A.

Poller L, van den Besselaar AM, Jespersen J, Tripodi A, Houghton D. Correction for lack of coincidence of normal and abnormal calibration slopes in ISI determination. *Thromb Haemost*. 1999;81:935-939.

Poller L, van den Besselaar AM, Jespersen J, Tripodi A, Houghton D. A comparison of linear and orthogonal regression analysis for local INR determination in ECAA coagulometer studies. *Br J Haematol*. 1998;102:910-917.

Poller L. Prothrombin time (chapter 6). In: Jespersen J, ed. *Laboratory Techniques in Thrombosis. A Manual*. 2nd ed. Dordrecht: Kluwer Academic Publishers; 2000.

Procedures for Validation of INR and Local Calibration of PT/INR Systems. Approved Guideline. Wayne, Pa: CLSI; 2005. Document H54-A.

Proctor RR, Rappaport SI. The partial thromboplastin time with kaolin: a simple screening test for first stage plasma clotting factor deficiencies. *Am J Clin Pathol*. 1961;36:212-219.

Rapaport SI, Vermylen J, Hoylaerts M, et al. The multiple facets of the partial thromboplastin time APTT. *Thromb Haemost*. 2004;2(912):2250-2259.

White II GC. The partial thromboplastin time: defining an era in coagulation. *J Thromb Haemost*. 2003;1:2267-2270.

Thrombin Time

Jim RT, Goldfein S. Hageman trait (Hageman factor deficiency). *Am J Med*. 1957;23:824-831.

Love JE, Ferrell C, Chandler WL. Monitoring direct thrombin inhibitors with a plasma diluted thrombin time. *Thromb Haemost*. 2007;98(1):234-242.

Penner JA. Experience with a thrombin clotting time assay for measuring heparin activity. *Am J Clin Pathol*. 1974;61:645-653.

Walenga JM, Hoppensteadt D, Koza M, Wallock M, Pifarre R, Fareed J. Laboratory assays for the evaluation of recombinant hirudin. *Haemostasis*. 1991;21(suppl 1):49-63.

Fibrinogen Assays

Bargnoux AS, Dupuy AM, Biron-Andreani C, Schved JF, Cristol JP. Immunonephelometric determination of fibrinogen on citrated or heparinized plasma: comparison with functional Clauss method. *Clin Lab*. 2005;51(5-6):285-288.

Clauss A. Rapid physiological coagulation method in determination of fibrinogen. *Acta Haematol.* 1957;17(4):237-46.

Llamas P, Santos AB, Outeirino J, Soto C, Tomas JF. Diagnostic utility of comparing fibrinogen Clauss and prothrombin time derived method. *Thromb Res.* 2004;114(1):73-74.

Nieuwenhuizen W. Biochemistry and measurement of fibrinogen. *Eur Heart J.* 1995;16(suppl A):6-10;

Palareti G, Maccaferri M. Specific assays of hemostasis proteins: fibrinogen. *Ric Clin Lab.* 1990;20(2):167-176.

Factor Assays

Al-Sharif FZ, Aljurf MD, Al-Momen AM, et al. Clinical and laboratory features of congenital factor XIII deficiency. *Saudi Med J.* 2002;23:552-554.

Assays of Von Willebrand Factor Antigen and Ristocetin Cofactor Activity. Approved Guideline. Wayne, Pa: NCCLS (now CLSI); 2002. Document H51-A.

Barrowcliffe TW. Monitoring haemophilia severity and treatment: new or old laboratory tests. *Haemophilia.* 2004;10(suppl 4):109-114.

Dempfle CE, Harenberg J, Hochreuter K, Heene DL. Microtiter assay for measurement of factor XIII activity in plasma. *J Lab Clin Med.* 1992;119:522-528.

Determination of Factor Coagulant Activities. Approved Guideline. Wayne, Pa: NCCLS (now CLSI); 1997. Document H48-A.

Ingerslev J, Jankowski MA, Weston SB, Charles LA, ReFacto Field Study Participants. Collaborative field study on the utility of a BDD factor VIII concentrate standard in the estimation of BDDr Factor VIII:C activity in hemophilic plasma using one-stage clotting assays. *J Thromb Haemost.* 2004;2:623-628.

Mixing Studies and Inhibitors

Chang SH, Tillema V, Scherr D. A "Percent Correction" formula for evaluation of mixing studies. *Am J Clin Pathol.* 2002;117:62-73.

Exner T. Diagnostic methodologies for circulating anticoagulants. *Thromb Haemost.* 1995;74(1):338-344.

Haya S, Moret A, Cid AR, et al. Inhibitors in haemophilia A: current management and open issues. *Haemophilia.* 2007;13(suppl 5):52-60.

Kaczor DA, Bickford NN, Triplett DA. Evaluation of different mixing study reagents and dilution effect in lupus anticoagulant testing. *Am J Clin Pathol.* 1991;95:408-411.

Kasper CK. Laboratory tests for factor VIII inhibitors, their variation, significance, and interpretation. *Blood Coagul Fibrinolysis.* 1991;2(suppl 1):7-10.

Ledford-Kraemer M. All mixed up about mixing studies. *CLOT-ED.* 2004;3:2-6.

Rooney AM. The Taipan snake venom time: a new test for lupus anticoagulant. *J Clin Pathol.* 1994;47:497-501.

Sahud M. Laboratory diagnosis of inhibitors. *Semin Thromb Hemost.* 2000;26:195-203.

Thom J, Ivey L, Eikeboom J. Normal plasma mixing studies in the laboratory diagnosis of lupus anticoagulant. *J Thromb Haemost.* 2003;1:2689-2691.

Triplett DA, Brandt J. Laboratory identification of the lupus anticoagulant. *Br J Haematol.* 1989;73:139-142.

Establishing Reference Ranges and Validating Coagulation Laboratory Testing

Andrew M, Paes B, Johnson M. Development of hemostatic system in neonate and young infant. *Am J Pediatr Hematol Oncol.* 1990;12:95-104.

Andrew M, Paes B, Milner R, et al. Development of the human coagulation system in the full-term infant. *Blood.* 1987;70:165-172.

Andrew M, Paes B, Milner R, et al. Development of the human coagulation system in the healthy premature infant. *Blood.* 1988;72:1651-1657.

Andrew M, Vegh P, Johnston M, Bowker J, Ofosu F, Mitchell L. Maturation of the hemostatic system during childhood. *Blood.* 1992;80:1998-2005.

Collection, Transport and Processing of Blood Specimens for Testing Plasma-Based Coagulation Assays and Molecular Hemostasis Assays. Approved Guideline. 5th ed. Wayne, Pa: CLSI; 2008: Document H21-A5.

How to Define and Determine Reference Intervals. Approved Guideline. 2nd ed. Wayne, Pa: NCCLS (now CLSI); 2000. Document C28-A2.

Laga AC, Cheves TA, Sweeney JD. The effect of specimen hemolysis on coagulation test results. *Am J Clin Pathol.* 2005;126:748-755.

Quality Control

Ye JJ, Ingels SC, Parvin CA. Performance evaluation and planning for patient-based quality control procedures. *Am J Clin Pathol.* 2000;113:240-248.

Bland JM, Altman DG. Statistical methods for assessing agreement between two methods of clinical assessment. *Lancet.* 1986;1(8476):307-310.

Results Reporting

Koepke JA, McLaren CE, Wijetunga A, Houwen B. A method to examine the need for duplicate testing of common coagulation tests. *Am J Clin Pathol.* 1994;102:242-247.

Quality Assurance

Hertzberg MS, Mammen J, McCraw A, Nair SC, Srivastava A. Achieving and maintaining quality in the laboratory. *Haemophilia*. 2006;12(suppl 3):61-67.

Meijer P, Haverkate F. External quality assessment and the laboratory diagnosis of thrombophilia. *Semin Thromb Hemost*. 2005;31:59-65.

van den Besselaar AM, Bertina RM. Multicentre calibration of secondary reference material for thromboplastin, rabbit, plain, coded CRM 149R. *Thromb Haemost*. 1991;65: 263-267.

CHAPTER 7

Platelet Testing

Kandice Kottke-Marchant, MD, PhD

Patient History

Evaluation of a patient for platelet disorders should start with a thorough medication and clinical history together with testing to exclude coagulation protein and fibrinolytic disorders, followed by an assessment of platelet number and morphology, culminating with studies of platelet function.

Clinical History for Patients with a Bleeding Diathesis

A thorough clinical and family bleeding history should be obtained before beginning a laboratory evaluation of platelet function. This is covered in more detail in chapter 5. The history should include an assessment of the duration, pattern, and severity of bleeding problems, including whether the bleeding is spontaneous or is associated with trauma or surgery. A lifelong bleeding diathesis may suggest a congenital thrombocytopenia or platelet dysfunction, but an onset in adulthood does not necessarily exclude a congenital problem. In assessing whether a true hemorrhagic disorder exists, it is often helpful to assess if the bleeding is out of proportion to the degree of trauma, or whether blood transfusions were required for relatively minor surgical procedures, such as tooth extractions.

Platelet-mediated bleeding disorders result in microvascular bleeding, which typically manifests as a mucocutaneous bleeding pattern. Ecchymosis, petechiae, purpura, epistaxis, and gingival bleeding are commonly observed. This pattern is in contrast to that observed with coagulation protein disorders, where deep tissue bleeding and hemarthroses are more common. Von Willebrand disease, an abnormality of von Willebrand factor (vWF), has bleeding symptoms very similar to platelet dysfunction, and evaluation for von Willebrand disease should be included in the initial evaluation of a possible platelet disorder. Bleeding diatheses due to vascular malformations

may give a bleeding pattern similar to platelet disorders, but the pattern is often more focal than diffuse. Acquired purpuras, such as seen with disseminated intravascular coagulation (DIC), vasculitis, or infections, usually can be distinguished from platelet dysfunction, as platelet disorders typically cause bleeding from mucous membranes ("wet" purpura), while vascular purpura is usually confined to the skin ("dry" purpura).

Many drugs, as well as foods such as garlic or caffeine, can affect platelet function, so a complete drug and dietary history should be obtained. It is important to remember that aspirin, an irreversible inhibitor of platelet function, is an ingredient of many over-the-counter and prescription medications, such as cold or flu remedies (Table 7-1). The clinical history should include investigation of systemic disorders, such as renal disease, hepatic failure, connective tissue disorders, myeloproliferative disorders, myelodysplastic disorders, malignancy, and cardiovascular disease, as platelet dysfunction is associated with many of these diseases. Additionally, some specific clinical features, such as albinism, deafness, nephritis, and susceptibility to infections, may help in the differential diagnosis of the inherited platelet disorders.

Clinical History for Patients with Thrombosis

Thrombosis can involve the venous and/or arterial vascular systems. Venous thrombosis, such as deep vein thrombosis or pulmonary embolism, is rarely associated with platelet dysfunction. On the other hand, arterial thrombosis may have platelet activation as a mediating step but usually is also underscored by vascular disease, such as atherosclerosis and vascular inflammation, or an antiphospholipid antibody syndrome. Heparin-induced thrombocytopenia usually presents 5 to 14 days post heparin therapy, so eliciting a careful history of heparin exposure is necessary to identify this disorder. Other platelet disorders associat-

Table 7-1. Drugs that Affect Platelet Function

Drug Classification	Examples
Nonsteroidal anti-inflammatory agents	Aspirin: irreversible cyclooxygenase inhibitor Ibuprofen, indomethacin, naproxen, sulindac: reversible cyclooxygenase inhibitors
Thienopyridines	Clopidogrel, ticlopidine: $P2Y_{12}$ antagonists
Glycoprotein IIb/IIIa receptor antagonists	Abciximab, tirofiban, eptifibatide
Drugs that increase cyclic adenosine monophosphate (cAMP) levels	Dipyridamole, cilostazole, prostacyclin, iloprost
Antibiotics	Penicillin, cephalosporins, nitrofuantoin, miconazole
Volume expanders	Dextran, hydroxyethyl starch
Anticoagulants	Heparin
Antifibrinolytic agents	Streptokinase, tissue plasminogen activator, urokinase, ε-aminocaproic acid
Cardiovascular drugs	Nitroglycerin, isosorbide dinitrate, propranolol, nitroprusside, nifedipine, verapamil, diltiazem, quinidine
Psychotropic drugs	Imiprimine, amitriptyline, nortryptaline, chlorpromazine, promethazine, flufenazine, trifluoperzaine, haloperidol
Anesthetics	Dibucaine, tetracaine, metycaine, cyclaine, butacaine, nepercaine, procaine, cocaine, plaquenil, halothane
Chemotherapeutic agents	Mithramycin, daunorubicin, BCNU
Antihistamines	—
Herbal supplements	Dae-Jo-Whan, Geiji-Bokryung-Hwan (GBH), Ganoderma lucidum (Ling-Zhi)
Foods	Caffeine, flavinoids (chocolate, red wine, green tea), antioxidants (vitamin E, garlic, onion, gingko biloba, turmeric)

Sources: Bennett JS. Acquired platelet function defects. In: Gresele P, Page C, Fuster V, Vermylen J, eds. *Platelets in Thrombotic and Non-Thrombotic Disorders: Pathophysiology, Pharmacology and Therapeutics.* Cambridge University Press; 2002: 689-706.
Janossy K, Ball DR, Jefferson P. Platelet function and diet. *Anaesthesia.* 2006;61:508-509.

ed with thrombosis include the myeloproliferative disorders, so enquiring about other bone pain or bone marrow disorders and examining for hepatosplenomegaly may be helpful.

Platelet Count, Indices, and Morphology

A first step in the investigation of platelet disorders should be measurement of the platelet count, platelet indices, and review of peripheral smear morphology. This step helps in classifying the dis-

order in the setting of thrombocytopenia, thrombocytosis, or normal platelet count. It also helps to exclude other pathologies that may affect platelet number or function, such as the leukemias, myeloproliferative disorders, myelodysplastic disorders, or consumptive coagulopathies, such as DIC.

Sample Collection and Processing Considerations

Blood samples collected for platelet counting should be collected into an EDTA (ethylenediaminetetraacetic acid) anticoagulant, typically a purple-capped vacutainer tube. The sample should be mixed thoroughly after collection to prevent in vitro clotting. A peripherally collected specimen is ideal, but collection from indwelling catheters is acceptable, provided the flushing liquid is removed prior to sampling to avoid dilution. If multiple specimens are drawn at the same time, the light blue top tube (containing sodium citrate anticoagulant) for coagulation testing should be drawn before the EDTA tube to avoid carryover of the EDTA into the coagulation samples. The platelet count is usually stable for up to 24 hours postcollection. An air-dried, peripheral, Wright-stained smear can be made from the EDTA specimen for platelet morphologic analysis.

Test Performance and Interpretation

It is best to assess platelet morphology in the thin part of the smear where the erythrocytes have good morphology and are present in a thin monolayer, keeping in mind that the feathered edge or the lateral sides of the smear should be scanned for platelet clumps. On a properly prepared, Wright-stained blood smear, the platelets are of roughly uniform size (approximately 2-μm diameter), with abundant purple-staining granules (Figure 7-1). Large platelets, dumbbell-shaped platelets, and megakaryocyte fragments are unusual. The presence of more than a few large platelets suggests increased platelet turnover, myeloproliferative disorder, or a congenital macrothrombocytopenia. Degranulated, pale, gray platelets are an infrequent finding and suggest in vivo platelet activation or an alpha granule disorder.

The platelet count is generally between 150,000 to 400,000/μL of blood in normal individuals, although values lower than 150,000/μL can be adequate for hemostasis. True thrombocytopenia must be distinguished from spurious in vitro platelet clumping, known as pseudothrombocytopenia, an in vitro phenomenon not associated with disease, due to EDTA-dependent, cold-reacting platelet agglutinins or platelet binding to neutrophils (platelet satellitism). These agglutinins may be observed in patients with high immunoglobulin levels, anticardiolipin antibodies, or infections and usually only bind platelets when calcium is chelated, such as in an EDTA blood collection tube. A pseudothrombocytopenia associated with the glycoprotein IIb/IIIa (GPIIb/IIIa) antagonist drug abciximab has also been reported. Pseudothrombocytopenia can be detected by examining a peripheral smear, where

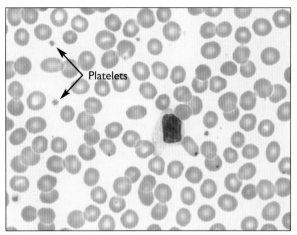

Figure 7-1. Platelets on a Wright-stained peripheral smear are small (approximately 2 to 3 μm) diameter and have a purple granular cytoplasm, reflective of the many alpha granules. (Original magnification X100.)

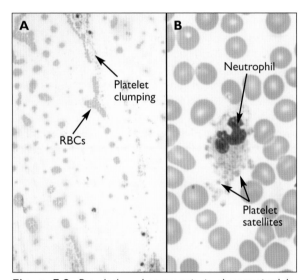

Figure 7-2. Pseudothrombocytopenia is characterized by clumping of platelets near the feathered edge of a peripheral blood smear (A). Platelet satellitism may also be observed, where platelets are bound to the membranes of neutrophils or monocytes (B). (A: Wright stain, original magnification X10; B: Wright stain; original magnification X100.)

large aggregates of platelets are observed, often around the feathered edge (Figure 7-2, A). Pseudothrombocytopenia is often accompanied by platelet satellitism, where platelets are bound to the cytoplasmic membranes of neutrophils, through interaction of P-selectin on platelets and P-selectin glycoprotein ligand-1 on neutrophils (Figure 7-2, B). A more accurate platelet count can be established by collecting the blood sample in either citrate or heparin anticoagulants or by collecting blood directly from a finger stick into a diluent. Giant platelets observed with macrothrombocytopenia syndromes also can give falsely low platelet counts, as the large platelets may be counted as leukocytes by automated cell counters.

Automated hematology analyzers that perform complete blood counts (CBCs) measure overall platelet count as well as the size of each platelet. The mean platelet volume (MPV) is an indication of platelet size. Normal MPV ranges are approximately 7 to 11 fL. The platelet distribution width (PDW) is a measure of the dispersion of platelet sizes and is analogous to the red cell distribution width (RDW). The MPV can be an indication of platelet turnover, as platelets newly released from the bone marrow are larger and tend to decrease in size with age in the circulation. In patients with rapid turnover, the platelets will, in general, be larger due to the larger size of newly produced platelets, and their PDW will be increased due to a mixture of large and small platelets. There is some evidence that elevated MPV may be associated with unfavorable outcomes in individuals with myocardial infarction, likely as a marker of platelet turnover.

True congenital macrothrombocytopenias usually have uniformly large platelets, with a very high MPV and normal PDW; often the platelets are at least twice normal size and may be as large as erythrocytes. However, in many patients with congenital macrothrombocytopenia, the platelet indices may not be available because they may "flag out" on automated hematology analyzers due to the low number of platelet events and the abnormal distribution. Newer techniques based on messenger ribonucleic acid (mRNA) detection in platelets (reticulated platelets) may also be helpful to indicate the rate of thrombopoiesis, as mRNA levels are high in newly formed platelets and decline progressively during blood circulation time.

Some platelet disorders can be associated with unique platelet and/or leukocyte morphology. In von Willebrand disease, Glanzmann thrombasthenia, and myeloproliferative disorders, the platelets have unremarkable morphologic features. Giant platelets are seen in Bernard-Soulier disease; other macrothrombocytopenia syndromes associated with myosin heavy chain gene defects (MYH9) additionally may have Dohle body-like neutrophil inclusions. Immunostains for non-muscle myosin heavy chain-A may be useful to show abnormal subcellular localization in neutrophils from patients with MYH9 mutations. In patients with Wiskott-Aldrich syndrome, the platelets may be

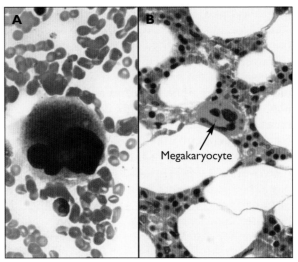

Figure 7-3. Megakaryocyte morphology on a bone marrow aspirate smear (A) and biopsy section (B). Megakaryocyte morphology is best observed on the aspirate smear, as the entire cell can be visualized. The biopsy sections are best for estimating megakaryocyte number and adequacy, but only a portion of the cell may be visualized due to the thin sections. (A: Wright stain, original magnification X50; B: Hematoxylin and eosin stain, original magnification X40.)

small. Platelets in the gray platelet syndrome, an alpha granule deficit, are characteristic for being pale, gray, and hypogranular on a Wright-stained blood smear. Some platelet storage pool disorders may have morphologically normal platelets by light microscopy but may have decreased alpha and/or dense granules by electron microscopy.

Bone Marrow Analysis

Sample Collection and Processing Considerations

Bone marrow analysis typically includes study of bone marrow aspirate smears or touch preparations, clot sections, and core biopsy material. Characteristically, a bone marrow sample is aspirated from the iliac crest with a needle. The marrow particles are placed on a glass coverslip or slide and then smeared to create monolayers of cells for examination. After air drying, the slides are Wright stained for microscopic analysis. A large-bore needle is used to sample a core of bone. If the aspirate specimen does not appear to have visible spicules, the core biopsy sample can be touched gently to a glass slide to prepare a touch preparation. The bone biopsy is then fixed with formalin or other fixative and processed by decalcification, followed by wax embedding, sectioning, and staining with hematoxylin and eosin. The residual aspirate specimen is allowed to clot, then fixed, embedded, and processed for microscopy.

Bone marrow analysis in the evaluation of platelet disorders is of use in the assessment of megakaryocyte number and morphology. Megakaryocytes are observed best in the spicular areas of the bone marrow aspirate, although they may be found throughout the smears. The aspirate smears are the preferred material to assess megakaryocyte nuclear and cytoplasmic morphology, since the entire cell is visible (Figure 7-3, A). Megakaryocyte number is best assessed on the biopsy sections, as the megakaryocytes can be evaluated in comparison with overall cellularity; megakaryocyte clustering and associated fibrosis can also be assessed. Caution should be used in evaluating megakaryocyte morphology on the biopsy sections, which are only 4 to 6 μm in thickness and often show only a portion of the entire cell because of the large size of the megakaryocytes (Figure 7-3, B).

Bone Marrow Interpretation

Examination of the bone marrow may be helpful in evaluating the etiology of both thrombocytopenic and thrombocytotic disorders, but it is not generally helpful in the evaluation of platelet dysfunction with a normal platelet count. In a patient with thrombocytosis, the bone marrow examination may be helpful to diagnose a myeloproliferative disorder or suggest a reactive process. Myeloproliferative disorders include chronic myelogenous leukemia (CML), polycythemia vera, essential thrombocythemia, and myelofibrosis. Bone marrow features that suggest a myeloproliferative process include megakaryocytic or hematopoietic hyperplasia, increased eosinophils and/or basophils, and marrow fibrosis. (Diagnostic details of myeloproliferative disorders are beyond the scope of this chapter; the reader is directed to the recent World Health Organization publication.) In a thrombocytopenic patient, the bone marrow examination is useful for determining the number of megakaryocytes; absence indicates dysfunctional marrow and decreased platelet production, while increased numbers suggests peripheral destruction with attempted bone marrow compensation. Bone marrow examination can also detect myelophthisic disorders, such as acute leukemia, lymphoma, or metastatic malignancy, which could explain a patient's thrombocytopenia.

Platelet Function Testing

Platelet function testing may be used to detect or characterize an intrinsic or acquired platelet dysfunction. Platelet function testing also has been utilized to assess and monitor the therapeutic effect of antiplatelet drugs, but the clinical utility of such testing is still debated. In the initial evaluation of platelets, it is desirable to perform a screening test to evaluate platelet function. However, caution must be exercised in interpreting platelet function testing, as abnormal results are often observed in thrombocytopenic patients; distinction of an abnormal result due to intrinsic platelet dysfunction versus thrombocytopenia alone may not be possible. Keep in mind that normal initial platelet function testing results do not completely exclude a platelet function disorder, nor are abnormal results necessarily diagnostic of a disorder.

The platelet functions measured by laboratory tests include adhesion, aggregation, activation, and granule release, although platelet aggregation studies are most commonly performed. More specialized tests, such as platelet reticulocyte and flow cytometry analysis, evaluate platelet surface glycoproteins and RNA content. Whole blood platelet function assays have been developed for use in the evaluation of platelet function. Most of these whole blood assays utilize small stand-alone devices that measure platelet function and can be used in laboratories that otherwise could not perform platelet function studies; some can be utilized in the near-patient setting. However, many of these devices are still in development or in the early stages of clinical implementation, and their performance has not been completely characterized for the detection of platelet function defects or monitoring of antiplatelet drugs. These devices include the PFA-100® System (Siemens Healthcare Diagnostics Inc, Tarrytown, New York), the VerifyNow™ (Ultegra) System (Accumetrics, San Diego, California), the Plateletworks® (Helena Laboratories, Beaumont, Texas), the Impact-R (Diamed, Switzerland), and the Hemostasis Analysis System (Hemodyne Inc, Richmond, Virginia). The thromboelastograph, while not a new technology, recently has been adapted for use in the clinical coagulation laboratory; it measures a combination of coagulation, platelet function, and fibrinolysis.

Bleeding Time

For nearly a century, the bleeding time was the only platelet function test available. The bleeding time is a test that is fraught with variability and involves the creation of a cut in the skin and measurement of the time it takes for bleeding to stop. Since its initial development by Duke in 1910, various attempts have been made to standardize the bleeding time test, including the current implementation of a blood-pressure device to standardize venous pressure and the use of a spring-loaded blade to create a standardize cut length and depth. The bleeding time result depends not only upon platelet number and function, but also on fibrinogen concentration, adequate vascular function, orientation and size of the incision, site of the incision, skin quality, skin temperature, operator technique, and patient cooperation.

Many laboratories have stopped performing the bleeding time completely due to its variability, poor reproducibility, and lack of correlation with intraoperative bleeding. The one benefit of the bleeding time over all other platelet function tests is that it requires the function of the normal vessel wall for hemostasis.

PFA-100

The PFA-100 System is a device that measures both platelet adhesion and aggregation in whole blood, using a high-shear testing system (Figure 7-4).

Specimen Considerations

Samples for PFA-100 testing should be collected using a peripheral venipuncture when possible. The PFA-100 utilizes a whole blood specimen collected into a light blue top vacutainer tube containing buffered trisodium citrate. Both 3.8% (0.129 M) and 3.2% (0.105 M) citrate can be used, but the reference ranges will vary between the anticoagulants. Since 3.2% citrate is the anticoagulant recommended for most coagulation assays, this anticoagulant would be preferred for the PFA-100 to facilitate sample collection consistency. It is recommended that each laboratory establish its own reference range using blood samples from normal individuals. Specimens should be kept at room temperature and transported to the laboratory in a timely fashion. Testing should be completed within 4 hours of phlebotomy. Transporting specimens for platelet function testing in a pneumatic tube should be avoided unless validated at an individual institution, as this may adversely affect platelet function, leading to falsely abnormal results. Guidelines for performing PFA-100

testing are being developed by the Clinical and Laboratory Standards Institute (CLSI).

Test Performance and Interpretation

The disposable cartridges contain a membrane with a central aperture (147-μm diameter) coated with aggregation agonists (collagen/epinephrine [EPI] or collagen/adenosine diphosphate [ADP]). A blood sample (either 800 or 900 μL) is pipetted into a sample cup. The blood is drawn out of the cup and passed through the aperture at a high shear rate (5000 to 6000 sec^{-1}), where the platelets adhere to the membrane, aggregate, and cause aperture occlusion (Figure 7-4). When the blood flow ceases, the instrument measures the "closure time," which is a reflection of platelet function.

A normal closure time indicates normal platelet function, while a prolonged closure time indicates decreased platelet function; a shortened closure time can be seen with an elevated platelet count or increased platelet function. Prolonged closure times can be seen with thrombocytopenic samples (<150,000 platelets/μL) or anemic samples (hematocrit <35%). Due to the limitations of the blood volume in the sample cup, the instrument can only measure closure times up to 300 seconds. Beyond that, the closure time is reported as >300 seconds.

The PFA-100 generates several error codes, such as "Insufficient Sample" when the entire patient sample has been expended, and "Maximum Syringe Travel Reached" when the machine syringe has completed its traverse before closure time is reached. If either of these two codes is reported together with an abnormally long closure time, the test result likely is valid. However, if either of these codes or the code "Flow Obstruction" is reported with a short closure time or one in the normal range, this indicates a problem with the sample or machine malfunction, and the result should not be reported. Duplicate testing is only recommended when the initial result shows a prolonged closure time. An algorithm for performing and reporting PFA-100 results is shown in Figure 7-5.

Good quality control must be maintained on the instrument, including daily electronic checks and vacuum checks, as well as validation of each new cartridge lot with a fresh normal sample. External proficiency testing for the PFA-100 has recently become available; this challenge utilizes normal donor blood drawn into special sample collection tubes.

The collagen/EPI cartridge is the primary screening cartridge; it detects platelet dysfunction

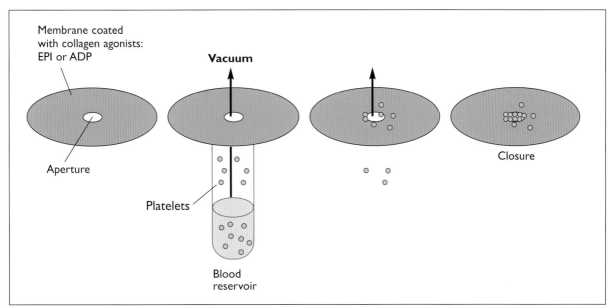

Figure 7-4. The Platelet Function Analyzer-100 (PFA-100), manufactured by Siemens, uses disposable cartridges that contain a membrane with a central aperture (147 μm) coated with aggregation agonists (collagen/epinephrine [EPI] or collagen/adenosine diphosphate [ADP]). A blood sample (either 800 or 900 μL) is pipetted into a sample cup. The blood is drawn out of the cup and passed through the aperture at a high shear rate (5000 to 6000 sec^{-1}), where the platelets adhere to the membrane, aggregate, and cause aperture occlusion. When the blood flow ceases, the instrument measures the "closure time," which is a reflection of platelet function.

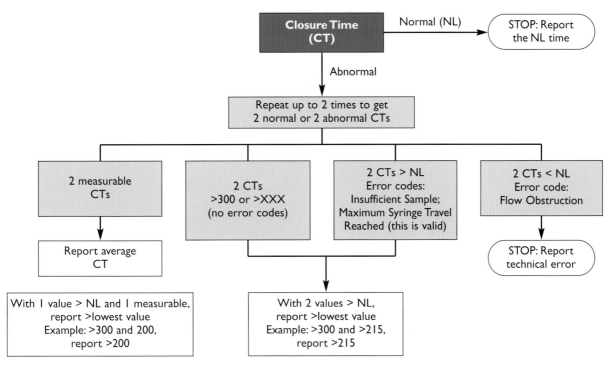

Figure 7-5. A laboratory testing algorithm for the Platelet Function Analyzer-100 (PFA-100). If the first closure time is normal, the result can be reported. However, if the first closure time is abnormal, the test should be repeated up to two more times to result in two abnormal or two normal closure times. If the sample volume is exhausted prior to closure being detected, the result may be reported as "greater than" the time of sample exhaustion. For example, a result of >255 seconds indicates that the sample was exhausted at 255 seconds and closure still had not occurred. If two or more "greater than" results are obtained, the result reported should be "greater than" the lowest value. For example, if the results are >300 seconds and >215 seconds, the result would be >215 seconds.

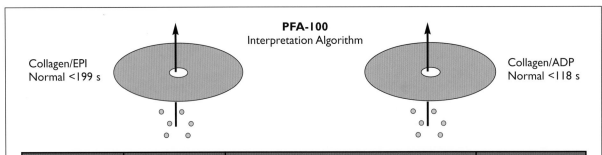

Collagen/EPI Result	Collagen/ADP Result	Interpretation	Reasons
Normal	Normal	Normal	Normal function
Abnormal	Normal	The closure time with the collagen/EPI cartridge is abnormal, but the time with the collagen/ADP cartridge is normal. This pattern suggests some abnormality of platelet function and is often seen with an aspirin-like drug effect.	Usually aspirin
Abnormal	Abnormal	The closure times with both the collagen/EPI cartridge and the collagen/ADP cartridge are abnormal. This pattern suggests an abnormality of platelet function, von Willebrand disease, or an antiplatelet drug effect.	Platelet dysfunction, drugs, von Willebrand disease
Normal	Abnormal	The closure time with the collagen/EPI cartridge is normal, and the closure time with the collagen/ADP cartridge is abnormal. This pattern is unusual but suggests an abnormality of platelet function, von Willebrand disease, or an antiplatelet drug effect.	Platelet dysfunction, drugs, von Willebrand disease

Figure 7-6. Laboratory reporting algorithm for the Platelet Function Analyzer-100 (PFA-100). The platelet testing device uses two cartridges, one with epinephrine (EPI) and collagen and the other with adenosine diphosphate (ADP) and collagen. Most patients with platelet dysfunction will have a prolonged closure time with both cartridges. However, an aspirin-like drug effect will specifically prolong the collagen/EPI closure time with a normal collagen/ADP closure time.

induced by intrinsic platelet defects, von Willebrand disease, or platelet-inhibiting agents. Aspirin-like drugs give a prolonged closure time with the collagen/EPI cartridge and a normal closure time with the collagen/ADP cartridge due to a high ADP concentration in the cartridge. See Figure 7-6 for a PFA-100 interpretation algorithm. Von Willebrand disease, intrinsic platelet dysfunction, and non-aspirin drugs characteristically give an abnormal closure time with both cartridges. In some studies, the PFA-100 has been shown to be more sensitive than the bleeding time, especially for the detection of type I von Willebrand disease. The PFA-100 is very sensitive to the GPIIb/IIIa antagonist drugs but is relatively insensitive to thienopyridine drugs, such as clopidogrel and ticlopidine. The PFA-100 may not be sensitive to all types of von Willebrand disease and platelet dysfunction. For example, in type 2N von Willebrand disease, in which the primary defect is a decreased factor VIII, the PFA-100 will give normal results. Additionally, the PFA-100 may not detect platelet storage pool disorders in some patients or some macrothrombocytopenia disorders. It has not yet been established whether the PFA-100 is useful in monitoring the therapeutic effect of antiplatelet drugs (see chapter 25 for a more complete discussion). The PFA-100 results can be affected by low platelet counts and low hematocrits, but are not affected by heparin or deficiencies of coagulation factors other than fibrinogen.

Platelet Aggregation

Platelet aggregation measures the ability of agonists to cause in vitro platelet activation and platelet-platelet binding. As such, platelet aggregation is often useful to distinguish intrinsic platelet disorders involving surface glycoproteins, signal transduction, and platelet granules.

Specimen Considerations

Platelet aggregation testing can be performed either in whole blood or using a suspension of platelets in plasma, termed platelet-rich plasma (PRP). Blood for platelet aggregation studies should be drawn into an anticoagulant solution of sodium citrate. Historically, platelet aggregation studies were performed using 3.8% sodium citrate, but with the rise in the use of 3.2% sodium citrate for other coagulation tests, the latter has become acceptable for platelet aggregation testing. Ideally, blood should be obtained from a peripheral venipuncture. The sample handling considerations for whole blood aggregation are similar to those listed for the PFA-100, since both are whole blood platelet function assays. Guidelines for performing platelet aggregation testing are under development by CLSI.

For optical platelet aggregation assays utilizing PRP, the sample should be kept at room temperature and transported to the laboratory expeditiously. Platelet aggregation testing should be completed within 4 hours of phlebotomy. The first step in sample processing requires the production of PRP by differential centrifugation of erythrocytes and leukocytes, resulting in a top suspension of platelets and plasma. The centrifugation speeds employed in laboratories vary widely, but conditions of $180g$ for 10 minutes at room temperature are typical. The PRP is removed from the top of the tube, and the sample is centrifuged again, typically $3000g$ for 15 minutes, to prepare platelet-poor plasma (PPP). These conditions usually result in a PPP platelet count of $<10,000/\mu L$.

The platelet count in the PRP should be measured, but it is usually slightly higher than the whole blood platelet count. Both PRP and PPP should then be "rested" at room temperature for approximately 30 minutes prior to platelet aggregation testing. Prior to testing, the platelet count in the PRP is often normalized to 200,000 to 250,000/μL by mixing appropriate ratios of PRP and PPP, although some recent studies have suggested that this practice may affect platelet aggregation results, and adjusting the platelet count

with saline may be more appropriate. With optical aggregation methodologies, PRP platelet counts $<100,000/\mu L$ may provide insufficiently turbid samples to provide reliable results. It may occasionally be necessary to test these samples, especially in the functional evaluation of patients with thrombocytopenia. For such samples, it may be helpful to adjust the platelet count of a normal sample to a similar low platelet count as a normal "thrombocytopenic" control. Another consideration in the preparation of PRP is macrothrombocytopenia; the large platelets may sediment with the leukocytes or erythrocytes during centrifugation, and the PRP may be nearly devoid of platelets when processed using standard techniques. Allowing these samples to sediment by gravity may be one way to procure adequate PRP for platelet aggregation testing. The optical aggregation assay may suffer from interference from hemolyzed, lipemic, and icteric plasma.

Test Performance and Interpretation

Platelet aggregation studies can be performed in whole blood by an impedance technique or in platelet-rich plasma by an optical, turbidimetric technique. Whole blood platelet aggregation can be combined with studies of dense granule adenosine triphosphate (ATP) release with a lumiaggregometer using a firefly luciferin-luciferase reagent system.

In the turbidimetric platelet aggregation assay, platelet aggregation is measured spectrophotometrically by the increase in light transmission after addition of an aggregation agonist in a stirred platelet sample. The agonists typically used include ADP, collagen, arachidonic acid, epinephrine, and occasionally thrombin receptor-activating peptide (TRAP) (Figure 7-7). For optical aggregation, the adequacy of the aggregation response is followed by quantifying the maximal percentage of aggregation or the slope of the aggregation curve. For whole blood aggregation, the extent of aggregation is determined by an impedance technique, with readout in ohms.

With use of a subthreshold concentration of agonist, there is typically a primary wave of aggregation, with subsequent disaggregation due to lack of granule release. Optimal platelet aggregation shows a biphasic pattern for the agonists ADP and epinephrine; the initial increase in aggregation is due to primary aggregation in response to activation of the glycoprotein IIb/IIIa platelet membrane receptor, while the second wave of

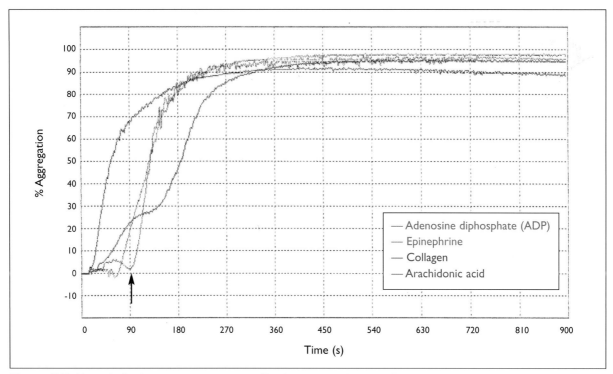

Figure 7-7. Normal platelet aggregation tracings. Platelet aggregation with 5 μM ADP (green), 100 μM epinephrine (red), 2 mg/mL collagen (blue), and 0.5 mg/mL arachidonic acid (black). At this concentration, ADP shows two waves of aggregation. After the initial wave of aggregation, platelet ADP release from dense granules stimulates the final and irreversible wave of aggregation. Collagen aggregation characteristically shows an initial shape change (arrow). Normal aggregation for all agonists is typically >70% aggregation, but laboratories should establish their own in-house reference ranges. (Optical platelet aggregation using a Helena PACKS-4 aggregometer.)

aggregation is the result of platelet degranulation with recruitment of additional platelet aggregates. Lack of a secondary wave suggests a platelet storage pool disorder caused either by reduced numbers of granules or defective release. If the platelet count in the PRP is too low (often below 150,000/μL), release of platelet granule contents may not produce a sufficiently high concentration in the sample to generate a secondary wave, generating a potentially falsely abnormal result. In these cases, it may be necessary to concentrate the platelets using slightly higher centrifugation speed, although this is technically difficult, and platelets should not be pelleted and resuspended. At higher agonist concentrations, the two waves merge into one wave of aggregation. Other agonists, such as arachidonic acid, thrombin receptor agonists, and collagen, usually show only a single wave of aggregation. Collagen characteristically shows an initial shape change prior to the wave of aggregation; this is seen as a transient increase in turbidity. Normal aggregation characteristically results in greater than 70% or 80% aggregation, but all laboratories should establish their own reference ranges for each agonist. Several studies

have shown that some normal individuals with no history of platelet-associated bleeding may not respond to concentrations of epinephrine typically used in aggregation assays. Some authors have removed nonresponders during calculation of reference ranges, but this limits the diagnostic specificity of epinephrine as an aggregating reagent.

Another important reagent used in the evaluation of platelet function by aggregation is the antibiotic ristocetin, which facilitates the binding of vWF to the glycoprotein Ib/IX/V complex (Figure 7-8). Ristocetin induced platelet aggregation (RIPA) evaluates aggregation upon the addition of both low and high concentrations of ristocetin, allowing detection of both increased and decreased sensitivity to ristocetin. A normal result requires the presence of both functional vWF *and* normal GPIb/IX/V, so RIPA can detect both von Willebrand disease and some platelet dysfunctions, such as Bernard Soulier syndrome.

Quality control for platelet aggregation is problematic, since no commercial controls are available. Obtaining normal results with a normal donor is the only feasible quality control method and ideally should be performed daily or whenev-

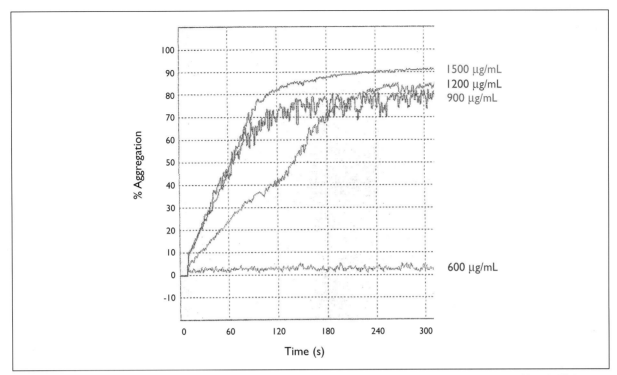

Figure 7-8. Platelet aggregation with several concentrations of ristocetin. Ristocetin stimulates a conformational change in von Willebrand factor, leading to aggregation through the glycoprotein Ib/IX/V complex. Note that aggregation is virtually absent at low ristocetin concentration (600 μg/mL), becoming progressively stronger until complete aggregation is reached somewhere between 1200 and 1500 μg/mL. (Optical platelet aggregation using a Helena PACKS-4 aggregometer.)

er patient samples are tested. An external platelet aggregation proficiency test has recently been made available; it uses a normal donor sample drawn into a special collection tube, with agonists provided.

Many factors can affect the platelet aggregation results, such as thrombocytopenia, thrombocytosis, processing technique, processing temperature, stirring rate, and processing time. In addition, clinicians ordering the tests should advise patients to discontinue, if possible, any medication, such as aspirin or nonsteroidal anti-inflammatory agents, which may interfere with the assessment of the test results.

A survey of platelet function testing techniques among North American Specialized Coagulation Laboratory Association (NASCOLA) member laboratories revealed that there is a wide variety in practice in the performance of platelet aggregation testing. The NASCOLA study revealed that the majority of laboratories reported results containing quantitative values (% aggregation and/or slope) and a qualitative interpretation. The majority of laboratories utilized the agonists collagen, arachidonic acid, epinephrine, and ristocetin; however, the final test concentrations of these ago-

nists varied widely, sometimes by several orders of magnitude.

There is little data in the literature to define the sensitivity and specificity of agonist concentration for the diagnosis of platelet disorders. However, the median concentrations of the agonists from this survey give some guidance for typical concentrations used in practice: ADP, 5 μM; collagen, 2.5 to 5.0 mg/mL; epinephrine, 10 to 18 μM; arachidonic acid, 0.5 to 1.6 mM; ristocetin low dose, 0.5 mg/mL; and ristocetin high dose, 1.2 to 1.25 mg/mL. The publication of a CLSI guideline for platelet aggregation testing should help to standardize the laboratory practice of platelet aggregation testing.

Plateletworks

Plateletworks is a new technology that is designed to determine platelet aggregation in fresh whole blood samples collected into EDTA or agonist-containing blood tubes. It utilizes a hematology cell analyzer to measure the platelet count in the EDTA tube and the ADP or collagen-containing tubes, in which the platelets have largely aggregated, and only the residual, nonaggregated platelets will be counted. The platelet aggregation is

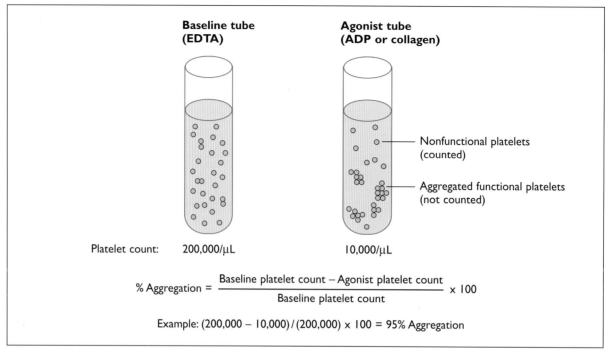

Baseline tube (EDTA)

Agonist tube (ADP or collagen)

Nonfunctional platelets (counted)

Aggregated functional platelets (not counted)

Platelet count: 200,000/μL   10,000/μL

$$\% \ Aggregation = \frac{Baseline \ platelet \ count - Agonist \ platelet \ count}{Baseline \ platelet \ count} \times 100$$

Example: (200,000 − 10,000) / (200,000) × 100 = 95% Aggregation

Figure 7-9. Plateletworks is designed to determine platelet aggregation in fresh whole blood samples collected into EDTA or agonist-containing blood tubes. It utilizes a hematology cell analyzer to measure the platelet count in the EDTA tube and the adenosine diphosphate (ADP) or collagen-containing tubes, in which the platelets have largely aggregated, and only the residual, nonaggregated platelets will be counted. The platelet aggregation is assessed by measuring the change in platelet count between the EDTA tube and the agonist tubes. The number of functional platelets can be calculated by subtracting the agonist platelet count from the EDTA platelet count. Likewise, the percent of platelet aggregation can be calculated by (EDTA platelet count − agonist platelet count)/(EDTA platelet count) × 100.

assessed by measuring the change in platelet count between the EDTA tube and the agonist tubes. This is the first platelet function test to simultaneously measure both platelet count and platelet aggregation. The current formulation of this test requires testing of the platelet count within 10 minutes in the ADP and collagen tubes, as platelets have been shown to disaggregate with prolonged incubation in the agonist tubes. This short time constraint may make the Plateletworks more applicable to a point-of-care application or to hospitals with rapid sample transportation systems. The number of functional platelets can be calculated by subtracting the agonist platelet count from the baseline EDTA platelet count. Likewise, the percent of platelet aggregation can be calculated by (baseline platelet count − agonist platelet count)/(baseline platelet count) x 100 (Figure 7-9). There is limited clinical experience with this technology, but it has been utilized to monitor the effect of GPIIb/IIIa antagonist antiplatelet drugs and also to monitor residual platelet function after open-heart surgery as a guide to platelet transfusion.

VerifyNow

VerifyNow (Ultegra) is a rapid, automated, turbidimetric whole blood assay designed to assess platelet aggregation, based on the ability of activated platelets to bind fibrinogen. Whole blood is directly sampled from a vacutainer tube into a disposable test cartridge. Fibrinogen-coated polystyrene microparticles agglutinate in whole blood in proportion to the number of available platelet GPIIb/IIIa receptors. The standard cartridge (VerifyNow IIb/IIIa Assay) uses TRAP agonist-activated platelets and fibrinogen-coated microparticles, and is designed to measure specifically the effect of GPIIb/IIIa antagonist drugs. The standard cartridge is not sensitive to drugs such as aspirin or the thienopyridines (clopidogrel and ticlopidine), and it is not designed to detect platelet functional disorders or von Willebrand disease. Use of cartridges with different agonists allows distinction of the effects of antiplatelet drugs, such as aspirin and clopidogrel. The VerifyNow Aspirin Assay is a cartridge that uses an arachidonic agonist to aid in detecting platelet dysfunction due to aspirin therapy. Sample results

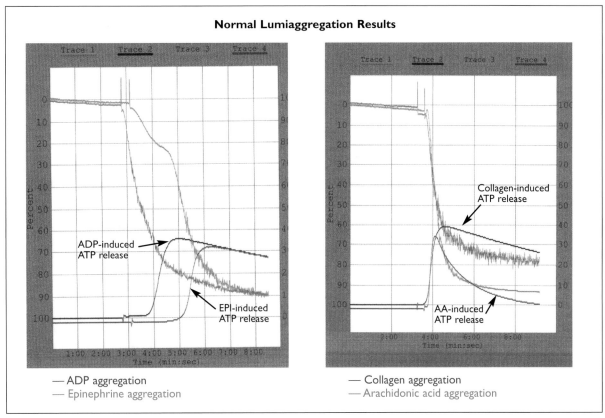

Normal Lumiaggregation Results

— ADP aggregation
— Epinephrine aggregation

— Collagen aggregation
— Arachidonic acid aggregation

Figure 7-10. Lumiaggregation. The platelet aggregation technique can be accompanied by a measure of dense granule release using a lumiaggregation technique based on the release of adenosine triphosphate (ATP) by dense granules during platelet aggregation. This diagram shows simultaneous measurement of platelet aggregation and release with adeno- sine diphosphate (ADP) and epinephrine (EPI) (left), and arachidonic acid (AA) and collagen (right). The aggregation curve with the agonists is shown at the top, and the corresponding ATP release from the dense granules is shown at the bottom. Note that the granule release corresponds to the secondary wave of aggregation.

are reported in aspirin reaction units (ARUs). A recommended reference range is ARU 620 to 672 in patients not taking aspirin, but this range should be verified by individual laboratories. An ADP-based cartridge (VerifyNow P2Y12 Assay) designed to detect the inhibition of platelet function by clopidogrel has been recently approved for clinical diagnostic testing.

The Ultegra assay uses a whole blood sample collected into 3.2% sodium citrate but requires a specific plastic blue top vacutainer tube that collects 2.0 mL of blood. The sample should rest 30 minutes after collection prior to assay, but it is stable for up to 4 hours after phlebotomy. The instrument may not report results in patients with a hematocrit less than 30%, but software is in development to extend testing to anemic patients. The results may be affected by a platelet count lower than 92,000/μL. Quality control includes running an electronic control daily. There are also two levels of wet control materials available.

Release Assays

Studies of alpha or dense granule release are helpful to discern alpha and dense granule storage pool disorders from platelet release disorders. However, laboratory techniques are limited and are not widely available.

Lumiaggregation

The platelet aggregation technique can be accompanied by a measure of dense granule release using a lumiaggregation technique. This capitalizes on the release of ATP by dense granules during platelet aggregation. A luciferin-luciferase enzyme reagent that is extracted from fireflies is added to the sample prior to aggregation. The high-energy phosphate bonds in ATP react with the enzyme, creating a luminescence. This can be quantified during aggregation to detect dense granule release. This technique has been adapted to both whole blood and optical aggregometers made by Chrono-log (Havertown, Pennsylvania)

(Figure 7-10). A modified lumiaggregation technique has been developed for the diagnosis of heparin-induced thrombocytopenia. Platelet granule release can also be measured by studying platelet ATP to ADP ratios.

Release Markers

The platelet alpha granules contain many proteins, including the platelet-specific proteins platelet factor 4 (PF4) and β-thromboglobulin (βTG). In the plasma of normal individuals, the level of these two proteins is low. However, during in vivo platelet activation or thrombosis, the platelets degranulate, and levels of these two proteins increase substantially. Research studies have shown increased levels of these release markers in patients with acute myocardial infarction and other thrombotic disorders. Enzyme-linked immunosorbant assays (ELISA) are available for both of these analytes, but these have not been widely used clinically due to stringent sample collection and processing requirements. These samples require collection in a special tube containing platelet inhibitors (citrate, theophylline, adenosine, dipyridamole), immediate chilling on ice, and rapid centrifugation to prevent spurious in vitro platelet release.

P-selectin

P-selectin is an intrinsic membrane glycoprotein located on the inner table of the alpha granule membrane. Upon platelet activation, P-selectin is expressed on the outer platelet surface, and some P-selectin is shed into the plasma. ELISA assays for soluble P-selectin are available but are fraught with the same limitations as the PF4 and βTG assays. Measurement of platelet-associated P-selectin is typically performed by flow cytometry (covered below).

Adhesion Assays

Platelet Adhesion Chambers

Many specialized experimental devices for studying platelet adhesion have been developed, but these are used largely in the research setting. The devices range from laminar parallel plate flow systems, to capillary flow systems, to rotating disc systems. These typically utilize video microscopy and a digital camera system to study the adhesion of platelets to the substrate and can be adapted to the study of whole blood, PRP, or washed platelets. Various substrate materials have been

tested, such as vWF, collagen, and artificial biomaterials used in cardiovascular devices. While a very powerful tool to study the details of platelet adhesion, these devices have not yet been adapted for clinical diagnostic laboratory use.

Impact

A device in development is the Impact, a modified cone and plate viscometer. It utilizes a small (130 µL) whole blood sample collected into 3.2% sodium citrate. The apparatus induces laminar flow to the sample with uniform shear stress (1800 sec^{-1}) between a disposable coverslip and a rotating polystyrene cone and measures shear-induced platelet adhesion and aggregation. Under these conditions, platelets adhere to the surface and form elongated aggregates aligned with the laminar flow lines. The platelets are stained, and the percentage of surface covered and the average size of the objects are determined via a charge couple display (CCD) camera that is integrated into the instrument. Results are available within 6 minutes. The instrument records several test parameters, including surface coverage, average size, and a distribution histogram of the adherent platelets.

Limited studies have been performed utilizing the Impact to detect platelet abnormalities, such as Glanzmann thrombasthenia, as well as afibrinogenemia and type 3 von Willebrand disease. Platelet adhesion to the polystyrene has been shown to be dependent upon plasma vWF and fibrinogen binding to the plates and on platelet glycoproteins Ib and IIb/IIIa. The Impact has also been shown to correlate with postoperative bleeding complications in cardiac surgery. A modification of this assay, using pre-exposure of the platelets to agonists arachidonic acid or ADP, allows it to measure inhibition of platelet aggregation by aspirin or clopidogrel. The Impact is available in Europe but not yet marketed in the USA.

Platelet Mechanical Assays

Hemostasis Analysis System

Historically, evaluation of clot retraction was based on the measurement of exuded serum from a clotted sample or the decrease in the size of a clot with time. The Hemostasis Analysis System is a new instrument that functions as a dynamic clot retractometer. It is available in Europe but is still under development in the USA. It is a modified mechanical tensile testing apparatus that measures platelet force development in a citrated whole blood sample, placed between two parallel plates; the sample is stimulated to clot by addition of agents such as thrombin. During clotting, the platelet cytoskeleton contracts, and the clot retraction results in the displacement of the moveable top plate, measured by a displacement transducer. The device measures both the tensile properties of the platelet cytoskeleton as well as the integrity of the GPIIb/IIIa receptor–fibrinogen linkage. One instrument parameter measured is platelet contractile force (PCF), the force produced by platelets during clot retraction. This parameter is sensitive to platelet number, platelet metabolic status, GPIIb/IIIa function, and antithrombin activities. Other parameters measured are clot elastic modulus (CEM) and thrombin generation time (TGT). Samples for measurement of PCF are reportedly stable for up to 10 days, although this remains to be independently verified. The Hemostasis Analysis System has been shown to detect decreased PCF during cardiopulmonary bypass surgery, with some platelet disorders, with acute uremia, and with GPIIb/IIIa antagonist drugs. Its role in the hemostasis testing laboratory has not yet been determined.

Whole Blood Viscoelastometry

A number of different techniques have been used to assess the viscosity and elasticity of clotted whole blood, including the Thrombelastograph® Hemostasis Analyzer (Haemonetics Corp, Braintree, Massachusetts), ROTEM® (Diagnostica Stago Inc, Parsippany, New Jersey), and the Sonoclot® Coagulation & Platelet Function Analyzer (Sienco Inc, Arvada, Colorado). Platelets increase both the dynamic viscosity and elasticity of clotted whole blood in a dose-dependent manner. Approximately 70% of the elastic shear modulus of normal clotting whole blood is directly proportional to the platelet count in the sample, with the remainder due to the fibrinogen concentration. Most of these assays are based either on the clotting of native whole blood, recalcification of sodium citrate anticoagulated whole blood, or recalcification plus activation with tissue factor or contact system activators like Celite or kaolin. All of these methods are dependent on platelet activation by thrombin generated in the sample, making these assays essentially insensitive to platelet inhibition by aspirin, other nonsteroidal anti-inflammatory medications, and the thienopyridines. These assays are sensitive to GPIIb/IIIa antagonists. They are insensitive to von Willebrand dis-

ease and other non-thrombin-dependent platelet abnormalities.

Recently, a modified Thromboelastograph assay called Platelet Mapping has been introduced to assess non-thrombin-based platelet activation. This assay uses heparin to block thrombin activity and a combination of reptilase and factor XIIIa to activate fibrinogen and cross-link it. Inhibition of platelets by nonsteroidal anti-inflammatory agents and thienopyridines can be assessed by adding arachidonic acid or ADP as agonists. Further studies are needed to assess the diagnostic accuracy of this method.

Flow Cytometry

Flow cytometry has been utilized to study platelet structure and function, but this technique is only employed in specialized centers. Flow cytometric analysis is based on the detection of cell surface proteins with fluorescently-labeled antibodies. The flow cytometer is an instrument in which cells in a buffer suspension pass through a laser beam. The scatter of light in the forward and side directions indicates cell size and granularity, respectively. A population of platelets can be electronically selected (gated) based on size and granularity (Figure 7-11). Coupled with fluorescently labeled antibodies to surface proteins, the expression of a panel of proteins can be analyzed for each platelet individually. Typically 10,000 or more cells are studied in each sample, making flow cytometry a powerful tool to study platelet function. Benefits of platelet flow cytometry include the ability to detect the activation state of circulating platelets, to study the reactivity of platelets to specific agonists, and to study platelet function in a very small sample with a relatively low platelet count.

Platelet flow cytometry can be used to detect the presence of typical platelet surface glycoproteins as well as decreased expression or deficiency of these glycoproteins. It has been used to detect the absence of GPIIb/IIIa receptors in patients with Glanzmann thrombasthenia and has been used to study deficiencies of glycoproteins Ia, Ib, IIb, IV, and IX.

Platelet activation leads to a conformational change in some surface receptors, such as GPIIb/IIIa, and increased surface expression of several markers, such as P-selectin from the alpha granules. During activation, there is redistribution of the phospholipid in the platelet membrane, with increased surface expression of phosphatidyl

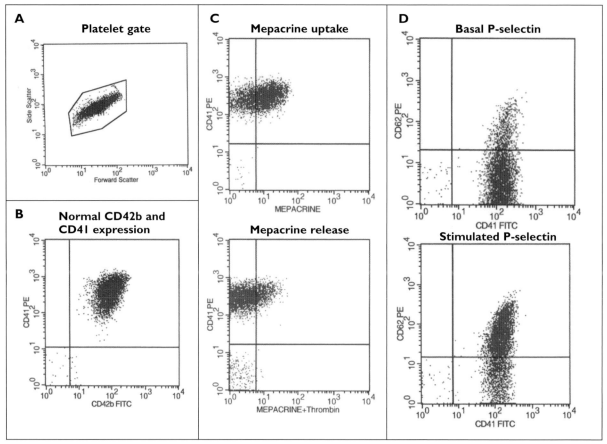

Figure 7-11. Platelet flow cytometry. Platelets can be studied by flow cytometric analysis to assess platelet surface receptors and functionality of platelet release and aggregation. The platelet population can be gated by its characteristic size and granularity in a whole blood specimen (Panel A). Using a combination of fluorescently labeled antibodies, surface expression of glycoproteins can be evaluated, as shown for CD41 (GPIIb/IIIa) and CD42b (GPIX) (Panel B). Platelet flow cytometry can be coupled with platelet agonists to study alpha and dense granule function. Mepacrine, a fluorescence compound that is taken up by platelet dense granules, can be used to study dense granule number and function (Panel C). Resting platelets should take up mepacrine; once stimulated with an agonist, the mepacrine fluorescence decreases. Resting platelets do not express high levels of the alpha granule protein P-selectin (Panel D). However, upon stimulation, most platelets degranulate and begin to express P-selectin on the plasma membrane surface. (Becton Dickinson, FACScan.)

serine. Additionally, there is increased binding of several proteins to surface receptors, such as thrombospondin and fibrinogen. With fibrinogen binding to GPIIb/IIIa, a new epitope can be detected, known as a ligand-induced binding site (LIBS). Thus, with the use of appropriate antibodies, the percentage of activated platelets in a specimen can be determined. Measurement of platelet activation by flow cytometry has been utilized to diagnose alpha and dense granule storage pool disorders and release/signaling disorders, where measurement of activation-dependent surface markers are measured before and after addition of a platelet agonist, such as TRAP or ADP. In a research setting, study of in vivo platelet activation has employed drawing a blood sample into a cocktail of platelet inhibitors to prevent further in vitro activation; this technique has been utilized to demonstrate increased in vivo platelet activation during acute myocardial infarction and coronary angioplasty.

Flow cytometric methods can be adapted to measure platelet function by using a combination of activation-dependent antibodies and gating techniques. For example, platelet aggregation can be measured by flow cytometry by coupling a GPIIb/IIIa antibody with a forward/side scatter gate that differentially captures the larger platelet aggregates. Measure of platelet microparticle release can be performed by gating on a low forward-scatter region (below that of resting platelets) and counting the number of cells in this region that express GPIIb/IIIa and an activation marker. Dense granules can be assessed using mepacrine, which is selectively taken up into the dense granules. Measure of mepacrine uptake corresponds to dense granule number, and mepacrine release after platelet stimulation can be used to assess platelet signaling and the release reaction. Platelet alpha granule release can be followed by probing platelet-bound P-selectin, which is newly expressed on the platelet surface following activation (Figure 7-11). Platelet procoagulant activity, due to formation of coagulation complexes on the phospholipid surface of activated platelets, can be studied using coagulation protein-specific antibodies, such as those to factor Va or Xa.

Another use of flow cytometry is in the detection of platelet-associated immunoglobulins, for the detection of platelet autoantibodies in patients with immune thrombocytopenic purpura (ITP) and drug-induced thrombocytopenias, which is sensitive, but not specific. This test can be made more specific for drug-induced antibodies by incubating the platelets in the presence of the drugs in question or by using activation-dependent tests, such as 51Cr release or 14C serotonin release. Antigen-capture assays, such as MAIPA (monoclonal antibody immobilization of platelet antigens), have improved specificity further by being able to detect antibody binding to specific platelet surface glycoproteins.

Platelet flow cytometry can be performed on whole blood samples collected into 3.2% sodium citrate if in vitro activation studies or aggregation will be performed. Alternatively, samples can be drawn into CTAD (citrate, theophylline, adenosine, and dipyridamole) to prevent in vitro activation. Platelets progressively activate during in vitro storage, so testing of platelet function by flow cytometry should be performed within 1 hour of phlebotomy. If only detection of surface glycoproteins is desired, as in the diagnosis of Glanzmann thrombasthenia, the samples may be used for up to 24 hours.

Platelet Turnover (Platelet Reticulocyte Analysis)

Platelets with increased RNA content (reticulated platelets) can be measured by flow cytometry using dyes that bind to RNA and DNA, such as thiazole orange, auramine O, and coriphosphine. Several different flow cytometric techniques for performing this analysis have been published, but there is currently no internationally standardized method. The published normal reference ranges vary considerably, from 1% to more than 16%

reticulated platelets. This variability is likely due to the different techniques employed. Reticulated platelet analysis has been studied as a diagnostic tool to evaluate whether thrombocytopenia is due to increased platelet destruction or decreased platelet production, as platelets newly released from bone marrow have increased RNA content. It is anticipated that implementation of reticulated platelet counts may help to avoid bone marrow examination in some individuals with thrombocytopenia. This assay has recently been automated on the CELL DYN-4000 (Abbott Diagnostics, Abbott Park, Illinois). It has also been adapted to the Sysmex XE-2100 (Sysmex Corp, Kobe, Japan), where a nuclei acid-specific dye is detected in platelets in the reticulocyte channel; the parameter is termed the immature platelet fraction (IPF). The IPF has been shown to be useful in the diagnosis of peripheral platelet consumption and as a guide to transfusion after hematopoietic stem cell transplantation.

Electron Microscopy

Electron microscopy may be utilized for the ultrastructural evaluation of platelets, particularly in patients with suspected storage pool disorders, showing a decrease or absence of the organelles (cytoplasmic dense bodies) storing adenine nucleotides, serotonin, and calcium. Giant platelet disorders also have characteristic electron microscopic findings.

Platelet Genetic Testing

Genetic testing for diagnosis of platelet disorders is not widely available. Some specialized research laboratories may perform gene sequencing in the diagnosis of Glanzmann thrombasthenia or Bernard-Soulier syndrome. Polymorphisms in some of the platelet surface glycoproteins, such as PLA-1/2 (HPA-1a/1b), that may be responsible for neonatal alloimmune thrombocytopenia have been studied by platelet flow cytometry, but a polymerase chain reaction (PCR)-based assay kit has been recently commercialized (GTI, Milwaukee, Wisconsin). Other studies of platelet genomics, such as single nucleotide polymorphism (SNP) analysis of surface glycoproteins for predicting cardiovascular risk, are limited to research studies at this time. Measurement of platelet transcriptome and proteome, which seek to measure the variety of mRNA and platelet species in the platelet, are in their infancy but hold clinical promise in the diagnosis of some platelet disorders.

ADAMTS-13 Assays

Thrombotic thrombocytopenic purpura (TTP) is a disorder that results from a deficiency of or an inhibitor directed against ADAMTS-13 (a disintegrin-like metalloprotease with thrombospondin type 1 repeats), a metalloproteinase that cleaves vWF. The initial assays developed to measure ADAMTS-13 activity were based upon its ability to cleave the high-molecular-weight multimers of vWF and utilized electrophoretic multimer analysis to identify the cleavage. These assays required a long incubation of the patient's plasma with vWF, followed by the multimer assay, which itself takes up to 2 days. Thus, these multimer-based assays could not be performed in a clinically relevant time frame, since the diagnosis of TTP usually has a high degree of clinical urgency. An adaptation of the ADAMTS-13 assay has used a collagen-binding assay instead of the multimer electrophoresis step, but this assay still requires a prolonged incubation of the patient's plasma with vWF.

It has been determined that ADAMTS-13 cleaves vWF at Y1605-M1606. Several rapid ADAMTS-13 assays are being developed that no longer rely on measuring the change in molecular weight of vWF, but are chromogenic or fluorometric substrate-based assays that directly measure the enzymatic activity of ADAMTS-13. Rapid fluorescence resonance energy transfer (FRET)-based assays are being developed and are nearing clinical availability in the USA. One such assay uses a peptide substrate that contains a 73-amino acid residue of vWF from D1596 to R1668. ELISA assays are also being developed that can detect the anti-ADAMTS-13 antibodies observed in some patients with TTP.

Summary

Laboratory testing for platelet function is more complex than plasma-based assays for coagulation proteins because of the cellular nature of platelets. Platelet testing has mainly been limited to large medical centers, but there has been significant technological development of newer platelet function assays that have brought some platelet function testing capabilities to smaller laboratories and point-of-care settings.

Suggested Reading

Platelet Count and Indices

Bain BJ. *A Beginner's Guide to Blood Cells: A Practical Guide.* 2nd ed. Oxford, UK: Blackwell; 2006.

Bowles KM, Cooke LJ, Richards EM, Baglin TP. Platelet size has diagnostic predictive value in patients with thrombocytopenia. *Clin Lab Haematol.* 2005;27:370-373.

Fiorin F, Steffan A. Pradella P, Bizzaro N, Potenze R, De Angelis V. IgG platelet antibodies in EDTA-dependent pseudothrombocytopenia bind to platelet membrane glycoprotein IIb. *Am J Clin Pathol.* 1998;110:178-183.

Huczek Z, Kochman J, Filipiak KJ, et al. Mean platelet volume on admission predicts impaired reperfusion and long-term mortality in acute myocardial infarction treated with primary percutaneous coronary intervention. *J Am Coll Cardiol.* 2005;46:284-290.

Kaito K, Otsubo H, Usui N, et al. Platelet size deviation width, platelet large cell ratio and mean platelet volume have sufficient sensitivity and specificity in the diagnosis of immune thrombocytopenia. *Br J Haematol.* 2005;128:698-702.

Lewis SM, Tatsumi N. Collection and handling of blood. In: Lewis SM, Bain BJ, Bates I, eds. *Dacie and Lewis Practical Haematology.* 10th ed. St Louis, Mo: Elsevier; 2006: 1-10.

Sane DC, Damaraju LV, Topol EJ, et al. Occurrence and clinical significance of pseudothrombocytopenia during abciximab therapy. *J Am Coll Cardiol.* 2000; 36:75-83.

Tubes and Additives for Venous Blood Specimen Collection. Approved Standard. 5th ed. Wayne, Pa: NCCLS (now CLSI); 2003. Document H01-A5.

Bone Marrow Evaluation

Bates I. Bone marrow biopsy. In: Lewis SM, Bain BJ, Bates I, eds. *Dacie and Lewis Practical Haematology.* 10th ed. St Louis, Mo: Elsevier; 2006: 115-131.

Jaffe ES, ed. *Pathology and Genetics. Tumours of Haematopoietic and Lymphoid Tissues. World Health Organization Classification of Tumours.* Lyon, France: IARC Press; 2003.

Evaluation of Platelet Function

General References

Harrison P. Platelet function analysis. *Blood Rev.* 2005;19:111-123.

Kottke-Marchant K, Corcoran G. The laboratory diagnosis of platelet disorders: an algorithmic approach. *Arch Pathol Lab Med.* 2002;126:133-146.

Michelson AD. Platelet function testing in cardiovascular diseases. *Circulation.* 2004;110:e489-e493.

Michelson AD. The clinical approach to disorders of platelet number and function. In: Michelson AD. *Platelets*. Amsterdam: Academic Press; 2002: 541-545.

Nicholson NS, Panzer-Knodle SG, Haas BS, et al. Assessment of platelet function assays. *Am Heart J*. 1998;135:S170-178.

PFA-100

Fressinaud E, Veyradier A, Truchaud F, et al. Screening for von Willebrand disease with a new analyzer using high shear stress: a study of 60 cases. *Blood*. 1998;91:1325-1331.

Hayward CPM, Harrison P, Cattaneo M, Ortel TL, Rao AK. Platelet function analyzer (PFA-100) closure time in the evaluation of platelet disorders and platelet function. *J Thromb Haemost*. 2006;4:312-319.

Kottke-Marchant K, Powers JB, Brooks L, Kundu S, Christie DJ. The effect of antiplatelet drugs, heparin and preanalytical variables on platelet function detected by the platelet function analyzer (PFA-100™). *Clin Appl Thromb Hemost*. 1999;5:122-130.

Kundu SK, Heilmann EJ, Sio R, Garcia C, Davidson RM, Ostgaard RA. Description of an in vitro platelet function analyzer - PFA-100. *Semin Thromb Hemost*. 1995; 21:106-112.

Mammen EF, Comp PC, Gosselin R, et al. PFA-100 system: a new method for assessment of platelet dysfunction. *Semin Thromb Hemost*. 1998; 24:195-202.

Bleeding Time

Brinkhous KM. W.W. Duke and his bleeding time test: a commentary on platelet function. *JAMA*. 1983;250: 1210-1214.

Burns ER, Lawrence C. Bleeding time: a guide to its diagnostic and clinical utility. *Arch Pathol Lab Med*. 1989;113:1219-1224.

Lind SE. The bleeding time does not predict surgical bleeding. *Blood*. 1991;77:2547-2552.

Aggregation

Born GVR. Aggregation of blood platelets by adenosine diphosphate and its reversal. *Nature*. 1962;194:927-929.

Cattaneo M, Lecchi A, Zighetti ML, Lussana F. Platelet aggregation studies: autologous platelet-poor plasma inhibits platelet aggregation when added to platelet-rich plasma to normalize platelet count. *Haematologica*. 2007;92(5):694-697.

Ghosh K, Nair S, Kulkarni B, Khare A, Shetty S, Mohnty D. Platelet function tests using platelet aggregometry: need for repetition of the test for diagnosis of defective platelet function. *Platelets*. 2003;14:351-354.

Harder S, Klikhardt U, Graff J, et al. In vitro dose response to different GP IIb/IIIa-antagonists: interlaboratory comparison of various platelet function tests. *Thromb Res*. 2001;102:39-48.

Machin SJ, Preston E, for the British Society for Haematology BCSH Haemostasis and Thrombosis Task Force. Guideline on platelet function testing. *J Clin Pathol*. 1988;41:322-1330.

Mani H, Luxembourg B, Klaffling C, Erbe M, Lindhoff-Last E. Use of native or platelet count adjusted platelet rich plasma for platelet aggregation measurements. *J Clin Pathol*. 2005;58:747-750.

Moffat KA, Ledford-Kraemer MR, Nichols WL, Hayward CPM. Variability in clinical laboratory practice in testing for disorders of platelet function. *Thromb Haemost*. 2005;93:549-553.

Mustard JF, Packham MA. Factors influencing platelet function: adhesion, release, and aggregation. *Pharmacol Rev*. 1970;22(2):97-187.

Nosal R. Antiplatelet and antileukocyte effects of cardiovascular, immunomodulatory and chemotherapeutic drugs. *Cardiovasc Hematol Agents Med Chem*. 2006;4:237-261.

Platelet Function Testing by Aggregometry. Proposed Guideline. Wayne, Pa: Clinical and Laboratory Standards Institute (CLSI); 2007. Document H58-P.

Triplett DA, Harms CS, Newhouse P, Clark C. *Platelet Function. Laboratory Evaluation and Clinical Application*. Chicago, Ill: ASCP Press; 1978.

Zhou L, Schmaier AH. Platelet aggregation testing in platelet-rich plasma: description of procedures with the aim to develop standards in the field. *Am J Clin Pathol*. 2005;123:172-193.

Ultegra

Buch AN, Singh S, Roy P, et al. Measuring aspirin resistance, clopidogrel responsiveness, and postprocedural markers of myonecrosis in patients undergoing percutaneous coronary intervention. *Am J Cardiol*. 2007;99:1518-1522.

Coller BS, Lang D, Scudder LE. Rapid and simple platelet function assay to assess GPIIb/IIIa receptor blockade. *Circulation*. 1997;95:860-867.

Wheeler GL, Braden GA, Steinhubl SR, et al. The Ultegra rapid platelet-function assay: comparison to standard platelet function assays in patients undergoing percutaneous coronary intervention with abciximab therapy. *Am Heart J*. 2002;143:602-611.

Plateletworks

Craft RM, Chavez JJ, Snider CC, Muenchen RA, Carroll RC. Comparison of modified Thromboelastograph and Plateletworks whole blood assays to optical platelet aggregation for monitoring reversal of clopidogrel inhibition in elective surgery patients. *J Lab Clin Med*. 2005;145:309-315.

Ostrowsky J, Foes J, Warchol M, Tsarovsky G, Blay J. Plateletworks platelet function test compared to the thromboelastograph for prediction of postoperative outcomes. *J Extra Corpor Technol.* 2004;36:149-152.

Sackett E, Nuttall GA, Ereth MH, Oliver WC Jr. Validation of the Plateletworks Point of Care Platelet Function Analyzer. *Anesth Analgesia.* 2000;Abstract S86.

White MM, Krishnan R, Kueter TJ, Jacoski MV, Jennings LK. The use of the point of case Helena ICHOR/ Plateletworks and the Accumetrics Ultegra RPFA for assessment of platelet function with GP IIb/IIIa antagonists. *J Thromb Thrombolysis.* 2004;18:163-169.

Platelet Activation Assays

Gurney D, Lip GYH, Blann AD. A reliable plasma marker of platelet activation: does it exist? *Am J Hematol.* 2002;70:139-144.

Stewart MW, Etches WS, Boshkov LK, Gordon PA. Heparin-induced thrombocytopenia: an improved method of detection based on lumi-aggregometry. *Br J Haematol.* 1995;91:173-177.

White MM, Foust JT, Mauer AM, Robertson JT, Jennings LK. Assessment of lumiaggregometry for research and clinical laboratories. *Thromb Haemost.* 1992; 67:572-577.

Platelet Adhesion Assays

Gupta AS, Wang S, Link E, et al. Glycocalyx-mimetic dextran-modified poly(vinyl amine) surfactant coating reduces platelet adhesion on medical-grade polycarbonate surface. *Biomaterials.* 2006;27:3084-3095.

Hainaud P, Brouland JP, Andre P, et al. Dissociation between fibrinogen and fibrin interaction with platelets in patients with different subtypes of Glanzmann's thrombasthenia: studies in an ex vivo perfusion chamber model. *Br J Haematol.* 2002; 119(4):998-1004

Impact

Lubetsky A, Jarrach R, Brill A, Marin G, Savion N, Varon D. Cone and Plate(let) parameters as predictors of bleeding in patients undergoing cardiac surgery. *Blood.* 2002;100:2720.

Savion N, Varon D. Impact - The Cone and Plate(let) Analyzer: testing platelet function and anti-platelet drug response. *J Pathophysiol Hemost Thromb.* 2006; 35:83-88.

Spectre G, Brill A, Gural A, et al. A new point-of-care method for monitoring anti-platelet therapy: application of the Cone and Plate(let) Analyzer. *Platelets.* 2005;16:293-299.

Hemostasis Analysis System

Carr ME. Development of platelet contractile force as a research and clinical measure of platelet function. *Cell Biochem Biophys.* 2003;38:55-78.

Carr ME. Measurement of platelet force: the Hemodyne 7 Hemostatic Analyzer. *Clin Lab Review.* 1995;9:312-320.

Carr ME, Martin EJ, Kuhn JG, Spiess BD. Onset of force development as a marker of thrombin generation in whole blood: the thrombin generation time (TGT). *J Thromb Haemost.* 2003;1:1977-1983.

Flow Cytometry

Abrams C, Shattil SJ. Immunological detection of activated platelets in clinical disorders. *Thromb Haemost.* 1991;65:467-175.

Catellier DJ, Aleksic N, Folsom AR, Boerwinkle E. Atherosclerosis Risk in Communities (ARIC) Carotid MRI Flow Cytometry Study of Monocyte and Platelet Markers: Intraindividual Variability and Reliability. *Clin Chem.* 2008 May 29 (Epub ahead of print).

Linden MD, Frelinger AL III, Barnard MR, Przyklenk K, Furman MI, Michelson AD. Application of flow cytometry to platelet disorders. *Semin Thromb Hemost.* 2004;30:501-512.

Wall JE, Buijs-Wilts M, Arnold JT, Wang W, White MM. A flow cytometric assay using mepacrine for study of uptake and release of platelet dense granule contents. *Br J Haematol.* 1995:89:380-385.

Platelet Turnover

Briggs C, Hart D, Kunka S, Oguni S, Machin SJ. Immature platelet fraction measurement: a future guide to platelet transfusion requirement after haematopoietic stem cell transplantation. *Transfus Med.* 2006;16:101-109.

Butkiewicz AM, Kemona H, Dymicka-Pickarska V, Matowicka-Karna J, Radziwon P, Lipska A. Platelet count, mean platelet volume and thrombocytopeoietic indices in healthy women and men. *Thromb Res.* 2006;118:199-204.

Robinson MSC, Machin S, Mackie IJ, Harrison P. Two colour analysis of reticulated platelets. *Clin Lab Haemotol.* 2000;22:211-213.

Salvagno GL, Montagnana M, Degan M, et al. Evaluation of platelet turnover by flow cytometry. *Platelets.* 2006;17:170-177.

ADAMTS-13

Kokame K, Nobe Y, Kokubo Y, Okayama A, Miyata T. FRETS-VWF73, a first flourogenic substrate for ADAMTS13 assay. *Br J Haematol.* 2005;129:93-100.

Uemura M, Fujimura Y, Matsumoto M, et al. Comprehensive analysis of ADAMTS13 in patients with liver cirrhosis. *Thromb Haemost.* 2008;99:1019-1029.

Fibrinolysis Testing

Wayne L. Chandler, MD

Clinical History Suggesting the Need for Fibrinolytic Testing

Fibrinolytic testing is most appropriate in patients with unexplained bleeding or arterial thrombosis starting at a relatively young age (typically less than age 50) after more routine coagulation studies have proven negative. With regard to fibrinolysis-related bleeding, a common finding is delayed bleeding, often associated with recurrent, excessive wound hematomas after surgery or trauma. Several acquired forms of fibrinolytic bleeding have been described, which will be discussed in more detail in chapter 13. A history of unexplained arterial thrombosis in the absence of other arterial risk factors, such as hypertension, diabetes, hyperlipidemia, smoking, etc, could prompt a fibrinolytic workup for fibrinolytic hypofunction. Initial fibrinolytic testing for unexplained bleeding and thrombosis is shown in Figure 8-1. For more details and interpretation of test results, see chapters 13 and 20.

Sample Collection and Processing for Fibrinolytic Testing

Two factors influence when samples for fibrinolytic assays should be drawn: (1) circadian rhythms in the fibrinolytic system and (2) the effect of an acute phase response on fibrinolysis. The fibrinolytic system shows a prominent circadian rhythm, with peak levels of tissue plasminogen activator (tPA) antigen, plasminogen activator inhibitor-1 (PAI-1) activity, and PAI-1 antigen, and nadir levels of tPA activity in the morning at approximately 0600 to 0800 h, with the reverse at 1800 to 2000 h (Figure 8-2). On average, PAI-1 levels are twice as high in the morning as they are in the afternoon or evening, but the difference between morning and evening may be up to 10-fold in some individuals. When possible, samples for fibrinolytic testing should be drawn in the morning at approximately 0800 h. As normal lev-

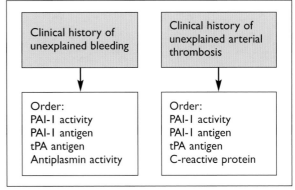

Figure 8-1. Recommended fibrinolytic tests for patients with a history of unexplained bleeding or arterial thrombosis.

els of fibrinolytic parameters may vary with time of day, the time the sample is drawn should be carefully noted. Clinical laboratories measuring and reporting results of tPA and PAI-1 assays should have reference ranges based on time of day, similar to ranges for other factors with a circadian rhythm, like cortisol. Acute phase responses due to infection, inflammation, cancer, and other processes result in increased production of PAI-1 and tPA. Levels of tPA and PAI-1 can rise in as little as 2 to 3 hours after an inflammatory stimulus such as surgery.

If the patient is being evaluated for increased risk of arterial thrombosis associated with elevated PAI-1 activity, the sample should be drawn when the patient is not experiencing a transient rise in PAI-1 due to an acute phase response. This can be difficult to assess clinically; therefore, we recommend drawing a C-reactive protein (CRP) along with fibrinolytic assays for assessment of arterial thrombotic risk. If the CRP is normal, it suggests that an acute phase response is not occurring, and the tPA and PAI-1 can be interpreted. If the CRP is elevated, the tPA and PAI-1 values may be transiently elevated as well and should not be used as an indicator of thrombotic risk.

Plasma anticoagulated with sodium citrate can be used for most fibrinolytic assays, with a few

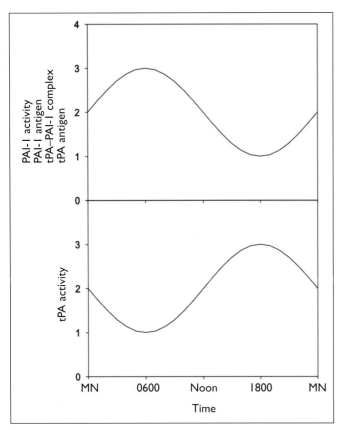

Figure 8-2. Circadian rhythms of the fibrinolytic system. Secretion of plasminogen activator inhibitor-1 (PAI-1) shows a circadian rhythm, with peak secretion in the morning and nadir secretion in the evening. Secretion of tissue plasminogen activator (tPA) is relatively constant during the day. Increased PAI-1 secretion in the morning leads to increased levels of PAI-1 activity, PAI-1 antigen, tPA–PAI-1 complex, and total tPA antigen in the morning (upper graph), but reduced levels of tPA activity (lower graph). The time the sample was obtained must be taken into account in interpreting fibrinolytic studies.

Table 8-1. Acceptable Sample Types for Fibrinolytic Assays

Assay	Sample Type
Antiplasmin activity	Citrate plasma
D-dimer antigen	Citrate plasma
Point-of-care D-dimer tests	Whole blood
Euglobulin lysis time	Citrate plasma
Lipoprotein (a)	EDTA plasma
PAI-1 activity	Citrate plasma
PAI-1 antigen	Citrate plasma
Plasmin-antiplasmin	Citrate plasma
Plasminogen activity	Citrate plasma
Plasminogen antigen	Citrate plasma
Thrombin-activatable fibrinolysis inhibitor (TAFI) antigen	Citrate plasma
Thrombo-Wellco test	Clot tube with thrombin activator and aprotinin
tPA activity	Acidified citrate plasma
tPA antigen	Citrate plasma
uPA activity	Citrate plasma
uPA antigen	Citrate plasma
Whole blood lysis time	Citrate whole blood

notable exceptions such as tPA activity. Table 8-1 shows the type of sample that should be obtained for each of the fibrinolytic assays described in this chapter. The in vivo concentration of active tPA in blood is a steady-state function of the rate of tPA secretion, the rate of tPA inhibition by PAI-1 (proportional to the PAI-1 activity), and the rate of tPA clearance by the liver. PAI-1 activity is typically in excess of tPA activity in blood. Therefore, if the reaction between tPA and PAI-1 is not blocked when the sample is drawn, essentially all tPA activity will be lost in the sample due to PAI-1 inhibition.

To accurately measure tPA activity, blood must be drawn into an acidified citrate tube, which is available commercially (Biopool® Stabilyte™, Trinity Biotech, Bray, Ireland). In the presence of elevated fibrinolytic activity (as occurs during thrombolytic therapy), plasminogen activation, fibrin degradation, and fibrinogen degradation can continue to occur in the sample after it is drawn. To stabilize fibrin degradation fragment levels in samples with potentially elevated fibrinolytic activity present, blood should be drawn into sampling tubes containing the fibrinolytic inhibitor aprotinin and the serine protease inhibitor D-Phe-Pro-Arg chloromethylketone (PPACK) in addition to citrate. These tubes are available commercially (Haematologic Technologies Inc, Essex Junction, Vermont) but are not typically stocked by most hospitals. If PAI-1 antigen is being measured, care must be taken not to activate platelets, resulting in PAI-1 release. Blood should be drawn into citrate anticoagulant and the

Table 8-2. Comparison of Clinical Utility of D-dimer Assays

D-dimer Assay	Useful in Evaluation of VTE	Useful in Evaluation of DIC
Qualitative latex agglutination assays	No	Yes
Semiquantitative and threshold assays	Yes	No
Quantitative enzyme immunoassays	Yes	Yes
Quantitative latex immunoassays	Yes	Yes

Abbreviations: VTE, venous thromboembolism; DIC, disseminated intravascular coagulation

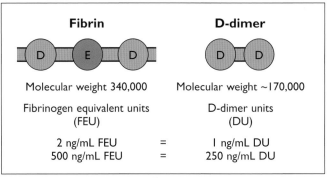

Figure 8-3. Comparison of different units used to report results for quantitative D-dimer assays: fibrinogen equivalent units (FEU) versus D-dimer units (DU).

sample maintained at room temperature until the platelets can be removed by centrifugation. Ethylenediaminetetraacetic acid (EDTA) can stimulate platelet release and should not be used as an anticoagulant for fibrinolytic studies.

D-Dimer and Fibrin Split Product Assays

Fibrinogen, fibrin, and their degradation products can be measured immunologically in plasma using specific antibodies. Degradation of fibrin by plasmin results in a variety of different fragments, from large pieces of polymerized fibrin to small fragments of a single fibrin molecule like D-dimer. The most common assay currently used to assess fibrin degradation is the D-dimer assay, a breakdown product consisting of two fibrin D-domains cross-linked by factor XIIIa and proteolytically released by plasmin. The D-dimer assay will be discussed first, followed by less common fibrin degradation product assays. Table 8-2 shows the different types of D-dimer assays currently available. While all of these assays are termed "D-dimer" tests, the reality is they detect a range of cross-linked fibrin degradation fragments that contain the D-dimer moiety.

The specificity of different tests for D-dimer itself versus DDE or other fibrin fragments is a function of the antibodies used to detect D-dimer. If purified fibrin degradation fragments prepared in vitro are used, variable results can be found

using different D-dimer assays. In contrast, in vivo, a wide range of fibrin degradation fragments is present in plasma, and most D-dimer assays appear to give similar results. One difficulty in comparing D-dimer results between assays is the different units used to report results by different companies (Figure 8-3).

A D-dimer fragment is approximately half the molecular weight of fibrin, though it varies somewhat depending of the method used to prepare the D-dimer standard in vitro. Some companies report D-dimer in D-dimer units (DU), assuming a molecular weight half that of fibrin. Other companies report in fibrinogen equivalent units (FEU), a measure of how many degraded fibrins this level of D-dimer represents. Since fibrin is approximately twice the size of D-dimer, 1 FEU represents twice as many ng/mL as 1 D-dimer unit (ie, 2 ng/mL FEU = 1 ng/mL DU). Another problem of reporting is whether nanograms or micrograms are reported, and whether milliliters or liters are used, giving a grand total of eight different reporting units. Based on College of American Pathologists surveys, the most common reporting units are FEU, either µg/mL or ng/mL. The type of unit generated has real clinical importance, as it will determine the threshold used for the evaluation of thrombosis and disseminated intravascular coagulation (DIC). Care must be taken to determine the type of units used for a particular assay and the diagnostic threshold for that method. The reference range for D-dimer in healthy subjects is typically less than about 500 ng/mL FEU, equivalent to less than 1.5 nmol/L degraded fibrin.

The first assays developed to measure D-dimer clinically were latex agglutination assays. Latex particles are coated with anti-human D-dimer antibodies. Plasma is added to the particles and the mixture rotated. Visually observed agglutination of the latex particles indicates the presence of D-dimer. Dilutions of patient plasma can be assayed to estimate the concentration of D-dimer in the sample. These assays are rapid and inexpensive but are relatively insensitive. These assays cannot reproducibly detect the difference between normal and modest elevations of D-dimer, are subject to observer bias as to what constitutes a positive test, and at best provide a qualitative indication of D-dimer level. They are not suitable for the evaluation of patients with suspected pulmonary embolism or proximal deep venous thrombosis, as they cannot reliably detect normal versus slight elevations of D-dimer.

Modified forms of the agglutination assay have been developed to detect a specific level or threshold of D-dimer in samples at a cutoff value for ruling out pulmonary embolism. These assays are designed to give only a positive or negative result and are typically used as a point-of-care test or near-patient test. Below the threshold, the assay is negative and the probability of pulmonary embolism is low; above the threshold, elevated D-dimer is present in the sample, indicating fibrin is present in the vascular system, but not the amount or location. Therefore, a positive result has low specificity and is not useful clinically in the diagnosis of thromboembolism or DIC.

One version of this assay is based on the use of a whole blood sample and an autologous red cell agglutination assay (SimpliRED™ D-dimer, AGEN Biomedical, Acacia Ridge, Australia). The assay utilizes a conjugate of two monoclonal antibodies, one that binds to the red blood cell surface (but itself does not cause agglutination) and another that is specific for D-dimer and D-dimer containing cross-linked fibrin degradation products. The conjugate will coat the red cells but will not cause agglutination in samples with normal D-dimer levels. If D-dimer containing cross-linked fibrin degradation fragments are present in blood samples at elevated levels, they will bind to the antibody conjugate on the red blood cells, causing cross-linking between cells, which results in visible agglutination.

Another version of the assay also requires a whole blood sample and utilizes a single test cartridge with colloidal gold particles coated with anti-human D-dimer antibodies (Clearview® Simplify™ D-dimer, AGEN Biomedical, Acacia Ridge, Australia). If the sample contains D-dimer antigen, this forms a complex with the antibody-gold conjugate. Aided by buffer, this complex flows along the strip in the aqueous phase until it is captured by a second D-dimer monoclonal antibody immobilized in the test zone. It produces a visible line if positive, much like over-the-counter pregnancy tests. These assays are a definite improvement over standard latex agglutination assays but may still suffer from observer bias due to the use of a visually determined endpoint. While still in use as point-of-care tests for rapid evaluation of threshold D-dimer levels, they have been replaced in many clinical settings by rapid, automated, quantitative D-dimer assays.

The first quantitative D-dimer assays were microtiter plate enzyme immunoassays. These assays were sensitive to low levels of D-dimer (detection limit approximately 30 ng/mL FEU) and have acceptable precision (coefficient of variation about 4% at normal plasma levels) for clinical evaluation of thrombosis, but were slow, requiring several hours to perform. This format has been updated, including one rapid automated version (VIDAS® D-Dimer Exclusion™, bioMérieux Inc, Durham, North Carolina) with similar detection limit and precision. Another version of a quantitative D-dimer assay is the latex immunoassay. This method is based on a fine suspension of antibody-coated latex particles. The rate of agglutination, monitored optically by an automated instrument, is proportional to the antigen concentration. These assays take approximately 15 minutes and have detection limits and precisions similar to enzyme immunoassays.

D-dimer latex immunoassays are available on most automated coagulation analyzers. Another advantage of all quantitative D-dimer assays is that they can also be used to assess high levels of D-dimer associated with DIC and other consumptive coagulopathies. Accurate measurement of high D-dimer levels is not possible using manual qualitative latex agglutination assays or semi-quantitative threshold-type assays. Current research is evaluating the use of D-dimer assays after the institution of therapy in patients with venous thrombosis as a measure of the efficacy of the therapy, so accurate and precise quantitation of D-dimer levels may be even more important in the future.

Most D-dimer assays depend on the use of monoclonal antibodies for specificity, making these assays susceptible to interference from human anti-mouse antibodies (HAMA). These antibodies can occur spontaneously in patients treated with mouse monoclonal antibodies therapeutically, and rarely in patients with monoclonal gammopathies such as multiple myeloma.

Rheumatoid factors may cross-react with a variety of polyclonal and monoclonal antibodies. Interference due to cross-reacting anti-immunoglobulins can appear as either a false-negative, if the assay antibody is prevented from binding, or a false-positive, if the anti-immunoglobulin cross-links assay antibodies, for example binding the coating and signal antibody in an enzyme linked immunoassay format. False increases in D-dimer can occur if the sample contains excessive levels of plasmin or plasminogen activators, which can lead to in vitro formation of fibrin fragments that can react in the D-dimer assay.

In addition to D-dimer assays, other methods for assessing fibrin degradation products are also available but used less often, with less direct clinical utility. The Thrombo-Wellco Test™ (Remel, Lenexa, Kansas) detects fibrin degradation products by first clotting all remaining fibrinogen in the sample using a snake venom, followed by detection of residual fibrinogen/fibrin fragments using a latex agglutination assay based on antifibrinogen antibodies. It suffers from a lack of both sensitivity and specificity. Only high levels of degradation fragments can be detected, so it is not a useful test for ruling out thrombosis. Factors that slow the formation or polymerization of fibrin in the preparation tube will leave residual fibrinogen in the sample, producing false-positive results.

Other fibrin degradation fragment assays are available with varying specificities to different fragments of fibrin. Most of these are latex agglutination assays with limited clinical utility compared to quantitative D-dimer assays. Some of these assays detect fibrinogen as well as fibrin degradation products; therefore, they will be elevated when fibrinogen is being actively degraded. This can occur in "primary" fibrinolysis syndromes, the most common of which is thrombolytic therapy with a nonfibrin-specific plasminogen activator like urokinase or streptokinase. Fibrin degradation product assays that detect fibrinogen degradation may show a greater elevation than D-dimer in cases of primary fibrinolysis versus DIC and other causes of secondary fibrinolysis. A downside of fibrin/fibrinogen degradation product assays is that they must be done on serum, as they will cross-react with residual fibrinogen in the sample, giving a false-positive result. This can occur if fibrinogen is not completely removed from the sample due to incomplete clotting (eg, heparin contamination in the blood).

Fibrinopeptide release assays have been used on a research basis to evaluate fibrinogen degradation by plasmin versus thrombin activation of fibrinogen. When thrombin activates fibrinogen, it releases two fibrinopeptide A molecules and two fibrinopeptide B molecules. Fibrinopeptide B consists of the first 14 residues of the amino-terminus of the beta-chain of fibrinogen (Bβ 1-14). Plasmin can also degrade fibrinogen, releasing a fragment that consists of the first 42 residues of the beta-chain (Bβ 1-42) or 28 residues (Bβ 15-42) if thrombin has already released fibrinopeptide B. When Bβ 1-42 levels are elevated to a greater extent than fibrinopeptide A or B levels, it suggests that falling fibrinogen levels may be due more to fibrinogen degradation by plasmin than fibrinogen activation by thrombin. These assays are seldom used clinically.

Plasminogen Assays

Plasminogen levels in plasma are usually measured by an activity assay based on addition of streptokinase to plasma, which binds to plasminogen, forming a complex with plasmin-like activity. A plasmin-sensitive chromogenic substrate is then added and the absorbance monitored at 405 nm. Streptokinase-plasminogen complex is not inhibited by α_2-antiplasmin. The slope of the absorbance change versus time curve is proportional to the activatable plasminogen concentration. Plasminogen assays are typically calibrated with pooled normal plasma or commercial calibrators and reported as percent of normal plasma. The reference range for plasminogen activity in healthy subjects is approximately 70% to 130% of normal, equivalent to 1.7 to 3.1 μmol/L, although laboratories performing these tests should establish in-house reference ranges.

Most plasminogen activity assays have a coefficient of variation of 3% to 4%. Positive interference in the assay can occur if other enzymes that can cleave the plasmin-sensitive chromogenic substrate are present in the sample (very uncommon). This can be evaluated by repeating the assay with the plasminogen activator (streptokinase) reagent removed, which should show no activity. Negative interference can occur if the sample contains factors that inhibit the plasminogen activator (such as high-titer antistreptokinase antibodies). In addition to the activity assay, plasminogen can also be measured antigenically using polyclonal or monoclonal antibodies in a variety of immunoassay formats.

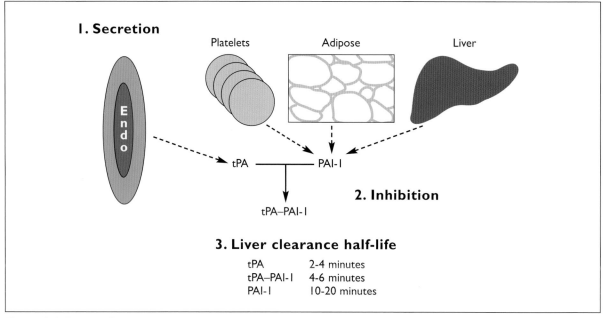

Figure 8-4. Factors affecting the levels of tissue plasminogen activator (tPA) and plasminogen activator inhibitor-1 (PAI-1) in plasma. tPA is secreted by vascular endothelial cells (Endo), PAI-1 is secreted by liver and adipose tissue and released from platelets. tPA and PAI-1 react to form the inactive tPA–PAI-1 complex. Active tPA is cleared by the liver faster than tPA–PAI-1 complex or active PAI-1. Total tPA antigen assays measure the sum of free tPA and tPA–PAI-1 complex. Total PAI-1 antigen assays measure the sum of free PAI-1 and tPA–PAI-1 complex.

Alpha 2-Antiplasmin Assays

Alpha 2-antiplasmin, also known as α_2-plasmin inhibitor, can be measured using a back-titration assay with purified plasmin. In this assay, an excess of plasmin is added to plasma and the mixture incubated to allow the plasmin to react to completion with α_2-antiplasmin. A plasmin-sensitive chromogenic substrate is then added to measure the amount of residual plasmin activity remaining. Percent of normal plasma concentration is plotted against residual plasmin activity to prepare a calibration curve. Alpha 2-antiplasmin assays are typically calibrated with pooled normal plasma or commercial calibrators and reported as percent of normal plasma. The reference range for α_2-antiplasmin activity in healthy subjects is approximately 80% to 120% of normal, equivalent to 0.8 to 1.2 μmol/L. Most α_2-antiplasmin activity assays have a coefficient of variation of about 4% to 5%. These assays are not truly specific for α_2-antiplasmin; they will detect anything in plasma that inhibits plasmin. A number of other proteins can inhibit plasmin slowly, including α_2-macroglobulin. Since α_2-antiplasmin is faster at inhibiting plasmin than these other proteins, the incubation time in the assay can be set to maximize the inhibition of plasmin by α_2-antiplasmin while limiting plasmin inhibition by other proteins.

Other synthetic inhibitors of plasmin (such as ε aminocaproic acid, tranexamic acid, aprotinin, or PPACK) can produce falsely elevated results in the assay if present in the sample. This may be of clinical importance, as fibrinolysis inhibitor drugs are often used during cardiopulmonary bypass for open heart surgery procedures. Falsely low results for α_2-antiplasmin can be seen if the sample contains other enzymes that can cleave the plasmin-sensitive chromogenic substrate, making residual plasmin activity appear increased.

Tissue Plasminogen Activator Assays

Blood contains both active tPA and active PAI-1, the principal inhibitor of tPA (Figure 8-4). tPA reacts with PAI-1, forming an inactive tPA–PAI-1 complex. Higher levels of PAI-1 are associated with an increased fraction of tPA–PAI-1 versus active tPA in blood. Since tPA–PAI-1 is cleared more slowly by the liver than active tPA, increased PAI-1 levels lead to higher total tPA levels but lower active tPA levels in blood. Active tPA, tPA–PAI-1 complex, and total tPA levels can be measured in plasma. In vivo, the level of active tPA in blood is a function of (1) the rate of tPA secretion; (2) the rate of tPA inhibition, which is a function of the PAI-1 concentration; and (3) the rate of tPA clearance by the liver. Even though the concentration of active PAI-1 is usually in molar excess compared to active tPA, in vivo PAI-1 never has the chance to completely inhibit all of the tPA because new tPA is always being released into the blood by the endothelium. If a sample of blood is drawn without blocking the reaction between PAI-1 and tPA, PAI-1 will continue to inhibit tPA ex vivo, resulting in little if any active tPA remaining. To accurately determine the concentration of active tPA present in vivo, the reaction between tPA and PAI-1 must be blocked by drawing the blood into a special acidified citrate solution.

Tissue plasminogen activator is released primarily from vascular endothelial cells, the majority of which are located in capillary beds. When a tourniquet is placed on the arm, it traps blood in the venous system; this allows the tPA being secreted by the capillary beds to concentrate in the venous blood, falsely increasing tPA levels and decreasing apparent active PAI-1 levels. Tourniquets should be left on the arm for the minimum time possible when drawing samples for all hemostatic studies, but particularly for fibrinolytic evaluations. In the past, venous occlusion tests were used in an attempt to evaluate the adequacy of tPA release from endothelium. Venous occlusion studies consisted of placing a blood pressure cuff on the arm, inflating to between arterial and venous blood pressure for 10 to 20 minutes (trapping venous blood in the arm), then drawing a blood sample. Venous occlusion traps tPA secreted by vascular endothelium in the arm in the venous blood. Healthy endothelium produces little if any PAI-1; therefore, venous occlusion results in elevated levels of tPA antigen and activity levels, but lower levels of PAI-1 activity, due to consumption by reacting with active tPA. PAI-1 antigen levels show little change during venous occlusion after correction for the hemoconcentration that occurs.

Based on the finding that high initial PAI-1 levels in venous blood were associated with a smaller increase in tPA activity during venous occlusion, venous occlusion studies were evaluated as a risk factor for thrombosis. Subsequent studies by Keber et al showed that venous occlusion studies do not stimulate tPA release; occlusion just traps tPA already being secreted. The final result of venous occlusion studies could be predicted based on basal tPA activity, tPA antigen and PAI-1 activity prior to occlusion. In the end, venous occlusion studies did not add diagnostic accuracy to assessment of thrombotic risk and have been largely abandoned.

Several different assays have been developed to measure active tPA in acidified plasma. All of them are based on active tPA converting plasminogen to plasmin, followed by determination of plasmin activity using a plasmin-sensitive chromogenic substrate. In the absence of fibrin or large fibrin degradation fragments, tPA converts plasminogen to plasmin very slowly. Therefore, all tPA activity assays include some form of tPA catalyst, typically a fragment of fibrin/fibrinogen or polylysine.

Exogenous plasminogen is added to assure the sample plasminogen concentration does not limit the rate of the reaction. The major problem to overcome in tPA activity assays is the presence of active PAI-1 and active α_2-antiplasmin, which can cause interference in the assay by inhibiting tPA and plasmin formed when the acid pH in the sample is neutralized to allow the active tPA in the sample to activate plasminogen to plasmin. Two approaches have been used to minimize or eliminate this problem. The optimal tPA activity assay is a bioimmunoassay based first on binding of active tPA in acidified plasma to a noninhibitory anti-tPA antibody bound to a microtiter plate, followed by washing the plasma containing PAI-1 and α_2-antiplasmin off the plate. This is followed by addition of the assay reagent containing a tPA stimulator, plasminogen, and plasmin-sensitive chromogenic substrate. Purified active tPA in buffer is used to calibrate the assay. An international standard for tPA activity and antigen is available from the National Institute for Biologic Standards and Controls in the United Kingdom. This standard is the basis for the calibration of most commercial assays.

A second, less sensitive and specific method for measuring tPA activity is to combine acidified plasma directly with buffer, plasminogen, tPA stimulator, and plasmin-sensitive chromogenic substrate. This version of the tPA activity assay is performed at relatively high dilution to reduce the effect of PAI-1 and α_2-antiplasmin in the assay, decreasing their interference, but not eliminating it. This type of diluted tPA activity assay suffers interference from high levels of active PAI-1 in the sample and from synthetic tPA and plasmin inhibitors (such as PPACK and aprotinin). Since essentially all interfering factors are removed by the initial tPA binding and washing step in the tPA activity bioimmunoassay, it is the preferred method. The reference range for active tPA in healthy subjects follows a circadian rhythm driven by changes in PAI-1 secretion, with peak active tPA in the evening about 1800 h (0.8 to 2.6 IU/mL equivalent to 18 to 58 picomole/L) and nadir levels in the morning about 0600 h (0.3 to 1.7 IU/mL equivalent to 7 to 38 picomole/L). The bioimmunoassay for active tPA has a coefficient of variation of about 5% at 0.1 IU/mL and an active tPA detection limit of about 0.01 IU/mL. The bioimmunoassay is not affected by interference from other plasminogen activators (streptokinase, urokinase), which can result in false elevations in dilution-based assays, or from other plasmin or tPA inhibitors in plasma, such as aprotinin or PPACK.

Enzyme immunoassays have been developed to measure essentially all forms of tPA, including free tPA, tPA–PAI-1 complex, and total tPA (free plus complexed tPA). The most common tPA antigen assays measure total tPA in plasma. Since PAI-1 activity is typically in excess compared to tPA activity, and since total tPA is usually measured in citrate plasma, which does not stop the continued ex vivo reaction between tPA and PAI-1, most of the tPA measured in total tPA antigen assays is in the form of tPA–PAI-1 complex. Total tPA antigen assays using both polyclonal and monoclonal antibodies have been described. Because most tPA in plasma is in the form of tPA–PAI-1 complex, which is cleared more slowly than active tPA, the reference range for total tPA in healthy subjects follows a circadian rhythm that is opposite to that seen for active tPA (Figure 8-2), again driven by changes in PAI-1 secretion. Peak total tPA levels occur in the morning about 0600 h (3 to 12 ng/mL equivalent to 46 to 180 picomole/L) and nadir levels in the evening about 1800 h (3 to 8 ng/mL equivalent to 46 to 120 picomole/L). Enzyme immunoassays for total tPA antigen have a coefficient of variation of about 6% at 10 ng/mL and a total tPA detection limit of about 1.0 ng/mL. tPA–PAI-1 complex immunoassays have been described. Results from these assays are similar to total tPA antigen results, particularly in patients with elevated PAI-1, which drives most of the tPA to tPA–PAI-1 complex. tPA immunoassays suffer from the same potential interferences related to rheumatoid factors and human anti-mouse antibodies if monoclonal antibodies are used in the assay.

Plasminogen Activator Inhibitor-1 Assays

In vivo PAI-1 is produced primarily by liver and abdominal adipose tissue. In tissue culture, endothelial cells also produce PAI-1, but there is little evidence healthy endothelium in vivo produces PAI-1. It is possible that endothelium produces small amounts of PAI-1 during an inflammatory response, but this is controversial. PAI-1 is also stored in platelet alpha granules and can be released from platelets if samples are improperly handled. PAI-1 is unstable at room temperature in plasma. To accurately measure PAI-1 activity and antigen, blood should be drawn into citrate anticoagulant, avoiding difficult or slow blood draws, which may release platelet PAI-1. The sample should be centrifuged for at least 30,000g-min (eg, 10 min at 3000g or 15 min at 2000g), and the platelet-free plasma either assayed immediately or frozen at -70°C. Several different forms of PAI-1 exist in plasma, including active PAI-1, tPA–PAI-1 complex, a small amount of urokinase plasminogen activator (uPA)–PAI-1 complex, and latent PAI-1, an inactive form of PAI-1 due to a conformational change in the molecule. PAI-1 activity can be measured in plasma using bioimmunoassays, complex formation assays, or back-titration methods similar to those described for α_2-antiplasmin.

The bioimmunoassay is the preferred method. In the bioimmunoassay, a microtiter plate is coated with active tPA. Plasma is added to the plate, which is incubated to allow active PAI-1 in the sample to bind to the active tPA on the plate, forming tPA–PAI-1 complex. Unbound PAI-1 and other proteins and drugs are then washed off and the bound PAI-1 quantified using enzyme linked anti-PAI-1 antibodies. In both the complex formation and back-titration assays, an excess of active tPA or active uPA is added to plasma and allowed to react with active PAI-1 in the sample. In complex formation assays, PAI-1 activity in the original plasma is estimated as the difference between tPA–PAI-1 or uPA–PAI-1 complex originally present and the amount after addition of exogenous plasminogen activator. tPA–PAI-1 or uPA–PAI-1 complex levels are measured using enzyme immunoassays. In the back-titration assay, excess active tPA or uPA is added to the plasma, and the difference between the amount added initially and the residual amount is equivalent to the amount of active PAI-1 in the sample. PAI-1 activity back-titration assays suffer from all the potential limitations of tPA activity assays if the dilution-type tPA assay is used to measure residual active tPA, including falsely elevated PAI-1 activity levels due to other factors in the plasma (such as aprotinin or PPACK) that inhibit the added plasminogen activator or plasmin generated in the assay. The reference range for active PAI-1 in healthy subjects follows a circadian rhythm driven by changes in PAI-1 secretion, with peak active PAI-1 in the morning about 0600 h (2 to 15 U/mL equivalent to 45 to 330 picomole/L) and nadir levels in the evening about 1800 h (2 to 9 U/mL equivalent to 45 to 200 picomole/L). The bioimmunoassay for active PAI-1 has a coefficient of variation of about 6% at 10 U/mL and an active PAI-1 detection limit of about 1 U/mL.

The most common antigenic assays for PAI-1 measure either total PAI-1 antigen (including active PAI-1, tPA–PAI-1 complex, and latent PAI-1) or tPA–PAI-1 complex. The major concern with total PAI-1 antigen assays is release of latent PAI-1 from platelets during blood sampling. Platelets contain PAI-1, which is released during platelet activation. Due to conformational inactivation of PAI-1 at 37°C, approximately 90% of the PAI-1 in platelets is the inactive latent form. If platelets are activated during the blood draw process due to shear forces, difficult draw, prolonged tourniquet times, or the wrong anticoagulant (such as EDTA), platelets will release substantial quantities of latent PAI-1 antigen, which has little effect on PAI-1 activity but produces a false elevation of plasma total PAI-1 antigen. The reference range for total PAI-1 in healthy subjects follows a circadian rhythm that parallels active PAI-1 driven by changes in PAI-1 secretion. Peak total PAI-1 levels occur in the morning about 0600 h (4 to 50 ng/mL equivalent to 77 to 1000 picomole/L) and nadir levels in the evening about 1800 h (4 to 20 ng/mL equivalent to 77 to 400 picomole/L). Enzyme immunoassays for total PAI-1 have a coefficient of variation of about 6% at 20 ng/mL and a total PAI-1 detection limit of about 1 ng/mL.

Global Tests of Fibrinolysis

The earliest tests of fibrinolytic activity were clot lysis times. In general, whole blood or plasma is clotted either by allowing unaltered whole blood to clot, recalcifying citrated blood, or activating citrated blood with tissue factor or a contact system activator plus calcium. The blood clot is then

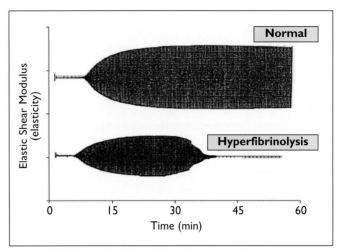

Figure 8-5. Example of hyperfibrinolysis detected by measurement of elastic shear modulus (thromboelastometry) on recalcified whole blood. The width of the curve is proportional to clot elasticity. The normal sample shows onset of clotting, increase in clot strength, followed by stable clot elasticity. The sample with increased fibrinolytic activity due to elevated tissue plasminogen activator levels shows formation then lysis of clot elasticity.

monitored for lysis over a period of minutes to hours. A host of individual variations in whole blood lysis time protocols have been described, but the most common are simple whole blood clot formation followed by lysis. Native whole blood from healthy subjects clots in less than an hour but can take days to lyse, making native whole blood lysis times clinically impractical as a routine assay. While whole blood clot lysis can be detected visually, the most common methods for determining lysis times measure clot viscoelasticity, roughly equivalent to clot strength. As blood clots, the elasticity and viscosity of the blood increases as platelets and fibrinogen are activated, forming a dense platelet-fibrin mesh. Since whole blood clot lysis is slow in healthy subjects, the percent fall in viscosity or elasticity after an hour is generally described rather than the time to total lysis of the clot. Clot viscoelasticity can be measured using a variety of different devices, including cone and plate viscometers, thromboelastometry (Figure 8-5), and dynamic viscosity measurements.

Whole blood clot lysis times are insensitive to small or even moderate changes in plasminogen activator levels. Typically they only show evidence of lysis when extreme elevations of plasminogen activator levels occur that overcome plasma and platelet released PAI-1. Even so, they have a clinical role to play in the evaluation of the acutely bleeding patient, particularly for unexplained bleeding during surgery. Whole blood lysis times can be used to detect major increases in the level of plasminogen activators in blood, as occurs during cardiopulmonary bypass, when increased tPA is released from endothelium, and during the anhepatic phase of liver transplantation, due to lack of tPA clearance by the liver. When plasminogen activator levels are elevated in the sample, whole blood clot lysis times may be short, on the order of 15 to 30 minutes.

Whole blood lysis time assays suffer from a variety of different sensitivity and specificity problems. Whole blood clot lysis times are not a measure of any single factor. An increased fibrinogen concentration prolongs the total time to lysis, as does an increased platelet count, which adds strength and the fibrinolytic inhibitor PAI-1 to the clot. Increased plasma PAI-1 or α_2-antiplasmin levels also slow the rate of clot lysis by inhibiting plasminogen activation and plasmin activity in the sample. Endogenous or exogenous elevation of plasminogen activators in the sample, including tPA, uPA, streptokinase, and others, accelerates plasminogen activation and shortens whole blood clot lysis times. Decreased levels of the fibrin cross-linking protein factor XIIIa may also shorten lysis times. In most cases, longer lysis times are generally correlated with increased levels of active PAI-1, while short lysis times are associated with increased levels of plasminogen activators. Due to the lack of specificity, this type of assay has largely been supplanted by more specific measures of fibrinolytic factors, except in the situation of rapid testing associated with open heart or liver transplant surgery.

Due to the prolonged lysis times seen with whole blood clot lysis, a modified method known as the euglobulin lysis time was developed. Briefly, proteins in plasma are precipitated using a combination of low ionic strength and acidic pH. The precipitate is removed, dissolved in physiologic buffer, and clotted. The time to lysis is then determined using any one of a variety of techniques, including the measurement of change in absorbance or viscoelasticity, as described for whole blood lysis times. tPA, PAI-1, plasminogen, and fibrinogen all precipitate as part of the euglobulin fraction. In contrast, α_2-antiplasmin is left in the supernatant and does not precipitate. Thus, the euglobulin preparation is essentially a crude method for eliminating α_2-antiplasmin. Since plasmin generated in the euglobulin lysis time is not inhibited by α_2-antiplasmin, the lysis times are shorter than whole blood lysis times. The percentage of original plasma tPA, PAI-1, plasminogen, and fibrinogen recovered in the euglobulin fraction is dependent on the pH and other conditions used to prepare the precipitate. There is no standardized euglobulin lysis time method; many different protocols have been described, producing highly variable results. Euglobulin lysis times in healthy subjects range from about 1 to 12 hours, with an average of about 5 to 8 hours. The euglobulin lysis time is positively correlated with PAI-1 levels (higher PAI-1 leads to longer lysis times) and inversely correlated with active plasminogen activator levels (shortened times). Euglobin lysis times can be measured optically, by viscoelastometry, or by using fibrin-coated plates. Euglobulin lysis times provide an overall measure of the rate of fibrinolysis, with the same specificity and sensitivity problems described for whole blood lysis times. In most situations, euglobulin lysis times have been replaced with specific measurements of fibrinolytic proteins.

Lipoprotein (a)

Lipoprotein (a) is a low-density lipoprotein-like particle containing an additional protein moiety called apolipoprotein (a), which is homologous to plasminogen. In distinction to plasminogen, lipoprotein (a) cannot develop enzymatic activity, so high levels of lipoprotein (a) are thought to competitively interfere with plasmin activity and have been shown to be a risk factor for arterial thrombosis. Lipoprotein (a) levels can be measured antigenically using a variety of techniques, including enzyme immunoassays and latex immunoassays. Lipoprotein (a) levels vary widely in the population, due primarily to genetic factors. Apolipoprotein (a) molecules show significant differences in length, due to variations in the number of kringle-4 like tandem repeats in the gene. Apolipoprotein (a) length can be evaluated phenotypically using immunoblotting or by gene sequencing. These assays require plasma from EDTA-anticoagulated whole blood.

Research Assays

A number of different research fibrinolytic assays are available for measuring other aspects of the fibrinolytic system. These assays are not commonly used for clinical evaluation of patients but play a role in understanding the regulation of fibrinolysis.

Enzyme immunoassays for measuring plasmin-antiplasmin (PAP) complexes have been described. These complexes have a half-life of approximately 5 hours in plasma and provide an indication of the rate and amount of plasmin formation that is occurring in vivo.

Urokinase plasminogen activator activity, uPA–PAI-1 complex, and total uPA antigen assays have been described based on methods similar to those discussed above for tPA activity, tPA–PAI-1 complex, and total tPA. Urokinase plasminogen activator is a trace protein in plasma thought to be due to leakage from uPA secretion in the kidney. There is little indication for uPA assays clinically.

Immunoassays have been developed to measure different forms of thrombin-activatable fibrinolysis inhibitor (TAFI), including the unactivated zymogen form of TAFI, activated TAFI, and proteolytically cleaved inactive TAFI. Some studies suggest that elevated levels of TAFI are a risk factor for venous thrombosis, but TAFI assays are not routinely measured clinically at this time.

Suggested Reading

Sample Collection and Processing for Fibrinolytic Testing

Kluft C, Verheijen JH. Leiden fibrinolysis working party: blood collection and handling procedures for assessment of tissue-type plasminogen activator (t-PA) and plasminogen activator inhibitor-1 (PAI-1). *Fibrinolysis*. 1990;4(suppl 2):155-161.

Macy EM, Meilahn EN, Declerck PJ, Tracy RP. Sample preparation for plasma measurement of plasminogen activator inhibitor-1 antigen in large population studies. *Arch Pathol Lab Med*. 1993;117:67-70.

D-dimer and Fibrin Split Product Assays

Amiral J, Minard F, Plassart V, Vissac AM, Chambrette B. Reactivity of D-dimer assays with the fibrinogen-fibrin split products generated by thrombolytic agents. *Blood Coagul Fibrinolysis*. 1990;1:525-530.

Bauer KA. Laboratory markers of coagulation and fibrinolysis. In: Colman RW, Marder VJ, Clowes AW, George JN, Goldhaber SZ, eds. *Hemostasis and Thrombosis: Basic Principles and Clinical Practice*. Philadelphia, Pa: Lippincott Williams & Wilkins; 2006: 851-873.

Dempfle CE. Validation, calibration, and specificity of quantitative D-dimer assays. *Semin Vasc Med*. 2005; 5:315-320.

Dempfle CE, Zips S, Ergül H, Heene DL. The Fibrin Assay Comparison Trial (FACT): evaluation of 23 quantitative D-dimer assays as basis for the development of D-dimer calibrators. *Thromb Haemost*. 2001;85:671-678.

Di Nisio M, Squizzato A, Rutjes AW, Buller HR, Zwinderman AH, Bossuyt PM. Diagnostic accuracy of D-dimer test for exclusion of venous thromboembolism: a systematic review. *J Thromb Haemost*. 2007; 5:296-304.

Plasminogen and Antiplasmin Assays

Friberger P, Knos M, Gustavsson S, Aurell L, Claeson G. Methods for determination of plasmin, antiplasmin and plasminogen by means of substrate S-2251. *Haemostasis*. 1978;7:138-145.

tPA and PAI-1 Assays

Alessi M, Juhan-Vague I, Declerck P, Collen D. Molecular forms of plasminogen activator inhibitor-1 (PAI-1) and tissue-type plasminogen activator (t-PA) in human plasma. *Thromb Res*. 1991;62:275-285.

Bos R, Revet M, Nieuwenhuizen W. A bio-immuno assay to determine free tissue-type plasminogen activator (t-PA) in Stabilyte Plasma. *Fibrinolysis*. 1994;8(suppl 2):163-165.

Chandler WL, Jascur ML, Henderson PJ. Measurement of different forms of tPA in plasma. *Clin Chem*. 2000;46:38-46.

Chandler WL, Trimble SL, Loo SC, Mornin D. Effect of PAI-1 levels on the molar concentrations of active tissue plasminogen activator (t-PA) and t-PA/PAI-1 complex in plasma. *Blood*. 1990;76:930-937.

Cheema SP, Webster NR, Dunn F, Bellamy MC. Mediators of fibrinolysis in orthotopic liver transplantation. *Clin Transplant*. 1996;10:24-27.

Declerck PJ, Moreau H, Jespersen J, Gram J, Kluft C. Multicenter evaluation of commercially available methods for the immunological determination of plasminogen activator inhibitor-1 (PAI-1). *Thromb Haemost*. 1993;70:858-863.

Dzik WH, Arkin CF, Jenkins RL, Stump DC. Fibrinolysis during liver transplantation in humans: role of tissue-type plasminogen activator. *Blood*. 1988;71:1090-1095.

Keber D, Blinc A, Fettich J. Increase of tissue plasminogen activator in limbs during venous occlusion: a simple haemodynamic model. *Thromb Haemost*. 1990;64:433-437.

Lucore CL, Sobel BE. Interactions of tissue-type plasminogen activator with plasma inhibitors and their pharmacologic implications. *Circulation*. 1988;77:660-669.

Mahmoud M, Gaffney PJ. Bioimmunoassay (BIA) of tissue plasminogen activator (t-PA) and its specific inhibitor (t-PA/INH). *Thromb Haemost.* 1985;53:356-359.

Rånby M, Sundell B, Nilsson TK. Blood collection in strong acidic citrate anticoagulant used in a study of dietary influence on basal tPA activity. *Thromb Haemost.* 1989;62:917-922.

Spiess BD. Thromboelastography and cardiopulmonary bypass. *Semin Thromb Hemost.* 1995;21(suppl 4):27-33.

Spiess BD, Wall MH, Gillies BS, Fitch JCK, Soltow LO, Chandler WL. A comparison of thromboelastography with heparinase or protamine sulfate added in-vitro during heparinized cardiopulmonary bypass. *Thromb Haemost.* 1997;78:820-826.

Wiman B, Andersson T, Hallqvist J, Christina R, Ahlbom A, deFaire U. Plasma levels of tissue plasminogen activator/plasminogen activator inhibitor-1 complex and von Willebrand factor are significant risk markers for recurrent myocardial infarction in the Stockholm Heart Epidemiology Program (SHEEP) study. *Arterioscl Thromb Vasc Biol.* 2000;20:2019-2023.

Other Fibrinolytic Assays

Lang T, Bauters A, Braun SL, Pötzsch B, von Pape KW, Kolde HJ, Lakner M. Multi-centre investigation on reference ranges for ROTEM thromboelastometry. *Blood Coagul Fibrinolysis.* 2005;16:301-310.

Liszka-Hackzell JJ, Ekback G. Analysis of the information content in Sonoclot data and reconstruction of coagulation test variables. *J Med Syst.* 2002;26:1-8.

Smith AA, Jacobson LJ, Miller BI, Hathaway WE, Manco-Johnson MJ. A new euglobulin clot lysis assay for global fibrinolysis. *Thromb Res.* 2003;112:329-337.

Urano T, Nishikawa T, Nagai N, Takada Y, Takada A. Amounts of tPA and PAI-1 in the euglobulin fraction obtained at different pH: their relation to the euglobulin clot lysis time. *Thromb Res.* 1997;88:75-80.

CHAPTER 9

Laboratory Aspects of Thrombophilia Testing

Charles Eby, MD

Introduction

The decision to order laboratory tests for identification of inherited and acquired thrombophilia risk factors should be preceded by an assessment of the potential benefit to each patient before performing a battery of expensive tests (see chapters 17 and 18) and awareness of the preanalytical variables that can affect the accuracy of hemostasis tests in general (see chapter 4) as well as those that are unique to thrombophilia testing. In order to maximize analytical accuracy and precision, assays should be calibrated with a certified standard, if available, and reference ranges should be determined from a sample of at least 40 individuals that reflect the population to be tested.

This chapter will critique frequently employed methods for performing thrombophilia tests that are currently offered by most North American reference laboratories. Analytes that are not routinely used for clinical decision making, or are considered to be research tools, will also be reviewed. Figure 9-1 provides an overview of the common tests included in evaluation of a thrombophilic patient.

Protein C

Protein C is a vitamin K-dependent serine protease zymogen that regulates thrombin generation by inactivating factors Va and VIIIa. In vivo, thrombin activation of protein C is accelerated when thrombin and protein C interact with their respective receptors, thrombomodulin (TM) and endothelial cell protein C receptor (EPCR), on the surface of endothelial cells. Activated protein C (APC) with its cofactor, protein S, degrades factor Va and factor VIIIa, reducing the rate of prothrombin activation (Figure 9-2).

Inherited heterozygous, or partial, protein C deficiency is a moderate risk factor for venous thromboembolism (VTE), while inherited severe protein C deficiency (homozygous or compound heterozygous) is associated with fetal central nervous system and retinal thrombotic events and neonatal purpura fulminans. There are two types of inherited protein C deficiency. Type I is characterized by decreased synthesis of fully functional protein (antigen = activity). Type II mutations translate into dysfunctional protein (antigen > activity). Approximately 85% of the protein C mutations submitted to an international registry are type I.

Acquired protein C deficiency is much more common than the inherited form and can be separated into decreased synthesis (hepatic disease, vitamin K deficiency, vitamin K antagonism [warfarin], and drug-induced decreased protein synthesis [L-asparaginase]), increased clearance (acute thrombosis, disseminated intravascular coagulation, acute medical illness, or trauma), and hemodilution (posthemorrhage resuscitation with crystalloid, postplasmapheresis).

Elevated protein C levels are of doubtful clinical significance and have been reported with advancing age (~4% increase/decade), nephrotic syndrome (rate of hepatic synthesis > rate of renal loss), pregnancy, oral contraceptive therapy, and postmenopausal hormone replacement therapy.

Protein C Functional Assays

The key component of commercial protein C functional assays is in vitro activation of plasma protein C by Agkistrodon contortrix contortrix (Southern Copperhead) snake venom (Protac®). Protein C activation is independent of calcium ion concentration and degree of protein C gamma carboxylation. The endpoint is either fibrin clot formation (clotting method) or hydrolysis of an APC-specific chromogenic substrate (spectrophotometric method).

Clotting Method

The clotting method is a two-stage phospholipid activated partial thromboplastin time (aPTT)-based or thromboplastin (prothrombin time [PT])-

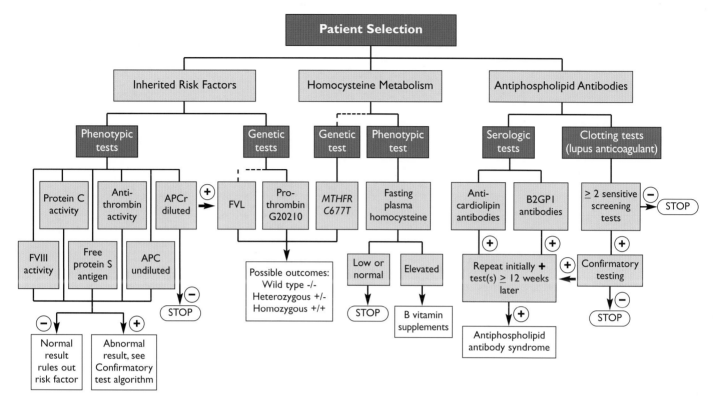

Figure 9-1. Thrombophilia risk factor screening. Laboratory testing for thrombophilic risk includes testing for inherited risk factors, homocysteine metabolism, and antiphospholipid antibodies. Initial screening tests should include phenotypic or functional assays for factor VIII, protein C, protein S, antithrombin, and activated protein C resistance (APCr). Further diagnostic algorithms for specific risk factor diagnosis are covered in chapter 19. Genetic testing for prothrombin G20210A mutation should be included, but genotyping for factor V Leiden (FVL) and methylenetetrahydrofolate reductase (MTHFR) should only be performed if the screening assays are abnormal. Testing for antiphospholipid antibodies should be included; these are discussed in chapter 22.

based coagulation assay. In the first stage, test plasma is incubated with Protac to completely activate protein C. In the second stage, an activator (aPTT, PT, or Russell viper venom reagent) is added. Time to detection of fibrin clot or rate of substrate hydrolysis is *inversely* proportional to plasma protein C activity.

Advantages of protein C clot-based assays include sensitivity for detection of both type I and II protein C deficiencies, and performance on most coagulation instruments regardless of fibrin clot detection method. Disadvantages include the potential to underestimate protein C activity if factor VIII is elevated or protein S is low in the test plasma, or the patient has factor V Leiden—conditions that may shorten the clotting time. Additionally, the undercarboxylated form of protein C, due to warfarin antagonism or vitamin K deficiency, is able to be activated by Protac but does not degrade Va and VIIIa, leading to shorter clotting times and lower protein C results compared to chromogenic spectrophotometric or antigen assays. Finally, overestimation of protein C

clotting activity may occur if test plasma contains heparin, a direct thrombin inhibitor (DTI) anticoagulant, a specific factor inhibitor, or a lupus anticoagulant—conditions that may prolong the clotting time. Dilution of test plasma with protein C-depleted plasma and a heparin neutralizing agent normalizes protein S, coagulation factors, and fibrinogen activities, and improves the accuracy of clot-based protein C activity assays, but does not totally eliminate interference from potent inhibitors and anticoagulants.

Spectrophotometric Method

The spectrophotometric method is a one-stage assay. Protac, the chromogenic substrate, and test plasma are combined, and either the rate or endpoint of hydrolysis of the substrate and release of an indicator, such as para-nitroanaline dye, is *directly* proportional to APC activity. This method is not affected by qualitative or quantitative variations of protein S or coagulation factors, anticoagulants, or inhibitors, and is more precise than the clot-based method. However, disadvantages

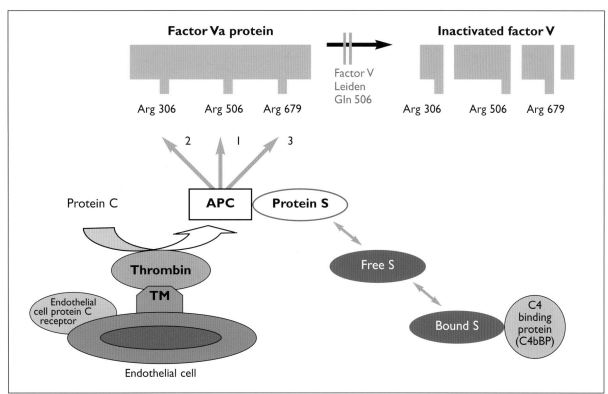

Figure 9-2. Several thrombophilic risk factors center around the inactivation of thrombin. When thrombin binds to the endothelial protein thrombomodulin (TM), its substrate specificity changes from fibrinogen to protein C, resulting in formation of the protease activated protein C (APC). APC, together with its cofactor, protein S, proteolytically inactivates factor Va by cleavage at three arginine residues in sequential order: 506, 306, and 679. APC also degrades factor VIIIa. Deficiencies of either protein C or protein S are known thrombophilic risk factors, due to decreased inactivation of factors Va and VIIIa and subsequent increased thrombin production. A common genetic mutation of factor V, with arginine 506 replaced by glutamine (factor V Leiden) is also a thrombotic risk factor by a similar mechanism.

include requirement for an instrument with spectrophotometer capabilities and insensitivity for type II mutations that do not alter rate of hydrolysis of synthetic substrate, but do reduce the rate of factor Va and/or VIIIa degradation. However, type II protein C mutations are uncommon. In addition, spontaneous hydrolysis of substrate may positively bias results based on endpoint methods. To avoid this artifact, a blank should be included with patient samples and the rate of hydrolysis monitored in the blank. Based on North American and European proficiency surveys, the spectrophotometric, or chromogenic, protein C activity assays are the more common method for protein C deficiency screening.

Protein C Antigen Assays

Originally, protein C antigen was measured with the "Laurell rocket" immunoelectorphoretic method. Under alkaline conditions, negatively charged protein C migrates toward the anode of an electric field, and polyclonal anti-protein C antibodies embedded in the agaraose gel remain stationary. Protein C-antibody complexes precipitate when equilibrium is achieved and are detected by staining the dried gel with a nonspecific protein stain, such as Coomassie Brilliant Blue. Dilutions of pooled normal plasma are run alongside test plasmas, and the logarithm of the distance from the application well to the peak of the precipitation arc is proportional to the logarithm of protein C concentration. While the method is simple and inexpensive, immunoelectrophoresis is labor intensive and time consuming, and is insensitive for detection of type II protein C deficiencies. In addition, carboxylated and undercarboxylated protein C migrate at different rates. The chelator ethylenediaminetetraacetic acid (EDTA) must be added to both the gel and electrophoresis buffer to obtain a single precipitation arc and accurate measurement of total protein C antigen.

Several commercial protein C antigen enzyme-linked immunosorbent assays (ELISA) are available, employing different combinations of capture and detection protein C antibodies with variable sensitivities for undercarboxylated protein C and

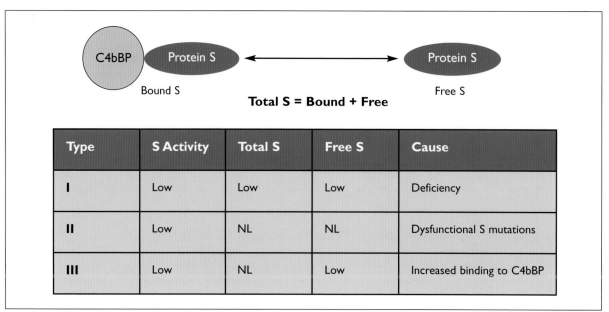

Type	S Activity	Total S	Free S	Cause
I	Low	Low	Low	Deficiency
II	Low	NL	NL	Dysfunctional S mutations
III	Low	NL	Low	Increased binding to C4bBP

Figure 9-3. Diagnosis of protein S deficiency. Protein S is a cofactor for protein C in the degradation of factors Va and VIIIa. Protein S is present bound to the complement regulatory protein C4b binding protein (C4bBP) and also free in plasma. Only free protein S can serve as a functional cofactor for protein C. Total protein S assays measure both bound and free protein S, while free protein S assays measure only the unbound form. Type I, II, and III protein S deficiencies can be diagnosed by a combination of activity and total and free protein S assays.

protein C/inhibitor complexes. Overall, method comparisons typically produce fairly concordant results. While all antigen assays are insensitive to protein C type II mutations and should not be used for screening purposes, they can help to discriminate between type I and type II mutations in patients with low protein C activity, as type I mutations will have decreased activity and antigen levels, while type II mutations will have only decreased activity. However, there is no evidence to indicate that type I and II deficiencies confer different risks for an initial or recurrent venous thromboembolic event or require different management. Therefore, protein C antigen testing is not recommended for routine thrombophilia evaluation. Confirmation of protein C deficiency by genetic testing is not currently feasible due to the variety of potential mutations.

Commercial and "homemade" protein C methods should be calibrated against the international protein C reference standard to improve interlaboratory precision. Protein C reference intervals typically range from ~70% to ~150%, but each laboratory should establish their own in-house reference ranges. The population utilized for establishing a reference range should not include people with liver disease, individuals taking warfarin-like drugs, or women using oral contraceptives or hormone replacement therapy, and ideally should reflect the age range of the expected patient population.

An algorithm for the diagnosis of congenital protein C deficiency and distinguishing it from acquired causes of protein C deficiency is presented in chapter 19.

Protein S

Protein S is a nonenzymatic cofactor that accelerates APC proteolysis of factors Va and VIIIa. Protein S synthesis occurs primarily in the liver and requires vitamin K-dependent gamma carboxylation to be fully functional. Approximately 60% of total protein S antigen (total protein S) is inactive, noncovalently bound to the beta chain of the complement regulatory C4b binding protein (C4bBP), and 40% is unbound (free protein S) and functional (protein S activity) (Figure 9-2).

Inherited heterozygous (partial) deficiency of protein S is generally considered to be a mild to moderate risk factor for venous thromboembolic events. Severe inherited protein S deficiency has been reported in infants with neonatal purpura fulminans.

Three types of heterozygous protein S deficiency are recognized (Figure 9-3). Type I deficiency consists of decreased synthesis of functional protein (concordant reductions of total protein S antigen, free protein S antigen, and protein S activity) and accounts for approximately 70% of reported mutations. Type II deficiency is due to protein S mutations that translate into dysfunctional protein (decreased protein S activity, normal free and total protein S antigen). Type III deficiency is defined by an increased binding affinity of protein S for C4bBP (protein S activity and free protein S antigen reduced, total protein S antigen normal). While type II is uncommon, it may be underdiagnosed due to incomplete laboratory evaluation. The existence of type III deficiency has been challenged, with examples of protein S deficient kindreds whose affected members possess a common protein S mutation but are a mixture of type I and III phenotypes. One explanation offered is an age-related increase in total and bound protein S, but not free protein S, due to increased C4bBP.

Similar to protein C, acquired protein S deficiency is much more common than the inherited form and can be grouped into three categories: (1) decreased synthesis (hepatic disease, vitamin K deficiency, vitamin K antagonism [warfarin], and drug-induced decreased protein synthesis [L-asparaginase]); (2) increased clearance (acute thrombosis, disseminated intravascular coagulation, acute medical illness, or trauma); and (3) hemodilution (posthemorrhage resuscitation with crystalloid, postplasmapheresis). In addition, protein S activity decreases with oral contraceptive and hormone replacement use, and during the second and third trimesters of pregnancy.

Protein S levels increase in postmenopausal women, but there is no age-associated increase in men. There is no known clinical significance for an elevated protein S level.

Protein S Functional Assays

Protein S Activity Assays

In protein S activity assays, prolongation of a fibrin clot endpoint is dependent upon the concentration of functional protein S in the test plasma. Protein S-depleted plasma and APC are added to dilutions of test plasma, and an activator is added (phospholipid, aPTT-based; thromboplastin, PT-based; or Russell viper venom-based). Advantages of clot-based protein S activity assays include sensitivity for all subtypes of protein S deficiency and compatibility with most coagulation instruments. The major disadvantage of functional protein S assays is poor specificity (false-positive results). Plasma from patients with activated protein C resistance due to factor V Leiden mutation may

shorten the clotting time in a protein S activity assay, leading to an underestimation of protein S activity. Elevated factor VIII and VII activities in test plasma can also produce shorter aPTT and PT clotting times, respectively, leading to reporting of falsely lower protein S activities. False "normal" protein S activities can occur if an anticoagulant such as heparin, specific factor inhibitor, or a lupus anticoagulant is present in the test plasma and prolongs the clotting time, leading to an overestimate of protein S activity. Falsely elevated protein S activity levels can be seen during therapy with a direct thrombin inhibitor.

Analytical inaccuracies have been reduced in some commercial assays by greater dilution of test plasma with protein S-deficient plasma and the addition of factor Va and heparin inhibitors. However, certain instrument-protein S activity reagent combinations are less sensitive than other instrument-reagent combinations and are more likely to produce false normal results, according to a recent review of North American Specialty Coagulation Laboratory Association (NASCOLA) proficiency test results. Finally, protein S activity assays are inherently imprecise. College of American Pathologists proficiency survey results consistently report coefficients of variation for protein S activity of 17% to 23%—typically double the interlaboratory variability reported for protein S antigen assays.

Protein S Antigenic Assays

Protein S concentration can be reported as either *total* (free + C4bBP bound) or *free* antigen. The accuracy of total protein S immunologic assays is dependent upon equivalent sensitivity of protein S antibodies for free and bound protein S (assay is independent of C4bBP concentration) or assay conditions that ensure an adequate shift of equilibrium to free protein S (extensive plasma dilution and incubation prior to exposure to capture antibody). The Laurell electroimmunoassay method can be problematic due to the potential for separate precipitation arcs of free and bound protein S, preventing accurate determination of total protein S concentration. Several ELISA methods are available, employing different combinations of polyclonal and monoclonal protein S antibodies to capture and detect protein S. Validation of any total protein S immunoassay should include evidence that addition of C4bBP to plasma does not reduce detection of total protein S due to antibody insensitivity to bound protein S. NASCOLA proficiency test results indicate that total protein S antigen testing is insensitive for detection of type I protein S deficiency and more likely to produce false normal results compared to free protein S antigen or protein S activity methods. While total protein S determination is necessary to identify type III deficiency, a genetic basis for this subtype is in doubt, as is its clinical significance.

The original and still "gold standard" method for determining free protein S antigen concentration involves precipitation of C4bBP-protein S with polyethylene glycol (PEG) and measurement of free protein S antigen in the supernatant. However, this method is both technically demanding and imprecise, and it has been supplanted by development of immunoassays employing monoclonal antibodies with specificity for free protein S. Most commercial ELISA methods use capture and detection monoclonal antibodies that recognize different epitopes on free protein S, while one kit uses immobilized C4bBP to capture free protein S and a monoclonal antibody for detection. In a fully automated immunoturbidimetric assay, increased turbidity due to agglutination of microparticles coated with anti-free protein S antibodies correlates with free protein S concentration. Ex vivo increases in free protein S concentration can result from dilution and prolonged storage at room temperature, necessitating careful attention to these preanalytical variables.

In general, free protein S antigen assays are more accurate and precise, compared to total protein S antigen and protein S activity assays, for detection of type I protein S deficiency. However, free protein S antigen assays are insensitive for detection of type II protein S (qualitative) deficiency. While the International Society of Thrombosis and Haemostasis subcommittee on thrombosis and the World Health Organization recommend free protein S as the preferred method to screen for protein S deficiency, there is no consensus among clinical laboratories, with some performing only free protein S antigen or protein S activity, and others using an algorithm starting with protein S activity and measuring free and total protein S antigen to "confirm" a low activity result. Commercial and "homemade" protein S methods should be calibrated against the international protein S reference standard to improve interlaboratory precision. Reported protein S antigen and activity reference intervals typically range from ~60% to ~150%, and separate ranges for premenopausal women who do or do not use oral contraceptives are rarely provided, complicating interpretation of "low values" for these individuals.

In general, abnormally low protein S levels based on currently available functional and antigen assays must be carefully scrutinized before labeling a patient with a genetic disorder, which may be difficult to revise. A guideline for diagnosis of protein S deficiency is included in chapter 19. Confirmation of protein S deficiency by genetic testing currently is not feasible due to the variety of potential mutations.

Antithrombin

Antithrombin inactivates factor Xa and thrombin (factor IIa), as well as other serine proteases. The rate of antithrombin (AT) inhibition of these activated serine protease enzymes is accelerated by glycosaminoglycan molecules expressed on the surface of endothelial cells, and by heparin when administered as an anticoagulant drug.

Inherited heterozygous deficiency of antithrombin is a major risk factor for venous thromboembolic events. Complete deficiency of antithrombin activity has not been reported and may not be compatible with life. There are two main types of inherited antithrombin deficiency. Type I mutations produce hyposynthesis of fully functional protein (antigen = activity). Type II mutations translate into dysfunctional protein (antigen > activity). Three type II subtypes are recognized: reactive site (RS) mutations that alter antithrombin interaction with Xa and thrombin active sites; heparin binding site (HBS) mutations that affect heparin binding to antithrombin; and pleotropic (PE) mutations with multiple effects.

When an antithrombin defect is diagnosed in patients with venous thromboembolic events, it is usually classified as a type I deficiency. The most common type II deficiency is a heparin binding defect. Both type I and type II antithrombin deficiencies have decreased heparin cofactor antithrombin activity, which is the most sensitive method for detecting antithrombin deficiency. In order to distinguish type I from II deficiencies, antithrombin antigen and antithrombin activity without heparin acceleration must be determined. However, these tests are performed in relatively few hospital or reference laboratories, making it likely that some type II antithrombin-deficient patients are incorrectly labeled as type I deficient due to incomplete laboratory evaluation. Clinical observations suggest that individuals who are heterozygotes for a type II heparin binding site antithrombin defect are at minimal or no increased risk for VTE. Therefore, it is important to fully characterize antithrombin defects in order to provide a more accurate assessment of thrombosis risk to patients and their family members.

Acquired antithrombin deficiency can occur in patients with nephrotic syndrome (antithrombin renal loss > hepatic synthesis), fulminant hepatic failure or end-stage liver disease, and disseminated intravascular coagulation. Pregnancy-associated changes in antithrombin are minimal. Drug-induced antithrombin deficiency can occur during unfractionated heparin infusions (~10% decrease in antithrombin activity), but not during low molecular heparin therapy. L-asparaginase, a chemotherapy drug used to treat acute lymphoblastic leukemia, reduces hepatic synthesis of antithrombin.

Antithrombin Assays

Antithrombin Functional Assays

Antithrombin inhibition of thrombin or factor Xa can be measured in the presence of heparin (heparin cofactor activity), or in the absence of heparin (progressive antithrombin activity), in a chromogenic assay. Bovine thrombin (factor IIa) or human factor Xa are preferred substrates since they are not inactivated by heparin cofactor II, another plasma protein with serine protease inhibitor activity. In the heparin cofactor activity assay, bovine thrombin (or human factor Xa) and heparin are added in excess to diluted test plasma, and the rate of enzyme inhibition is correlated with the antithrombin activity in the test plasma. Residual IIa or Xa activity is determined by hydrolysis of a synthetic chromogenic substrate which releases a para-nitroanaline dye when hydrolyzed by IIa or Xa, producing a color change that is quantified by measuring light absorption at a specific wavelength. The rate, or endpoint, of substrate hydrolysis is inversely related to the antithrombin activity in the test plasma.

The assay for progressive antithrombin activity is identical except for the absence of heparin. A low heparin cofactor activity result combined with a normal progressive antithrombin activity confirms a diagnosis of type II heparin binding site antithrombin deficiency. Advantages of the heparin cofactor activity assay include its sensitivity for all types of antithrombin deficiency and lack of interference from variations in coagulation factor activities, lupus anticoagulants, or heparin, since fibrin clot formation is not monitored. Disadvantages of the heparin cofactor activity assay include the requirement for an instrument with spectrophotometer capability and interference from high concentrations of hemoglobin, bilirubin, and trigylcerides. In addition, if bovine thrombin is the substrate in the assay, false-negative results may occur if the test plasma contains a direct thrombin inhibitor drug, which would bind to and neutralize thrombin and produce an overestimate of antithrombin activity.

Antithrombin Antigen Assay

There are multiple methods available for measuring antithrombin concentration, including Laurell electroimmunodiffusion, radial immunodiffusion, ELISA, and automated nephelometry. When combined with a low heparin cofactor activity assay, the antithrombin antigen result will discriminate between type I (low antigen) and type II (normal antigen) deficiency. However, antithrombin antigen assays are insensitive screening tests because the antigen level is normal in patients with type II antithrombin deficiency.

Antithrombin heparin cofactor activity reference intervals are typically in the 80% to 120% range and should be calibrated against the International Reference Standard. Unlike patients with heterozygous protein C and S deficiencies, there is essentially no overlap of antithrombin activities between deficient patients and reference groups. In thrombophilia testing proficiency surveys, antithrombin activity assays are more precise than protein C and S assays, with coefficients of variation typically approximately 10%.

Activated Protein C Resistance

In 1993, Dahlbäck observed a blunted prolongation of the aPTT when activated protein C was added to the plasmas of a subset of patients with thrombophilia, and he named the phenomenon activated protein C resistance. In 1994, Bertina and colleagues discovered the predominant molecular basis of APC resistance: substitution of an alanine for guanine at nucleotide 1691 changes arginine to glutamine at amino acid position 506 of factor V, eliminating one of three APC cleavage sites, which delays degradation of factor Va and impedes APC-mediated inhibition of thrombin generation. This mutation was named factor V Leiden (FVL), after the city in which it was discovered (Figure 9-2).

Activated Protein C Resistance Assays

First-generation APC resistance assays typically use undiluted patient plasma, purified APC, and aPTT activator (phospholipid and Ca^{++}).

The added APC degrades factors Va and VIIIa and prolongs the aPTT. An aPTT test with and without added APC is performed in parallel, and the results are expressed as a ratio: (aPTT+APC)/aPTT. For normal individuals, the ratio is typically >2, while factor V Leiden heterozygotes have a ratio <2, and homozygotes typically have a ratio <1.5. This first-generation APC resistance test can be performed on any manual or automated coagulation analyzer. However, the first-generation APC resistance test provides poor discrimination between FVL negative individuals and FVL carriers. To achieve 100% sensitivity, a high cut-off value for the APC resistance ratio is required, reducing the specificity of the assay to ~50%. In addition, a false-positive low APC ratio can result from interference due to conditions that prolong the aPTT, including acquired or congenital coagulopathies, lupus anticoagulants, and parenteral and oral anticoagulants. Therefore, an aPTT-based method using undiluted plasma should not be used to screen for APC resistance if the patient has an elevated baseline aPTT.

Several changes have been made to address the deficiencies of the original, first-generation APC resistance method. In the modified aPTT APC test, patient plasma is diluted to 25% (1:4) with factor V depleted plasma containing a heparin neutralizing substance, producing improved specificity for FVL without sacrificing sensitivity, and improving discrimination between heterozygous and homozygous FVL states. However, the modified aPTT APC resistance method is more expensive to perform and is insensitive for non-FVL causes of APC resistance that are associated with an increased risk for VTE (see "Global Tests of Hypercoaguability"), and a potent lupus anticoagulant or high concentrations of heparin and direct thrombin inhibitors may still produce false-positive results.

Alternative approaches to detecting APC resistance have been validated, including tissue factor and Russell viper venom (RVV)-activated clotting assays and a factor Xa chromogenic assay. In one commercial RVV-based assay, factor X is activated by Russell viper venom, with or without the addition of Agkistrodon contortrix contortrix (Southern Copperhead) venom, to activate endogenous protein C, and the resulting clotting times are expressed as a ratio: (RVV+APC)/RVV. Advantages of the RVV screening method for FVL include excellent discrimination of normal subjects from FVL carriers and less interference from lupus anticoagulants and heparin (polybrene is added to neutralize heparin). However, the assay also does not completely discriminate between heterozygous and homozygous FVL genotypes, and plasma from patients who are taking oral anticoagulants require predilution with factor V depleted plasma (1:4) in order to prevent false-positive results.

Residual platelets in plasma samples that have undergone freezing and thawing before performing APC resistance testing can produce lower ratios and could be a cause of false-positive results, particularly with the undiluted aPTT-based method. APC ratios can vary widely among healthy subjects, may not be normally distributed, and, given the high prevalence of FVL among Caucasians (3% to 7%), asymptomatic carriers may be included in a reference range population. The common practice of determining a cut-off for APC resistance based on the mean minus 2 standard deviations from a small reference group (n=20) may be inaccurate. Alternative approaches include selecting a larger number of controls and determining if log transformation of the APC ratios will produce a normal distribution. If not, a rank order method can be applied. Finally, all healthy subjects can be genotyped for FVL and carriers removed from the control group. In an effort to reduce interassay imprecision, some laboratories determine the APC ratio of a control pooled normal plasma and divide each patient APC ratio by the control APC ratio to obtain a normalized APC ratio. However, reporting normalized APC ratios does not eliminate the variation in APC ratios between laboratories using different reagents and methods.

Factor VIII Activity

Although no mutations analogous to factor V Leiden have been discovered in the factor VIII gene, multiple epidemiologic studies have confirmed that elevated factor VIII activity is a risk factor for both initial and recurrent idiopathic venous thromboembolism. In most studies, factor VIII was measured in a one-stage clotting assay, as described in chapter 6, with the potential for inaccurately low or elevated activities due to plasma variables such as a nonspecific inhibitor or elevated fibrinogen, respectively. However, Cristina et al (2004) found that when one-stage clotting and chromogenic factor VIII activities were compared in VTE patients who had stopped taking oral anticoagulation therapy, similar increased relative risks for recurrent VTE were obtained for patients with factor VIII activities greater than the 90th percentile compared to patients with factor VIII activities less than the 25th percentile (6.21 and 5.43 respectively), and intra- and interassay coefficients of variation were similar.

Factor VIII activity is higher in non-O blood group individuals and increases with advancing age. It is also elevated during pregnancy and malignancies, and is acutely elevated due to inflammatory, infectious, and traumatic states. Currently, there is uncertainty regarding how to integrate a particular patient's factor VIII activity into an overall assessment of the patient's thrombotic risk profile. Two important issues are timing of factor VIII activity measurement and determination of a laboratory specific cut-off. An elevated factor VIII activity during an acute VTE may not reflect a patient's steady state activity. Concurrently measuring other acute phase proteins (fibrinogen, C-reactive protein [CRP]) may help interpret an elevated factor VIII result. Selecting a cut-off value for factor VIII activity to be a VTE risk factor is particularly challenging. First, cut-off values in published studies were derived from unique populations and cannot be generally applied. Second, laboratory proficiency surveys confirm a high degree of interlaboratory variability in factor VIII activity measurement at high factor VIII levels, in part due to the many reagent and coagulation instrument combinations in use, making a single cut-off value impractical. Third, prospective studies validating the use of factor VIII activity to determine duration of oral anticoagulation therapy have not been completed.

Similar issues apply to measurement of factor IX and factor XI. While case-control studies confirm an association between elevated factor IX and factor XI antigen concentrations and venous thromboembolic events, clinical laboratories measure factor IX and factor XI activity by clotting assays, and there is insufficient current evidence to support routine measurement of these clotting factors in thrombophilia work-ups.

Fibrinogen Abnormalities

Congenital dysfunction of fibrinogen is uncommon, but some types can be associated with a tendency to thrombosis. The thrombotic risk may arise from several mechanisms, including rapid fibrin polymerization and decreased rate of degradation by plasmin. The assays employed to diagnose these disorders, including fibrinogen assay, thrombin time, reptilase time, and immunologic fibrinogen, are covered in chapters 6 and 15, and a diagnostic approach to these thrombophilic disorders is covered in chapter 20.

Lupus Anticoagulant

Acquired autoantibodies that recognize epitopes on selected proteins when bound to phospholipid surfaces can be identified by their in vitro behavior of prolonging a variety of clotting tests. The laboratory assays and diagnostic algorithms for antiphospholipid antibodies, which include lupus anticoagulants, are covered in chapter 22.

Factor V Leiden and Prothrombin G20210A Mutations

Shortly after the description of activated protein C resistance, a point mutation in factor V gene that substitutes glutamine for arginine at amino acid position 506, eliminating an APC cleavage site and causing APC resistance, was discovered and named factor V Leiden (FVL). The frequency of the mutation ranges from 2% to 15% in different regions of Europe and is much lower among indigenous peoples of Asia, Africa, and North and South America. The relative risk of VTE associated with FVL is estimated to be 5- to 10-fold for heterozygosity and as high as 50- to 100-fold for homozygosity. Due to its relatively high frequency among Caucasians, coinheritance of FVL with other inherited or acquired thrombophilia risk factors must be considered when evaluating patients with a strong personal and family history of venous thromboembolic events.

Additional mutations have been identified in the factor V gene, but they are not typically included in the laboratory evaluation of thromobphilia. Mutations that eliminate a different APC cleavage site of factor V at amino acid 306 (factor V Cambridge and factor V Hong Kong) are extremely rare, and the relative risk of venous thrombosis associated with these mutations is unknown. A group of nucleotide substitutions located in exons 13 and 16 of the factor V gene define a unique chromosome 1 allele known as factor V HR2 haplotype. The HR2 allele does not contain the FVL G1691A mutation, and its frequency is similar among European, east African, and Asian populations, supporting an earlier occurrence of the HR2 allele. None of the amino acid substitutions coded by the HR2 haplotype alters an APC cleavage site, yet the A4070G mutation is associated with mild

APC resistance via an unknown mechanism. Preliminary epidemiologic studies are inconsistent regarding the thrombotic risk of inheritance of the HR2 allele (frequency 12% in US Caucasians) or its co-inheritance with FVL. Until the publication of definitive data supporting a thrombotic risk for HR2, routine screening of thrombophilic patients for this allele is not indicated.

The average prothrombin activity is approximately 20% higher among patients with a history of VTE compared to a control population. Sequencing of the prothrombin gene identified a point mutation in the 3 prime non-coding region of factor II (G20210A) that correlated with higher prothrombin activity and an increased risk of VTE. Ribonucleic acid (RNA) expression and survival studies have shown that messenger RNA (mRNA) containing the point mutation is translated more efficiently than wild type prothrombin mRNA, providing a mechanism to explain the connection between the prothrombin gene G20210A mutation (PGM), elevated factor II, and increased risk of venous thrombosis. Similar to factor V Leiden, PGM is common among people of European ancestry, ranging from 2% to 5%, and is extremely rare in populations that have not acquired European alleles. The relative risk of venous thrombosis is estimated to be 5-fold for PGM heterozygosity. Reliable estimates for the relative risk associated with PGM homozygosity do not exist due to its rarity. Recently, a C20209T mutation has been described in African-American and some Middle Eastern populations, involving the DNA nucleotide base immediately upstream from the 20210 location. While its clinical significance has not been determined, the patients identified have presented with venous thrombosis and stroke.

The FVL mutation was originally identified using primers and the polymerase chain reaction (PCR) to amplify the region of interest surrounding nucleotide 1691. The FVL mutation (G to A) eliminates a MnlI endonuclease cleavage site, which can be detected on agarose gel electrophoresis as a loss of a digest fragment. Since the PGM does not create or destroy an endonuclease cleavage site, the original detection method used a PCR primer to introduce a second mutation that created a HindIII cleavage site in the presence of the G20210A PGM, which yielded an extra fragment on gel electrophoresis. While some molecular diagnostic laboratories continue to use these methods, alternative in-house and commercial methods have been developed to automate and simplify the detection of FVL and PGM, including real-time PCR instruments, Invader® (Third Wave Technologies Inc, Madison, Wisconsin) and INFINITI® (AutoGenomics Inc, Carlsbad, California) platforms. While genetic testing is not affected by most of the preanalytical and analytical variables that can affect the accuracy and precision of functional and antigenic assays, genetic testing is vulnerable to contamination from other sources of DNA, which may result in false-positive or false-negative results, and to ambiguous output, whether direct DNA sequencing or indirect detection of a mutation. Stringent quality control of all steps is necessary to ensure accurate molecular test results.

Plasma Homocysteine Concentration

Demethylation of the essential amino acid methionine produces homocysteine, which may be converted to cysteine and excreted in the urine or remethylated to methionine (Figure 9-4). Three vitamins are necessary: cofactors (B_6 and B_{12}) or substrate (folate) for the enzymatic elimination (vitamin B_6) and remethylation (vitamin B_{12} and folate) of homocysteine. Homozygous deficiency of any of the key enzymes involved in homocysteine degradation or remethylation produces severe hyperhomocysteinemia, homocystinuria, mental retardation, and premature arterial and venous thromboembolic events. Mild to moderate hyperhomocysteinemia due to a partial deficiency of certain enzymes is associated with an increased risk for myocardial infarction, stroke, peripheral arterial ischemia, and VTE. Risk factors for acquired hyperhomocysteinemia include vitamin B_6, B_{12}, and folate deficiencies and renal failure.

Plasma homocysteine circulates in various oxidized and protein-bound forms. Prior to analysis, homocysteine (tHcy) is reduced to produce a homogeneous, unbound, sulfhydryl form of homocysteine. Initially, various chromatography methods were employed to measure homocysteine, but they were not suitable for routine use in clinical laboratories. In the 1990s, two commercial immunoassays were developed: a fully automated fluorescence polarization immunoassay (FPIA) and an enzyme immunoassay (EIA). Both assays are based on conversion of homocysteine to S-adenosyl-homocysteine (SAH), which competes with either fluoresceinated SAH (FPIA) or immobilized SAH (EIA) for binding to an anti-SAH antibody. Method comparison studies confirm FPIA is

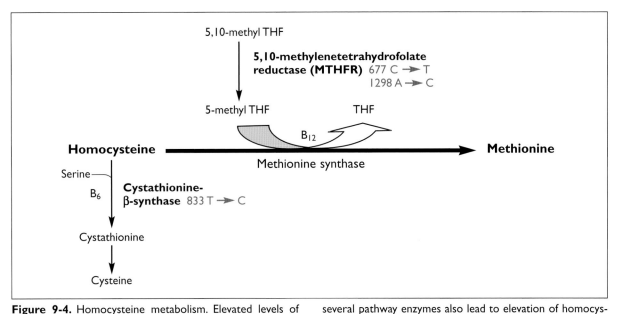

Figure 9-4. Homocysteine metabolism. Elevated levels of plasma homocysteine have been associated with an increased thrombotic risk. Homocysteine is a precursor amino acid in the synthesis of methionine and cysteine. High levels of homocysteine may be observed in nutritional deficiency of vitamins B_6, B_{12}, and folate. However, genetic mutations of several pathway enzymes also lead to elevation of homocysteine, with variable thrombotic risk. The homozygous 833 T to C mutation of cystathionine beta synthase leads to congenital homocystinuria, while the 677 C to T and 1298 A to C mutations of the MTHFR enzyme lead to a modest or no increase in homocysteine levels respectively.

the more precise assay, but interlaboratory agreement is not ideal due to the lack of a certified reference material.

Two preanalytical variables can affect the accuracy of tHcy measurements. Ingestion of a methione-rich high-protein meal may raise tHcy 10% to 15% after 6 to 8 hours. Ex vivo release of tHcy from red cells is time and temperature dependent, and can be minimized by collection of anticoagulated blood that is either promptly centrifuged and plasma poured off or whole blood that is stored on ice until processed.

Two mutations in the 5,10-methylenetetrahydrofolate reductase (MTHFR) gene, *C677T* and *A1298C*, reduce homocysteine remethylation. Among Caucasians, 10% to 12% are 677 TT homozygous, which reduces enzyme activity to 35% and increases plasma total homocysteine (tHcy) concentrations by 25% compared to wild type. Individuals who are heterozygous (CT) for the 677 mutation do not have elevated tHcy. A recent meta analysis of epidemiologic studies supported a weak association for both hyperhomocysteinemia and *MTHFR C677T* with VTE risk. However, studies conducted in North America did not confirm an association between the *MTHFR C677T* mutation and VTE risk, probably due to greater fortification of foods with B vitamins as compared with other regions of the world.

Homozygosity for *A1298C* occurs in ~10% of Caucasians, reduces enzyme activity to ~60%, and is not associated with elevated tHcy or increased risk for venous thromboembolic events. Genotyping for *MTHFR C677T* and *A1298C* can be done by PCR amplification of the appropriate gene loci and gel electrophoresis after HinfI and MboII restriction digests respectively. Commercial instruments that employ novel methods for detecting allele-specific nucleotide substitutions are also available.

Presently, screening thrombophilic patients for hyperhomocysteinemia, but not for *MTHFR C677T* or *A1298C* mutations, is recommended by many hemostasis experts, and most patients with elevated tHcy respond to folate supplementation alone or to a combination of vitamin B_6, B_{12}, and folate.

Global Tests of Hypercoagulability

Endogenous Thrombin Potential

PT-based and aPTT-based clotting tests measure the lag time from the addition of an activator to the initial burst of thrombin and resultant fibrin formation. Measuring the peak rate and cumulative amount of thrombin generation in addition to the lag time identifies patterns of endogenous thrombin potential (ETP) that correlate with various hemorrhagic and hypercoagulable conditions. While the concept and application of extended monitoring of in vitro thrombin generation were presented more than 50 years ago, recent advances in methodology and data analysis are leading to investigations of ETP as a screening test for inherited hypercoaguable conditions, including FVL; prothrombin 20210 mutation; deficiencies of protein S, antithrombin, and protein C; and to provide insights into the prothrombotic mechanisms of oral contraceptives, lupus anticoagulants, and elevated intrinsic pathway factor activities.

While ETP can not be used for thrombophilia evaluation or anticoagulant drug monitoring at this time due to insufficient prospective management studies, it has a promising future. The basic method involves addition of thromboplastin to citrated platelet-poor plasma and continuous measurement of thrombin generation by monitoring thrombin-dependent hydrolysis of a fluorescently-labeled synthetic substrate. Modifications of the assay include variations in the concentration of tissue factor and phospholipid, substitution of platelet-rich plasma for phospholipid, and comparison of ETP results with and without activated protein C or thrombomodulin to screen for FVL and protein C deficiency, respectively. Preanalytical variables primarily involve platelet activation during collection and transport of citrated whole blood when ETP is performed on platelet-rich plasma, and lysis of residual platelets when ETP is performed on frozen-thawed platelet-poor plasma.

Undiluted Activated Protein C Resistance

The original APC resistance method, performed with undiluted patient plasma, is sensitive for detecting FVL, but nonspecific, with a false-positive rate of ~15%. Nevertheless, patients who are APC resistant and FVL negative are at increased risk for venous thrombosis. Multiple inherited and acquired variables can produce APC resistance: elevated factor VIII, pregnancy, oral contraceptives, and factor V HR2. Therefore, an undiluted APC resistance test could be considered as a screening test for thrombophilic risk factors affecting the APC pathway. However, most clinical and laboratory hemostasis experts do not include this test in their thrombophilia algorithms, relying on more specific assays for APC pathway thrombotic risk assessment.

Activated Partial Thromboplastin Time

Since the activated partial thromboplastin time (aPTT) is a global clotting test, it is reasonable to assume that elevations of coagulation proteins (factors VIII, IX, XI, and fibrinogen), individually associated with an increased risk for VTE, would produce a shortening of the aPTT, making the aPTT a potential screening test for selected hypercoaguable risk factors. In one case-control study by Tripodi et al (2004), aPTT ratios (subject aPTT / normal pooled plasma aPTT) from patients with a history of a VTE were compared to healthy control subjects. The median aPTT ratio was significantly lower for the VTE group (0.97) compared to the control group (1.00). Eleven percent of the VTE group had an aPTT ratio lower than the fifth percentile value of the control group (0.87), for an odds ratio of 2.4. After accounting for differences in ABO blood type and FVIII activities between the two groups, the odds ratio for an aPTT ratio below the fifth percentile of the control group remained elevated (2.1), indicating that other factors are involved in producing shortened aPTTs in patients with a history of a VTE. Limitations to using an aPTT ratio as an inexpensive screening test for selected elevated coagulation factor levels include the poor sensitivity and specificity of the aPTT ratio, due to the marked overlap between VTE and control subjects, and evaluation of only one reagent-coagulation instrument combination. Furthermore, the aPTT ratio is influenced by anticoagulant therapy such as heparin, DTIs, or warfarin.

Activation and Degradation Peptides

Activation of several coagulation factors releases peptide fragments that can be measured with sensitive immunoassays, providing insights into the in vivo coagulation status of patients potentially at increased risk for venous thromboembolic events. Commercial immunoassays have been developed to detect F_{1+2}, a peptide released from prothrombin when activated by factor Xa; fibrinopeptide A (FPA), a 16 amino terminus peptide released from the alpha chains of fibrinogen by thrombin during conversion to fibrin; and thrombin-antithrombin (TAT) complexes formed by the inactivation of thrombin by antithrombin. When compared to control subjects, some protein C-deficient patients in stable health and not anticoagulated have higher concentrations of F_{1+2} but similar concentrations of FPA, suggesting that some patients with inherited deficiencies of coagulation-regulating factors have increased basal rates of thrombin generation, and they could be tipped toward excess fibrin generation by acquired prothrombotic states. Similar results have been obtained from asymptomatic patients with protein S deficiency, factor V Leiden, and prothrombin G20210A mutations, but interestingly not from patients with antithrombin deficiency. F_{1+2} levels in protein C and protein S deficient patients who are therapeutically anticoagulated with a vitamin K antagonist like warfarin are not significantly higher when compared to a control group, which is consistent with the efficacy of oral anticoagulation therapy for prevention of recurrent VTE. Elevated concentrations of TAT complexes have also been reported for asymptomatic carriers of FVL and patients with partial deficiencies of protein C and protein S. However, the ranges of TAT levels from asymptomatic carriers of thrombophilic risk factors overlap substantially with the ranges for control populations, limiting the predictive value of results for individual at-risk patients. Elevations of FPA were originally applied to detection of acute venous and arterial thromboembolic events, but sensitive quantitative D-dimer assays have replaced its use in routine clinical situations.

Immunoassays for F_{1+2}, TAT, and FPA share a common vulnerability to ex vivo activation of thrombin, which can produce false-positive elevations in the measured peptide or protein complex. To minimize this source of preanalytical error, phlebotomy technique must be meticulous and minimally traumatic, and thrombin activation within the collection tube stringently inhibited by using combinations of specific and nonspecific thrombin inhibitors rather than simply sodium citrate. Presently, these assays should be considered for research use and not applied to routine assessment of patients' risk for thromboembolic events.

Conclusion

As presented in Figure 9-1, there is a broad menu of tests that can potentially be performed as part of a laboratory evaluation of a patient for underlying thrombophilia risk factors. In this chapter, the major thrombophilia test methods were reviewed; classifications of deficiencies were defined for protein C, protein S, and antithrombin; and important preanalytical and analytical sources of inaccurate results were discussed. In chapter 19, algorithms are presented to provide guidance for using the tests and methodologies presented in this chapter to appropriately evaluate patients with venous thromboembolic events.

Suggested Reading

Thrombophilia Testing Overview

Dahlback B. Advances in understanding pathogenic mechanisms of thrombophilic disorders. *Blood.* 2008;112:19-27.

Jespersen J, Bertina RM, Haverkate F, eds. *Laboratory Techniques in Thrombosis. A Manual.* Second Revised Edition of the ECAT Assay Procedures. New York: Springer; 2000.

Mackie I, Cooper P, Kitchen S. Quality assurance issues and interpretation of assays. *Semin Hematol.* 2007; 44:114-125.

Moll S. Thrombophilias: practical implications and testing caveats. *J Thromb Thrombolysis.* 2006;21:7-15.

Rowan RM, van Assendelft OW, Preston FE, eds. *Advanced Laboratory Methods in Haematology.* London: Arnold; 2002.

Protein C

Boyer C, Rothschild C, Wolf M, et al. A new method for the estimation of protein C by ELISA. *Thromb Haemost.* 1984;36:579-589.

Kottke-Marchant K, Comp P. Laboratory issues in diagnosing abnormalities of protein C, thrombomodulin, and endothelial cell protein C receptor. *Arch Pathol Lab Med.* 2002;126:1337-1348.

Marlar RA, Adcock DM. Clinical evaluation of protein C: a comparative review of antigenic and functional assays. *Hum Pathol.* 1989;20:1040-1047.

Richards SM, Olson T, Keyes LD. Quantitative and functional assays compared for determination of zymogen and activated human protein C. *Clin Chem.* 1990;36:1892-1896.

Protein S

Faioni EM. Reliable estimates of plasma protein S levels: are we getting any closer? *Thromb Haemost.* 2001;86:1139-1140.

Goodwin AJ, Rosendaal FR, Kottke-Marchant K, Bovill EG. A review of the technical, diagnostic, and epidemiologic considerations for protein S assays. *Arch Pathol Lab Med.* 2002;126:1349-1366.

Lane DA, Mannucci PM, Bauer KA, et al. Inherited thrombophilia: part 1. *Thromb Haemost.* 1996;76:651-652.

Lane DA, Mannucci PM, Bauer KA, et al. Inherited thrombophilia: part 2. *Thromb Haemost.* 1996;76: 824-834.

Perrson KE, Hillarp A, Dahlback B. Analytical considerations for free protein S assays in protein S deficiency. *Thromb Haemost.* 2001;86:1144-1147.

Van Cott EM, Ledford-Kraemer M, Meijer P, Nichols WL, Johnson SM, Peerschke EIB. An analysis of North American Specialized Coagulation Laboratory Association proficiency testing. *Am J Clin Pathol.* 2005;123:778-785.

Antithrombin

Goodnight SH, Schaeffer JL, Sheth K. Measurement of antithrombin III in normal and pathologic states using chromogenic substrate S-2238: comparison with immunoelectrophoretic and factor Xa inhibition assays. *Am J Clin Pathol.* 1980;73:639-647.

Hortin GL, Tollefsen DM, Santoro SA. Assessment of interference by heparin cofactor II in the DuPont ACA antithrombin-III assay. *Am J Clin Pathol.* 1988;89:515-517.

Kottke-Marchant K, Duncan A. Antithrombin deficiency: issues in laboratory diagnosis. *Arch Pathol Lab Med.* 2002;126:1326-1336.

Odegard OR, Abilgaard U. Antithrombin III: critical review of assay methods: significance of variations in health and disease. *Haemostasis.* 1978;7:127-134.

Activated Protein C Resistance

Dahlback B, Carlsson M, Svensson PJ. Familial thrombophilia due to a previously unrecognized mechanism characterized by poor anticoagulant response to activated protein C: prediction of a cofactor to activated protein C. *Proc Natl Acad Sci USA.* 1993; 90:1004-1008.

de Ronde H, Bertina RM. Laboratory diagnosis of APC-resistance: a critical evaluation of the test and the development of diagnostic criteria. *Thromb Haemost.* 1994;72(6):880-886.

Favaloro EJ, Mirochnik O, McDonald D. Functional activated protein C resistance assays: correlation with factor V DNA analysis is better with RVVT- than APTT-based assays. *Br J Biomed Sci.* 1999;56:23-33.

Tripodi A, Negri B, Bertina RM, Mannucci PM. Screening for the FV:Q506 mutation: evaluation of thirteen plasma-based methods for their diagnostic efficacy in comparison with DNA analysis. *Thromb Haemost.* 1997;11(3):436-439.

Factor VIII

Cristina L, Benilde C, Michela C, Mirella F, Giuliana G, Gualtiero P. High plasma levels of factor VIII and risk of recurrence of venous thromboembolism. *Br J Haematol.* 2004;124:504-510.

Koster T, Blann AD, Briet E, Vandenbroucke JP, Rosendaal FR. Role of clotting factor VIII in effect of von Willebrand factor on occurrence of deep-vein thrombosis. *Lancet.* 1995;345:152-155.

Kraaijenhagen RA, in't Anker PS, Koopman MM, Reitsma PH, et al. High plasma concentration of factor VIII is a major risk factor for venous thromboembolism. *Thromb Haemost.* 2000;83:5-9.

Kyrle PA, Minar E, Hirschl M, et al. High plasma levels of factor VIII and the risk of recurrent venous thromboembolism. *New Engl J Med.* 2000;343:457-462.

Factor V Leiden and HR2

Bertina RM, Koeleman BP, Koster T, et al. Mutation in blood coagulation factor V associated with resistance to activated protein C. *Nature.* 1994;369:64-67.

Benson JM, Ellingsen D, El-Jamil M, et al. Factor V Leiden and factor V R2 allele: high-throughput analysis and association with venous thromboembolism. *Thromb Haemost.* 2001:86:1188-1192.

Faioni EM, Franchi F, Bucciarelli P, et al. Coinheritance of the HR2 haplotype in the factor V gene confers an increased risk of venous thromboembolism to carriers of factor V R506Q (factor V Leiden). *Blood.* 1999;94:3062-3066.

Prothrombin 20210 Mutation

Danckwardt S, Hartmann K, Katz B, et al. The prothrombin 20209 C—>T mutation in Jewish-Moroccan Caucasians: molecular analysis of gain-of-function of 3' end processing. *J Thromb Haemost.* 2006;4(5):1078-1085.

Poort SR, Rosendaal FR, Reitsma PH, Bertina RM. A common genetic variation in the 3'-untranslated region of the prothrombin gene is associated with elevated plasma prothrombin levels and an increase in venous thrombosis. *Blood.* 1996;88(10):3698-3703.

Rosendaal FR, Doggen CJ, Zivelin A, et al. Geographic distribution of the 20210 G to A prothrombin variant. *Thromb Haemost.* 1998:79:706-708.

Warshawsky I, Hren C, Sercia L, Shadrach B, Deitcher SR, Newton E, Kottke-Marchant K. Detection of a novel point mutation of the prothrombin gene at position 20209. *Diagn Mol Pathol.* 2002;11(3):152-156.

Homocysteine

Heijer MD, Lewington S, Clarke R. Homocysteine, MTHFR and risk of venous thrombosis: a meta-analysis of published epidemiological studies. *J Thromb Haemost.* 2004:3:292-299.

Refsum H, Smith AD, Ueland PM, et al. Facts and recommendations about total homocysteine determinations: an expert opinion. *Clin Chem.* 2004;50:3-32.

Tripodi A, Chantarangkul V, Lombardi R, Lecchi A, Mannucci PM, Cattaneo M. Multicenter study of homocysteine measurement: performance characteristics of different methods, influence of standards on interlaboratory agreement of results. *Thromb Haemost.* 2001:85:291-295.

Global Tests of Hypercoaguability

Baglan T. The measurement and application of thrombin generation. *Br J Haematol.* 2005;130:653-661.

Chanarangkul V, Clerici M, Bressi C, Tripodi A. Standardization of the endogenous thrombin potential measurement: how to minimize the effect of residual platelets in stored plasma. *Br J Haematol.* 2004;124:355-357.

Rodeghiero F, Tosetto A. Activated protein C resistance and factor V Leiden mutation are independent risk factors for venous thromboembolism. *Ann Intern Med.* 1999;130:643-650.

Tripodi A, Chantarangkul V, Martinelli I, Bucciarelli P, Mannucci PM. A shortened activated partial thromboplastin time is associated with the risk of venous thromboembolism. *Blood.* 2004:104(12):3631-3634.

Activation and Degradation Peptides

Bauer KA, Barzegar S, Rosenberg RD. Influence of anticoagulants used for blood collection on plasma prothrombin fragment F1+2 measurements. *Thromb Res.* 1991;63:617-628.

Bauer KA, Broekmans AW, Bertina RM, et al. Hemostatic enzyme generation in the blood of patients with hereditary protein C deficiency. *Blood.* 1988;71:1418-1426.

Bauer KA, Humphries S, Smillie B, et al. Prothrombin activation is increased among asymptomatic carriers of the prothrombin G20210A and factor V Arg506Gin mutations. *Thromb Haemost.* 2000;84:396-400.

Martinelli I, Bottasso B, Duca F, et al. Heightened thrombin generation in individuals with resistance to activated protein C. *Thromb Haemost.* 1996;75(5):703-705.

Omote M, Asakura H, Takamichi S, et al. Changes in molecular markers of hemostatic and fibrinolytic activation under various sampling conditions using vacuum tube samples from healthy volunteers. *Thromb Res.* 2008 Jun 23. [Epub ahead of print]

Simioni P, Scarano L, Gavasso S, et al. Prothrombin fragment 1+2 and thrombin-antithrombin complex levels in patients with inherited APC resistance due to factor V Leiden mutation. *Br J Hematol.* 1996;92:435-441.

Section 3

Algorithmic Approach to Bleeding Disorders

Elizabeth M. Van Cott, MD, Editor

Chapter 10. Abnormal Activated Partial Thromboplastin Time

Chapter 11. Prolonged Prothrombin Time

Chapter 12. Normal Prothrombin Time and Activated Partial Thromboplastin Time

Chapter 13. Fibrinolytic Bleeding Disorders

Chapter 14. Platelet Disorders

Chapter 15. Abnormal Thrombin Time

Chapter 16. Von Willebrand Disease

Abnormal Activated Partial Thromboplastin Time

John D. Olson, MD, PhD

Introduction

Since its initial description in 1953, the partial thromboplastin time (PTT) has been a valuable tool for the monitoring of heparin therapy and the evaluation of patients with a suspected bleeding disorder. Methods for automating the reading of the endpoint for both the prothrombin time (PT) and the activated partial thromboplastin time (aPTT) occurred in the 1960s and the early 1970s. Simultaneously, manufacturers began providing reagents for the performance of the test. In the 1970s, the PTT method changed. An activator, such as kaolin, celite, micronized silica, or ellagic acid, was added to the test to activate contact factors and change the PTT to the aPTT. This is the method that prevails today. Wide variability continues in the responsiveness of the reagents used for the aPTT in all of its clinical applications.

Clinical Etiologies

As depicted in Table 10-1, there are six common causes for a prolonged aPTT. The clinically important issues are described briefly below. Of interest, each of these different etiologic categories for a prolonged aPTT requires a different action on the part of the clinician who must respond to the abnormal test result.

Table 10-1. Common Etiologies of an Isolated Prolongation of the aPTT

- ❑ Medication: heparin and direct thrombin inhibitors
- ❑ Coagulation factor deficiency associated with hemorrhage
- ❑ Coagulation factor deficiency of little or no clinical significance
- ❑ Nonspecific coagulation factor inhibitor (lupus anticoagulant)
- ❑ Specific coagulation factor inhibitor
- ❑ Spurious or preanalytic errors

Case Study

A 42-year-old woman, the victim of a motor vehicle accident, presented to the emergency department. At the time of presentation, she was conscious. Following an immediate imaging evaluation, she was taken to the operating room for the open reduction of multiple fractures of the left hip and femur. She received four units of packed red blood cells intraoperatively.

In the postoperative period, following admission to the surgical intensive care unit, her vital signs were stable, and the following results of laboratory studies were obtained:

Test	Value	Reference Range
PT (s)	12.5	10-13
aPTT (s)	70	25-36
Thrombin time (s)	>150	18-22
Fibrinogen (mg/dL)	145	130-330
Hemoglobin (g/dL)	11	12-15
Hematocrit (%)	34	36-44
WBC (per μL)	9000	4000-10,000
Platelet count (per μL)	145,000	150,000-450,000

The laboratory findings led to the evaluation of the prolonged aPTT. At the beginning of the testing algorithm for determining the etiology of an abnormal aPTT, the thrombin clotting time was performed and found to be greater than 150 seconds (reference range = 18 to 22 seconds).

What are some possible causes for the prolonged aPTT?

Medications

The aPTT is the most common test used for monitoring therapy with unfractionated heparin. In addition, unfractionated heparin can contaminate coagulation specimens for coagulation testing that are drawn inappropriately from arterial or central venous lines. When the aPTT is used to monitor heparin therapy, it is of value in guiding dose

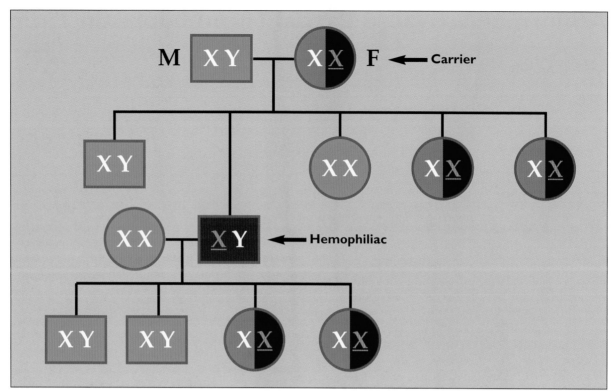

Figure 10-1. Diagram of a family pedigree for hemophilia A. Since hemophilia A (factor VIII deficiency) is an X-linked disorder, females are carriers of the disorder since they have two X chromosomes. A female carrier will pass hemophilia to approximately 50% of male offspring, while approximately 50% of female offspring will be carriers. All female offspring of hemophiliac males are obligate factor VIII carriers, while all male offspring will not be affected.

adjustments, as detailed in chapter 24. When heparin contaminates the specimen, it can lead to unnecessary and costly additional laboratory work, and, if not detected promptly, it can obfuscate the interpretation of those tests. Also, even though the test cannot be used to monitor the concentration of low-molecular-weight heparin, it will commonly cause prolongation of the aPTT.

Although heparins are by far the most common of the medications causing a prolongation in the aPTT at the present time, other medications in clinical use (hirudin analogues, eg, lepirudin or bivalirudin, and argatroban) directly inhibit thrombin and prolong the aPTT. Other less-frequent causes may include hydroxy-ethyl starch, hematin, and Suramin. In addition, Taularidine, an additive in some intravenous medications, can also cause a long aPTT. Early exclusion of a medication as an etiology for a prolonged aPTT is important and may require the use of the thrombin clotting time or the addition of heparinase to the specimen for the correct interpretation of the result. See chapter 6 for a laboratory algorithm for detecting a medication effect on the aPTT.

Coagulation Factor Deficiency Associated with Significant Hemorrhage

Inherited deficiencies of not only the most commonly deficient coagulation factors, factor VIII (in hemophilia A and von Willebrand disease), factor IX, or factor XI, but also involving other coagulation factors, can lead to a significant prolongation of the aPTT. In addition, acquired deficiencies of multiple coagulation factors, as might be seen in liver disease or consumption coagulopathy, may also prolong the aPTT. The latter acquired causes are frequently accompanied by a prolongation in the PT (see chapter 12 for more information regarding prolongation of both the PT and the aPTT). When factors are severely deficient, action on the part of the clinician to replace deficient coagulation factors or, in the case of acquired deficiencies, to treat the underlying etiology of the factor deficiency is often required.

Von Willebrand Disease

The most common of the inherited bleeding disorders is von Willebrand disease. It is primarily a platelet function disorder; however, the coagulant factor VIII molecule circulates bound to von

Willebrand factor (vWF), which serves to stabilize the factor VIII in the circulation. Thus, patients with von Willebrand disease have reduced factor VIII and can therefore have a prolonged aPTT. The degree of aPTT prolongation is proportional to the severity of the deficiency of factor VIII, and in many cases of von Willebrand disease, the level of factor VIII will not be sufficient to cause an abnormal aPTT, or the prolongation may be mild. Von Willebrand disease is discussed in detail in chapter 16.

Factor VIII Deficiency

Factor VIII deficiency (hemophilia A or classic hemophilia) is a bleeding disorder that is inherited as a sex-linked recessive trait, affecting only boys and men, but transmitted only by women. Figure 10-1 illustrates a typical pedigree in hemophilia A. The disease occurs at the rate of about 1 in 10,000 of the population, and spontaneous mutations may be seen in up to 30% of cases. Over 100 mutations in the factor VIII gene have been described as a cause for hemophilia A, with inversions of intron 22 accounting for approximately 50% of mutations.

The severity of the bleeding disorder is directly related to the level of factor VIII in the patient, as is severity of the prolongation of the aPTT. Table 10-2 depicts the relationship of factor level to clinical bleeding. In very mild cases, the aPTT may be normal or only very slightly prolonged, depending on the sensitivity of the reagent used for the test. In contrast, patients with severe disease will have a markedly prolonged aPTT, approaching 90 seconds or more. Although occasional patients with very mild factor VIII deficiency may have a normal aPTT, the test is a very sensitive test for detecting patients with inherited factor VIII deficiency. In contrast, the aPTT does not parallel the level of factor VIII closely enough to monitor therapy. Assaying the concentration of factor VIII in the blood must be done in order to monitor therapy. Factor VIII deficiency is treated with concentrates of factor VIII, either extracted from plasma or via recombinant technology. Patients with mild or moderate disease may respond to the stimulation of the release of factor VIII following treatment with nasal or parenteral DDAVP (desmopressin).

Factor VIII levels in female hemophilia carriers are typically near 50% of normal, with normal levels of vWF. The ratio of factor VIII/vWF antigen is less than 0.5 in more than 90% of hemophilia carriers; however, genetic testing is still recommended to establish the carrier state. A more severe, iso-lated decrease in factor VIII levels in a woman could indicate lyonization, hemizygous chromosome X seen in Turner syndrome, or type 2N von Willebrand disease.

Factor IX Deficiency

Like factor VIII deficiency, factor IX deficiency (Christmas disease or hemophilia B) is also inherited as a sex-linked recessive trait. All of the principles regarding factor VIII that are outlined above also apply to factor IX deficiency; however, it is less frequent, occurring at a rate of about 1 in 25,000 males. The balance of the issues regarding the aPTT and factor assays are managed in factor IX deficiency as they are in factor VIII deficiency. Factor IX deficiency is treated with concentrates that are prepared via extraction from plasma or recombinant technology.

Factor XI Deficiency

Factor XI deficiency (hemophilia C) is an autosomally inherited disorder that is rare in most ethnic groups, occurring at a rate of approximately 1 case in 1,000,000 population. The exception to this is the unusually high prevalence in Ashkenazi Jews, in whom the mutation of the factor XI gene may be as high as 13%. Following von Willebrand disease, factor XI deficiency is the second most common inherited bleeding disorder among women; however, in Ashkenazi Jewish women, it may be the most common. In contrast to factor VIII and IX deficiencies, the manifestation of bleeding is variable, often being milder than predicted by the factor level but, in contrast, also occasionally being more severe than expected. Deficiency of factor XI causes an isolated prolongation of the aPTT, and the diagnosis and monitoring is dependent on assay of factor XI. There is no commercially available concentrate of factor XI in the USA, leaving fresh frozen plasma as the only option for treatment of hemorrhage or prophylaxis prior to procedures.

Table 10-2. Severity of Hemophilia

Severity	Factor VIII (%)	Spontaneous Hemorrhage	Traumatic Hemorrhage
Very mild	> 20	Never	Rare
Mild	5-20	Rare	Common
Moderate	2-5	Unusual	Always
Severe	< 2	Common	Always

Other Inherited and Acquired Deficiencies

Although unusual, it is possible for the multiple deficiencies of factors seen in disseminated intravascular coagulation or liver disease to present with prolongation of only the aPTT; however, the PT is often equally or more sensitive to these deficiencies. This means that there is more frequently an isolated prolongation of the PT or prolongation of both the PT and aPTT. This is similarly true of the inherited deficiencies of fibrinogen and the factors X, V, and prothrombin, which can affect both tests.

Coagulation Factor Deficiencies Not Associated with Bleeding

As students, most of us were taught about the intrinsic and extrinsic activation of the common pathway of coagulation. This approach to thinking about the mechanism of thrombin generation grows from the activation processes that occur in vitro. Activation of coagulation in vivo proceeds differently, involving considerable cross-activation between the "intrinsic" and "extrinsic" systems. Because of this difference between the in vivo and in vitro activation processes, patients may have significant deficiencies of coagulation factors involved in contact activation (factor XII, prekallikrein, and high-molecular-weight kininogen [HMWK]) that lead to a markedly prolonged aPTT (often greater than 100 seconds) but have no associated bleeding diathesis. In general, these deficiencies are of importance only because they cause an abnormal test and could potentially mask other deficiencies that may cause problems. In general, a prolonged aPTT caused by deficiency of factor XII, prekallikrein, or HMWK is of no clinical consequence.

While factor XII (Hageman factor) deficiency is not associated with a bleeding risk, it has been reported to be associated with an increased risk of venous thrombosis. Indeed, the initial patient described to have factor XII deficiency died as a result of a pulmonary embolism. Severe factor XII deficiency is a rare cause of a markedly prolonged aPTT, but mildly decreased factor XII is a not uncommon finding that is typically associated with a mildly prolonged aPTT. An association between prekallikrein (Fletcher factor) deficiency or high-molecular-weight kininogen (Williams trait) deficiencies and venous thromboembolism is also speculated. Prekallikrein deficiency shows a prevalence in a wide distribution of racial groups and may not be more common in African Americans, as originally described. Laboratory diagnosis of HMWK deficiency may be hampered due to limited availability of HMWK-deficient plasma necessary to perform the factor assay.

Lupus Anticoagulant

Lupus anticoagulants are a type of antiphospholipid antibodies that can lead to a significant prolongation of the aPTT, which usually fails to correct in a mixing study. Demonstration that this anticoagulant activity is dependent on phospholipid confirms the diagnosis of a lupus-type anticoagulant. When identified, the patient may have an increased risk for a hypercoagulable state. If a lupus anticoagulant occurs in patients with thrombosis, more prolonged and possibly increased intensity of anticoagulation may be required. Lupus anticoagulants are discussed in detail in chapter 22.

Specific Coagulation Factor Inhibitors

Patients with hemophilia and severe factor deficiency can develop alloantibodies that recognize the deficient factor when it is transfused. In addition, nonhemophilic patients may spontaneously develop autoantibodies that recognize and rapidly clear their own coagulation factor—a condition referred to as acquired hemophilia. Recognition of a specific inhibitor by the clinician and laboratory is critical because the resulting hemorrhage may be life threatening and the management of the patient extremely difficult.

Alloantibodies in Hemophilic Patients

Among hemophilic patients with severe factor VIII or factor IX deficiency, 10% to 15% will develop alloantibodies to therapeutically infused coagulation factor. The presence of these antibodies has little effect on the already prolonged aPTT of the patient but will change the result of the mixing study, indicating the presence of an inhibitor. If incubated mixing studies are performed, the results will often show a delayed acting inhibitor, with progressive inhibition with time. Following the demonstration of the inhibitor, a Bethesda unit (BU) assay would be done in order to quantify the level of the inhibitor (see chapter 6). Patients are classified based on the value of the number of Bethesda units and the anamnestic response (or lack thereof) of the BU to exposure to factor VIII.

Low responders/low titer are at ≤5 BU, while high responders/high titer are >5 BU. The distinction matters with the design of treatment, the low responders being treated initially with high doses of factor VIII in an attempt to neutralize the antibody and achieve recoverable factor VIII in the plasma for hemostasis. High responders and low responders who fail factor VIII infusion receive concentrates that will activate coagulation without the need for factor VIII, referred to as bypassing activity. Examples include activated prothrombin complex concentrates and activated factor VII. Inducing tolerance to factor VIII treatment by continuous exposure to low or high doses of factor VIII concentrates has met with variable success.

Autoantibodies to Factor VIII and Other Factors

Rarely (0.2 to 0.5 per 1,000,000 in the population per year) autoantibodies can arise spontaneously to coagulation factors, leading to significant bleeding. Although autoantibodies to all factors have been described, those recognizing factor VIII are by far the most common. Of cases with anti-factor VIII autoantibodies, approximately 50% arise with no recognizable underlying associated illness and are seen more frequently in the elderly population. Of the balance, 10% to 15% are associated with pregnancy, most often in the first pregnancy and arising within 2 to 3 months postpartum. The rest of the cases are associated with autoimmune disease, malignancy, allergy (often to medication), and dermatologic disorders. The clinical diagnosis and management of acute hemorrhage are as described above for the alloantibodies. Of interest, the kinetics of autoantibody binding to factor VIII may be different than alloantibodies, allowing for some residual factor VIII to assay in the plasma and variable results in determining the number of Bethesda units.

In contrast to alloantibodies, autoantibodies sometimes respond to immune suppression to induce remission. In the cases associated with pregnancy, as many as three-fourths of patients will have spontaneous remission or require only corticosteroid therapy, with a mortality of only 5%. The other cases are more difficult to treat, often requiring more aggressive immunosuppressive therapy and with spontaneous remission in only one-third of cases. The mortality rate is as high as 20%. In general, the higher the BU titer at diagnosis, the more resistant the patient will be to treatment.

Spurious Causes of an Elevated aPTT

It is important to remember that there are several preanalytic (spurious) causes for an abnormal aPTT. Common preanalytic problems include:

❑ Unusually high hematocrit (prolongs the aPTT and PT) or low hematocrit (shortens the aPTT).

❑ Altering the concentration of citrate; for example, underfilling the blood collection tube (prolongs the aPTT), overfilling the tube (shortens the aPTT), or incorrectly preparing a citrate tube for a patient with high hematocrit.

❑ Incorrect timing of the collection.

❑ Contamination of the specimen with EDTA; for example, if a purple top (EDTA) tube is collected prior to a blue top (citrate) tube.

❑ Formation of clots in the specimen (shortens the aPTT).

❑ Delay in the transport and processing of the specimen (prolongs the aPTT).

❑ Inappropriate temperature of transport or storage.

❑ Collection through or above an intravenous line, leading to dilution or contamination.

Any of the above etiologies can cause an abnormal aPTT. Thus, it is important that when a new reagent or instrument is put into production to perform the aPTT, the new method is evaluated to determine how effective it is in distinguishing these etiologies. Therefore, the validation of a new aPTT method will require the following determinations:

❑ New reference range in the apparently healthy population: testing of a minimum of 20 (preferably more than 30) apparently healthy individuals.

❑ Sensitivity to heparin and possibly other drugs: testing 30 specimens from patients receiving unfractionated heparin, comparing to the heparin assay or, using the CuSum method, comparing to the existing reagent used in the laboratory, as described in chapter 6.

❑ Sensitivity to a factor deficiency, particularly factors VIII and IX: testing serial dilution of pooled normal plasma in factor deficient plasma, comparing the factor concentration at the upper limits of the reference range.

❑ Sensitivity of the reagent to a lupus-type anticoagulant: testing several saved, known lupus anticoagulants to test the sensitivity of the new reagent.

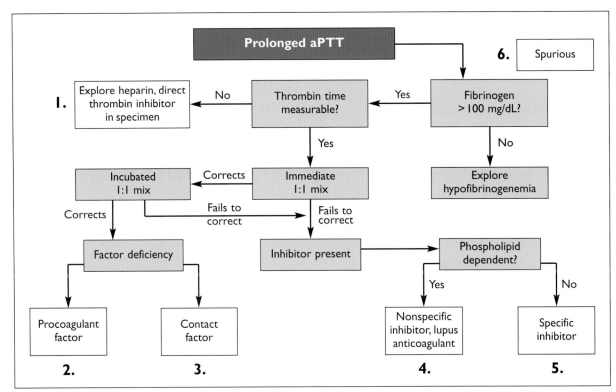

Figure 10-2. Algorithm for evaluation of the prolonged activated partial thromboplastin time (aPTT). A stepwise approach to evaluating the prolonged aPTT is shown. It is not the only clinically acceptable algorithm, but it is one approach that may be helpful for sorting the six potential clinical etiologies.

❑ Performance of the aPTT reagent in evaluating the mixing study in factor deficient and anticoagulant-containing plasmas.

Laboratory Testing

An algorithm for the evaluation of the prolonged aPTT is presented in Figure 10-2 and discussed below. In most cases, by the time the evaluation of a prolonged aPTT begins, the value of the fibrinogen will be known. If it is not, it is often useful to determine that there is sufficient fibrinogen to support the test. It is frustrating to go through an entire evaluation of a prolonged aPTT only to find that the patient suffers from hypofibrinogenemia.

If the fibrinogen concentration is sufficient, an evaluation to exclude heparin or other medications is useful. The use of a sensitive thrombin time (control of 18 to 20 seconds) can be very helpful. When using a sensitive thrombin time, if there is sufficient heparin or direct thrombin inhibitor in the specimen to prolong the aPTT beyond the reference range only 1 second, the thrombin time will be greater than 150 seconds in the majority of cases. In the case study presented at the beginning of the chapter, the value of the thrombin time

raised the possibility of heparin contamination of the specimen. Consultation with the nursing staff in the intensive care unit confirmed that the specimen had been drawn from an arterial line that was "kept open" with heparin. Repeat coagulation studies performed on a specimen collected from a peripheral vein revealed an aPTT within the reference range. Therefore, the finding of a measurable thrombin time excludes heparin as the etiology of a significantly prolonged aPTT. If the thrombin time is measurable, then a mixing study is useful. For details regarding performance of the mixing study, see chapter 6. A few points are worth emphasizing here.

Mixing studies on a very mildly prolonged aPTT (less than 3 to 5 seconds above the reference range) will frequently be confusing and difficult to interpret. An approach to the evaluation of the mildly prolonged aPTT can be seen in the algorithm in Figure 10-3.

Normal pooled plasma (>20 normal donors) or a commercial reagent plasma (frozen, not lyophilized) should be used for mixing with the patient's plasma and should have a known, near 100% concentration of coagulation factors, particularly factors V and VIII.

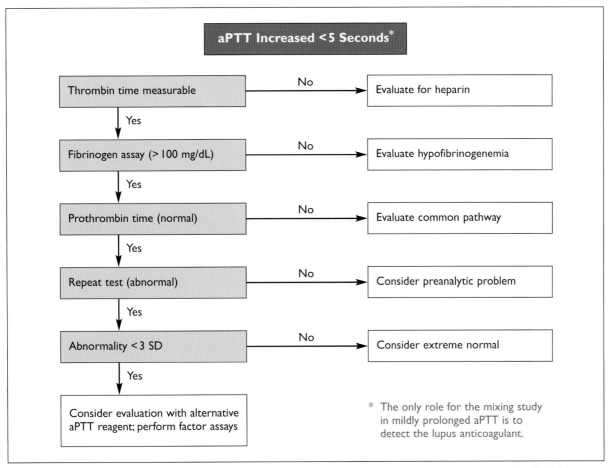

Figure 10-3. Algorithm for evaluation of the slight prolongation of the activated partial thromboplastin time (aPTT). A stepwise approach to evaluating the slightly prolonged aPTT is shown. The lupus anticoagulant causes the aPTT 1:1 mix to be more prolonged than the initial aPTT, a phenomenon that is seen more often with kaolin reagents.

The value of the mixing study primarily is to determine whether there is an inhibitor present. Interpretation of the control result and the result of test assays for time dependency of known inhibitors can be very difficult. The approach to the mixing study described here is one of mixing an equal volume of the patient plasma and the control plasma, and immediately performing the aPTT. If the resulting aPTT fails to correct to the reference range, an inhibitor is likely to be present. If the value corrects into the reference range, it is then necessary to perform an incubated mix to determine if a time-dependent inhibitor is present. To perform such a mix (Figure 10-4), an equal volume of patient plasma and control plasma are mixed (tube 1), and the control plasma (tube 2) and the patient plasma (tube 3) are all incubated at 37°C for 1 to 2 hours, separately. At the end of the incubation, the aPTT is measured, and the mixed specimen (tube 1, the test) and mixture of the patient and control (tube 4, the control), which is

composed of equal portions of tubes 3 and 2 that were incubated separately then mixed immediately prior to performing the aPTT, are compared. If the two mixed specimens (test and control) have aPTTs that are equal, an interpretation of correction is made, and an evaluation for factor deficiency would proceed.

The clinical history in the patient will determine whether one would initially evaluate contact activation factors or procoagulant factors. If there is a prolongation of the test specimen, when compared to the control specimen, of 10% to 15%, it is interpreted that the mix failed to correct and an inhibitor is present. The next step is to determine if there is phospholipid dependence in the inhibitor, as discussed in chapter 22. A variety of tests are available to determine phospholipid dependence of inhibition, more than one of which should be available for the evaluation, because none are 100% sensitive or specific to the presence of a lupus-type anticoagulant. If phospholipid

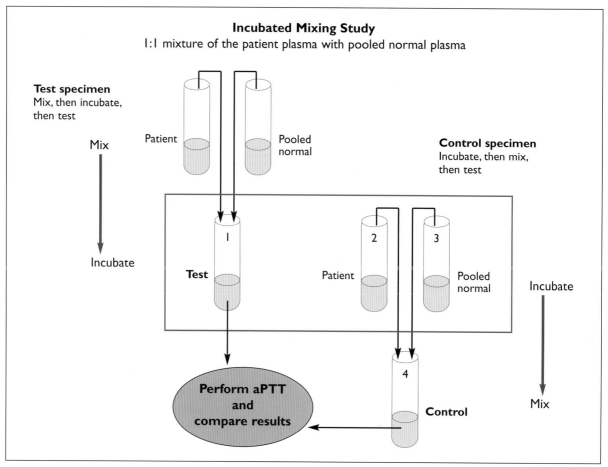

Figure 10-4. To perform an incubated mixing study, an equal volume of patient plasma and control plasma are mixed (tube 1), and the control plasma (tube 2) and the patient plasma (tube 3) are all incubated at 37°C for 1 to 2 hours, separately. At the end of the incubation, the activated partial thromboplastin time (aPTT) is measured, and the mixed specimen (tube 1, the test) and mixture of the patient and control (tube 4, the control), which is composed of equal portions of tubes 3 and 2 that were incubated separately then mixed immediately prior to performing the aPTT, are compared.

dependence is determined and the patient's clinical history is consistent, the diagnosis of a nonspecific (lupus-type) anticoagulant is made. If there is no phospholipid dependence, the patient should be evaluated for the presence of a specific coagulation inhibitor. The initial step in testing for a specific coagulation inhibitor is to perform assays for factors VIII, IX, XI, and XII.

It is important to remember that the purpose of the incubated mix is to detect an inhibitor, not to determine the etiology (increase the sensitivity of the test, not the specificity). It is a common misconception that the finding of a time-dependent inhibitor indicates a specific inhibitor. It is true that most specific inhibitors are time dependent (approximately 85% to 90%) and that most lupus anticoagulants are not (only about 10%). However, the prevalence of lupus anticoagulants and the rarity of specific inhibitors mean that most time-dependent inhibitors are seen with lupus anticoagulants.

Accelerated aPTT

The majority of abnormal aPTT results seen in the laboratory are prolonged; however, there are unusual situations in which the aPTT may be accelerated or shorter than the reference range. When the patient's blood (specimen) is partially activated, and some activated factors are present and the process of activation is interrupted by addition of citrate, the subsequent aPTT can be shorter than expected because of the partial activation. It has a "head start" of a sort. There are three situations in which this may occur:

Spurious. If the phlebotomist has difficulty with the collection of the specimen, such as overfilling a tube or with a difficult venipuncture or unusual trauma, some activation of coagulation may occur. Often the result will be a clotted specimen, but on occasion, the activation will not be complete and the resulting coagulation studies will be accelerated.

Treatment with activated clotting factors. Treatment with activated clotting factors can shorten coagulation times in the laboratory. Most patients receiving these preparations have a long aPTT prior to treatment, but on occasion, when they do not, the treatment may cause both the PT and aPTT to be accelerated.

Consumptive coagulopathy. Early in the development of disseminated intravascular coagulation, before coagulation factors are consumed, activated coagulation factors can circulate in the blood. The activation process can then be stopped or significantly slowed by the collection into citrate. When the process is reactivated by the addition of calcium to the plasma, the activation has a "head start" and the resulting clotting time can be accelerated.

Summary

The brief case study (clinical history) presented at the beginning of the chapter provides an introduction to the laboratory evaluation of the prolonged aPTT. The particular clinical situation depicted is relatively common. Despite efforts to flush central lines to free them of heparin, heparin contamination of coagulation specimens drawn from central lines continues to be a difficulty.

Although simplified, the two algorithms provide an approach to thinking about the evaluation of prolonged aPTT. Regardless of the approach that is used, it is important to determine the etiology of the prolonged aPTT in order to know the potential impact on the patient and the intervention required of the clinician. As one performs the evaluation, it is important to continually keep in mind:

- ❏ that in addition to unfractionated heparin and low-molecular-weight heparin, other medications, such as direct thrombin inhibitors, will prolong the aPTT;
- ❏ the nuances of the mixing study and its importance in the interpretation of a prolonged aPTT;
- ❏ the difficulties of assigning the phospholipid dependence of inhibitors; and
- ❏ the critical importance of interpretation of the laboratory result in view of the patient's clinical history and findings.

Finally, although rare, there are clinical conditions of which the laboratory and clinician should be aware that can arise that will accelerate the aPTT.

Suggested Reading

General References and Review

Olson JD, Addressing clinical etiologies of a prolonged aPTT. *CAP Today.* 1999 Sep;13(9):28,30,32.

Clinical Etiologies
Medications

Olson JD, Arkin CF, Brandt JT, et al. Laboratory monitoring of unfractionated heparin therapy. *Arch Path Lab Med.* 1998;122(9):782-798.

Coagulation Factor Deficiency
Associated with Significant Hemorrhage

Chee YL, Greaves M. Role of coagulation testing in predicting bleeding risk. *Hematol J.* 2003;4(6):373-378.

Collins P, McCartney N, Davies R, et al. A population based, unselected consecutive cohort of patients with acquired hemophilia A. *Br J Haemotol.* 2004;124:86.

Graw J, Brackmann HH, Oldenburg J, Schneppenheim R, Spannagl M, Schwaab R. Haemophilia A: from mutation analysis to new therapies. *Nat Rev Genet.* 2005;6:488-501.

Kessler CM, Mariani G. Clinical manifestations and therapy of the hemophilias. In: Coleman RW, et al, eds. *Hemostasis and Thrombosis: Basic Principles and Clinical Practice.* 5th ed. Philadelphia, Pa: Lippincott, Williams & Wilkins; 2006: 887.

Lillicrap DP, White BN, Holden JJ, et al. Carrier detection in the hemophiliac. *Am J Hematol.* 1987;26:285-296.

Verbruggen B, Meijer P, Novakova I, Van Heerde W. Diagnosis of factor VIII deficiency. *Haemophilia.* 2008;14(suppl 3):76-82.

Coagulation Factor Deficiencies
Not Associated with Bleeding

Fuhrer G, Gallimore MJ, Heller W, Hoffmeister HE. FXII. *Blut.* 1990;61:258-266.

Girolami A, De Marco L, Dal Bo Zanon R, Patrassi G, Cappellato MG. Rarer quantitative and qualitative abnormalities of coagulation. *Clin Haematol.* 1985;14:385-411.

Girolami A, Randi ML, Gavasso S, Lombardi AM, Spiezia F. The occasional venous thromboses seen in patients with severe (homozygous) FXII deficiency are probably due to associated risk factors: a study of prevalence in 21 patients and review of the literature. *J Thromb Thrombolysis.* 2004;17:139-143.

Kitchens CS. The contact system. *Arch Pathol Lab Med.* 2002;126(11):1382-1386.

Lupus Anticoagulant

Brandt JT, Barna LK, Triplett DA. Laboratory identification of lupus anticoagulants: results of the Second International Workshop for Identification of Lupus Anticoagulants. On behalf of the Subcommittee on Lupus Anticoagulants/Antiphospholipid Antibodies of the ISTH. *Thromb Haemost.* 1995;74:1597-1603.

Clyne LP, White PF. Time dependency of lupuslike anticoagulants. *Arch Intern Med.* 1988;148:1060-1063.

Dlott JS, Triplett DA. Diagnosing antiphospholipid antibody syndrome. *CAP Today.* 1999 Mar;13(3):84-88.

Specific Coagulation Factor Inhibitors

Astermark J. Overview of inhibitors. *Semin Hematol.* 2006;43(2 suppl 4):S3-7.

Hay CR, Brown S, Collins PW, Keeling DM, Liesner R. The diagnosis and management of factor VIII and IX inhibitors: a guideline from the United Kingdom Haemophilia Centre Doctors Organisation. *Br J Haematol.* 2006;133:591-605.

Kessler CM, Mariani G. Acquired disorders of coagulation: the immune coagulopathies. In: Coleman RW, et al, eds. *Hemostasis and Thrombosis: Basic Principles and Clinical Practice.* 5th ed. Philadelphia, Pa: Lippincott, Williams & Wilkins; 2006: 1061.

Sahud MA. Factor VIII inhibitors: laboratory diagnosis of inhibitors. *Semin Thromb Hemost.* 2000;26(2):195-203.

Spurious Causes of an Elevated aPTT

Koepke JA, Rodgers JL, Ollivier MJ. Pre-instrumental variables in coagulation testing. *Am J Clin Pathol.* 1975;64:591-596.

Laboratory Testing Algorithm

Kaczor DA, Bickford NN, Triplett DA. Evaluation of different mixing study reagents and dilution effect in lupus anticoagulant testing. *Am J Clin Pathol.* 1991;95:408-411.

Wight J, Paisley S. The epidemiology of inhibitors in hemophilia A: a systematic review. *Haemophilia.* 2003;4:418.

CHAPTER 11

Prolonged Prothrombin Time

Mark T. Cunningham, MD
Chad R. Rund, DO

Introduction

A prolonged prothrombin time (PT) with or without a prolonged activated partial thromboplastin time (aPTT) is a common coagulation test abnormality. There are several potential etiologies. A prolonged PT with a normal aPTT is seen in factor VII deficiency, factor VII inhibitor, warfarin therapy, mild forms of liver disease, vitamin K deficiency, and disseminated intravascular coagulation (DIC). A prolonged PT with a prolonged aPTT is seen in deficiencies or inhibitors of single common pathway factors (fibrinogen, factors II, V, and X), combined factor V and factor VIII deficiency, hereditary deficiency of vitamin K-dependent coagulation factors, warfarin anticoagulation, superwarfarin poisoning, heparin therapy, direct thrombin inhibitor therapy, dilutional coagulopathy, lupus anticoagulant, dysfibrinogenemia, and severe forms of liver disease, vitamin K deficiency, and DIC.

This chapter will discuss the entities in the above differential diagnosis and will present an algorithm for sequential test selection that can be used to make the correct diagnosis.

Differential Diagnosis of a Corrected Mixing Study

The differential diagnosis of a prolonged PT with a normal aPTT is summarized in Table 11-1. The differential diagnosis of a prolonged PT and prolonged aPTT is summarized in Table 11-2.

The 1:1 mixing study of the PT and aPTT is the first step in the evaluation of a prolonged PT with or without a prolonged aPTT. The purpose is to distinguish between a factor deficiency and a factor inhibitor. Full correction of the mixing study

Table 11-2. Differential Diagnosis of a Prolonged PT and Prolonged aPTT

Corrected 1:1 Mixing Studies (PT, aPTT)
Dilutional coagulopathy
Disseminated intravascular coagulation
Drugs
Warfarin
Other vitamin K antagonists
Factor deficiency
Factor II deficiency
Factor V deficiency
Factor V and factor VIII deficiency, combined
Factor X deficiency
Fibrinogen deficiency and dysfibrinogenemia
Vitamin K-dependent coagulation factor deficiency, hereditary
Liver disease (severe)
Toxins
Superwarfarin poisoning
Vitamin K deficiency (severe)

Noncorrected 1:1 Mixing Studies (PT, aPTT)
Drugs
Direct thrombin inhibitors (argatroban, bivalirudin, lepirudin)
Heparin
Factor inhibitor
Factor II inhibitor
Factor V inhibitor
Factor X inhibitor
Lupus anticoagulant

Table 11-1. Differential Diagnosis of a Prolonged PT and Normal aPTT

Corrected 1:1 Mixing Study
Disseminated intravascular coagulation
Drugs
Warfarin
Other vitamin K antagonists
Factor deficiency
Factor VII deficiency
Mild to moderate decrease in factor II, V, X, or fibrinogen
Liver disease
Vitamin K deficiency

Noncorrected 1:1 Mixing Study
Factor inhibitor
Factor VII inhibitor

Case Studies

Prolonged PT with Normal aPTT

A 55-year-old man with a history of alcohol abuse and malnutrition presented with altered mental status and rectal bleeding. Laboratory testing showed the following:

Test	Value	Reference Range
Hemoglobin (g/dL)	11.0	14-18
WBC (per μL)	9000	4000-10,000
Platelet count (per μL)	210,000	150,000-450,000
Aspartate aminotransferase (U/L)	52	3-70
Alanine aminotransferase (U/L)	35	3-78
Gamma glutamyltransferase (U/L)	260	5-75
PT (s)	18	11-13
aPTT (s)	30	27-33
PT 1:1 mix (s)		
Immediate, 0 min	11	11-13
Incubated, 60 min	13	11-13
Factor V activity (%)	85	50-150
Factor VII activity (%)	20	50-150

The patient was given 4 mg of vitamin K subcutaneously. Four hours later, the PT was repeated and showed full correction into the reference range. What is the most likely cause for the prolonged PT?

Answer: Vitamin K deficiency. The diagnosis is made by demonstrating a prolonged PT with or without a prolonged aPTT, full correction of the 1:1 mixing study for the PT, decreased vitamin K-dependent factor VII, normal vitamin K-independent factor V, and correction of the PT after administration of vitamin K. Additional laboratory testing would most likely show decreased levels of other vitamin K-dependent factors as well (II, IX, and X).

Prolonged PT with Prolonged aPTT

A 64-year-old woman developed hematuria 3 days after a coronary artery bypass graft procedure as well as persistent bleeding from surgical wounds. The surgeon used bovine thrombin (fibrin glue) to maintain hemostasis during the operation. Laboratory testing showed the following:

Test	Value	Reference Range
Hemoglobin (g/dL)	13.0	12-16
WBC (per μL)	8000	4000-10,000
Platelet count (per μL)	330,000	150,000-450,000
PT (s)	20	11-13
aPTT (s)	48	27-33
Thrombin time (s)	19	14-21
PT 1:1 mix (s)		
Immediate, 0 min	17	11-13
Incubated, 60 min	19	11-13
aPTT 1:1 mix (s)		
Immediate, 0 min	42	27-33
Incubated, 60 min	46	27-33
Factor II activity (%)	90	50-150
Factor V activity (%)	<1	50-150
Factor VII activity (%)	87	50-150
Factor X activity (%)	105	50-150

What is the most likely cause for the prolonged PT and aPTT?

Answer: Factor V inhibitor. The use of bovine thrombin during surgery is a major risk factor for the development of a factor V inhibitor, as most bovine thrombin preparations also contain bovine factor V. The diagnosis is made by demonstrating a prolonged PT, prolonged aPTT, noncorrection of the 1:1 mixing studies for the PT and aPTT, and isolated deficiency of factor V. A Bethesda assay showed a factor V inhibitor titer of 30 Bethesda units.

indicates a single factor deficiency or multiple factor deficiencies. Noncorrection of the mixing study indicates a factor inhibitor. Occasionally, depending on the reagent, the presence of multiple factor deficiencies can lead to noncorrection of the PT mixing study. The performance of the mixing study is covered in more detail in chapter 6.

Dilutional Coagulopathy

Dilutional coagulopathy occurs in bleeding patients following massive transfusion with packed red blood cells and crystalloid (eg, normal saline). This usually occurs in the setting of trauma or surgery and manifests as continued or worsening bleeding despite transfusion. Packed red blood cells and crystalloid are devoid of coagulation factors; therefore, patients develop greater depletion of coagulation factors as they receive greater volumes of transfused product.

The diagnosis is made in the correct clinical setting by demonstrating a prolonged PT, prolonged aPTT, full correction of the 1:1 mixing study, and normal or only mildly increased D-dimer or fibrin degradation product (FDP) levels. Plasma fibrinogen may also be decreased, and levels below 100 mg/dL are sufficient to cause bleeding and prolong the PT and aPTT. Dilutional coagulopathy is typically treated with fresh frozen plasma (FFP) to normalize the PT and aPTT, and cryoprecipitate to increase the fibrinogen to 100 mg/dL or above.

Disseminated Intravascular Coagulation

Disseminated intravascular coagulation can manifest as a prolonged PT or a prolonged PT and aPTT. The mechanism of prolongation involves increased utilization (consumption) of multiple coagulation factors that exceeds the synthetic capacity of the liver. The D-dimer test is a screening test with a high sensitivity for DIC. Plasma or serum FDPs can also be tested but are less sensitive than D-dimer. DIC is covered in more detail in chapter 13.

Drugs

Warfarin

Warfarin and other vitamin K antagonists cause prolongation of the PT with a normal aPTT at therapeutic doses (international normalized ratio [INR] = 2.0 to 3.0). Prolongation of the aPTT in addition to the PT can sometimes occur when the INR is greater than 1.5, but the elevation of the aPTT may vary between aPTT reagents. Laboratory monitoring of warfarin therapy is covered in more detail in chapter 23.

The diagnosis of warfarin as the cause of a prolonged PT with or without a prolonged aPTT can usually be made by verifying, through medical record review or discussion with the patient's physician, that the patient is receiving warfarin therapy. Unintended warfarin ingestion can be diagnosed by demonstrating full correction of the 1:1 mixing study of the PT and aPTT, decreased levels of vitamin K-dependent factors (II, VII, IX, and X), normal levels of vitamin K independent factors (V, VIII), and increased plasma levels of warfarin as determined by high performance liquid chromatography (HPLC).

Warfarin over-anticoagulation is treated according to guidelines established in 2008 by the American College of Chest Physicians (Table 11-3). Treatment modalities include warfarin dose adjustment, vitamin K (oral or intravenous), FFP, prothrombin complex concentrate, and recombinant factor VIIa. The specific modalities used are based on the magnitude of the INR and the severity of bleeding.

Herbal medications may be able to interact with warfarin to increase the risk for bleeding. Case studies have reported bleeding complications in association with danshen, devil's claw, garlic, ginkgo biloba, ginseng, herbal tea, and guilinggao.

The mechanism could theoretically be through anticoagulant effects (with increased PT/INR) or through antiplatelet effects (without increasing the PT/INR). There are no reports of herbal medications by themselves (in the absence of warfarin) causing prolongation of the PT or aPTT. Although randomized studies have yet to show a definitive link between herbal medications and bleeding risk while on warfarin, physicians and laboratorians should be aware of the potential because of warfarin's narrow therapeutic index.

Factor Deficiency

Factor II Deficiency

Factor II (prothrombin) deficiency is usually inherited as an autosomal recessive bleeding disorder, although it can be acquired. Inherited deficiency is very rare (prevalence 1 in 200,000) and is caused by mutations in the factor II gene on chromosome 11 that result in decreased synthesis or impaired function of the protein product.

Acquired isolated factor II deficiency is associated with lupus anticoagulant and anecdotally associated with lymphoma and *Mycoplasma pneumoniae* infection. The mechanism of lupus anticoagulant-associated factor II deficiency involves non-neutralizing autoantibodies against factor II that cause increased clearance of the protein and subsequent factor II deficiency. This topic is discussed in more detail in chapter 22.

The diagnosis of factor II deficiency is made by demonstrating a prolonged PT, prolonged aPTT, normal thrombin time, full correction of the 1:1 mixing study for the PT and aPTT, and an isolated decrease in factor II activity levels. Testing for prothrombin antigen levels is not widely available and usually is not necessary to establish a diagnosis. An inherited etiology is supported by the presence of a childhood history of bleeding, positive family history, and, if necessary, mutation analysis of the factor II gene. Heterozygous individuals usually have factor II activity levels near 50%. Homozygotes or compound heterozygotes typically have factor II levels less than 30%, with bleeding complications more likely when the factor II level is less than 10%.

Factor II deficiency typically is treated with FFP for bleeding episodes and surgery. The use of prothrombin complex concentrates, provided the product contains sufficient prothrombin, is an alternative treatment. The minimum factor II activity level recommended for surgery is 40% to 50%. Factor II has a biological half-life of approxi-

Table 11-3. Management of Warfarin When the International Normalized Ratio (INR) is Nontherapeutic; Based on 2008 American College of Chest Physicians Guidelines

INR	Bleeding	Treatment
Above therapeutic, but <5.0	Not significant	(1) Lower or omit warfarin dose (2) Monitor INR more frequently (3) Resume warfarin at lower dose when INR is therapeutic (4) If only minimally above therapeutic range, no dose reductions may be required
≥5.0 to <9.0	Not significant	(1) Omit one or two warfarin doses (2) Monitor INR more frequently (3) Resume warfarin at lower dose when INR is therapeutic (4) If patient has increased risk of bleeding, then omit a warfarin dose and give vitamin K1 (1 to 2.5 mg orally) (5) If patient needs urgent surgery, then omit a warfarin dose and give vitamin K1 (≤5 mg orally) to reduce INR in 24 hours. If still high, then give additional vitamin K1 (1 to 2 mg orally)
≥9.0	Not significant	(1) Hold warfarin (2) Give vitamin K1 (2.5 to 5 mg orally) to decrease INR over 24 to 48 hours (3) Monitor INR more frequently and use additional vitamin K if necessary (4) Resume warfarin at lower dose when INR is therapeutic
Elevated	Serious	(1) Hold warfarin (2) Give vitamin K1 (10 mg by slow intravenous infusion) (3) Supplement with fresh frozen plasma, prothrombin complex concentrate, or recombinant factor VIIa, based on urgency (4) Repeat vitamin K1 dose every 12 hours as needed
Elevated	Life threatening	(1) Hold warfarin (2) Give prothrombin complex concentrate, fresh frozen plasma, or recombinant factor VIIa (3) Supplement with vitamin K1 (10 mg by slow intravenous infusion) (4) Repeat (2) and (3) as needed

Source: Ansell J, Hirsh J, Hylek E, et al. Pharmacology and Management of the Vitamin K Antagonists: American College of Chest Physicians Evidence-Based Clinical Practice Guidelines (8th Edition). Chest. 2008;133:160-198.

mately 73 hours; therefore, the factor II activity level should be monitored once every other day (trough level).

Factor V Deficiency

Factor V deficiency is an autosomal recessive bleeding disorder and is very rare (prevalence unknown; studies limited to case reports and case series). It is caused by mutations in the factor V gene on chromosome 1 that result in decreased synthesis or impaired function of the protein product. Severe factor V deficiency also results in a decrease in platelet alpha granule-associated factor V.

The diagnosis of factor V deficiency is made by demonstrating a prolonged PT, prolonged aPTT, normal thrombin time, full correction of the 1:1 mixing study for the PT and aPTT, and an isolated decrease in factor V activity levels. Factor V levels are typically in the range of 0% to 17%.

Other causes of low factor V, including combined inherited factor V and factor VIII deficiency, DIC, and liver disease, should be excluded.

Factor V deficiency is treated with FFP for bleeding episodes and surgery. The minimum factor V activity level recommended for surgery is 10% to 30%. Factor V has a biological half-life of approximately 12 to 36 hours; therefore, the factor

V activity level should be monitored once every 12 hours (trough level). Recombinant factor VIIa has also been successfully used. The use of platelet transfusions may also be helpful as a source of factor V replacement.

Factor V and Factor VIII Deficiency, Combined

Combined factor V and factor VIII deficiency is an autosomal recessive bleeding disorder caused by mutations in either the LMAN1 (lectin, mannose-binding 1) gene on chromosome 18 or the MCFD2 (multiple combined factor deficiency 2) gene on chromosome 2. Each gene encodes a distinct protein that localizes to the endoplasmic reticulum-Golgi intermediate complex (ERGIC-53, now called LMAN1) and functions in the intracellular trafficking of factor V and factor VIII. Mutations result in defective secretion of factors V and VIII and subsequent deficiency of these proteins in the blood. The disorder is particularly prevalent in Middle Eastern and Sephardic Jews, with a prevalence of 1 in 100,000.

The diagnosis is made by demonstrating a prolonged PT, prolonged aPTT, full correction of the 1:1 mixing study for the PT and aPTT, decreased factor V and factor VIII activity levels, and normal levels of other coagulation factors. The factor V and factor VIII levels typically range from 5% to 30%. Genotyping for this disorder is available in some specialized laboratories.

Combined factor V and factor VIII deficiency is treated with FFP for bleeding episodes and surgery. The minimum factor V and factor VIII activity levels recommended for surgery are 10% to 30%, and 30% to 100%, respectively. The factor V and factor VIII activity level should be monitored once every 12 hours (trough level) based on biological half-lives of 12 to 36 hours and 13 hours, respectively.

Factor VII Deficiency

Factor VII deficiency is an autosomal recessive bleeding disorder and is very rare (prevalence approximately 1 in 500,000; studies limited to case reports and case series). It is caused by mutations in the factor VII gene on chromosome 13 that result in decreased synthesis or impaired function of the protein product.

The diagnosis of factor VII deficiency is made by demonstrating a prolonged PT, normal aPTT, normal thrombin time, full correction of the 1:1 mixing study for the PT, and decreased factor VII activity levels. The level of other coagulation factors is normal. Factor VII levels are typically in the range of 0% to 20%.

Other causes of low factor VII, including DIC, liver disease, vitamin K deficiency, and warfarin therapy, should excluded.

Factor VII deficiency is treated with FFP for bleeding episodes and surgery. The minimum factor VII activity level recommended for surgery is 10% to 20%. Factor VII has a biological half-life of approximately 5 hours; therefore, the factor VII activity level should be monitored once every 4 to 6 hours (trough level). Prothrombin complex concentrate and recombinant factor VIIa have also been successfully used.

Factor X Deficiency

Factor X deficiency is inherited as an autosomal recessive bleeding disorder, and it can also be acquired. Inherited deficiency is very rare (prevalence 1 in 200,000) and is caused by mutations in the factor X gene on chromosome 13 that result in decreased synthesis or impaired function of the protein product.

Acquired isolated factor X deficiency is associated with AL (light chain) amyloidosis and anecdotally associated with malignancy (stomach carcinoma, acute myeloid leukemia) and valproic acid therapy. The mechanism of amyloidosis-associated factor X deficiency is a combination of factor X sequestration by amyloid fibrils and factor X dysfunction due to abnormal protein glycosylation.

The diagnosis of factor X deficiency is made by demonstrating a prolonged PT, prolonged aPTT, full correction of the 1:1 mixing study for the PT and aPTT, and decreased factor X activity levels. Factor X levels are typically in the range of 0% to 26%. The Russell viper venom time will be prolonged, as the venom activates factor X, but the thrombin time will be normal. An inherited etiology is supported by the presence of a childhood history of bleeding, positive family history, and, if necessary, mutation analysis of the factor X gene. Bleeding symptoms are typically more severe with a factor X level less than 15%.

Other causes of low factor X, including DIC, liver disease, vitamin K deficiency, and warfarin therapy, should be excluded.

Factor X deficiency is treated with FFP for bleeding episodes and surgery. The minimum factor X activity level recommended for surgery is 10% to 40%. Factor X has a biological half-life of approximately 33 hours; therefore, the factor X activity level should be monitored once every day (trough level). Prothrombin complex concentrate and recombinant factor VIIa have also been successfully used. Treatment of amyloid-associated acquired factor X deficiency is problematic because of the shortened half-life of factor X. Fresh frozen plasma is not usually helpful in these patients, but plasma exchange, splenectomy, factor VIIa and prothrombin complex concentrates are potential alternatives.

Fibrinogen Deficiency

Fibrinogen deficiency is a group of inherited disorders caused by mutations in the fibrinogen Aα (FGA), Bβ (FGB), or γ (FGG) genes on chromosome 4. There are three subtypes of fibrinogen deficiency: afibrinogenemia, hypofibrinogenemia, and dysfibrinogenemia.

Afibrinogenemia and hypofibrinogenemia. Afibrinogenemia is characterized by absent to trace fibrinogen antigen levels in plasma. Hypofibrinogenemia is characterized by decreased but more than trace fibrinogen antigen levels in plasma. These subtypes are autosomal recessive, have a prevalence of 1 in 1,000,000, and cause bleeding.

The diagnosis of afibrinogenemia/hypofibrinogenemia is made by demonstrating a prolonged PT, prolonged aPTT, full correction of the 1:1 mixing study of the PT and aPTT, normal D-dimer, decreased fibrinogen activity and antigen levels, and a normal fibrinogen activity to fibrinogen antigen ratio. The thrombin time is also prolonged in proportion to the decrease in fibrinogen level, but the thrombin time does not correct after heparin neutralization. Fibrinogen levels typically range from 0.5 to 1.5 mg/dL in afibrinogenemia, and 20 to 80 mg/dL in hypofibrinogenemia.

Other causes of low fibrinogen, including DIC, drug effect (L-asparaginase, valproic acid), and liver disease, should be excluded.

Afibrinogenemia and hypofibrinogenemia are treated with cryoprecipitate for bleeding episodes and surgery. The minimum fibrinogen activity level recommended for surgery is 100 mg/dL. Fibrinogen has a biological half-life of approximately 99 hours; therefore, the fibrinogen activity level should be monitored once every 3 to 4 days (trough level).

Dysfibrinogenemia. Dysfibrinogenemia is characterized by a low fibrinogen activity to fibrinogen antigen ratio in plasma. This disorder is usually autosomal dominant, causes bleeding and/or thrombosis, and has a prevalence of 0.8% in patients with a history of venous thrombosis. Screening coagulation tests sometimes show a prolonged PT with or without a prolonged aPTT. The diagnostic approach for dysfibrinogenemia is discussed in detail in chapter 15.

Dysfibrinogenemia associated with bleeding has been successfully treated with cryoprecipitate. Thrombosis has been successfully treated with warfarin, the combination of cryoprecipitate and argatroban, and fondaparinux. In one case, fondaparinux was effective after failure of low-molecular-weight heparin (enoxaparin). Laboratory monitoring may not be required; however, one may consider using the thrombin time or fibrinogen activity to antigen ratio to monitor cryoprecipitate therapy.

Vitamin K-Dependent Coagulation Factor Deficiency, Hereditary

Hereditary vitamin K-dependent coagulation factor deficiency is an autosomal recessive bleeding disorder, very rare (prevalence unknown; studies limited to case reports), and characterized by decreased levels of the procoagulant factors II, VII, IX, and X, and anticoagulant proteins C, S, and Z. It is caused by mutations in the genes for either the γ-carboxylase or the vitamin K epoxide reductase (VKORC1). The combined action of these two enzymes is necessary for the synthesis of gamma-carboxyglutamic acid (Gla) residues on all vitamin K-dependent proteins.

The disease presents in the newborn period and childhood. Clinical manifestations are variable with respect to bleeding location/severity, responsiveness to vitamin K replacement, and the presence of skeletal abnormalities.

The diagnosis is made by demonstrating a prolonged PT, prolonged aPTT, full correction of the 1:1 mixing study for the PT and aPTT, decreased activity levels of all vitamin K-dependent factors (II, VII, IX, X, protein C, protein S), normal activity levels of factor V and VIII, and identification of a mutation within the γ-carboxylase gene or the VKORC1 gene. Gene analysis will require the expertise of a highly specialized laboratory. Other causes of vitamin K-dependent factor deficiency, including vitamin K deficiency, warfarin therapy, and superwarfarin poisoning, should be excluded.

Treatment approaches include vitamin K replacement and FFP. The therapeutic goal is to normalize the PT and aPTT. For some patients, vitamin K replacement is sufficient to normalize these tests.

Liver Disease

The coagulopathy of liver disease can manifest as a prolonged PT in mild disease, and a prolonged PT and aPTT in severe disease. The mechanism involves impaired hepatocellular function, leading to decreased synthesis and secretion of all coagulation factors except factor VIII (which is synthesized by endothelial cells). The diagnosis is made by demonstrating decreased synthetic liver function tests (total protein and albumin), elevated hepatocellular liver function tests (ie, aspartate aminotransferase, alanine aminotransferase), prolonged PT with or without a prolonged aPTT, correction of the 1:1 mixing study for the PT and aPTT, decreased levels of vitamin K-dependent factors (II, VII, IX, X), decreased vitamin K-independent factor V, and normal or elevated factor VIII. Patients with severe liver disease (eg, end-stage cirrhosis) may not have elevated hepatocellular liver function tests, so a liver biopsy might be needed to document liver disease.

Factor deficiency due to liver disease is treated with vitamin K and FFP. The goal is to normalize the PT and aPTT. The underlying liver disease should also be treated to prevent further hepatocellular damage and impaired coagulation factor synthesis.

Superwarfarin Poisoning

Superwarfarins are chemical derivatives of warfarin designed to be fat-soluble long-acting anticoagulants with 100 times greater potency. They are most commonly used as rodenticides and are not intended for human consumption. Examples of superwarfarin rodenticides include the 4-hydroxycoumarin derivatives (brodifacoum, bromadialone, and difenacoum) and the indanedione derivatives (diphacinone, chlorphacinone, and pindone). The most common cause of poisoning is brodifacoum, which is used in the commercial product D-Con®.

Superwarfarin poisoning is associated with accidental ingestion (particularly in children), suicide attempt, industrial exposure, Munchausen syndrome, or drug abuse. Most cases (90%) occur in children. Clinical manifestations of bleeding are numerous. Fatal bleeding appears to be very rare (about 1 in 10,000 cases).

The diagnosis is made by demonstrating a prolonged PT, prolonged aPTT, full correction of the 1:1 mixing study for the PT and aPTT, markedly decreased levels of the vitamin K-dependent coagulation factors (II, VII, IX, X), normal levels of vitamin K-independent factors (V, VIII), and elevated blood levels of superwarfarin as measured by high pressure liquid chromatography (HPLC). It is typical to observe PT values >100 seconds, with INR >9, and factor levels <5%, with rapid rebound of the abnormalities after short-term therapy. The turnaround time of superwarfarin blood levels can be several days to weeks; therefore, a presumptive diagnosis can be made based on clinical suspicion and the results of the PT, aPTT, mixing studies, and factor levels.

Superwarfarin poisoning should be suspected when the above test abnormalities occur in the absence of warfarin therapy, or when test abnormalities persist or recur despite treatment with standard doses of vitamin K or FFP.

Treatment includes vitamin K and FFP, with the goal of normalizing the PT and aPTT. Prolonged vitamin K replacement of several weeks often is required due to the long elimination half-life of superwarfarins.

Vitamin K Deficiency

Newborns

Bleeding in the newborn due to vitamin K deficiency (ie, hemorrhagic disease of the newborn) has an incidence of 0.01% to 1.5% in the absence of vitamin K prophylaxis. Newborns are sensitive to developing vitamin K deficiency because they lack intestinal microbial flora that synthesize vitamin K. Anatomic locations of bleeding include cranial subperiosteum (cephalohematoma), umbilicus, intracranial, intrathoracic, gastrointestinal tract, urogenital tract, circumcision, and needlestick sites.

The disorder is classified into three types based on age of onset: early (< 24 hours), classical (2 to 7 days), and late (1 week to 6 months). The early form (very rare) is precipitated by the maternal use of certain medications, such as anticonvulsants (carbamazepine, phenytoin, barbiturates), antibiotics (cephalosporins, rifampin, isoniazid), and vitamin K antagonists (warfarin). Bleeding can be prevented by discontinuing or replacing the offending medication prior to delivery.

The classical form (most common) is caused by breast feeding (breast milk has inadequate vitamin K) or inadequate formula milk intake. Bleeding can be prevented by a single prophylactic dose of vitamin K after birth or by adequate feeding with formula milk.

The late form (very rare) is caused by inadequate vitamin K intake (ie, breast feeding or inadequate formula milk intake) combined with undiagnosed hepatobiliary disease. Bleeding can be prevented by administering 3 doses of prophylactic vitamin K.

Prophylactic vitamin K is administered in 3 oral doses of 2 mg each. The first dose is given within 24 hours of birth after 1 or 2 feedings (1 mL syrup diluted with 9 mL water). The second dose is given at 1 week of age. The third dose is given at 1 month of age but can be omitted if the PT is normal.

Adults

Adults may develop vitamin K deficiency due to decreased dietary intake of vitamin K (malnutrition), drugs that inhibit vitamin K reductases (antibiotics containing an N-methyl-thiotetrazole side chain), impaired bile salt synthesis/secretion (liver disease, biliary tract disease), and impaired intestinal absorption of vitamin K/bile salt complexes (celiac disease, Crohn's disease, etc). Bleeding manifestations are highly variable.

Vitamin K deficiency can manifest as a prolonged PT in mild disease, and a prolonged PT and aPTT in severe disease. The mechanism of the coagulopathy is based on the special role of vitamin K in protein synthesis. Vitamin K is a cofactor for the hepatocellular carboxylase enzyme that catalyzes posttranslational carboxylation of glutamic acid residues to gamma-carboxyglutamic acid (Gla residues) on the N-terminal regions of vitamin K-dependent coagulation proteins (factors II, VII, IX, X, protein C, and protein S). Gla residues are required for calcium-dependent binding of these proteins to phospholipid surfaces, which is essential to their function in vivo as well as in vitro for the PT and aPTT.

The diagnosis is made by demonstrating a prolonged PT with or without a prolonged aPTT, full correction of the 1:1 mixing study for the PT and aPTT, decreased levels of vitamin K-dependent factors (II, VII, IX, and X), normal levels of vitamin K-independent factors (V, VIII), and normalization of the PT and aPTT within 2 to 6 hours after subcutaneous administration of 1 to 5 mg of vitamin K. Complex cases may require demonstration of increased plasma levels of PIVKA (proteins induced by vitamin K absence), such as noncarboxylated factor II (PIVKA-II), or decreased serum levels of vitamin K.

Differential Diagnosis of a Noncorrected Mixing Study

Drugs

Direct Thrombin Inhibitors

Direct thrombin inhibitors (DTIs) are typically used to treat patients with heparin-induced thrombocytopenia with thrombosis. Examples of this class of anticoagulant include argatroban, bivalirudin, and lepirudin. They are discussed in detail in chapter 24. DTIs cause prolongation of the PT and aPTT at therapeutic doses, with greater effects on the aPTT. Their mechanism of action involves binding to the active site of thrombin and inhibition of proteolytic activity toward fibrinogen. The PT and aPTT are prolonged because these tests are dependent on the conversion of fibrinogen to fibrin as the endpoint of the clotting time.

The diagnosis of DTI effect on the PT/aPTT can usually be made by clinical correlation through medical record review or discussion with the patient's physician. The 1:1 mixing study of the PT and aPTT will show noncorrection. All clot-based factor activity assays will show increasing activity upon increasing specimen dilution (nonlinearity), consistent with dilution of an inhibitor. The thrombin time is usually markedly prolonged during DTI therapy.

There are few reports describing the treatment of bleeding due to DTI overdose. No antidotes for DTIs are available. One study reported successful treatment of argatroban overdose with FFP.

Heparin

Unfractionated heparin (UFH), low-molecular-weight heparin (LMWH), and the active site pentasaccharide of heparin (fondaparinux) can prolong the PT and aPTT. This is especially prevalent in hospitalized patients who have blood obtained from venous access sites that are maintained with heparin, even when the venous access is saline flushed prior to phlebotomy.

For each of these heparins, the degree of prolongation is greater for the aPTT than the PT. Unfractionated heparin has a greater effect on the aPTT than LMWH, and LMWH has a greater effect than fondaparinux. The use of heparin-neutralizing agents in commercial thromboplastin reagents has helped to minimize the prolongation of the PT.

The mechanism by which heparins prolong the aPTT and PT is by forming a binary complex with antithrombin, producing a conformational change in antithrombin that gives it potent inhibitory activity against multiple serine proteases of the coagulation cascade, including factor VIIa. The most potent inhibitory activity is against factor Xa in the case of LMWH and fondaparinux.

The thrombin time is a very sensitive screening test for UFH contamination, showing significant prolongation at therapeutic doses. UFH can be confirmed by adding a heparin-neutralizing agent (heparinase, polybrene, etc) to the patient plasma to show normalization of the PT or with a normal reptilase time. When heparin is present, the aPTT and thrombin time prolongations can be normalized by heparin neutralization, but PT prolongations that are due to heparin often do not normalize. This might be due to the increase in tissue factor pathway inhibitor (TFPI) that occurs during heparin therapy.

Protamine sulfate is the antidote for bleeding due to UFH therapy. The dose is 1 mg of protamine sulfate to neutralize 100 units of circulating heparin. The quantity of circulating heparin can be estimated from the total amount of heparin administered, taking into account its elimination half-life of 1 hour. The aPTT should be checked immediately after the protamine dose.

Recombinant factor VIIa has been found to be effective as a treatment for bleeding due to LMWH or fondaparinux.

Factor Inhibitors

Factor II Inhibitor

Factor II inhibitors can be divided into three classes based on their effect on the PT, aPTT, and 1:1 mixing study for the PT and aPTT. The first class, which is the most common, produces a prolonged PT, prolonged aPTT, and corrected 1:1 mixing studies for the PT and aPTT. These inhibitors are non-neutralizing antibodies that increase the clearance rate of prothrombin and are usually associated with lupus anticoagulant secondary to systemic lupus erythematosus. Factor II activity levels are decreased. This topic is discussed in more detail in chapter 22.

The second class is very rare, with three cases reported in the literature. They produce a prolonged PT, prolonged aPTT, and noncorrected 1:1 mixing study for the PT and aPTT. These inhibitors are antibodies that neutralize either prothrombin's procoagulant function (associated with bleeding) or thrombin's secondary binding site

function (associated with recurrent arterial thrombosis).

The third class of factor II inhibitor produces a normal PT and normal aPTT. These inhibitors are directed against thrombin (but not prothrombin). They are often associated with the use of bovine thrombin or fibrin glue and are typically antibodies against bovine thrombin that have little cross-reactivity with human thrombin. These are usually neutralizing antibodies that inhibit the catalytic function of bovine thrombin and produce a prolonged thrombin time when bovine thrombin (but not human thrombin) is used as the reagent. Sometimes the antibody does cross-react with human thrombin and prolongs the thrombin time, even when human thrombin is used, and can also prolong the PT and aPTT. This topic is also discussed in chapter 15.

There are very rare reports describing the treatment of factor II inhibitors. One case was successfully treated by immunosuppression, using a combination of prednisolone and azathioprine. Treatment should be monitored by periodic measurement of factor II activity and factor II inhibitor titers.

Factor V Inhibitor

Factor V inhibitors are the most common type of inhibitor of the common coagulation pathway. More than 150 cases have been reported. Most (two-thirds) are associated with the use of bovine thrombin or fibrin glue as an aid in surgical hemostasis. These preparations are contaminated with trace amounts of bovine factor V and can induce the formation of antibodies that cross-react with human factor V.

Bleeding and mortality rates are dependent on whether the inhibitor develops spontaneously or in response to bovine thrombin. Bleeding rates are 72% and 33% in the spontaneous and bovine thrombin-associated inhibitors, respectively. Mortality rates are 17% and 6% in the spontaneous and bovine thrombin-associated inhibitors, respectively.

Factor V levels and inhibitor titers are also dependant on the etiology of the factor V inhibitor. The mean factor V levels are 4.5% and 9.6% in the spontaneous and bovine thrombin-associated inhibitors, respectively. The mean inhibitor titers are 114 Bethesda units and 17.2 Bethesda units in the spontaneous and bovine thrombin-associated inhibitors, respectively.

The inhibitors are of IgG isotype and inhibit factor V function by binding to the phospholipid binding region (C2 domain) of the factor V light chain.

The diagnosis is made by demonstrating a prolonged PT, prolonged aPTT, noncorrection of the 1:1 mixing study of the PT and aPTT, and an isolated decrease in levels of factor V activity. Sometimes the mixing study shows a time-dependent inhibitor (full correction at time zero, but noncorrection after incubation). The factor V inhibitor is quantified by the Bethesda assay.

Factor V inhibitors are most successfully treated by immunosuppression, using intravenous immunoglobulin (87% response) and glucocorticoids (74% response). Other successful treatments include cyclosporine A, cyclophosphamide, chemotherapy, plasmapheresis, and immunoadsorption by column chromatography. Platelets and prothrombin complex concentrate have a lower response rate (35% or less). Factor replacement therapy with FFP, cryoprecipitate, and recombinant factor VIIa are not effective alone. Treatment should be monitored by periodic measurement of factor V activity and factor V inhibitor titers.

Factor VII Inhibitor

Factor VII inhibitors are rare, with seven cases reported in the literature. Bleeding occurs in 71% of cases, and the mortality from bleeding is 14%. Factor VII activity levels range from < 1% to 38%. Limited data are available on inhibitor titers; one case reported 1 Bethesda unit.

The inhibitors are of IgG isotype and inhibit factor VII function by binding to the Gla domain of the factor VII light chain.

The diagnosis is made by demonstrating a prolonged PT, normal aPTT, noncorrection of the 1:1 mixing study of the PT, and decreased levels of factor VII activity. The factor VII inhibitor is quantified by the Bethesda assay.

There are very rare reports describing the treatment of factor VII inhibitors. Acute bleeding has been successfully treated with plasma exchange alone or the combination of tranexamic acid and methylprednisolone. Factor VII inhibitor production has been successfully treated with methylprednisolone alone or the combination of prednisolone and azathioprine. Treatment should be monitored by periodic measurement of factor VII activity and factor VII inhibitor titers.

Factor X Inhibitor

Factor X inhibitors are rare, with six cases reported in the literature. Bleeding occurs in 67% of cases, and no fatalities have been reported. Factor X activity levels range from 4% to 20%. Limited data are available on inhibitor titers; one case reported 5.4 Bethesda units.

The inhibitors are of IgG isotype and inhibit factor X function by binding to the Gla domain of the factor X light chain.

The diagnosis is made by demonstrating a prolonged PT, prolonged aPTT, noncorrection of the 1:1 mixing study of the PT and aPTT, and decreased levels of factor X activity. The factor X inhibitor is quantified by the Bethesda assay.

There are very rare reports describing the treatment of factor X inhibitors. Acute bleeding has been successfully treated with methylprednisolone, ε-aminocaproic acid, and prothrombin complex concentrate. Factor X inhibitor production has been successfully treated with methylprednisolone or intravenous immunoglobulin. Treatment should be monitored by periodic measurement of factor X activity and factor X inhibitor titers.

Lupus Anticoagulant

Lupus anticoagulants are inhibitors of phospholipid-dependent coagulation tests and are clinically associated with arterial and venous thrombosis and recurrent miscarriage. They usually prolong the aPTT alone, and sometimes prolong both the aPTT and PT. Lupus anticoagulants inhibit these tests by either interacting with anionic phospholipids in the reagent or by interacting with phospholipid binding proteins in patient plasma (such as beta 2-glycoprotein I, prothrombin). If the PT is prolonged, then one should consider the possibility of antiprothrombin antibodies, which cause increased clearance of prothrombin, leading to severe prothrombin deficiency and increased bleeding risk. Lupus anticoagulant is covered in more detail in chapter 22.

Algorithm

A laboratory testing algorithm for the evaluation of a prolonged PT with or without a prolonged aPTT is shown in Figure 11-1. Testing is directed toward evaluating the differential diagnosis just presented, using a scheme that proceeds in a stepwise fashion from simple to more complex testing. This algorithm is just one of many possible strategies that could be devised.

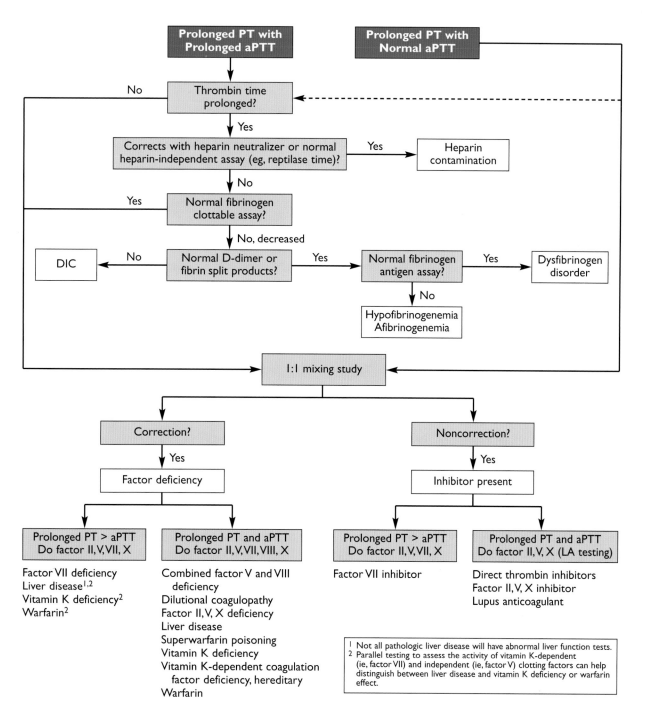

Figure 11-1. The investigation of a prolonged PT with or without a prolonged aPTT. Step 1: Evaluate for heparin contamination (prolonged thrombin time, correction with heparin neutralizer, normal reptilase time). Step 2: Evaluate for disseminated intravascular coagulation (DIC) (low fibrinogen activity, high FDP [fibrin degradation products]/D-dimer). Step 3: Evaluate for hypofibrinogenemia (low fibrinogen activity, normal FDP/D-dimer). Step 4: Evaluate for dysfibrinogenemia (low fibrinogen activity/fibrinogen antigen ratio). Step 5: Evaluate for factor deficiency (corrected 1:1 mixing study) versus factor inhibitor (noncorrected 1:1 mixing study).

If factor deficiency is suspected based on a corrected 1:1 mixing study, then proceed to specific factor levels to distin-

guish between an isolated factor deficiency (II, V, VII, X) versus multiple factor deficiency (combined factor V and VIII deficiency, DIC, dilutional coagulopathy, liver disease, superwarfarin poisoning, vitamin K deficiency, vitamin K-dependent coagulation factor deficiency, and warfarin). Medical history, response to vitamin K treatment, and further laboratory testing (liver function tests, HPLC analysis for superwarfarin) can help sort out the causes of multiple factor deficiency.

If an inhibitor is suspected based on a noncorrected 1:1 mixing study, then proceed to specific factor levels to distinguish between an isolated factor inhibitor (II, V, VII, X) versus an inhibitor that interferes with multiple coagulation factor assays (direct thrombin inhibitor, lupus anticoagulant [LA]).

Suggested Reading

Corrected Mixing Study

Dilutional Coagulopathy

Leslie SD, Toy PT. Laboratory hemostatic abnormalities in massively transfused patients given red blood cells and crystalloid. *Am J Clin Pathol.* 1991;96:770-773.

Murray DJ, Pennel BJ, Weinstein SL, Olson JD. Packed red cells in acute blood loss: dilutional coagulopathy as a cause of surgical bleeding. *Anesth Analg.* 1995;80:336-342.

Disseminated Intravascular Coagulation

Feinstein DI, Marder VJ, Colman RW. Consumptive thrombohemorrhagic disorders. In: Colman RW, Hirsh J, Marder VJ, Clowes AW, George JN, eds. *Hemostasis and Thrombosis: Basic Principles and Clinical Practice.* 4th ed. Philadelphia, Pa: Lippincott Williams & Wilkins; 2001: 1197-1233.

Horan JT, Francis CW. Fibrin degradation products, fibrin monomer and soluble fibrin in disseminated intravascular coagulation. *Semin Thromb Hemost.* 2001;27:657-666.

Lehman CM, Wilson LW, Rodgers GM. Analytic validation and clinical evaluation of the STA LIATEST immunoturbidimetric D-dimer assay for the diagnosis of disseminated intravascular coagulation. *Am J Clin Pathol.* 2004;122:178-184.

Taylor FB, Jr., Toh CH, Hoots WK, Wada H, Levi M. Towards definition, clinical and laboratory criteria, and a scoring system for disseminated intravascular coagulation. *Thromb Haemost.* 2001;86:1327-1330.

Wada H, Gabazza EC, Asakura H, et al. Comparison of diagnostic criteria for disseminated intravascular coagulation (DIC): diagnostic criteria of the International Society of Thrombosis and Hemostasis and of the Japanese Ministry of Health and Welfare for overt DIC. *Am J Hematol.* 2003;74:17-22.

Drugs: Warfarin

Ansell J, Hirsh J, Hylek E, et al. Pharmacology and Management of the Vitamin K Antagonists: American College of Chest Physicians Evidence-Based Clinical Practice Guidelines (8th Edition) *Chest.* 2008;133:160-198.

Fairwheather RB, Ansell J, van den Besselaar AMPH, et al. College of American Pathologists Conference XXXI on Laboratory Monitoring of Anticoagulant Therapy: laboratory monitoring of oral anticoagulant therapy. *Arch Pathol Lab Med.* 1998;122:768-781.

Heck AM, DeWitt BA, Lukes AL. Potential interactions between alternative therapies and warfarin. *Am J Health Syst Pharm.* 2000;57:1221-1227.

Holbrook AM, Pereira JA, Labiris R, et al. Systematic overview of warfarin and its drug and food interactions. *Arch Intern Med.* 2005;165:1095-1106.

Samuels N. Herbal remedies and anticoagulant therapy. *Thromb Haemost.* 2005;93:3-7.

Factor Deficiency

Factor II Deficiency

Akhavan S, Mannucci PM, Lak M, et al. Identification and three-dimensional structural analysis of nine novel mutations in patients with prothrombin deficiency. *Thromb Haemost.* 2000;84:989-997.

Bajaj SP, Rapaport SI, Fierer DS, Herbst KD, Schwartz DB. A mechanism for the hypoprothrombinemia of the acquired hypoprothrombinemia-lupus anticoagulant syndrome. *Blood.* 1983;61:684-692.

Collazos J, Egurbide MV, Atucha K, Esteban P, de Miguel J. Transient acquired factor II deficiency with Mycoplasma pneumoniae infection. *J Infect Dis.* 1991;164:434-435.

Lee ES, Hibsman BK, Liebman HA. Acquired bleeding disorder in a patient with malignant lymphoma: antibody-mediated prothrombin deficiency. *Cancer.* 2001;91:636-641.

Mammen EF. Factor II abnormalities. *Thromb Haemost.* 1983;9:13-16.

Peyvandi F, Mannucci PM. Rare coagulation disorders. *Thromb Haemost.* 1999;82:1207-1214.

Rouvier J, Collen D, Swart ACW, Verstraete M. Prothrombin metabolism in healthy subjects and in two patients with congenital hypoprothrombinemia. In: *Prothrombin.* Leiden, The Netherlands: Leiden University Press; 1983: 167-182.

Factor V Deficiency

Girolami A, Simioni P, Scarano L, Girolami B, Marchiori A. Hemorrhagic and thrombotic disorders due to factor V deficiencies and abnormalities: an updated classification. *Blood Rev.* 1998;12:45-51.

Lak BT, Sharifian R, Peyvandi F, Mannucci PM. Symptoms of inherited factor V deficiency in 35 Iranian patients. *Br J Haematol.* 1998;103:1067-1069.

Mellinger EJ, Duckert F. Major surgery in a subject with factor V deficiency. *Thromb Diath Haemorrh.* 1971;25:438-446.

Webster WP, Roberts HR, Penick GD. Hemostasis in factor V deficiency. *Am J Med Sci.* 1964;248:194-202.

Factor V Deficiency and Factor VIII Deficiency, Combined

Faioni EM, Fontana G, Carpani G, et al. Review of clinical, biochemical and genetic aspects of combined factor V and factor VIII deficiency, and report of a new affected family. *Thromb Res.* 2003;112:269-271.

Levine PH. Clinical manifestations and therapy of hemophilias A and B. In: Colman RW, Hirsh J, Marder VJ, Salzman EW, eds. *Hemostasis and Thrombosis: Basic Principles and Clinical Practice.* Philadelphia, Pa: JB Lippincott; 1987: 97-111.

Mellinger EJ, Duckert F. Major surgery in a subject with factor V deficiency. *Thromb Diath Haemorrh.* 1971;25:438-446.

Seligsohn U, Zivelin A, Zwang E. Combined factor V and factor VIII deficiency among non-Ashkenazi Jews. *N Engl J Med.* 1982;307:1191-1195.

Webster WP, Roberts HR, Penick GD. Hemostasis in factor V deficiency. *Am J Med Sci.* 1964;248:194-202.

Weiss AE, Webster WP, Strike LE, Brinkhous KM. Survival of transfused factor VIII in hemophilic patients treated with epsilon aminocaproic acid. *Transfusion.* 1976;16:209-214.

Zhang B, McGee B, Yamaoka JS, et al. Combined deficiency of factor V and factor VIII is due to mutations in either LMAN1 or MCFD2. *Blood.* 2006;107:1903-1907.

Zhang B, Spreafico M, Zheng, C, et al. Genotype-phenotype correlation in combined deficiency of factor V and factor VIII. *Blood.* 2008;111(12):5592-5600.

Factor VII Deficiency

Cooper DN, Millar DS, Wacey A, Banner DW, Tuddenham EG. Inherited factor VII deficiency: molecular genetics and pathophysiology. *Thromb Haemost.* 1997;78:151-160.

Marder V, Schulman R. Clinical aspects of congenital factor VII deficiency. *Am J Med.* 1964;37:182-194.

Mariani G, LoCoco L, Bernardi F, Pinotti M. Molecular and clinical aspects of factor VII deficiency. *Blood Coagul Fibrinolysis.* 1998;9(suppl 1): S83-S88.

Factor X Deficiency

Cooper DN, Millar DS, Wacey A, Pemberton S, Tuddenham EG. Inherited factor X deficiency: molecular genetics and pathophysiology. *Thromb Haemost.* 1997;78:161-172.

Peyvandi F, Mannucci PM, Lak M, et al. Congenital factor X deficiency: spectrum of bleeding symptoms in 32 Iranian patients. *Br J Haematol.* 1998;102:626-628.

Roberts HP, Lechler E, Webster WP, Penick GD. Survival of transfused factor X in patients with Stuart disease. *Thromb Diath Haemorrh.* 1968;13:884-893.

Fibrinogen Deficiency

Al-Fawaz IM, Gader AM. Severe congenital dysfibrinogenemia (fibrinogen-Riyadh): a family study. *Acta Haematol.* 1992;88:194-197.

Al-Mondhiry H, Ehmann WC. Congenital afibrinogenemia. *Am J Hematol.* 1994;46:343-347.

Collen D, Tytgat GN, Clacys H, Piessens R. Metabolism and distribution of fibrinogen. *Br J Haematol.* 1972;22:681-700.

Cunningham MT, Brandt JT, Laposata M, Olson JD. Laboratory diagnosis of dysfibrinogenemia. *Arch Pathol Lab Med.* 2002;126:499-505.

Ganti AK, Vose JM, Haire WD. Hematopoietic stem cell transplantation for Hodgkin's disease in a patient with dysfibrinogenemia and thrombosis. *J Thromb Thrombolysis.* 2007;23:155-158.

Haverkate F, Samama M. Familial dysfibrinogenemia and thrombophilia: report on a study of the SSC Subcommittee on Fibrinogen. *Thromb Haemost.* 1995;73:151-161.

Mammen EF. Fibrinogen abnormalities. *Semin Thromb Hemost.* 1983;9:1-9.

Wulf GG, Unterhalt M, Buchwald A, Zenker D, Kreuzer H, Hiddemann W. Hypercoagulability in a patient with hypodysfibrinogenemia: implications for clinical management. *Acta Haematol.* 1999;101:209-212.

Vitamin K Dependent Coagulation Factor Deficiency, Hereditary

Darghouth D, Hallgren, Shtofman RL, et al. Compound heterozygosity of novel missense mutations in the gamma-glutamyl-carboxylase gene causes hereditary combined vitamin K-dependent coagulation factor deficiency. *Blood.* 2006;108:1925-1931.

Liver Disease

Mammen EF. Coagulation defects in liver disease. *Med Clin North Am.* 1994;78:545-554.

Superwarfarin Poisoning

Chua JD, Friedberg WR. Superwarfarin poisoning. *Arch Intern Med.* 1998;158:1929-1932.

Katona B, Wason S. Superwarfarin poisoning. *J Emerg Med.* 1989;7:627-631.

Laposata M, Van Cott EM, Lev MH. Case records of the Massachusetts General Hospital: case 1-2007. A 40-year-old woman with epistaxis, hematemesis, and altered mental status. *N Engl J Med.* 2007:356:174-182.

LaRosa FG, Clarke SH, Lefkowitz JB. Brodifacoum intoxication with marijuana smoking. *Arch Pathol Lab Med.* 1997;121:67-69.

Routh CR, Triplett DA, Murphy MJ, Felice LJ, Sadowski JA, Bovill EG. Superwarfarin ingestion and detection. *Am J Hematol.* 1991;36:50-54.

Spahr JE, Maul JS, Rodgers GM. Superwarfarin poisoning: a report of two cases and review of the literature. *Am J Hematol.* 2007;82(7):656-660.

Vitamin K Deficiency

Hathaway WE. Vitamin K deficiency. *Southeast Asian J Trop Med Public Health.* 1993;24(suppl 1):5-9.

Suzuki S, Iwata G, Sutor AH. Vitamin K deficiency during the perinatal and infantile period. *Semin Thromb Hemost.* 2001;27:93-98.

Vermeer C, Hamulyak K. Pathophysiology of vitamin K-deficiency and oral anticoagulants. *Thromb Haemost.* 1991;66:153-159.

Noncorrected Mixing Study

Drugs

Direct Thrombin Inhibitors

Gosselin RC, Dager WE, King JH, et al. Effect of direct thrombin inhibitors, bivalirudin, lepirudin, and argatroban, on prothrombin time and INR values. *Am J Clin Pathol.* 2004;121:593-599.

Gosselin RC, King JH, Janatpour KA, Dager WE, Larkin EC, Owings JT. Comparing direct thrombin inhibitors using aPTT, ecarin clotting times, and thrombin inhibitor management testing. *Ann Pharmacother.* 2004;38:1383-1388.

Warkentin TE, Greinacher A, Craven S, Dewar L, Sheppard JA, Ofosu FA. Differences in the clinically effective molar concentrations of four direct thrombin inhibitors explain their variable prothrombin time prolongation. *Thromb Haemost.* 2005;94:958-964.

Yee AJ, Kuter DJ. Successful recovery after an overdose of argatroban. *Ann Pharmacother.* 2006;40:336-339.

Heparin

Berggvist D. Review of fondaparinux sodium injection for the prevention of venous thromboembolism in patients undergoing surgery. *Vasc Health Risk Manag.* 2006;2:365-370.

Firozvi K, Deveras RA, Kessler CM. Reversal of low-molecular-weight heparin induced bleeding with pre-existing hypercoagulable states with human recombinant factor VII concentrate. *Am J Hematol.* 2006;81:582-589.

Laposata M, Green D, Van Cott EM, Barrowcliffe TW, Goodnight SH, Sosilik RC. College of American Pathologists Conference XXXI on Laboratory Monitoring of Anticoagulant Therapy: the clinical use and laboratory monitoring of low-molecular-weight heparin, danaparoid, hirudin and related compounds, and argatroban. *Arch Pathol Lab Med.* 1998;122:799-807.

Olson JD, Arkin CF, Brandt JT, Cunningham MT, Giles A, Koepke JA, Witte DL. College of American Pathologists Conference XXXI on Laboratory Monitoring of Anticoagulant Therapy: laboratory monitoring of unfractionated heparin therapy. *Arch Pathol Lab Med.* 1998;122:782-798.

Smorgorzewska A, Brandt JT, Chandler WL, et al. Effect of fondaparinux on coagulation assays: results of College of American Pathologists proficiency testing. *Arch Pathol Lab Med.* 2006;130:1605-1611.

Factor Inhibitors

Factor II Inhibitor

Costa JM, Fiessinger JN, Capron L, Aiach M. Partial characterization of an autoantibody recognizing the secondary binding site(s) of thrombin in a patient with recurrent spontaneous arterial thrombosis. *Thromb Haemost.* 1992;67:193-199.

Madoiwa S, Nakamura Y, Mimuro J, et al. Autoantibody against prothrombin aberrantly alters the proenzyme to facilitate formation of a complex with its physiological inhibitor antithrombin III without thrombin conversion. *Blood.* 2001;97:3783-3789.

Scully MF, Ellis V, Kakkar VV, Savidge GF, Williams YF, Sterndale H. An acquired coagulation inhibitor to factor II. *Br J Haematol.* 1982;50:655-664.

Factor V Inhibitor

Favaloro EJ, Posen J, Ramakrishna R, et al. Factor V inhibitors: rare or not so uncommon? A multi-laboratory investigation. *Blood Coagul Fibrinolysis.* 2004;15:637-647.

Izumi T, Kim SW, Greist A, et al. Fine-mapping of inhibitory anti-factor V antibodies using factor V C2 domain mutants: identification of two antigenic epitopes involved in phospholipid binding. *Thromb Haemoast.* 2001;85:1048-1054.

Knobl P, Lechner K. Acquired factor V inhibitors. *Bailliere's Clin Haematol.* 1998;11:305-318.

Streiff MB, Ness PM. Acquired FV inhibitors: a needless iatrogenic complication of bovine thrombin exposure. *Transfusion.* 2002;42:18-26.

Factor VII Inhibitor

Aguilar C, Lucia JF, Hernandez P. A case of an inhibitor autoantibody to coagulation factor VII. *Haemophilia*. 2003;9:119-120.

Campbell E, Sanal S, Mattson J, et al. Factor VII inhibitor. *Am J Med*. 1980;68:962-964.

Delmer A, Horellou MH, Andreau G, et al. Life-threatening intracranial bleeding associated with the presence of an antifactor VII autoantibody. *Blood*. 1989;74:229-232.

Ingerslev J, Christiansen K, Sorensen B. International Registry on Factor VII Deficiency (IRF7) Steering Committee. Inhibitor to factor VII in severe factor VII deficiency: detection and course of inhibitor response. *J Thromb Haemost*. 2005;3:799-800.

Kamikubo Y, Miyamoto S, Iwasa A, Ishii M, Okajima K. Purification and characterization of factor VII inhibitor found in a patient with life threatening bleeding. *Thromb Haemost*. 2000;83:60-64.

Mehta J, Singhal S, Mehta BC. Factor VII inhibitor. *J Assoc Physicians India*. 1992;40:44.

Ndimbie OK, Raman BK, Saeed SM. Lupus anticoagulant associated with specific inhibition of factor VII in a patient with AIDS. *Am J Clin Pathol*. 1989;91:491-493.

Okajima K, Ishii M. Life-threatening bleeding in a case of autoantibody-induced factor VII deficiency. *Int J Hematol*. 1999;69:129-132.

Factor X Inhibitor

Lankiewicz MW, Bell WR. A unique circulating inhibitor with specificity for coagulation factor X. *Am J Med*. 1992;93:343-346.

Matsunaga AT, Shafer FE. An acquired inhibitor to factor X in a pediatric patient with extensive burns. *J Pediatr Hematol Oncol*. 1996;18:223-226.

Mulhare PE, Tracy PB, Golden EA, Branda RF, Bovill EG. A case of acquired factor X deficiency with in vivo and in vitro evidence of inhibitory activity directed against factor X. *Am J Clin Pathol*. 1991;96:196-200.

Ness PM, Hymas PG, Gesme D, Perkins HA. An unusual factor X inhibitor in leprosy. *Am J Hematol*. 1980;8:397-402.

Rao LV, Zivelin A, Iturbe I, Rapaport SI. Antibody-induced acute factor X deficiency: clinical manifestations and properties of the antibody. *Thromb Haemost*. 1994;72:363-371.

Lupus Anticoagulant

See chapter 22, "Suggested Reading."

CHAPTER 12

Normal Prothrombin Time and Activated Partial Thromboplastin Time

Timothy E. Hayes, DVM, MD

Introduction

When confronted with an acutely bleeding patient, the most pressing question posed by the clinician is often, "What do I do to make the bleeding stop?" This question is particularly important because the clinician can select from multiple therapeutic options, including procedures (cauterization, surgical intervention, etc), medications (desmopressin acetate [DDAVP], antifibrinolytics, etc), and transfusion of blood components (platelets, frozen plasma, cryoprecipitate, and specific clotting factors, such as concentrates of factor VIII, factor IX, von Willebrand factor [vWF], and now even recombinant factor VIIa). The most appropriate clinical response depends upon accurate identification of the underlying cause for the bleeding. The prothrombin time (PT) and the activated partial thromboplastin time (aPTT) are often used in the initial evaluation of a bleeding patient to specifically assess the potential role of defective secondary hemostasis as an underlying cause of the bleeding. Not infrequently, both test results are within normal limits. This chapter will discuss an approach to this clinical setting.

Clinical Evaluation

Defects in secondary hemostasis are common causes for increased bleeding risk. Examples include liver disease, vitamin K deficiency, dilutional coagulopathies, disseminated intravascular coagulation (DIC), inherited factor deficiencies, and medications, such as heparins, warfarins, and the direct thrombin inhibitors (DTIs). The PT and aPTT are frequently used to screen for these types of defects and, in general, are well suited for such use. In particular, these assays are widely available, easily and rapidly performed, relatively inexpensive, and fairly sensitive for the more common defects in secondary hemostasis. It is not uncommon for these two assays to be within normal limits in a patient who is reported to be actively

bleeding or to have a personal or family history of a bleeding tendency.

Before undertaking an exhaustive and expensive laboratory evaluation for a hemorrhagic disorder, it is good practice to have some communication between the laboratory and the patient's clinician. Good communication and an adequate understanding of clinical and laboratory hemostasis are instrumental in making appropriate decisions about the next steps, both in terms of ensuring a good patient outcome as well as appropriate use of medical resources.

As a starting point, it is not unreasonable to assess the significance of the bleeding in question. In some cases, the bleeding is determined to be within normal limits (for example, a certain amount of chest tube drainage is not unusual after some forms of cardiac surgery) and therefore may not warrant an extensive laboratory evaluation. An experienced clinician is in the best position to determine the "normal" amount of bleeding for the clinical setting. At the same time, it may be worthwhile discussing the likelihood that the bleeding is due to something other than a coagulopathy (eg, inadequate surgical hemostasis).

If the bleeding is thought to be clinically significant and an underlying coagulopathy cannot be excluded, then further laboratory evaluation is indicated. The patient's clinical information, especially the personal/family history and current medications, are important in determining the type and extent of the testing to be performed (see chapter 5).

False-Negative (Normal) PT and aPTT

Although the PT and aPTT are fairly sensitive for detecting most of the common defects in secondary hemostasis, they do have their limitations. Causes of false-negative results may be broadly divided into four categories: patient specific, pre-

Case Study

A 28-year-old woman delivered a term infant by vaginal delivery. There was little bleeding during delivery, and the patient was discharged 1 day postpartum. At day 3 postpartum, the patient developed extensive vaginal bleeding, with a hemoglobin of 9 g/dL, requiring hospitalization and transfusion. The patient related a similar delayed bleeding episode after a tonsillectomy at the age of 7 years, but no coagulation evaluation had been performed at that time.

At the time of admission, the PT was 10.0 seconds (reference range 9.9-13.0) and the aPTT was 28.9 seconds (reference range 24.6-32.8) with a normal platelet count. After the initial coagulation study results were obtained, repeat testing of PT and aPTT was performed, with similar normal results. Further testing showed a normal Clauss fibrinogen of 325 mg/dL (reference range 200-

400), a normal D-dimer of 280 ng/mL fibrinogen equivalent units (FEU) (normal level <500), and a normal platelet function screen with the PFA-100 (collagen/epinephrine closure time 175 seconds [normal level <198] and collagen/adenosine diphosphate [ADP] closure time 115 seconds [normal level <118]). These results do not suggest a coagulation protein disorder in the common, intrinsic, or extrinsic pathways, nor do they suggest a fibrinolytic or platelet disorder. For this reason, a qualitative factor XIII urea solubility assay was performed, which was abnormal. Subsequent quantitative factor XIII assay sent to a reference laboratory showed a decreased factor XIII activity of 3%, and a diagnosis of factor XIII deficiency was rendered. This case illustrates that significant clinical bleeding can be seen with normal PT and aPTT results in the setting of a factor XIII deficiency because factor XIII is responsible for cross-linking fibrin.

analytic, analytic, and postanalytic (see the algorithm presented in Figure 12-1).

Patient-Specific Causes

Clinically significant defects in secondary hemostasis may not be detectable by the PT and aPTT in several settings. Therapeutic interventions, such as transfusion of plasma-containing blood components (including cryoprecipitate, prothrombin complex concentrates, factor concentrates, etc) and administration of medications (including vitamin K, DDAVP, etc) may transiently correct the underlying coagulation defect. Pregnancy and the use of oral contraceptives (ethinyl estradiol, mestranol) may result in elevations of several procoagulant factors, including factors VII, VIII, XII, von Willebrand factor (vWF), and fibrinogen. Factor XI may decrease or increase during pregnancy. An acute phase process resulting from inflammatory and stressful states (including strenuous exercise, high altitude, and mental stress) may be associated with elevated levels of fibrinogen, vWF, and factor VIII. Laboratory specimens drawn during these clinical settings may yield normal test results.

Preanalytic Causes

There are numerous variables in the preanalytic phase of testing that may result in a falsely normal (false-negative) PT and/or aPTT. Erroneous results obviously may arise from drawing blood from the wrong patient. The phlebotomy process

may also introduce errors. A specimen obtained via a traumatic or repeated venipuncture effort may introduce tissue juice (thromboplastin) contamination, which may shorten the aPTT, especially if there is also a slow flow of blood into the tube. Shortening of the PT and/or aPTT may also be seen with excessive mixing of the specimen, hemolysis, and chilling of the specimen. All of these events may bring an otherwise abnormal (prolonged) PT/aPTT result down into the normal range (false-negative result). When in doubt, the PT and aPTT should be repeated on a freshly collected specimen. See chapter 4 for further discussion.

Analytic Causes

The sensitivity of the PT and aPTT to detect clinically significant factor deficiencies varies with the reagent used. For example, the sensitivity of the PT reagent is highly dependent upon the source of thromboplastin, while the phospholipid composition and the type of activator influence the sensitivity of the aPTT reagent. The sensitivity of aPTT reagents to factor deficiencies varies not only between manufacturers, but also between lots of reagents from the same manufacturer. These concerns may be somewhat mitigated by understanding these reagent-specific issues and carefully establishing the assay reference range. Some of the clinically significant mild hemophilias and von Willebrand disease may have an aPTT result that is just a few seconds above the upper limit of the reference range. A poorly generated reference range therefore may not discern these patients from normal. See chapter 6 for a more in-depth discussion.

Postanalytic Causes

Postanalytic errors might include transcription errors, calculation errors, and oral miscommunication of test results. For example, it is critical that the correct international sensitivity index (ISI), geometric mean of the reference range, and calculations are used in generating the international normalized ratio (INR). This is becoming more important as many laboratories have completely replaced PT reporting with the INR—even for non-warfarin patients.

Additional Testing

If it is determined that the patient with a normal PT and aPTT has a compelling bleeding history that may be due to a coagulopathy, and the above issues have been addressed, then it may be of value to consider additional testing, as outlined below.

Primary Hemostasis Defects

Defects in primary hemostasis, such as von Willebrand disease and platelet dysfunction, are a common cause of bleeding and should therefore be considered during the evaluation of a hemorrhagic disorder. Although von Willebrand disease typically presents with an elevated aPTT, patients with some subtypes of von Willebrand disease present with bleeding despite a normal PT and aPTT. See chapters 14 and 16 for further information.

Secondary Hemostasis Defects

Factor Assays

The aPTT is used to detect coagulation abnormalities of the intrinsic pathway. It is more sensitive to deficiencies of the "proximal" part of the pathway (factors XII, XI, IX, and VIII) than to the "distal" part (factor II and fibrinogen). Therefore, it is not unreasonable to include a fibrinogen level (factor I) in the initial evaluation of a bleeding patient with a normal PT and aPTT.

The hereditary hemophilias include deficiencies of factors VIII (hemophilia A), IX (hemophilia B), and XI (hemophilia C). Most aPTT reagents will yield prolonged (abnormal) test results when factor VIII or IX drop below 30% to 40% of normal,

but some reagents have failed to detect mild hemophilia levels as low as 25% (eg, a normal aPTT). Most patients with heterozygous factor XI deficiency will be asymptomatic, yet some will bleed, especially after trauma. These patients may have a normal aPTT.

The family history may be helpful but may also be misleading. For example, the lack of a family history for a bleeding disorder does not exclude a hereditary factor deficiency. As many as one-third of cases of hemophilia A and B may be due to recent mutations—most commonly being first detected in the carrier female—with no previous family bleeding history. Therefore, specific factor assays should be performed when there is a strong clinical suspicion of mild hemophilia, even if the aPTT is normal.

Factor XIII Deficiency

Factor XIII (FXIII) is a cysteine enzyme (transglutaminase) that is activated by thrombin. The active enzyme covalently cross-links fibrin molecules to each other and cross-links α_2-antiplasmin to fibrin. This cross-linking converts the loose fibrin polymer into a highly organized structure with increased tensile strength that imparts clot stability and resistance to fibrinolysis. FXIII also participates in wound healing, cell migration, and clot retraction.

FXIII deficiency may be either inherited or acquired. Bleeding is usually not significant unless FXIII deficiency is severe (levels are below 3%). The inherited form is an autosomal recessive disorder and is highly heterogeneous from a molecular perspective. It is relatively rare, with an estimated frequency of 1 in 2,000,000. Bleeding is variable, but usually severe. Umbilical site bleeding within a few days of birth occurs in roughly 80% of cases. Delayed bleeding after mild trauma is typical. Intracranial bleeding may be spontaneous or occur after mild trauma. Subcutaneous and muscle hematomas are common. Delayed bleeding and menorrhagia are seen. Hemarthrosis is uncommon. Pregnancies may be lost due to placental bleeding. Poor wound healing occurs in about 14% of patients. Some patients present with a milder form of bleeding. It has been reported that heterozygotes may also have bleeding symptoms.

An acquired deficiency may rarely be seen with FXIII inhibitors, which are typically IgG antibodies. These antibodies may arise after exposure to certain drugs, including penicillin, phenytoin, valproate, and isoniazid. These antibodies may be

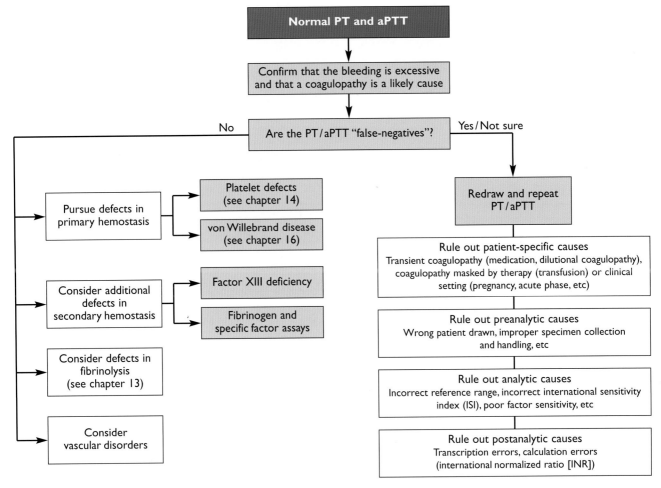

Figure 12-1. An algorithmic approach to the diagnosis of a bleeding etiology in a patient with a normal prothrombin time (PT) and activated partial thromboplastin time (aPTT). Initial evaluation should confirm that the bleeding is truly excessive and that the PT and aPTT results are not "false-negative" results due to recent transfusion therapy or drawing blood from the wrong patient. If a coagulopathy is likely, further directed workup would seek to rule out platelet dysfunction, von Willebrand disease, fibrinolysis abnormalities, as well as factor XIII and fibrinogen abnormalities. Negative workup for all of these would prompt a clinical evaluation to exclude vascular disorders.

associated with a severe deficiency of FXIII and significant bleeding. Acquired deficiency of FXIII may also be associated with certain malignancies (leukemia, plasmacytoma), sepsis, and DIC. The FXIII deficiency associated with these clinical conditions is usually mild and of minimal clinical significance.

The routine clotting assays (PT, aPTT, thrombin time, and fibrinogen) are normal even in severe FXIII deficiency. The qualitative screening assay is the clot solubility test. Patient plasma is incubated with thrombin to form a clot. This clot is then suspended in a 5M urea solution or 1% monochloroacetic acid. With severe FXIII deficiency (level < 3% of normal), normal fibrin cross-linking has not stabilized the fibrin clot and it dissolves within 60 minutes. An abnormal screening test result should be followed with mixing studies with normal plasma to rule out an inhibitor and further confirmed by a quantitative assay (functional and/or immunologic assays).

Fibrinolysis Defects

Defects in fibrinolysis may be associated with significant bleeding and a normal PT and aPTT. Hereditary forms are rare. Acquired forms are more common and are typically associated with thrombolytic therapy or DIC. See chapter 13 for further information.

Vascular Disorders

Certain vascular disorders, such as vasculitis, connective tissue disorders, vascular malformations, senile purpura and steroid-induced purpura may be associated with a bleeding tendency and a normal PT and aPTT. A variety of conditions can be association with a bruising tendency, including Ehlers-Danlos syndrome, hereditary hemorrhagic telangiectasia, osteogenesis imperfecta, Cushing syndrome, vitamin C deficiency (scurvy), Marfan syndrome, Fabry disease, and others. A complete discussion of these vascular disorders is beyond the scope of this book; however, the coagulation laboratory's role in the diagnosis of these disorders may include testing to exclude platelet disorders, DIC, or immune thrombocytopenic purpura (ITP), which may present as microvascular hemorrhage. Other pathology laboratories may be involved in skin biopsy and studies to exclude cutaneous vasculitis, such as testing for complement, antinuclear antibodies and antineutrophilic cytoplasmic antibody (ANCA).

Algorithm

An algorithmic approach to evaluating the setting of bleeding with a normal PT and aPTT is shown in Figure 12-1. This algorithm is just one of many possible strategies that might be used.

Suggested Reading

Approach to the Bleeding Patient

Clark P, Brennand J, Conkie JA, McCall F, Greer IA, Walker ID. Activated protein C sensitivity, protein C, protein S and coagulation in normal pregnancy. *Thromb Haemost.* 1998;79:1166-1170.

Dzik WH. Component therapy before bedside procedures. In: Mintz PD, ed. *Transfusion Therapy: Clinical Principles and Practice.* 2nd ed. Bethesda, Md: AABB Press; 2005: 1-26.

Goodnight SH, Hathaway WE. Screening tests in hemostasis. In: Goodnight SH, Hathaway WE. *Disorders of Hemostasis and Thrombosis: A Clinical Guide.* 2nd ed. New York: McGraw-Hill; 2001: 41-51.

Konkle BA. Clinical approach to the bleeding patient. In: Colman RW, Clowes AW, Godhaber SZ, Marder VJ, George JN, eds. *Hemostasis and Thrombosis: Basic Principles and Clinical Practice.* 5th ed. Philadelphia, Pa: Lippincott Williams and Wilkins; 2006: 1147-1158.

Schafer AI. Approach to bleeding. In: Loscalzo J, Schafer AI, eds. *Thrombosis and Hemorrhage.* 3rd ed. Philadelphia, Pa: Lippincott Williams and Wilkins; 2003: 315-329.

Factor XIII Deficiency

Anwar R, Miloszewski KJA. Factor XIII deficiency. *Br J Haematol.* 1999;107:468-484.

Dargaud Y, de Mazancourt P, Rugeri L, et al. An unusual clinical presentation of factor XIII deficiency and issues relating to the monitoring of factor XIII replacement therapy. *Blood Coagul Fibrinolysis.* 2008;19(5):447-452.

Francis JL. The detection and measurement of factor XIII activity: a review. *Med Lab Sci.* 1980;37:137-147.

Greenberg CS, Sane DC, Lai TS. Factor XIII and fibrin stabilization. In: Colman RW, Clowes AW, Goldhaber SZ, Marder VJ, George JN, eds. *Hemostasis and Thrombosis: Basic Principles and Clinical Practice.* 5th ed. Philadelphia, Pa: Lippincott Williams and Wilkins; 2006: 317-334.

Seitz R, Duckert F, Lopaciuk S, et al. ETRO Working Party on factor XIII questionnaire on congenital factor XIII deficiency in Europe: status and perspectives. *Semin Thromb Hemost.* 1996;22:415-418.

Vascular Disorders

Baxter BT. Heritable diseases of the blood vessels. *Cardiovasc Pathol.* 2004;14:185-188.

Myllyharju J, Kivirikko KI. Collagens and collagen-related diseases. *Ann Med.* 2001;33:4-6.

Zumberg M, Kitchens CS. Purpura and other hematovascular disorders. In: Kitchens CS, Alving BM, Kessler CM, eds. *Consultative Hemostasis and Thrombosis.* 2nd ed. Philadelphia, Pa; Saunders; 2007: 159-182.

CHAPTER 13

Fibrinolytic Bleeding Disorders

Wayne L. Chandler, MD

Clinical Evaluation of Fibrinolytic Bleeding Disorders

Clinical evaluation of fibrinolysis should be considered in patients with a history of bleeding and negative results on the more common coagulation assays. Increased bleeding due to fibrinolytic abnormalities is relatively rare but does occur. Mild to moderate delayed bleeding after surgery, trauma, or childbirth, often with wound hematomas, is commonly seen in congenital deficiencies of fibrinolytic inhibitors. Figure 13-1 shows an algorithm for the evaluation of the most common congenital fibrinolytic bleeding disorders. Acquired causes of fibrinolytic bleeding disorders include cirrhosis, liver transplantation, cardiopulmonary bypass, and treatment of thrombosis with exogenous plasminogen activators. Acquired bleeding due to elevated levels of plasminogen activators can be rapidly diagnosed using a whole blood clot lysis time monitored by changes in clot viscoelasticity (thromboelastometry). This type of testing is performed in many centers during open heart and liver transplantation procedures.

Plasminogen Activator Inhibitor-1 Deficiency

Plasminogen activator inhibitor-1 (PAI-1) deficiency is a rare congenital disorder with increased fibrinolysis leading to clinical bleeding symptoms. The clinically significant disorder is usually homozygous, as described by Fay et al (1997) in a family with complete deficiency of plasma PAI-1 activity in seven individuals. In this family, the deficiency was due to a homozygous dinucleotide insertion within exon 4 of the PAI-1 gene, resulting in a premature stop codon and a truncated nonfunctional PAI-1 protein. PAI-1 antigen was undetectable in plasma samples from four of the patients with the homozygous mutation, indicating complete deficiency of PAI-1. Transmission of

Case Study

Postoperative Bleeding and PAI-1 Deficiency

A 45-year-old woman is being evaluated for bleeding risk prior to elective hysterectomy for painful uterine fibroids. The patient has a history of excessive bleeding and wound hematomas requiring transfusion after cholecystectomy 8 years ago. The patient also has a history of delayed bleeding after the birth of each of her three children. The bleeding was severe enough to require postpartum transfusion and evacuation of retained blood in the uterus. The patient had no history of unusual bleeding in her parents, siblings, or children. The patient was on no medications. Routine hemostatic evaluation, including PT, aPTT, fibrinogen, platelet count, and von Willebrand disease testing, were all normal. Evaluation of her fibrinolytic system produced the following results: PAI-1 activity, undetectable; PAI-1 antigen, undetectable; tPA antigen, reduced. The patient was diagnosed with PAI-1 deficiency and treated with a fibrinolytic inhibitor during surgery, which for the first time was uneventful, without clinically significant postoperative bleeding.

homozygous PAI-1 deficiency is autosomal recessive. Clinically, the patients reported easy bruising, menorrhagia, and moderate to severe delayed bleeding after surgery or minor trauma, often associated with excessive wound hematomas. Bleeding episodes included joint and intracranial bleeding associated with minor trauma. It is interesting to note that excessive bleeding was not reported in 19 individuals heterozygous for the PAI-1 mutation. Another case report described complete lack of PAI-1 activity and antigen in a 36-year-old patient with lifelong epistaxis and bleeding after minor surgery, suggesting complete deficiency of PAI-1. Other studies have reported PAI-1

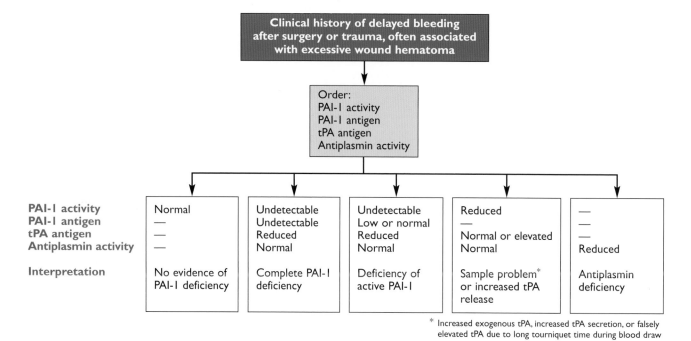

Figure 13-1. Recommended fibrinolytic testing for unexplained bleeding disorder.

Abbreviations: PAI-1, plasminogen activator inhibitor-1; tPA, tissue plasminogen activator.

deficiency in plasma but not platelets; low PAI-1 activity but normal PAI-1 antigen, suggesting a functional defect in the PAI-1 molecule; and reduced but not absent PAI-1 activity and antigen in patients with unexplained bleeding.

It seems clear that patients with complete deficiency of PAI-1 are at risk of bleeding. Patients may also be at risk if they have a complete absence of PAI-1 activity even if PAI-1 antigen is present. It is less clear whether low or low-normal PAI-1 levels are truly associated with a generalized increase in bleeding risk. Prospective studies of patients with low but not absent PAI-1 are needed to determine at what level partial PAI-1 deficiency becomes clinically significant.

Laboratory assessment of PAI-1 deficiency should include measurement of PAI-1 activity, PAI-1 antigen, and either total tissue plasminogen activator (tPA) antigen or tPA–PAI-1 complex. Table 13-1 shows typical tPA and PAI-1 results from patients with elevated PAI-1, normal PAI-1, PAI-1 deficiency, and prolonged venous occlusion. PAI-1 deficiency is likely if PAI-1 activity is essentially undetectable (below detection limit for the assay used), PAI-1 antigen is either absent or low, and tPA antigen or tPA–PAI-1 complex are low. Absence of both PAI-1 activity and antigen with low tPA antigen indicates complete PAI-1 deficiency. Absence of PAI-1 activity with low or normal PAI-1 antigen and low tPA antigen indicates dysfunctional PAI-1. When PAI-1 activity is low in

vivo, total tPA antigen is also low. This is due to more rapid hepatic clearance of active unbound tPA versus inactive tPA–PAI-1 complex. When PAI-1 activity is absent, tPA–PAI-1 complex will also be absent, and the majority of the tPA will be in the active unbound form. Since active unbound tPA is cleared faster than tPA–PAI-1 complex, the total amount of tPA in the blood in patients with PAI-1 deficiency is reduced. It is critical in diagnosing PAI-1 deficiency to demonstrate a low tPA antigen or tPA–PAI-1 complex in addition to a low PAI-1 activity.

The level of active tPA is similar in patients with PAI-1 deficiency and healthy subjects. In PAI-1 deficiency, more rapid hepatic clearance of active tPA is offset by the lack of active tPA inhibition by PAI-1. So, if circulating active tPA levels in patients with PAI-1 deficiency are similar to healthy subjects, why do PAI-1 deficient patients bleed? PAI-1 binds to the clot surface and is very active at inhibiting tPA trying to lyse the clot. It is thought that bleeding in PAI-1 deficiency is primarily due to lack of tPA inhibition at the clot, even though the level of circulating active tPA in blood is similar to healthy subjects.

A falsely low PAI-1 activity can occur in patients with elevated tPA in the sample due to venous trapping of tPA when the tourniquet is applied for an extended period during the blood draw. Low PAI-1 activity with elevated tPA can also be due to increased release of tPA in vivo,

Table 13-1. Effect of PAI-1 on tPA Levels in Plasma

Assay	High PAI-1	Normal* PAI-1	PAI-1 Deficiency	Venous Occlusion
PAI-1 activity (AU/mL)	50	2.0-15	<1	<1
PAI-1 antigen (ng/mL)	100	4-50	<1	25
tPA activity (IU/mL)	0.1	0.3-1.7	1.2	8
tPA antigen (ng/mL)	25	3-12	1.7	20

PAI-1, plasminogen activator inhibitor-1; tPA, tissue plasminogen activator.
*Values shown represent 0800 h reference range for healthy subjects.

reduced clearance of tPA, or administration of exogenous tPA (thrombolytic therapy). Acquired increases in tPA will be discussed later in this chapter.

Alpha 2-Antiplasmin Deficiency

Both homozygous and heterozygous α_2-antiplasmin deficiencies associated with bleeding have been described. Homozygous α_2-antiplasmin deficiency is a rare event; the prevalence is unknown. Transmission of homozygous α_2-antiplasmin deficiency is autosomal recessive. Patients typically have moderate to severe bleeding starting in childhood, including umbilical bleeding, easy bruising, hemarthrosis, gingival bleeding, and prolonged hemorrhage after minor trauma or dental procedures. Rebleeding from wounds and recurrent wound hematomas are often described. Intramedullary hematoma in the diaphyses of long bones has been described in several cases.

The risk of bleeding with heterozygous α_2-antiplasmin deficiency is more controversial. In heterozygous deficiency, α_2-antiplasmin levels are typically 35% to 71% of normal. Approximately 22% of heterozygotes reported a mild bleeding disorder similar to PAI-1 deficiency (delayed bleeding after surgery or trauma, wound hematomas). The majority of heterozygous subjects discovered as part of familial studies, though, have reported no bleeding episodes.

Acquired Causes of Fibrinolytic Bleeding

Acquired causes of increased tPA include increased tPA secretion associated with cardiopulmonary bypass and decreased tPA clearance asso-

Case Study

Tissue Plasminogen Activator Release During Renal Transplantation

A 62-year-old man was undergoing cadaveric renal transplantation secondary to diabetes associated renal failure. The surgery was proceeding normally when the surgeons noted increasing blood loss in the operative field. The anesthesiologist reported new onset bleeding from all intravenous and intra-arterial lines. Blood loss accelerated rapidly, and emergent red blood cell transfusions were started. Coagulation studies performed shortly after the bleeding started showed a normal platelet count, elevated fibrinogen, and normal prothrombin time. Rapid blood loss continued. Whole blood was sent for thromboelastometry. The sample showed normal onset of clotting times (consistent with normal prothrombin time), normal rate of development of clot strength (consistent with normal platelet count and fibrinogen), but rapid lysis of the clot, similar to that shown in Figure 8-5. A diagnosis of accelerated fibrinolysis was made, and the patient was treated with epsilon-amino-caproic acid. The excessive bleeding rapidly diminished. Blood samples drawn at the time of the increased bleeding subsequently showed elevated levels of tPA antigen. The cause of the increased tPA release was never determined.

ciated with advanced cirrhosis and liver transplantation. Cardiopulmonary bypass is a strong stimulator of tPA release, probably due to activation of the contact system and increased kallikrein generation by the artificial surface of the bypass circuit. Kallikrein then binds to the vascular endothelium, where it converts kininogen into bradykinin, which in turn stimulates tPA release from the endothelial cell. Within minutes after starting cardiopulmonary bypass, there is a rapid rise in active and total tPA levels and a drop in active PAI-1 levels. This rise in active tPA is associated with increased plasmin generation, increased fibrin degradation, and intraoperative and postoperative bleeding. Accelerated fibrinolysis during cardiopulmonary bypass is often treated with aprotinin, a nonspecific serine protease inhibitor that binds trypsin, plasmin, and kallikrein, or specific fibrinolytic inhibitors like epsilon-aminocaproic acid. Increased release of tPA can occur in situations other than cardiopulmonary bypass. Massive tPA release has been associated with electric shock, complicated labor, and disseminated intravascular coagulation.

Another cause of accelerated fibrinolysis is reduced hepatic clearance of tPA. The rate of tPA clearance is directly proportional to hepatic blood flow. In normal individuals, approximately 25% of cardiac output flows to the liver, which clears approximately 50% of the tPA from blood in a single pass. This rapid clearance gives tPA a short half-life, on the order of 2 to 4 minutes. In patients with cirrhosis, tPA levels rise as hepatic blood flow declines. Cirrhotic patients may be at increased risk of bleeding during surgery due to elevated tPA levels. In addition, since the liver is a source of PAI-1 production, if liver synthetic function is also failing, fibrinolysis may be accelerated due to both increased tPA levels and decreased PAI-1. During the anhepatic phase of liver transplantation, this can be particularly severe. Little if any tPA clearance occurs once hepatic blood flow is stopped and liver PAI-1 production is eliminated. tPA levels can rise 10-fold above baseline, and enhanced bleeding due to uncontrolled fibrinolysis can be severe. Once the new liver is connected to the blood flow, tPA levels rapidly drop back to normal within about 10 to 20 minutes.

One relatively rapid method for detecting clinically significant increases in fibrinolytic activity is the use of whole blood clot lysis times, monitored using some type of viscoelastic measurement (such as thromboelastometry). In healthy subjects, the whole blood clot lysis time can be on the order of several days, making the measurement too slow for practical use. In contrast, in subjects with major elevations of tPA due to increased release (as happens during cardiopulmonary bypass) or lack of clearance due to advanced cirrhosis or the anhepatic phase of liver transplantation, whole blood clot lysis times can be as short as 20 to 30 minutes.

When this is the case, bleeding is likely to occur due to rapid lysis of developing hemostatic clots.

Disseminated Intravascular Coagulation

Under normal conditions, activation of hemostasis is limited to the site of the wound. Little if any coagulation activation occurs in the flowing blood. Any activated coagulation factors released from the wound site are rapidly inhibited downstream. Thrombin, factor Xa, factor IXa, and factor XIa in the blood are rapidly inhibited by antithrombin bound to endothelial heparan sulfates, whereas factors Va and VIIIa are destroyed by activated protein C. Some disorders result in a major increase in the release or exposure of hemostatic activating factors, resulting in activation of platelets and coagulation throughout the vascular system. This is termed disseminated intravascular coagulation (DIC). Table 13-2 provides a list of potential causes of DIC. Figure 13-2 illustrates the

Case Study

DIC Associated with Gastric Carcinoma

A 65-year-old man was admitted for back pain, which he treated with 10 aspirin tablets per day. On admission, he was found to have a large retroperitoneal bleed into his psoas muscle, epistaxis, and gastrointestinal bleeding. Laboratory studies showed:

Test	Value	Reference Range
Platelet count (per μL)	34,000	150,000-400,000
Fibrinogen (mg/dL)	163	150-400
PT (s)	8.5	10.7-15.6
aPTT (s)	39	22-35
D-dimer (ng/mL FEU)	45,000	<600
Hematocrit (%)	31	36-45

The patient showed reduced platelets, normal fibrinogen, and prolonged PT and aPTT, consistent with decreased coagulation factor levels, and elevated D-dimer, suggesting increased fibrin generation and consumption of factors. The patient continued bleeding after admission. He was supported with transfusion of red blood cells, platelets, and fresh frozen plasma. Further workup demonstrated a metastatic gastric carcinoma. After resection of the primary tumor, the consumptive coagulopathy decreased in intensity. The patient subsequently died secondary to metastatic disease.

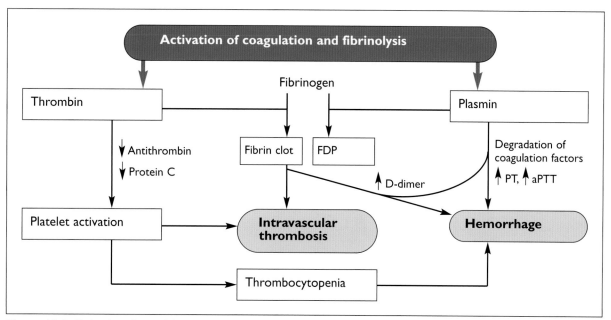

Figure 13-2. Pathophysiology of disseminated intravascular coagulation (DIC). Activation of the coagulation system leads to activation of thrombin and conversion of fibrinogen to fibrin, with activation of platelets and formation of microvascular thrombi. Simultaneous fibrinolytic activation leads to degradation of the fibrin clot and fibrinogen, along with degradation of other coagulation factors. This, in opposition, leads to a consumptive coagulopathy and hemorrhage. The balance between coagulation activation and fibrinolysis can dictate which clinical symptoms prevail, thrombosis or bleeding. Typical coagulation testing abnormalities seen in DIC include a decrease in fibrinogen with elevation of fibrin degradation products (FDP) and D-dimer, thrombocytopenia, elevated prothrombin time (PT) and activated partial thromboplastin time (aPTT), decreased coagulation factor levels, and decreased antithrombin and protein C.

pathophysiology of DIC. Factors that initiate disseminated coagulation in the blood include tissue factor released from damaged tissues through trauma, ischemia, or other causes; upregulation of tissue factor on cells; release or exposure of anionic phospholipids; and release of tumor-associated coagulation activating factors.

To understand DIC it is critical to remember that DIC is not a primary disorder; it is a secondary response of the hemostatic system to disseminated activation of coagulation, platelets, and the fibrinolytic system. DIC cannot be stopped until the underlying cause of the disseminated activation is diagnosed and treated. For example, some types of leukemia, particularly acute promyelocytic leukemia, can result in severe acute DIC due to release of coagulation activating factors by the leukemic cells as they die in the blood stream. Until the leukemia is diagnosed and treated, the DIC will continue. Symptomatic therapy can help reduce bleeding and thrombosis during DIC but will not stop it.

Disseminated intravascular coagulation is not a single disorder, but a spectrum of disease dependent on the amount and rate of release of coagulation activating factors into the blood. Slow release

of a small amount of coagulation-activating material into the blood, as occurs with moderate trauma and some slow-growing cancers, may be compensated for by normal hemostatic regulatory mechanisms. Steady release of an increasing

Table 13-2. Disorders Associated with Disseminated Intravascular Coagulation (DIC)

❏ Infections leading to sepsis

❏ Acute obstetrical complications (eg, amniotic fluid embolism)

❏ Preeclampsia and eclampsia

❏ Severe trauma, burns, or massive tissue necrosis

❏ Shock and heat stroke

❏ Venoms (snakes, spiders)

❏ Acute hemolysis (eg, drowning)

❏ Neoplasia (solid tumors, leukemias)

❏ Liver disease (eg, fulminant hepatic necrosis)

❏ Immunologic disorders (eg, heparin-induced thrombocytopenia)

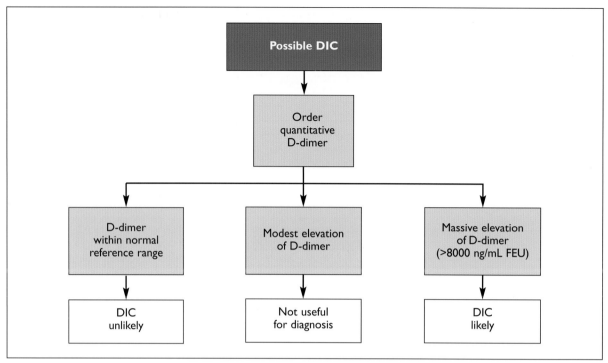

Figure 13-3. Clinical use of D-dimer assay to assess disseminated intravascular coagulation (DIC).

amount of coagulation activating factors (often seen in advanced adenocarcinomas) leads to chronic DIC with increased consumption of coagulation factors, thus the term consumptive coagulopathy. If the rate of release of coagulation activating factors increases slowly, there may be a compensatory increase in platelet and coagulation factor production. When the rate of release of coagulation activating factors increases beyond a certain point, the body cannot compensate for the increased loss of factors, leading to a drop in platelet count, fibrinogen, and coagulation factor levels. Eventually patients may acutely decompensate with rapid drops in hemostatic factor levels, leading to acute DIC. Acute DIC may also start rapidly due to a catastrophic insult, such as severe trauma, burn, obstetrical complication, shock, or overwhelming sepsis.

The clinical and laboratory presentation of DIC is dependent upon whether the insult to the hemostatic system is chronic versus acute, mild versus severe, localized versus systemic, thrombotic versus hemorrhagic, and whether coagulation activation or fibrinolytic activation predominates. Localized coagulation activation and consumption can occur within aortic aneurysms, large hemangiomas, and with injury to specific organs, such as hyperacute renal allograft rejection.

Laboratory diagnosis of DIC can be difficult due to the highly variable nature of inciting causes and clinical presentations. In general, for the more severe forms of DIC, diagnosis is based on low or dropping platelet count, low or dropping fibrinogen, prolongation of the prothorombin time (PT) or activated partial thromboplastin time (aPTT), and evidence of increased coagulation activation, usually by measuring markers of fibrin formation and degradation, including measurement of fibrin degradation products (FDPs), D-dimer, and soluble fibrin monomers. Platelet counts drop and FDP/D-dimer levels rise in most cases of acute severe DIC. Falling fibrinogen and prolongation of clotting tests occurs less often.

During severe DIC, thrombin is generated throughout the vascular system, leading to systemic formation and deposition of fibrin. Ongoing thrombin formation leads to an increase in thrombin inactivation by antithrombin and thrombomodulin, leading to a decrease in antithrombin and protein C levels. Concurrently, fibrinolysis is activated, leading to plasmin formation and degradation of both fibrinogen and fibrin. Under normal circumstances, fibrin is formed at the site of a wound and remains there polymerized as part of the clot. During DIC, fibrin is generated throughout the vascular system, leading to increased amounts of circulating fibrin monomers

Table 13-3. Comparison of Disseminated Intravascular Coagulation (DIC) Scoring Systems

	International Society on Thrombosis and Haemostasis	Japanese Ministry of Health and Welfare
Underlying disease	Essential	1 point
Bleeding	0 point	HM(-), 1 point HM(+), 0 point
Thrombotic organ failure	0 point	1 point
Prothrombin time (PT)	Prolongation ≥3 s, 1 point ≥6 s, 2 points	Ratio ≥1.25, 1 point ≥1.67, 2 points
Fibrinogen	≤1.0 g/L, 1 point	≤1.5 g/L, 1 point ≤1.0 g/L, 2 points
Fibrin degradation products (FDP) or D-dimer	Moderate increase, 2 points Strong increase, 3 points	FDP ≥10 μg/mL, 1 point FDP ≥20 μg/mL, 2 points FDP ≥40 μg/mL, 3 points
Platelet count (×10³/μL)	≤100, 1 point ≤50, 2 points	HM(+), 0 point HM(-), ≤120, 1 point ≤80, 2 points ≤50, 3 points
Total points, overt DIC	≥5 points	HM(+) ≥4 points HM(-) ≥7 points

HM(+), patients with hematopoietic malignancy; HM(-), patients without hematopoietic malignancy. Modified from Wada H, et al. Comparison of diagnostic criteria for disseminated intravascular coagulation. *Am J Hematol.* 2003;74:17-22. Copyright 2003 Wiley-Liss Inc. Reprinted with permission of Wiley-Liss, Inc, a subsidiary of John Wiley & Sons, Inc.

or small oligomers complexed with several fibrinogen molecules in the blood. This is termed soluble fibrin. Fibrinogen, soluble fibrin, and fibrin deposited on vessel walls are degraded by the fibrinolytic system, leading to a complex mixture of intact and degraded fibrin and fibrinogen. In vivo studies indicate that the majority of fibrinogen/fibrin products formed in the blood during DIC are larger in size than fibrinogen, indicating a complex of one or more fibrinogen/fibrin molecules. True D-dimer, a small fragment about half the molecular weight of fibrinogen, makes up only a small part of the degradation fragments present in vivo during DIC. Most D-dimer and FDP assays detect a range of different fibrin fragment sizes.

D-dimer and FDP assays are the single most useful tools in evaluating DIC. Rapid quantitative D-dimer assays are routinely available on most automated coagulation analyzers. Many hospitalized patients have a low platelet count and low fibrinogen due to causes other than DIC. If the D-dimer is normal, it essentially rules out DIC

(Figure 13-3). If the D-dimer level is modestly elevated, 1000 to 4000 ng/mL FEU (fibrinogen equivalent units), a modest amount of intravascular hemostatic activation is occurring, but symptomatic DIC is unlikely. If the D-dimer level is in the range of 8000 to 10,000 ng/mL FEU or more, DIC is increasingly likely. A large intravascular thrombus, such as extensive venous thromboembolism, may also give high D-dimer levels but seldom produces D-dimer levels above 8000 ng/mL FEU. Use of D-dimer to evaluate DIC requires a rapid, precise, quantitative D-dimer assay. Semiquantitative or threshold assays are not as useful. These assays may help rule out DIC in some cases but have a low positive predictive value.

Another use of quantitative D-dimer assays is monitoring the progress of DIC once it is diagnosed. Decreasing D-dimer levels are a good indication that the therapy is working and the DIC improving. While latex agglutination assays for FDP can be useful in the initial diagnosis, their qualitative nature makes them less helpful for monitoring the progress of DIC. Soluble fibrin monomer assays are potentially useful in the diagnosis and monitoring of DIC, but currently no rapid, quantitative, soluble fibrin monomer assays are approved for use, although they are under development.

Accurate diagnosis of DIC can be difficult due to the many causes and presentations that occur clinically. To aid in the diagnosis of DIC, scoring systems have been developed by the International Society on Thrombosis and Haemostasis and the Japanese Ministry of Health and Welfare. Table 13-3 shows a comparison of the scoring systems. In acute severe cases, there is good agreement by both scoring systems that DIC is occurring. In more borderline cases, studies have shown the two systems disagree due to different weights given to each factor in the score.

Suggested Reading

PAI-1 Deficiency

Dieval J, Nguyen G, Gross S, Delobel J, Kruithof EK. A lifelong bleeding disorder associated with a deficiency of plasminogen activator inhibitor type 1. *Blood.* 1991;77:528-532.

Fay WP, Shapiro AD, Shih JL, Schleef RR, Ginsberg D. Complete deficiency of plasminogen activator inhibitor type 1 due to frameshift mutation. *N Engl J Med.* 1992;327:1729-1733.

Fay WP, Parker AC, Condrey LR, Shapiro AD. Human plasminogen activator inhibitor-1 (PAI-1) deficiency: characterization of a large kindred with a null mutation in the PAI-1 gene. *Blood.* 1997;90:204-208.

Minowa H, Takahashi Y, Tanaka T, et al. Four cases of bleeding diathesis in children due to congenital plasminogen activator inhibitor-1 deficiency. *Haemostasis.* 1999;29:286-291.

Lee MH, Vosburgh E, Anderson K, McDonagh J. Deficiency of plasma plasminogen activator inhibitor 1 results in hyperfibrinolytic bleeding. *Blood.* 1993;81:2357-2362.

Repine T, Osswald M. Menorrhagia due to a qualitative deficiency of plasminogen activator inhibitor-1: case report and literature review. *Clin Appl Thromb Hemost.* 2004;10:293-296.

Schleef RR, Higgins DL, Pillemer E, Levitt LJ. Bleeding diathesis due to decreased functional activity of type 1 plasminogen activator inhibitor. *J Clin Invest.* 1989;83:1747-1752.

Takahashi Y, Tanaka T, Minowa H, et al. Hereditary partial deficiency of plasminogen activator inhibitor-1 associated with a lifelong bleeding tendency. *Int J Hematol.* 1996;64:61-68.

Zhang ZY, Wang ZY, Dong NZ, Bai X, Zhang W, Ruan CG. A case of deficiency of plasma plasminogen activator inhibitor-1 related to Ala15Thr mutation in its signal peptide. *Blood Coagul Fibrinolysis.* 2005;16:79-84.

Antiplasmin Deficiency

Favier R, Aoki N, de Moerloose P. Congenital alpha(2)-plasmin inhibitor deficiencies: a review. *Br J Haematol.* 2001;114:4-10.

Harish VC, Zhang L, Huff JD, et al. Isolated antiplasmin deficiency presenting as a spontaneous bleeding disorder in a 63-year-old man. *Blood Coagul Fibrinolysis.* 2006;17(8):673-675.

Saito H. Alpha 2-plasmin inhibitor and its deficiency states. *J Lab Clin Med.* 1988;112:671-678.

Acquired Causes of Fibrinolytic Bleeding

Bennett B, Croll A, Ferguson K, Booth NA. Complexing of tissue plasminogen activator with PAI-1, a2-macroglobulin, and C1-inhibitor: studies in patients with defibrination and a fibrinolytic state after electroshock or complicated labor. *Blood.* 1990;75:671-676.

Chandler WL, Alessi MC, Aillaud MF, Henderson P, Vague P, Juhan-Vague I. Clearance of TPA and TPA/PAI-1 complex: relationship to elevated TPA antigen in patients with high PAI-1 activity levels. *Circulation.* 1997;96:761-768.

Chandler WL, Fitch JCK, Wall MH, et al. Individual variations in the fibrinolytic response during and after cardiopulmonary bypass. *Thromb Haemost.* 1995;74:1293-1297.

Chandler WL, Velan T. Secretion of tissue plasminogen activator and plasminogen activator inhibitor 1 during cardiopulmonary bypass. *Thromb Res.* 2004; 112:185-192.

Chandler WL, Velan T. Plasmin generation and D-dimer formation during cardiopulmonary bypass. *Blood Coagul Fibrinolysis.* 2004;15:583-591

Eisses MJ, Seidel K, Aldea GS, Chandler WL. Reducing hemostatic activation during cardiopulmonary bypass: a combined approach. *Anesth Analg.* 2004;98:1208-1216.

Gram J, Janetzko T, Jespersen J, Bruhn H. Enhanced effective fibrinolysis following the neutralization of heparin in open heart surgery increases the risk of post-surgical bleeding. *Thromb Haemost.* 1990;63:241-245.

Hersch SL, Kunelis T, Francis RB Jr. The pathogenesis of accelerated fibrinolysis in liver cirrhosis: a critical role for tissue plasminogen activator inhibitor. *Blood*. 1987;69:1315-1319.

Holloway DS, Summaria L, Sandesara J, Vagher JP, Alexander JC, Caprini JA. Decreased platelet number and function and increased fibrinolysis contribute to postoperative bleeding in cardiopulmonary bypass patients. *Thromb Haemost*. 1988;59:62-67.

Huber K, Kirchheimer C, Korninger C, Binder BR. Hepatic synthesis and clearance of components of the fibrinolytic system in healthy volunteers and in patients with different stages of liver cirrhosis. *Thromb Res*. 1991;62:491-500.

Ray MJ, Marsh NA, Hawson GAT. Relationship of fibrinolysis and platelet function to bleeding after cardiopulmonary bypass. *Blood Coagul Fibrinolysis*. 1994;5:679-685.

Spiess BD, Vocelka C, Cochran RP, Soltow L, Chandler WL. Heparin coated bypass circuits (Carmeda) suppresses the release of tissue plasminogen activator during coronary artery bypass graft surgery. *J Cardiothor Vasc Anesth*. 1998;12:299-304.

Disseminated Intravascular Coagulation

Marder VJ, Feinstein DI, Colman RW, Levi M. Consumptive thrombohemorrhagic disorders. In: Colman RW, Marder VJ, Clowes AW, George JN, Goldhaber SZ, eds. *Hemostasis and Thrombosis: Basic Principles and Clinical Practice*. Philadelphia, Pa: Lippincott Williams and Wilkins; 2006: 1571-1600.

Horan JT, Francis CW. Fibrin degradation products, fibrin monomer and soluble fibrin in disseminated intravascular coagulation. *Semin Thromb Hemost*. 2001;27:657-666.

Lehman CM, Wilson LW, Rodgers GM. Analytic validation and clinical evaluation of the STA LIATEST immunoturbidimetric D-dimer assay for the diagnosis of disseminated intravascular coagulation. *Am J Clin Pathol*. 2004;122:178-184.

Taylor FB Jr, Toh CH, Hoots WK, Wada H, Levi M. Towards definition, clinical and laboratory criteria, and a scoring system for disseminated intravascular coagulation. *Thromb Haemost*. 2001;86:1327-1330.

Wada H, Gabazza EC, Asakura H, et al. Comparison of diagnostic criteria for disseminated intravascular coagulation (DIC): diagnostic criteria of the International Society of Thrombosis and Hemostasis and of the Japanese Ministry of Health and Welfare for overt DIC. *Am J Hematol*. 2003;74:17-22.

CHAPTER 14

Platelet Disorders

Kandice Kottke-Marchant, MD, PhD

Introduction

The laboratory evaluation of hemorrhagic platelet disorders usually begins with measurement of platelet count and platelet morphology, followed by platelet functional assays. Due to this typical approach, platelet disorders in this chapter have been divided into several categories, based upon whether they are associated with a *normal, increased,* or *decreased* platelet count. Some categories are then further subdivided by platelet morphology, with giant platelets, normal sized platelets, or small platelets.

In contrast, thrombotic disorders are usually multifactorial, entailing many diverse pathologic states, including vascular abnormalities and thrombophilias as well as platelets, so the platelet-associated thrombotic disorders are discussed together at the end of the review.

Platelet Dysfunction with Normal Platelet Count (Qualitative Platelet Disorders)

The laboratory finding of platelet dysfunction with a normal platelet count usually indicates a qualitative platelet disorder. These disorders would be considered in a patient with a bleeding diathesis despite a normal prothrombin time (PT), activated partial thromboplastin time (aPTT), and normal platelet count. An algorithmic approach to these disorders is shown in Figure 14-3. Tests for von Willebrand disease would also be normal. As

Case Study

A 45-year-old woman with a lifelong history of ecchymosis and epistaxis was evaluated for a hemostatic functional disorder prior to a cholecystectomy. Her initial laboratory studies showed a normal PT of 10.5 seconds (reference range 9.9-13.0 s) and a normal aPTT of 31.6 seconds (reference range 24.5-34.0 s). Her platelet count was normal at 210,000/μL with a normal mean platelet volume (MPV) of 9.9 fL. A peripheral smear showed a normal number of platelets and normal platelet morphology. Because of the history of bleeding, a platelet function screen with the PFA-100 was performed, together with testing for von Willebrand disease. The PFA-100 was abnormal, with >300 second closure time with the collagen/epinephrine cartridge and >300 second closure time with the collagen/ADP (adenosine diphosphate) cartridge. The von Willebrand antigen was normal at 140% (reference range 45%-150%), with a normal ristocetin cofactor of 128% (reference range 41%-223%) and factor VIII activity of 135% (reference range 57%-199%).

These results are unlikely to reflect a coagulation protein disorder or von Willebrand disease, as the PT and aPTT are normal and all of the von Willebrand assays are also normal. Despite a normal platelet count, the abnormal PFA-100 suggests a platelet function defect, but this could be due to either an antiplatelet drug effect or an intrinsic platelet disorder. A careful drug history did not reveal any medication that could potentially affect platelet function, so additional platelet testing was performed. Platelet aggregation studies showed an absence of aggregation with the agonists ADP, epinephrine, collagen, and arachidonic acid (Figure 14-1, A), but normal aggregation with ristocetin (Figure 14-1, B). Platelet flow cytometry was performed, which showed absent surface expression of glycoprotein (GP) IIb/IIIa (CD41 and CD61), but normal expression of glycoprotein Ib (CD42b) (Figure 14-2). Taken together, the results are diagnostic for Glanzmann thrombasthenia, a congenital deficiency of the GPIIb/IIIa fibrinogen receptor on the platelet surface.

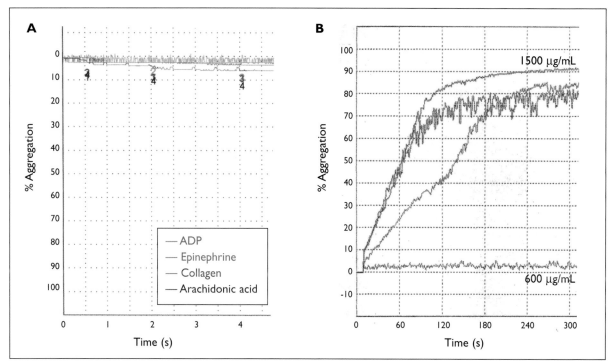

Figure 14-1. A. Platelet aggregation in the patient showed an absent aggregation response to adenosine diphosphate (ADP), epinephrine, collagen, and arachidonic acid. These agonists all lead to platelet aggregation mediated by the glycoprotein (GP) IIb/IIIa fibrinogen receptor. B. In contrast, aggregation to ristocetin was normal. Ristocetin induces a conformational change in von Willebrand factor and facilitates aggregation through the GPIb/IX/V complex. (Helena PACKS-4 Optical Aggregometer.)

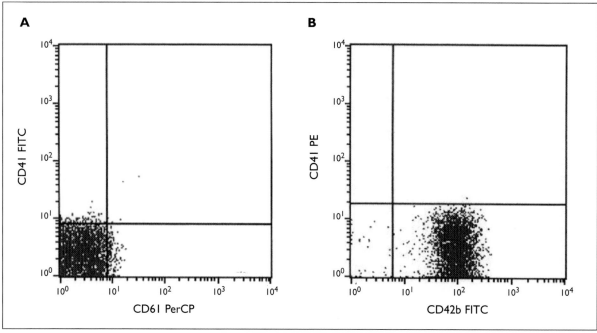

Figure 14-2. A. Platelet flow cytometry showed absent expression of both CD41 and CD61, indicative of a lack of surface expression of both the glycoprotein (GP) IIb (α_{IIb}) and GPIIIa (β_3) protein chains. This finding is diagnostic for Glanzmann thrombasthenia, when seen in conjunction with the platelet aggregation studies in Figure 14-1. B. Although CD41 is absent, the platelets still express the GPIb receptor, CD42b. (Becton Dickinson FACScan.)

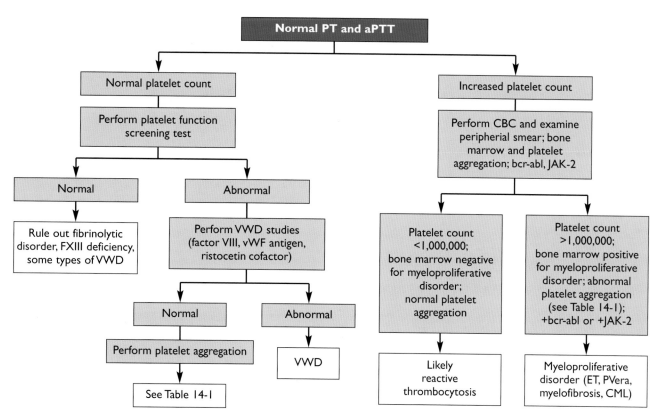

Figure 14-3. Diagnostic algorithm for distinguishing platelet abnormalities with a normal or increased platelet count. In patients with a normal platelet count, coagulopathies, such as von Willebrand disease (VWD) and fibrinolytic dysfunction, should be excluded. In patients with an elevated platelet count, myeloproliferative processes should be distinguished from a reactive thrombocytosis. Abbreviations: CML, chronic myelogenous leukemia; ET, essential thrombocytopenia; PVera, polycythemia vera.

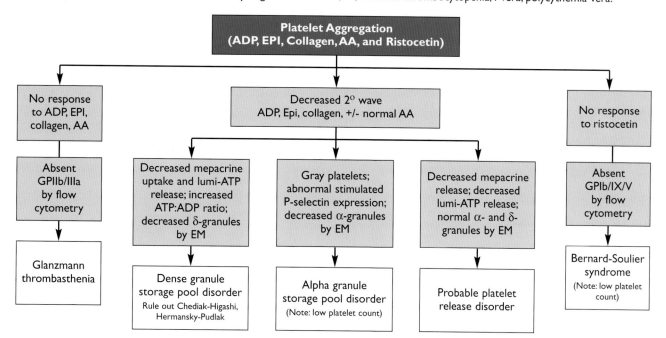

Figure 14-4. Platelet aggregation studies can be useful to distinguish between many acquired platelet disorders. A lack of response to adenosine diphosphate (ADP), epinephrine (EPI), collagen, and arachidonic acid (AA), with normal aggregation to ristocetin, is characteristic for Glanzmann thrombasthenia. Conversely, lack of aggregation response to ristocetin suggests Bernard-Soulier syndrome. Other ancillary tests can help to distinguish between the platelet granule storage pool disorders and platelet release defects. Abbreviations: ATP, adenosine triphosphate; EM, electron microscopy; GP, glycoprotein.

Table 14-1. Association of Platelet Aggregation Results with Platelet Disorders

Disorder	ADP 1o	2o	AA	EPI	Collagen	Ristocetin	Other Studies
Storage pool disorders							
Dense granule platelet storage pool disorder (δ-SPD)	N	↓	N	N or ↓	N or ↓	N	Decreased ATP release by lumiaggregometry Decreased dense granule mepacrine uptake by flow cytometry Decreased dense granules by TEM Abnormally high ATP/ADP ratio Acquired SPD " exhausted platelets" (CPB, DIC, TTP, HUS, MPD) Albinism in Hermansky-Pudlak and Chediak-Higashi Infections, small platelets seen with Wiskott-Aldrich
Alpha (α)-SPD Gray platelet syndrome	Var	Var	N	N	Var ↓	N	Pale platelets on smear Decreased α-granules by TEM Decreased P-selectin in some patients, but high constitutive expression in others
Surface glycoprotein disorders							
Bernard-Soulier syndrome	N	N	N	N	N	↓ or Abs	Macrothrombocytopenia Deficiency of GPIb/IX/V (1 or more) by flow cytometry
Glanzmann thrombasthenia	↓ or Abs	↓ or Abs	↓ or Abs	↓ or Abs	↓ or Abs	N	Deficiency of GPIIb and/or GPIIIa by flow cytometry Mutations of GPIIb and/or GPIIIa
Platelet-type VWD	N	N	N	N	N	↑	Loss of high-molecular-weight vWF multimers Gain of function mutation of GPIb
VWD, other	N	N	N	N	N	↓, N or ↑	FVIII:C, vWF:Ag, vWF:RiCof, vWF multimers
ADP receptor	↓	↓	N	N or ↓	N or ↓	N	Gene defects of P2Y₁₂ or P2Y₁
GPIa/IIa or GPVI disorder	N	N	N	N	↓	N	Associated mutations of GPIa/IIa or GPVI genes Decreased adhesion to collagen
Activation disorders							
Signal transduction disorders	N	↓	Var ↓	Var ↓	Var ↓	N	Decreased granule release with normal numbers of granules Receptor defects may show decreased aggregation to EPI, collagen only Decreased G-protein activation, phospholipase C activation, calcium mobilization, pleckstrin or tyrosine phosphorylation
Scott syndrome	N	N	N	N	N	N	Defective platelet procoagulant activity (ie, PF3) Defective microparticle formation
Defects of thromboxane synthesis or aspirin-like drug	N	↓↓	↓ or Abs	↓	↓ or Abs	N	Normal aggregation with prostaglandin G2 seen with aspirin or cyclooxygenase deficiency Decreased/absent prostaglandin G2 aggregation with thromboxane synthetase deficiency

Abbreviations: AA, arachidonic acid; Abs, absent; ADP, adenosine diphosphate; ATP, adenosine triphosphate; BUN, blood urea nitrogen; CPB, cardiopulmonary bypass; DIC, disseminated intravascular coagulation; EPI, epinephrine; GP, glycoprotein; HUS, hemolytic uremic syndrome; MPD, myeloproliferative disorder; N, normal; PF3, platelet factor 3; RiCof, ristocetin cofactor; SPD, storage pool disorder; TEM, transmission electron microscopy; TTP, thrombotic thrombocytopenic purpura; VWD, von Willebrand disease; vWF, von Willebrand factor.

Table 14-1. Association of Platelet Aggregation Results with Platelet Disorders (continued)

Disorder	ADP 1o	2o	AA	EPI	Collagen	Ristocetin	Other Studies
Other disorders with platelet dysfunction							
Myeloproliferative disorder	N	N	N	↓ or Abs delayed lag	N	N	Other abnormalities: α or δ-SPD, cyclooxygenase abnormality; other surface GP derangements; decreased or increased aggregation to ADP, collagen; spontaneous aggregation
Uremia	N	↓	↓	N / ↓	N / ↓	N	Abnormal creatinine, BUN Decreased PF3
Drug effects							
Thienopyridines	↓	Abs	N / ↓	N or ↓	N or ↓	N	History of clopidogrel or ticlopidine therapy
GPIIb/IIIa antagonists	↓ or Abs	↓ or Abs	↓ or Abs	↓ or Abs	↓ or Abs	N	History of treatment: abciximab, tirofiban, or eptifibatide Increased receptor occupancy by flow cytometry

highlighted in the case study, the platelet function screening test would be abnormal, and platelet aggregation studies would be performed to distinguish the disorders (Figure 14-4 and Table 14-1). This would be followed by more specific tests, if required. Most drug-induced platelet dysfunction will also demonstrate abnormal platelet function with a normal platelet count, so it is extremely important to take a careful drug history. Platelet aggregation abnormalities typically found with antiplatelet drugs, such as aspirin, glycoprotein IIb/IIIa (GPIIb/IIIa) antagonists, or the thienopyridines, can be found in Table 14-1. These are also discussed in chapter 25.

Glycoprotein Disorders

Glanzmann Thrombasthenia (GPIIb/IIIa Deficiency)

Glanzmann thrombasthenia is a congenital deficiency or dysfunction of GPIIb/IIIa (α_{IIb}/β_3 integrin), the fibrinogen receptor responsible for mediating platelet aggregation. It is an autosomal recessive disorder that manifests in lifelong mucocutaneous bleeding, with purpura, epistaxis, gingival bleeding, and menorrhagia as common features. Glanzmann thrombasthenia can be classified according to the amount of GPIIb/IIIa: type I, 0% to 5% of normal; type II, 6% to 20% of normal; and variant disease, 50% to 100% of normal with abnormal fibrinogen binding.

The genes for both α_{IIb} and β_3 are located at chromosome 17q21-23, with the α_{IIb} gene having 30 exons and the β_3 gene having 15 exons. Over 40 different mutations and partial deletions of both GPIIb and GPIIIa have been described. Because of the large number of mutations identified, many patients with Glanzmann thrombasthenia are compound heterozygotes. Often, mutations of one subunit prevent the formation of the entire complex on the platelet surface, and it is common to detect neither GPIIb nor GPIIIa on the surface of the platelet.

The complete blood count (CBC) in individuals with Glanzmann thrombasthenia is usually normal, with normal platelet morphology on the peripheral smear. In patients with Glanzmann thrombasthenia, the bleeding time and Platelet Function Analyzer-100 (PFA-100) will be abnormal (many laboratories no longer perform the bleeding time due to lack of sensitivity and specificity for platelet disorders). Absent or decreased aggregation response will be seen upon addition of adenosine diphosphate (ADP), collagen, epinephrine, and arachidonic acid aggregating agents, whereas the ristocetin-induced aggregation is normal (Figure 14-1). This finding is virtually diagnostic of Glanzmann thrombasthenia, but the disorder can be confirmed by platelet flow cytometry or crossed immunoelectrophoresis of platelet membrane proteins (Table 14-1 and Figure 14-2). Most patients with type I Glanzmann thrombasthenia will have no surface expression of either GPIIb or GPIIIa. Patients with type II Glanzmann thrombasthenia may have decreased, but detectable, surface expression of both proteins. Distinguishing patients with variant Glanzmann thrombasthenia from heterozygous carriers may be difficult, as both may have slightly decreased levels of the glycoprotein IIb/IIIa complex on the platelet surface. Flow cytometric analysis does not usually provide a quantitative measure of surface protein expression, but some quantitative assays are now available that employ calibration beads. Additional laboratory studies in patients with Glanzmann thrombasthenia will show decreased platelet-associated fibrinogen, defective fibrinogen binding to platelets (studied by flow cytometry in specialized laboratories), and decreased clot retraction.

Genetic diagnosis of Glanzmann thrombasthenia can be performed by direct sequencing of polymerase chain reaction (PCR) products from individual exons. However, these diagnostic tests are not widely available, even in research laboratories, because of the complexity of the gene and the large number of known mutations.

Glanzmann thrombasthenia has distinctive laboratory features and usually can be definitively diagnosed. However, afibrinogenemia, a rare deficiency of fibrinogen, can present with similar initial platelet aggregation results with absent response to ADP, epinephrine, collagen, and arachidonic acid. The aggregation defect in afibrinogenemia is restored with addition of fibrinogen to the specimen. Fibrinogen levels are markedly reduced or undetectable. Additionally, patients with afibrinogenemia have extremely prolonged fibrinogen-dependent clotting tests, such as PT and aPTT. Flow cytometric analysis of platelets in afibrinogenemia is normal. The GPIIb/IIIa antagonist drugs, such as abciximab or tirofiban, induce a Glanzmann-like inhibition of platelet function as part of their potent antiplatelet mechanism, but these are easily distinguished by the history of intravenous medication therapy.

The bleeding symptoms in Glanzmann thrombasthenia are significant, and prophylactic therapy with platelet transfusions is recommended prior to major surgical procedures. Desmopressin acetate (DDAVP) leads to increased blood levels of von Willebrand factor (vWF) and does not directly affect the platelets, but has been shown to lessen bleeding complications in Glanzmann thrombasthenia. Antifibrinolytic therapies, such as ε aminocaproic acid or tranexamic acid, have been shown to abrogate the bleeding complications, likely due to clot stabilization. There are some recent studies that have demonstrated success of recombinant factor VII (rFVIIa) in treatment of Glanzmann thrombasthenia.

Glycoprotein Ib/IX/V (Bernard-Soulier Syndrome)

Deficiency of the glycoprotein Ib/IX/V complex is accompanied by thrombocytopenia with large platelets, so this entity is discussed below under macrothrombocytopenia disorders.

Platelet-Type von Willebrand Disease

Most types of von Willebrand disease are due to abnormalities of the vWF protein, but platelet-type von Willebrand disease (also called pseudo-von Willebrand disease) is due to a molecular abnormality of the vWF receptor on platelet surfaces, GPIb/IX/V. This is an autosomal dominant inherited disorder in which an unusual gain-of-function abnormality of GPIb leads to increased binding of platelet GPIb/IX/V to vWF, with loss of high-molecular-weight vWF multimers from the plasma, resulting in a bleeding diathesis.

Platelet-type von Willebrand disease is due to a molecular defect of GPIb (Gly233Val, Gly233Ser, or Met239Val), leading to increased binding of vWF to the platelet surface. The affected amino acid residues are in a region of GPIb that adopts a β sheet conformation upon binding to vWF, and the mutations are thought to stabilize this binding.

The platelet morphology in platelet-type von Willebrand disease is usually normal, but some patients with giant platelet morphology have been described. Platelet-type von Willebrand disease closely resembles Type 2B von Willebrand disease, which is due to a gain-of-function mutation in the A2 domain of vWF, also leading to clearance of high-molecular-weight multimers from plasma. The two disorders can be distinguished by DNA analysis of exon 28 of the *VWF* gene, in which the mutations responsible for type 2B von Willebrand disease reside. This *VWF* sequencing is available at some large reference laboratories. Von Willebrand factor analysis is similar in the two disorders, with the findings of low to normal vWF and factor VIII levels and decreased functional ristocetin cofactor activity. Ristocetin aggregation shows an increased dose response in both disorders, and plasma vWF multimer analysis shows loss of high-molecular-weight forms. However, addition of cryoprecipitate or disialyated (asialo) vWF with its high-molecular-weight vWF forms causes spontaneous aggregation in platelets from individuals with platelet-type von Willebrand disease, but not with type 2B von Willebrand disease. See chapter 16 for a complete discussion of von Willebrand disease diagnosis.

Glycoprotein IV

Glycoprotein IV (GPIV), also known as CD36, is a protein on the platelet surface that is postulated to be an adhesive receptor for collagen and thrombospondin. It may also serve as a scavenger receptor for oxidized lipids. It is deficient in approximately 4% to 7% of healthy donors in the Japanese population and 7% to 10% of normal individuals in a sub-Saharan African population. The molecular basis of GPIV deficiency is a polymorphism in codon 90, leading to a Ser132Pro shift. A clinical association with GPIV deficiency has not been elucidated specifically.

Collagen Receptor Disorders (GPIa/IIa and GPVI)

The integrin $\alpha_2\beta_1$ (GPIa/IIa) is a platelet surface receptor for collagen that also is present on many other cell types. Rare patients with deficiency of this receptor and selective absence of platelet aggregation to collagen have been described. These patients have also been described to have decreased adhesion to vascular subendothelium, poor activation, and decreased spreading. An acquired deficiency of GPIa/IIa has been described in isolated cases of myeloproliferative disorders. Genetic polymorphisms of GPIa/IIa that lead to dysfunction have been described. A decreased platelet surface density of the receptor may correlate with decreased platelet adhesion to collagen and an increased bleeding risk in association with von Willebrand disease.

Recently, deficiency of GPVI has been described in Japanese subjects in conjunction with a mild bleeding diathesis and lack of collagen-induced platelet aggregation. A very rare patient with an antibody directed against GPVI that interfered with collagen-induced aggregation has been described.

Adenosine Diphosphate Receptor Abnormalities

As the primary receptor for adenosine diphosphate (ADP), genetic or functional variations in the $P2Y_{12}$ ADP receptor at chromosome 3q24-25 could have a marked effect on platelet function. Little is known about the pathology of the primary platelet ADP receptors. However, a rare congenital platelet defect characterized by severe impairment of platelet response to ADP has been reported, associated with a significant bleeding disorder and 2-bp deletion (CA) at codon 240 of the $P2Y_{12}$ gene. The deletion results in a frameshift for 28 residues before introducing a premature stop codon. In individuals with the congenital disorder, even high doses of ADP have been reported to induce minimal and rapidly reversible platelet aggregation, together with a markedly reduced binding of ADP to platelets and a failure of ADP to lower cytosolic cyclic adenosine monophosphate (cAMP). Other rare mutations of both the $P2Y_{12}$ and P2X1 genes that are associated with bleeding symptoms have been identified. Recently, a $P2Y_{12}$ H2 haplotype has been associated with peripheral arterial disease and variability in ADP-induced platelet aggregation.

Platelet Release Defects

Storage Pool Disorders (Alpha and Dense Granule Disorders)

Abnormalities of platelet secretion can be due to either deficiency of platelet granules or defects in the signal transduction events that regulate secretion or aggregation. Platelet storage pool disorders (SPDs) are the result of a deficiency of granules (alpha and/or dense granules). Platelet storage pool disorders can be congenital or acquired, but the molecular defect underlying most types is unknown. Dense granule storage pool disorders (δ-SPD) can be seen as a singular clinical entity or as part of other hereditary disorders, such as Chediak-Higashi syndrome, Hermansky-Pudlak syndrome, thrombocytopenia with absent radii (TAR syndrome), or Wiskott-Aldrich syndrome. Between 10% to 18% of patients with congenital platelet dysfunction have storage pool disorders. The alpha storage pool disorder (α-SPD) leading to the gray platelet syndrome is discussed below under macrothrombocytopenias, as this disorder leads to the production of decreased numbers of large platelets devoid of alpha granules, giving them a ghostly gray color on the peripheral smear. A rare α/δ-SPD that has features of both disorders has been described.

Patients with δ-SPD have a bleeding diathesis of variable severity, usually characterized by mucocutaneous bleeding. In more severely affected patients, perioperative bleeding is also observed. There have been rare familial associations with primary pulmonary hypertension and development of acute myelogenous leukemia. Therapy often is not necessary if the bleeding symptoms are mild, but preoperative platelet transfusions or DDAVP therapy may be required in some patients.

The platelet count and CBC indices are normal in patients with δ-SPD, and the peripheral smear

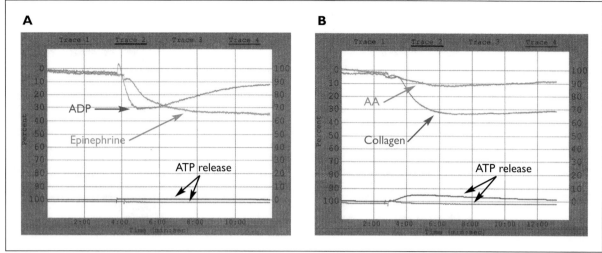

Figure 14-5. Lumiaggregation can be useful in the diagnosis of dense granule storage pool disorder (δ-SPD), as the luciferin-luciferase enzyme can detect granule adenosine triphosphate (ATP) release by luminescence. In δ-SPD, there is typically a primary wave of aggregation without a sec- ondary wave. The ATP release is decreased to absent. A. Agonists: adenosine diphosphate (ADP) at 5 µM and epinephrine at 10 µM. B. Agonists: collagen 0.2 mg/mL and arachidonic acid (AA) 0.5 mg/mL. (Chronolog Lumi-Aggregometer.)

shows platelets with apparently normal granule staining, as the purple granules usually visualized on the Wright stain are largely alpha granules, which are present in normal numbers in this disorder. Bone marrow evaluation is not usually helpful, as megakaryocytes are present in normal number with normal morphology. A decreased aggregation response to ADP, epinephrine, arachidonic acid, and collagen is often seen with δ-SPD (Figure 14-5). Aggregation studies with ADP and epinephrine characteristically show only a primary wave of aggregation, with an absent to decreased secondary wave, but aggregation studies with these agonists may be normal in up to 25% of patients. Aggregation with arachidonic acid is often normal, as well. The bleeding time and PFA-100 may be abnormal in some but not all patients. Decreased adenosine triphosphate (ATP) release is observed by lumiaggregometry (Figure 14-5), and decreased mepacrine dense granule uptake is observed by flow cytometry (Figure 14-6). Agonist-stimulated mepacrine release from dense granules is usually not impaired.

Ultrastructural abnormalities in these disorders usually show decreased dense granules, although dense granules are not usually absent (Figure 14-7, A). Often the dense granules lack their characteristic electron dense core and appear empty. The dense granule membrane protein, granulophysin (CD63), is usually expressed in normal amounts upon platelet activation in isolated δ-SPD. The activation-dependent expression of granulophysin

can be measured by platelet flow cytometry. Due to decreased platelet ADP content, an increased platelet ATP to ADP ratio is often seen with δ-SPD. α-SPD (gray platelet syndrome) has decreased alpha granules and is usually considered a macrothrombocytopenia.

Acquired platelet storage pool disorders can be seen with underlying myeloproliferative disorders where the platelet degranulation is defective as a result of the disease. Circulating "exhausted" platelets simulating SPD can be observed in clinical scenarios where there is ongoing in vivo platelet activation, such as cardiopulmonary bypass, with implantation of ventricular assist devices or total artificial hearts, and in disseminated intravascular coagulation and thrombotic thrombocytopenic purpura/hemolytic uremic syndrome (Figure 14-7, C). The isolated δ-SPD can be distinguished from these disorders by lack of specific clinical associations, coagulation test abnormalities, or thrombosis.

Chediak-Higashi Syndrome. Patients with Chediak-Higashi syndrome have a δ-SPD, but they also characteristically have neutropenia, with recurrent infections and oculocutaneous albinism. Neutrophils in patients with Chediak-Higashi syndrome have characteristic peroxidase-positive cytoplasmic granules. Patients with Chediak-Higashi syndrome have mutations of the CHS1 gene, located at 1q42-44. The protein encoded, lysosomal trafficking regulator (LYST), is involved in lipid-related protein trafficking and lysosome

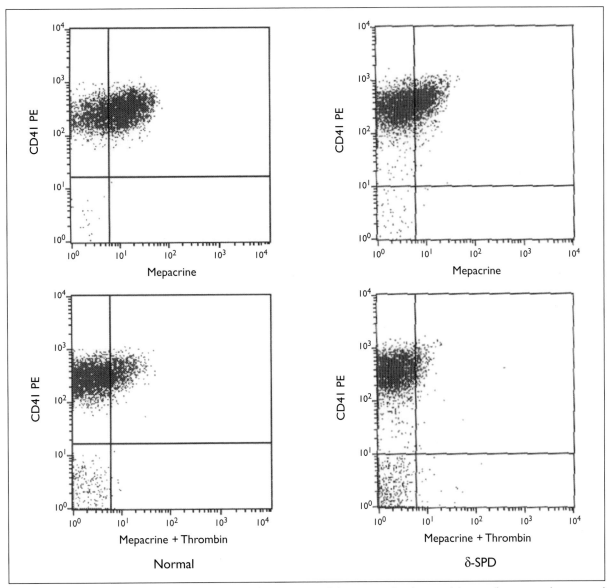

Figure 14-6. Flow cytometry can be helpful in the diagnosis of dense granule storage pool disorder (δ-SPD). Mepacrine is a fluorescent compound that is taken up into dense granules. Mepacrine uptake is decreased in patients with low numbers of dense granules (upper right panel) compared to normal (upper left panel). After stimulation with a platelet agonist (lower panels), mepacrine release is normal in both normal and δ-SPD patients. (Becton Dickinson FACScan.)

function. Several mutations of the gene have been described, but definitive genetic diagnosis is not widely available.

Hermansky-Pudlak Syndrome. Hermansky-Pudlak syndrome is a well-defined, autosomal recessive syndrome with oculocutaneous albinism; variable hypopigmentation of hair, skin, and irides; and pulmonary fibrosis. The syndrome is prevalent in Puerto Rico. Hermansky-Pudlak syndrome is a disorder of subcellular organelles in many tissues, including dense granules of platelets. There is defective trafficking and sorting of dense granule membrane proteins. Seven differ-

ent subtypes of the syndrome, due to seven different genetic mutations, have been described. Type I Hermansky-Pudlak is the most severe form of the syndrome and is due to a mutation of the Hermansky-Pudlak syndrome-1 (HPS-1) gene located at 10q23.1-13.3. This is the form of the syndrome most commonly observed in northwest Puerto Rico, where it has been described in 1 in 800 individuals. The other six subtypes of Hermansky-Pudlak have been described in relatively few individuals, but the gene defects in these disorders have recently been elucidated due to studies of associated storage pool disorders in

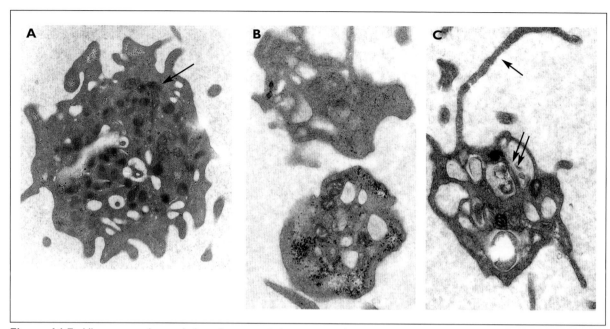

Figure 14-7. Ultrastructural morphology by transmission electron microscopy of platelets in dense granule storage pool disorder (δ-SPD), alpha storage pool disorder (α-SPD), and exhausted platelets. A. In δ-SPD, there are normal alpha granules (arrow), but a lack of dense granules. B. In α-SPD, alpha granules are not observed. C. In exhausted platelets that have undergone either in vivo or ex vivo release, neither alpha nor dense granules are observed, but the platelets have prominent filipodia (arrow) and expanded open canalicular system membranes (double arrow). (Original magnification X30,000.)

mice. Hermansky-Pudlak syndrome can be distinguished from isolated δ-SPD by the associated oculocutaneous albinism and its prevalence in the Puerto Rican population. Platelets in Hermansky-Pudlak and Chediak-Higashi syndromes have very low levels of the δ-granule membrane protein granulophysin, in contrast with normal levels in isolated δ-SPD. This suggests that the latter two disorders have defective granule packaging, while isolated δ-SPDs are due to inability to effectively store dense granule contents. The ATP to ADP ratio in platelets is elevated.

Signal Transduction Disorders

In addition to the storage pool disorders, platelet release defects can be seen with disorders of platelet signal transduction. Because platelet signal transduction is a complex and incompletely understood process, these are a poorly defined group of disorders. However, they may constitute a significant percentage of patients with abnormal secondary wave of aggregation and decreased granule release in whom alpha and dense granules are not deficient. These disorders include defects of the platelet cyclooxygenase and phospholipase C pathways, including phospholipase C activation, calcium mobilization, pleckstrin phosphorylation, and tyrosine phosphorylation.

Abnormalities of the GTP-binding proteins that link surface receptors and intracellular enzymes, such as $G\alpha_q$, $G\alpha_{i2}$, and $G\alpha_s$, have been described but are not well characterized clinically. Defects of the receptors for platelet agonists, such as ADP, collagen, thromboxane A2, and epinephrine, lead to abnormal signal transduction but have been included in the discussion of surface glycoprotein abnormalities, above. Defects of thromboxane A2 synthesis have been described, including defective liberation of arachidonic acid from the platelet membrane by phospholipase A2, cyclooxygenase enzyme deficiency, or thromboxane synthase deficiency.

In general, patients with signaling defects show decreased primary aggregation and decreased granule release without granule deficiency. In contrast to patients with δ-SPD, flow cytometric studies will show normal mepacrine uptake into dense granules, but decreased agonist-induced release. Identification of the exact defect requires detailed biochemical and genetic studies that are not available in most laboratories.

The rare individual with cyclooxygenase or thromboxane synthase deficiency will display an aspirin-like defect in aggregation despite no history of aspirin therapy. Platelet aggregation studies show a marked inhibition of platelet aggregation with arachidonic acid, depressed aggregation with collagen, and slightly decreased aggregation to ADP; ristocetin aggregation is usually normal. They also have defective production of thromboxane B2. Patients with cyclooxygenase deficiency additionally have defective production of prostacyclin (PGI_2) by endothelium, whereas this activity is normal with thromboxane synthase defects. True cyclooxygenase deficiency can be diagnosed by decreased cyclooxygenase levels on a Western blot of a platelet lysate, but patients with a dysfunctional protein mutant may have normal antigenic levels. Distinguishing these disorders from surreptitious aspirin or salicylate use requires a careful history, as many over-the-counter drug preparations contain salicylates. Serum salicylate levels may be helpful, but they have a short plasma half-life as they are rapidly cleared from the plasma and covalently bound to the cyclooxygenase enzyme. The finding of acetylated platelet cyclooxygenase could definitively identify salicylate use, but this is a research assay only performed in specialized centers.

Disorders of Platelet Procoagulant Activity (Scott Syndrome)

Platelets play an important procoagulant role, with assembly of coagulation complexes on activated platelet membranes that are rich in phosphatidyl serine. Scott syndrome is a rare, congenital, platelet functional disorder caused by defective "flip" of anionic phospholipids, such as phosphatidyl serine, to the outer table of the platelet membrane during platelet activation. These patients have normal platelet aggregation studies but have abnormal platelet procoagulant activity (platelet factor 3), decreased membrane binding of activated coagulation factors, and decreased microparticle formation.

Platelet Disorders with Thrombocytosis

Thrombocytosis is defined as an elevated platelet count higher than the laboratory's normal reference range. An elevated platelet count alone is not protective against bleeding; patients with thrombocytosis may be asymptomatic or display a wide range of symptoms, including both bleeding and thrombosis. Thrombocytosis is almost never congenital, and laboratory evaluation should be aimed at elucidating the underlying cause of the thrombocytosis. In patients with thrombocytosis,

the differential diagnosis is primarily between a reactive thrombocytosis and a clonal myeloproliferative process (essential thrombocythemia, chronic myelogenous leukemia, polycythemia vera, and myelofibrosis), or myelodysplastic process (5q-). The evaluation should include a CBC and peripheral blood smear. In addition, bone marrow evaluation, iron studies, cytogenetic study, and platelet aggregation study may be helpful. An algorithmic approach to the diagnosis of an increased platelet count is shown in Figure 14-3.

Reactive Thrombocytosis

Reactive, or secondary, thrombocytosis can be associated with many clinical disorders, such as iron deficiency, inflammatory and infectious diseases, and malignancies such as carcinomas or lymphomas, or with smoking or exercise. It can also be observed as a rebound thrombocytosis following splenectomy, or during treatment for immune thrombocytopenic purpura, pernicious anemia, or after cessation of myelosuppressive drugs.

Patients with reactive thrombocytosis have an elevated platelet count, but the degree of elevation is variable. The platelet count is typically less than $1,000,000/\mu L$, but a platelet count over $1,000,000/\mu L$ does not exclude a reactive thrombocytosis. A peripheral smear usually shows an increased number of platelets, with normal morphology. The mean platelet volume may be increased in patients with a rebound thrombocytosis due to production of new, larger platelets. The degree of thrombocytosis often may increase or decrease in conjunction with the leukocyte count in inflammatory disorders and malignancy, signaling that it is an acute phase reactant. Bone marrow examination in patients with a reactive thrombocytosis shows increased numbers of megakaryocytes, but other specific morphologic and molecular features associated with myeloproliferative or myelodysplastic disorders are not identified.

Laboratory studies in reactive thrombocytoses are usually not diagnostic. The platelet aggregation studies are typically normal. In contrast to myeloproliferative disorders, cytogenetic studies do not show a clonal process, bcr-abl is not rearranged, and Janus kinase 2 (JAK-2) mutations are not observed. Laboratory results may be suggestive of an acute phase response, with elevated C-reactive protein or interleukin 6. Thrombocytosis often can accompany iron deficiency; evaluation of serum iron studies together with the characteristic microcytic, hypochromic erythrocyte indices on the CBC can usually diagnose the condition and obviate the need for a bone marrow evaluation.

Myeloproliferative Disorders Associated with Thrombocytosis

Myeloproliferative disorders associated with thrombocytosis include essential thrombocythemia, chronic myelogenous leukemia, myelofibrosis, and polycythemia vera. Chronic myelogenous leukemia is associated with a t(9;22) chromosomal translocation associated with production of a bcr-abl fusion protein and is characterized by leukocytosis with eosinophilia and basophilia. A JAK-2 mutation is common in polycythemia vera, essential thrombocythemia, and myelofibrosis. In polycythemia vera, there is trilineage hematopoietic hyperplasia and increased erythrocyte mass. Essential thrombocythemia (ET) is characterized by thrombocytosis with little marrow fibrosis and is often associated with a JAK-2 mutation and expression of phospho-STAT5 protein. Myelofibrosis starts as a trilineage hematopoietic hyperplasia with prominent megakaryocytic hyperplasia, with transition to increasing marrow fibrosis and extramedullary hematopoiesis. The detailed clinical, bone marrow morphologic, pathogenic, and genetic features of these diseases are beyond the scope of this publication; the reader is referred to recent reviews for more information.

Of the myeloproliferative disorders, essential thrombocythemia and myelofibrosis are most often associated with thrombocytosis, although an elevated platelet count can be a feature of all myeloproliferative disorders. There is a great variability in the degree of thrombocytosis associated with myeloproliferative disorders, but often the platelet count is greater than 1,000,000/μL. In contrast, patients with reactive thrombocytoses typically have platelet counts less than 1,000,000/μL, but there is a great deal of overlap and platelet count alone cannot distinguish between these categories. In general, the peripheral smear will show a mixture of platelet sizes and platelet morphologies. Occasional degranulated giant platelets or megakaryocyte fragments may be observed. For diagnosis of myeloproliferative disorders, characteristic features of a specific disease can be discerned by examination of the peripheral blood smear and bone marrow, and by performing cytogenetic and molecular studies.

Platelet functional abnormalities have been described in all myeloproliferative disorders, but they have been best studied in essential thrombocythemia. Platelet aggregation studies in patients with myeloproliferative disorders are often abnormal, but the patterns observed are neither specific nor diagnostic. Essential thrombocythemia is often associated with an isolated defect in epinephrine-induced aggregation (Table 14-1). The decreased epinephrine-induced aggregation is thought to be due to down-regulation of α2-adrenergic receptors. Other patterns of platelet dysfunction with myeloproliferative disorders include decreased platelet aggregation to ADP or collagen, dense-granule storage pool pattern, abnormalities of the arachidonic acid pathway, and decreased receptors for fibrinogen, von Willebrand factor, or prostaglandin D2. Additionally, some patients may show evidence of circulating activated platelets or increased aggregation with various agonists, while others may have spontaneous in vitro aggregation without added agonists. In the clinical evaluation of patients with myeloproliferative disorders, it is important to remember that both bleeding and thrombosis can be observed in these patients, and that the results of the platelet functional tests will not necessarily distinguish whether the patient is at risk for bleeding or thrombosis. Bleeding complications are more frequently observed in patients with a platelet count greater than 1,000,000/μL, which may be due to an acquired deficiency of von Willebrand factor by adsorption to the platelets.

Abnormal platelet aggregation, when observed, may help to distinguish myeloproliferative from reactive thrombocytosis, but it is neither diagnostic nor specific. A decreased aggregation response to epinephrine may be observed in some normal donors and is also observed in patients with a congenital defect of the α2-adrenergic receptors, but these patients usually have a normal platelet count. In general, patients with myeloproliferative disorders have a higher platelet distribution width (PDW) than reactive thrombocytosis, but there is a great deal of overlap. The mean platelet volume (MPV) usually does not help distinguish between these two classes of disorders.

Platelet Disorders with Thrombocytopenia

Disorders in which the platelet count is decreased can be congenital or acquired, but they have been

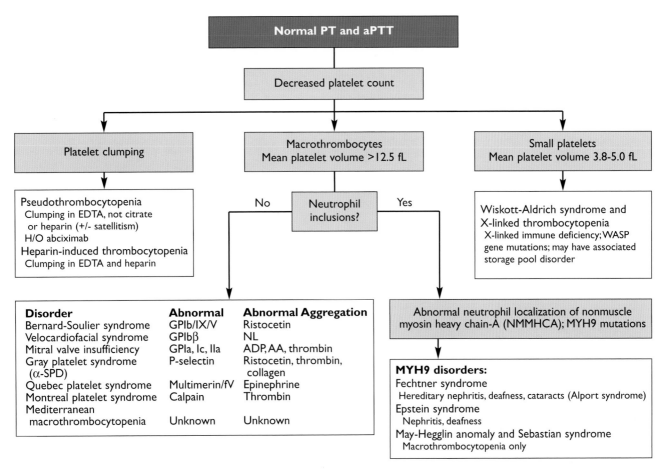

Figure 14-8. Diagnostic algorithm for distinction of platelet disorders with thrombocytopenia with increased or decreased platelet size. In the evaluation of thrombocytopenia, a pseudothrombocytopenia should be excluded. Platelet disorders with large platelets are usually congenital and include disorders with mutations of the myosin heavy chain gene (MYH9 disorders) and other surface glycoprotein abnormalities, such as a defect in GPIb/IX/V in Bernard-Soulier syndrome. Platelet disorders with small platelets are unusual and are observed in the Wiskott-Aldrich and X-linked thrombocytopenia syndromes. Abbreviations: AA, arachidonic acid; ADP, adenosine diphosphate; WASP, Wiskott-Aldrich syndrome protein gene.

grouped by platelet size in this discussion, as platelet size is one of the first distinguishing features appreciated by the evaluating pathologist (Figure 14-8).

Thrombocytopenia with Increased Platelet Size

The macrothrombocytopenia disorders are very rare and are all congenital in origin. Most are inherited in an autosomal dominant fashion and arise from congenital defects in platelet production by megakaryocyte or demarcation membrane systems, although the structural or genetic abnormalities are known in only a few disorders. Due to increased platelet turnover, some patients with acquired platelet destruction from disorders such as idiopathic thrombocytopenic purpura may have high MPVs due to the rapid release of new

platelets; but, in general, platelets in the macrothrombocytopenia disorders are much larger and more uniform in size.

Macrothrombocytopenias with Neutrophilic Inclusions (MYH9 Disorders)

Until recently, several macrothrombocytopenia disorders with neutrophil inclusions were recognized, including May-Hegglin anomaly, Sebastian syndrome, Fechtner syndrome, and Epstein syndrome. Genetic studies have established that these represent variable expression of a single disorder, characterized by mutations in the MYH9 gene (chromosome 22q12-13) that encodes the nonmuscle myosin heavy chain IIA protein (NMMHC-IIA). Clinically, these disorders can be characterized by a combination of symptoms and laboratory features that include a mild bleeding diathesis. Fechtner syndrome represents the

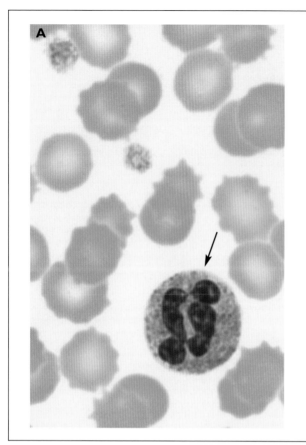
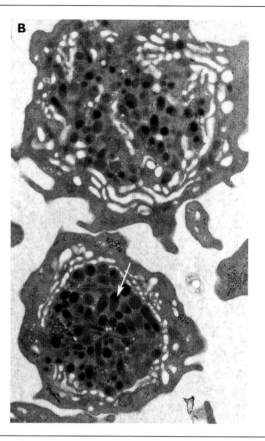

Figure 14-9. Platelet morphology in May-Hegglin, one of the congenital macrothrombocytopenia disorders due to mutations of the MYH9 gene. A. On the peripheral smear, there is thrombocytopenia with giant platelets. Dohle-like bodies are seen as blue to purple submembrane inclusions in neutrophils (arrow; Wright stain, original magnification X100). B. By electron microscopy, the platelets are large, with abundant alpha granules (arrow; original magnification X30,000).

severe end of the MYH9 spectrum, with the findings of nephritis, sensorineural hearing loss, and ocular abnormalities together with macrothrombocytopenia and neutrophil inclusions. Epstein syndrome differs from Fechtner syndrome by the lack of cataracts. May-Hegglin anomaly and Sebastian syndrome lack the kidney, ocular, and neural findings, but have macrothrombocytopenia and distinct types of neutrophil inclusions. The bleeding symptoms in the MYH9 disorders are usually mild and are not often a source of significant morbidity or mortality. The thrombocytopenia may develop in utero and is a rare cause of fetal or neonatal intracranial hemorrhage. The morbidity in these disorders is more related to the severity of the renal abnormalities.

The MYH9-related macrothrombocytopenia disorders are characterized by a decreased platelet count, with large platelets and the presence of neutrophilic inclusions, which can be observed on a peripheral smear (Figure 14-9, A). The Dohle-like bodies are blue, spindle-shaped inclusions in the periphery of the neutrophil cytoplasm. The thrombocytopenia is usually moderate, with platelet counts of 60,000 to 100,000/μL common, and the MPV is approximately 12.5 fL, but is often much larger. Some caution should be exercised in interpreting complete blood counts performed on automated instruments in these patients, as some of the platelets are so large they are outside of the "platelet gate" and may be counted as leukocytes. In general, the peripheral smear will show a uniform population of large, normally granulated platelets. Bone marrow analysis is not generally indicated but will show increased megakaryocytes, often surrounded by the production of large platelets.

In normal individuals, nonmuscle myosin heavy chain-A (NMMHCA) is distributed diffusely in the cytoplasm of neutrophils and platelets. Immunostains with an antibody against NMMHCA will show abnormal subcellular localization in the MYH9 disorders, with focal NMMHCA staining in association with the neutrophil inclusions.

Electron microscopic evaluation of the inclusions shows some distinctions: in May-Hegglin anomaly, the inclusions lack a limiting membrane, are free of specific granules, and consist of an amorphous cluster of ribosomes oriented along parallel microfilaments, whereas in Sebastian syndrome, the inclusions are composed of highly dispersed filaments and few ribosomes. Electron microscopic study of the platelets typically shows very large platelets with abundant alpha granules (Figure 14-9, B).

Laboratory platelet studies will usually show normal platelet aggregation and a normal bleeding time, attesting to the increased functionality of the larger platelets. Due to the large size of the platelets, centrifugation speeds for preparation of platelet-rich plasma may have to be modified or many platelets may sediment with the leukocytes or erythrocytes. Platelet surface glycoproteins are usually normal. Genetic analysis is not routinely available for MYH9, but 22 distinct mutations have been described, with R702C and E1841K being the most commonly observed. There is no genotype-phenotype relationship among the mutations identified.

Bernard-Soulier Syndrome

Bernard-Soulier syndrome is a congenital deficiency of the platelet glycoprotein Ibα/Ibβ/IX/V receptor, the surface receptor responsible for mediating vWF-induced platelet adhesion and aggregation. Although it is a disorder of a platelet surface glycoprotein, Bernard-Soulier syndrome has been classified in this chapter as a macrothrombocytopenia disorder because the initial pathologist's findings will typically be thrombocytopenia with an elevated MPV. Most of the Bernard-Soulier genetic defects are due to mutations of the GPIbα gene, but may also be due to defects of the GPIbβ or GPIX genes. Because the GPIb is expressed on the demarcation membrane system in the megakaryocytes, it is postulated that GPIb plays a role during megakaryopoiesis and maintenance of platelet size. The disorder is inherited as an incompletely recessive autosomal trait with severe lifelong mucocutaneous bleeding that often is out of proportion to the often-mild thrombocytopenia.

Individuals with Bernard-Soulier syndrome typically have moderately severe thrombocytopenia (40,000 to 100,000/μL), with uniformly large, granulated platelets observed on the peripheral smear. Platelets occasionally are the size of erythrocytes or lymphocytes, so automated platelet counts are often inaccurate. No neutrophil inclusions are present. This disorder has diagnostic surface glycoprotein abnormalities, so bone marrow analysis is not usually indicated for diagnosis.

Normal platelet aggregation is noted with exposure to ADP, collagen, epinephrine, and arachidonic acid, but aggregation is characteristically absent with the addition of ristocetin or botrocetin (Figures 14-4 and 14-10; Table 14-1). Aggregation induced by alpha-thrombin may also be abnormal. Caution should be exercised in preparation of platelet-rich plasma for aggregation studies in patients with Bernard-Soulier syndrome, as the platelets may sediment during centrifugation. Use of whole blood aggregation may be one way to circumvent this problem because centrifugation is not required prior to testing. Adhesion of platelets to subendothelium or immobilized vWF is markedly reduced at all shear rates in patients with Bernard-Soulier syndrome; this finding may have direct clinical consequences. Patients with Bernard-Soulier syndrome will also have an abnormal bleeding time and abnormal PFA-100 test results.

The genes for GPIbα and GPIbβ map to chromosomes 17 and 22, respectively. The genes encoding GPIX and GPV are both on chromosome 3. Over 30 different genetic causes of Bernard-Soulier syndrome have been identified, most frequently due to mutations of GPIbα with truncation, frameshift, or nonsense mutations leading to loss of the extracellular protein domain. More rarely, mutations of GPIbβ and GPIX have been described. A unique mutation of the GATA binding site of the GPIbβ promoter that leads to decreased GPIbβ transcription has been described. Although the genetic defect usually affects only one protein, the entire GPIb/IX/V complex is not expressed on the platelet surface if there is a deficiency of the GPIbα, GPIbβ, or GPIX chains. So-called "variant" Bernard-Soulier syndrome is due to qualitative genetic defects leading to a nonfunctional GPIb/IX/V complex, with at least partial expression on the platelet surface. The glycoprotein abnormality can be confirmed with flow cytometry or crossed immunoelectrophoresis, where a combined lack of GPIbα, GPIbβ, and GPIX is identified. Heterozygotes may have intermediate GPIb/IX/V content by flow cytometry. The flow cytometry results can be normal if the molecular defect affects function, but not production.

Defects in the GPIb gene, located on chromosome 22, are also associated with the more exten-

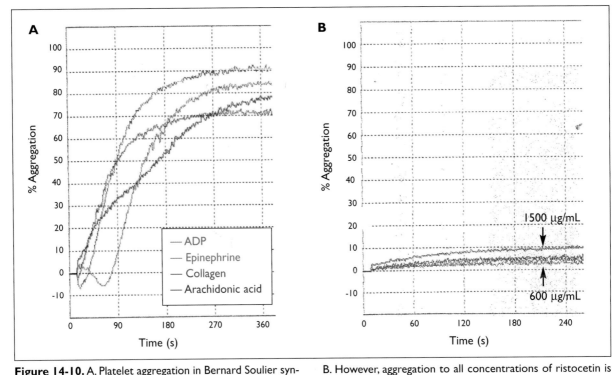

Figure 14-10. A. Platelet aggregation in Bernard Soulier syndrome is normal for adenosine diphosphate (ADP), epinephrine, collagen, and arachidonic acid. B. However, aggregation to all concentrations of ristocetin is markedly diminished to absent. (Helena PACKS-4 Optical Aggregometer.)

sive genetic defects observed in the velocardiofacial or Di George syndrome associated with a deletion in chromosome 22 (22q11.2). However, the GPIbβ gene defect in these patients is usually heterozygous, and bleeding symptoms are uncommon. These patients may have associated velopharyngeal insufficiency, conotruncal heart disease, immune deficiency, and learning disabilities together with an abnormality of GPIbβ.

Bernard-Soulier syndrome, although rare, has characteristic surface glycoprotein abnormalities and is not often confused with other platelet disorders after a laboratory analysis. However, the thrombocytopenia in Bernard-Soulier syndrome presenting in childhood can be mistaken for immune thrombocytopenic purpura (ITP), and patients with Bernard-Soulier syndrome have been known to undergo unnecessary splenectomy. The lack of neutrophil inclusions distinguishes Bernard-Soulier syndrome from the MYH9 disorders. Additional laboratory studies show normal vWF antigen and ristocetin cofactor activity to distinguish Bernard-Soulier syndrome from von Willebrand disease. Although some subtypes of von Willebrand disease may show decreased ristocetin aggregation, this can be distinguished from Bernard-Soulier syndrome because the abnormality corrects with the addition of normal plasma in

von Willebrand disease, and tests for von Willebrand factor are abnormal in von Willebrand disease.

Patients with Bernard-Soulier syndrome are subject to a lifelong bleeding diathesis and can be at increased risk for bleeding during surgical procedures. Patients do not usually require therapy on a routine basis, but preoperative therapy is usually prudent. There is no uniform therapeutic approach to these patients, but DDAVP and platelet transfusions are the most common. Due to risk of development of platelet refractoriness, the use of leukocyte-reduced platelets is advised. There is little experience with the use of recombinant factor VIIa.

The Gray Platelet Syndrome (Alpha Granule Storage Pool Disorder)

The gray platelet syndrome is a congenital alpha granule storage pool disorder (α-SPD) that is also considered a macrothrombocytopenia disorder. The disorder is characterized by mild, lifelong bleeding symptoms. Some patients may have marrow fibrosis, pulmonary fibrosis, and splenomegaly. Other storage pool disorders involving dense granules are discussed in the previous section, "Platelet Dysfunction with Normal Platelet Count."

Individuals with gray platelet syndrome have a variable thrombocytopenia and large (mean 13 fL), gray-appearing platelets on the peripheral blood smear due to decreased α -granules. Platelets may appear vacuolated. Bone marrow examination often shows reticulin fibrosis of the bone marrow, which is thought to be due to inability of megakaryocytes to store platelet-derived growth factor. It is unclear whether the pathophysiology of the gray platelet syndrome is due to premature release of α-granules from the cell or due to abnormal signal transduction or calcium flux.

Platelet aggregation studies may be normal for ADP and epinephrine but are often abnormal for thrombin and collagen (Table 14-1). Gray platelets are principally deficient in soluble proteins normally contained in alpha granules, such as platelet factor 4, β-thromboglobulin, vWF, thrombospondin, and platelet-derived growth factor. In contrast, the alpha granule membrane proteins are normal, suggesting that the disorder is primarily one of defective targeting and packaging of proteins into platelet α-granules. Flow cytometry studies have shown increased surface P-selectin, but decreased alpha granule P-selectin. By electron microscopy, platelets are enlarged and alpha granules are not present; however, dense granules, mitochondria, and lysosomes appear normal (Figure 14-7, B). Immunogold staining by electron microscopy may show P-selectin localization in the open canalicular system.

Pale platelets can also be seen with ongoing platelet activation and circulating "exhausted" platelets in patients with disseminated intravascular coagulation (DIC) or with implanted cardiac devices, but in these patients there will be a mixture of normal and pale platelets, and routine coagulation test results will be abnormal.

A related disorder is the Quebec syndrome, which is caused by a deficiency of multimerin, a high-molecular-weight multimeric protein that is stored as a complex with factor V in α-granules. Aggregation studies show a decreased response to epinephrine. Platelets have normal α-granule ultrastructural morphology but demonstrate protease-mediated degradation of α-granule proteins.

Thrombocytopenia with Decreased Platelet Size

Wiskott-Aldrich Syndrome and X-Linked Thrombocytopenia

Wiskott-Aldrich syndrome (WAS) and X-linked thrombocytopenia (XLT) are related X-linked recessive disorders caused by a defect in the Wiskott-Aldrich syndrome protein (WASP) gene. These disorders are characterized by immune deficiency, recurrent infections, eczema, and thrombocytopenia with small platelets and a platelet storage pool disorder. Affected patients may often present at birth with petechiae, bruising, and bloody diarrhea.

The WASP gene is expressed exclusively in hematopoietic stem cells; the WAS protein (WASp) is involved in signal transduction and adapter protein function and also is known to regulate actin filament assembly. WASp is involved in lymphoid and myeloid development through its ability to facilitate nuclear translocation of nuclear factor κB. Over 150 mutations of the WASP gene on the X chromosome have been identified. Affected individuals will have absent immunologic responses to polysaccharide antigens and progressive decline in T lymphocyte function. The XLT form involves defects in exon 2 of the WASP gene and is accompanied by a minimal immunodeficiency. These patients have an increased risk of bleeding due to small platelet mass and platelet dysfunction, even at modestly low platelet counts. The thrombocytopenia often responds to splenectomy, with a return in both platelet count and size, but postsplenectomy sepsis is a risk. A milder form without thrombocytopenia, X-linked neutropenia, occurs with WASP mutations that affect the Cdc42-binding site. Female carriers are asymptomatic.

The peripheral smear shows a decreased number of uniformly small platelets, and often the MPV is low (approximately half normal size). Bone marrow evaluation shows normal megakaryocyte number and morphology. Platelet dysfunction is severe; the platelets are unable to aggregate to ADP, epinephrine, and collagen. Platelets have decreased dense granules, and a storage-pool–like pattern is often seen. Lymphocytes are deficient in CD43 (sialophorin).

Thrombocytopenia with Normal Platelet Size

A thorough clinical history and bone marrow examination may be helpful in differentiating the underlying causes in thrombocytopenic platelet disorders with normal platelet morphology and size. This group of disorders includes both congenital and acquired thrombocytopenias that may be due to either decreased platelet production or increased platelet destruction (Figure 14-11). The

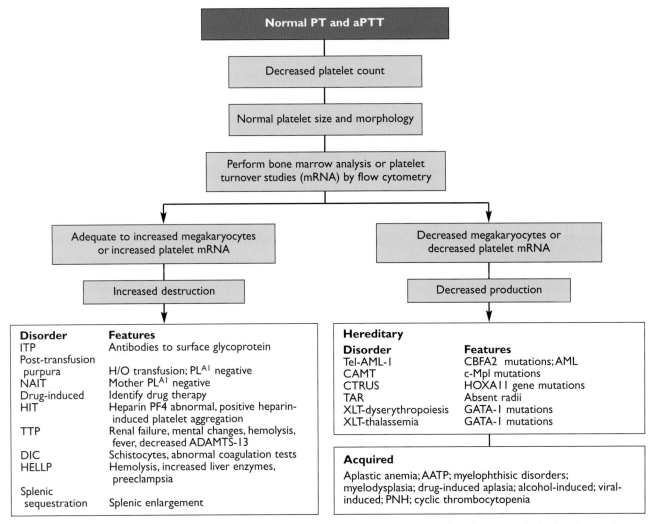

Figure 14-11. Algorithm for diagnosis of thrombocytopenic disorders associated with increased platelet destruction or decreased production. Abbreviations: see Appendix A.

number of megakaryocytes observed in the bone marrow specimen can help to distinguish between these etiologies, but analysis of platelet turnover by mRNA analysis may also be helpful.

The finding of adequate or increased megakaryocytes on the bone marrow or increased reticulated platelets in the face of peripheral thrombocytopenia suggests peripheral platelet destruction. Platelet functional tests usually are not helpful in differentiating between the entities in this class of disorders, as most functional studies will give abnormal results simply due to the low platelet number. The overall MPV is usually normal with destructive thrombocytopenia, but there is usually a range of platelet size with increased platelet distribution width (PDW) and many large platelets are seen, indicating the rapid platelet turnover. These disorders are invariably acquired, and an

underlying abnormality should be sought. In general, the clinical scenario is the most helpful in classifying these disorders.

Peripheral Platelet Destruction

Immune Destructive Thrombocytopenias

Immune thrombocytopenic purpura (ITP) is a disorder characterized by platelet sensitization with autoantibodies, leading to platelet destruction in the reticuloendothelial system and thrombocytopenia. It is defined clinically as thrombocytopenia in the presence of a normal hemoglobin level, normal WBC count and differential, and in the absence of hepatosplenomegaly, lymphadenopathy, and abnormalities of the radii or other underlying disease. Patients usually present with an acute onset of bleeding symptoms and marked thrombocytopenia.

Adult ITP has an incidence of 1 in 100,000, with a 2-3 to 1 female to male predominance. Childhood ITP has a peak age incidence of 2 to 4 years, with girls and boys affected equally. ITP is often associated with viral infections and may have a genetic predisposition, but the etiologic mechanism has not been elucidated. Patient prognosis in ITP is usually good, but severe bleeding complications can be seen when the platelet count is less than 10,000 to 20,000/μL. Acute treatment is usually aimed at raising the platelet count by using immunoglobulin therapy (IVIG or anti-D), steroids, vincristine, or plasmapheresis. In children, approximately 50% do not need further therapy, but only 10% to 30% of adults have a prolonged response to initial therapy. Long-term therapy, if necessary, may entail splenectomy. Up to 40% of patients may fail splenectomy; optimal treatment of these refractory patients is the subject of ongoing studies.

ITP is a diagnosis of exclusion, as the pathologic features are nondiagnostic. The platelet count is decreased and may be markedly decreased in patients with acute ITP. Peripheral smears often show a mixture of normal and larger platelets, with the MPV being normal to slightly increased. Erythrocyte and granulocyte counts and indices should be normal. Bone marrow examination may not be necessary if the patient presents with an acute isolated thrombocytopenia, with no family history of bleeding or thrombocytopenia. However, for patients with other abnormalities of the peripheral smear or complicated clinical features, bone marrow evaluation may be helpful. In uncomplicated ITP, the bone marrow usually demonstrates normal to increased numbers of megakaryocytes, with occasional megakaryocyte clustering. Megakaryocyte morphology is usually normal. Other pathologic changes in granulopoiesis or erythropoiesis should not be observed.

Autoantibodies to platelet surface glycoproteins can be detected by flow cytometry or immunoassay, although these findings may be neither sensitive nor specific. Immunoassays are available to identify the specific platelet surface glycoprotein target, such as GPIIb/IIIa or GPIb, but these assays cannot reliably distinguish ITP from nonimmune thrombocytopenia or myelodysplasia. Platelet functional tests usually are not helpful in diagnosing ITP, as the low platelet count alone can give abnormal functional test results. In rare patients with ITP due to anti-GPIIb/IIIa anti-

bodies, platelet flow cytometry may detect a Glanzmann thrombasthenia-like phenotype.

Since specific diagnostic tests for ITP are not available, the diagnostic approach usually involves excluding other causes of thrombocytopenia, such as congenital thrombocytopenias, and other causes of destructive thrombocytopenia, such as hypersplenism, heparin-induced thrombocytopenia (HIT), thrombotic thrombocytopenic purpura (TTP), and DIC. A family history of thrombocytopenia and the presence of uniformly large platelets on the peripheral smear suggest a congenital thrombocytopenia. For patients with prior or current exposure to heparin, with a characteristic decline in platelet count with or without thrombosis, heparin-induced thrombocytopenia tests, such as anti-platelet factor 4 enzyme-linked immunosorbent assay (ELISA) or serotonin release studies, are helpful to exclude HIT.

Thrombotic thrombocytopenic purpura is associated with a characteristic clinical pentad of renal failure, mental status changes, thrombosis, and hemolytic anemia in addition to the thrombocytopenia. Assays for vWF-cleaving metalloprotease (ADAMTS-13, a disintegrin and metalloprotease with thrombospondin-1-like domains) may be helpful in excluding TTP. A normal RBC reticulocyte count or normal serum lactic dehydrogenase (LDH) in ITP may also be helpful to exclude TTP. Disseminated intravascular coagulation is an intravascular consumptive coagulopathy often associated with infections, malignancy, or obstetric complications. It is usually accompanied by changes in coagulation assays, such as elevation of the PT, aPTT, and D-dimer, and decrease in fibrinogen, antithrombin, and protein C. Examination of the peripheral smear in both TTP and DIC should show schistocytes (RBC fragments) in addition to the thrombocytopenia. Testing for antinuclear antibody and anti-DNA antibody may be helpful to investigate systemic lupus erythematosus, which could indicate secondary ITP.

Alloimmune thrombocytopenias include post-transfusion purpura and neonatal alloimmune thrombocytopenia. Post-transfusion purpura is a rare alloimmune disorder in which immune destruction of both transfused and recipient platelets is seen, usually beginning 5 to 12 days post-transfusion. Patients often have low-frequency platelet antigens, and are frequently PL^{A1} (HPA-1a) negative (PL^{A2} homozygous or HPA-1b). Patients often are multiparous women who have been exposed to PL^{A1} during pregnancy and often are HLA-DR3-positive. Anti-PL^{A1} antibodies can be detected both in plasma and attached to the platelet surface.

Neonatal alloimmune thrombocytopenia (NAIT) is a severe thrombocytopenia that occurs shortly after birth. This syndrome is due to maternal immunization against fetal platelet antigens and occurs when the mother lacks a platelet antigen present on the baby's platelets; it is the platelet equivalent of hemolytic disease of the newborn. Maternal antibodies cross the placenta and cause the severe thrombocytopenia, which can result in intracranial hemorrhage. Several platelet antigen systems may be responsible for this scenario, but often the mother is PL^{A1} negative (PL^{A2} homozygote). Interestingly, this same PL^{A2} genetic polymorphism may impart a risk for coronary artery disease. The diagnosis is usually made by platelet surface glycoprotein phenotyping by ELISA or by molecular genetic studies.

Drug-induced thrombocytopenias due to immunologic and nonimmunologic platelet destruction can be seen with many drugs. Common causes of drug-induced autoimmune thrombocytopenia include quinidine, quinine, heparin, sulfonamide drugs, and gold salts. Various mechanisms have been implicated, with the drug acting as either a direct or indirect hapten. For example, drugs such as penicillin can bind directly to platelet surface glycoproteins and act as a direct hapten. On the other hand, drugs such as quinidine, quinine, and sulfonamides bind to platelet surface glycoproteins and induce a conformational change or result in a compound epitope that is immunogenic. Target glycoproteins implicated include GPIIb/IIIa and GPIb/IX/V. The development of immune complex formation and thrombocytopenia is characteristic of heparin-induced thrombocytopenia, discussed in chapter 21. Binding of drugs, such as eptifibatide or tirofiban, to the GPIIb/IIIa fibrinogen receptor may create a neoepitope (ligand-induced binding site, or LIBS) that is immunogenic. Other GPIIb/IIIa antagonists, such as abciximab, a chimeric monoclonal antibody to the fibrinogen receptor (Fab' fragment), may induce thrombocytopenia by an alloimmune mechanism. However, acute thrombocytopenia with abciximab can occur at the first exposure to the drug, and pre-existing antibodies and nonimmunologic mechanisms have been postulated. Pseudothrombocytopenia has also been described with abciximab due to in vitro platelet clumping associated with calcium-chelating EDTA anticoagulants.

The clinical features of drug-induced thrombocytopenia include thrombocytopenia within hours to days of drug ingestion. The thrombocytopenia, if severe, can lead to a bleeding diathesis characterized by mucosal bleeding, petechial hemorrhage, and purpura. In patients suspected of having immune drug-associated thrombocytopenia, the offending drug can be identified by discontinuation of medications, if medically feasible.

Immune drug-associated thrombocytopenia is usually diagnosed by a documented history of drug therapy with drugs known to cause thrombocytopenia, the time course of the thrombocytopenia, and the presence of drug-associated platelet antibodies. Other than thrombocytopenia on the peripheral smear, there are no characteristic pathologic features of this disorder. Drug-induced thrombocytopenias can be diagnosed by detecting the presence of platelet-associated antibody by flow cytometry or immunoassay, although this is a nonspecific finding that can also be seen with infections and autoimmune disorders, such as ITP. The drug dependence of the antibody binding can be demonstrated by incubating platelets with patient plasma in the presence of the drug. Conversely, patients with drug-induced thrombocytopenia may have negative testing and a presumptive diagnosis made on the basis of drug history.

Heparin induced thrombocytopenia (HIT) is a distinctive drug-induced thrombocytopenia associated with heparin therapy, in which antibodies are formed to heparin/platelet factor 4 complexes. It is described in chapter 21.

Nonimmune Destructive Thrombocytopenias

The major causes of nonimmune platelet destruction are disseminated intravascular coagulation and thrombotic thrombocytopenic purpura. However, nonimmune platelet destruction also may be seen with hematopoietic growth factors, tumor necrosis factor α/interferon γ, and interleukin-2. DDAVP therapy occasionally has been associated with development of thrombocytopenia in type 2B and platelet-type von Willebrand disease, due to rapid release of high-affinity vWF leading to platelet clumping. Thrombocytopenia can also be associated with use of porcine factor VIII in patients with factor VIII inhibitors.

Disseminated intravascular coagulation (DIC) occurs with an intravascular stimulation of the coagulation and fibrinolytic systems, leading to intravascular coagulation activation, fibrin formation, platelet activation, thrombocytopenia, fibrinolytic activation, and clotting factor consumption. DIC is not a discrete disease entity but is initiated by other disease processes that produce coagulation-activating or fibrinolytic-activating factors. Thrombocytopenia is a hallmark of DIC; the diagnostic features are covered in detail in chapter 13.

Thrombotic thrombocytopenic purpura (TTP) is an uncommon microangiopathy that results from an abnormality of a vWF-cleaving metalloproteinase (ADAMTS-13). A deficiency in ADAMTS-13, an acquired antibody to the metalloproteinase, or a mutation of *VWF* affecting the ADAMTS-13 cleavage site results in endothelial-associated accumulation of high-molecular-weight von Willebrand factor, leading to diffuse thrombus formation in small vessels and a decline of circulating platelets. Affected patients will show characteristic clinical symptoms of fever, thrombocytopenia, and microangiopathic hemolytic anemia accompanied by multiorgan failure, often manifesting as renal failure and mental status changes. The erythrocyte fragmentation likely occurs as red cells are sheared by fluid turbulence in areas of platelet aggregates.

Individuals with deficiency of ADAMTS-13 have a familial form of TTP that often presents in infancy or childhood and may recur as "chronic relapsing TTP," also known as the Upshaw-Schulman syndrome. Many genetic mutations of the ADAMTS-13 gene have been identified throughout the gene sequence. An idiopathic form of TTP that is seen in adulthood is thought to be due to autoantibodies directed against ADAMTS-13. Pathogenic antibodies that have been identified are typically directed against the cysteine-rich/spacer domain of ADAMTS-13.

Thrombotic thrombocytopenic purpura is associated with a high mortality due to the multiorgan failure. Supportive therapy has typically employed daily plasma exchange with fresh frozen plasma, and the majority of patients now survive the initial episode. However, relapse of TTP may be seen in 30% to 60% of patients, with relapse most frequent in the first month. Newer therapies based on infusion of recombinant ADAMTS-13 or rituximab to suppress the immune response have shown promise.

A clinically similar thrombotic microangiopathy can be observed with drugs such as cyclosporine, tacrolimus (FK506), quinine, ticlopidine, and clopidogrel, and after allogeneic organ transplantation. Hemolytic-uremic syndrome (HUS) is a related disorder that is often observed in children and associated with Shiga toxin-producing strains of *Escherichia coli* (serotype 0157:H7). Familial types of HUS may be due to a defect in a complement regulatory protein (plasma factor H or membrane-cofactor protein CD46).

Von Willebrand factor (vWF), a large multimeric protein produced by endothelial cells and megakaryocytes, is involved in platelet adhesion to sites of vascular injury and also serves as a carrier protein for coagulation factor VIII. vWF typically exists in plasma in a range of molecular sizes, but it is initially released from endothelial cells and tethered to the luminal cell membrane in an ultra-high-molecular-weight form via binding to the transmembrane protein P-selectin. A domain in the A2 domain of vWF is responsible for binding to platelet GPIb/IX/V, and the larger vWF multimers more avidly bind to platelets. Tethered vWF is rapidly processed into smaller fragments released into the plasma due to cleavage of Tyr842-843Met bonds by the ADAMTS-13 metalloprotease. ADAMTS-13 is the thirteenth member of a family of metalloproteases; it is a 190,000 molecular weight glycoprotein encoded on chromosome 9q34 that is produced primarily in the liver. In individuals with a deficiency of ADAMTS-13, the ultra-high-molecular-weight vWF remains attached to endothelial cells, leading to platelet adhesion, aggregation, and microvascular thrombosis. It is thought that these long vWF/platelet aggregates can detach and embolize, further facilitating microvascular thrombosis (Figure 14-12).

In TTP, a review of the CBC will highlight a normocytic anemia with profound thrombocytopenia; platelet counts are frequently below 20,000/μL. The reticulocyte count, red cell distribution width (RDW), and MPV are often increased, indicating the increased turnover of both erythrocytes and platelets. Morphologic evaluation of the peripheral blood smear reveals erythrocyte polychromasia and anisocytosis with prominent schistocytes (Figure 14-13). Bone marrow evaluation is typically not necessary in establishing a diagnosis of TTP but may be helpful to exclude other causes of thrombocytopenia. Bone marrow morphology is typically nonspecific and shows normal to increased numbers of megakaryocytes. Small vessels may show platelet thrombi, but these are not always observed. Autopsy reflects the pathophysiology of the disease, with platelet-rich thrombi plugging the microvasculature of the brain, heart, pancreas, spleen, adrenal, and kidneys, but rarely involving the liver or lungs, in distinction to DIC. These thrombi are also rich in vWF, if studied by

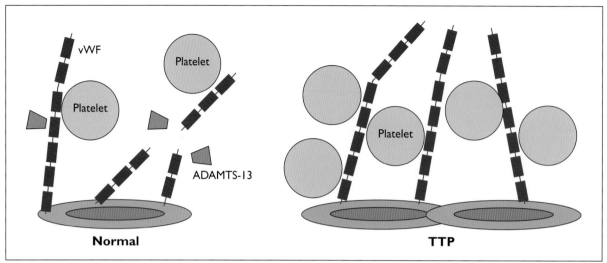

Figure 14-12. Pathophysiology of thrombotic thrombocytopenic purpura (TTP). Von Willebrand factor (vWF) typically is initially released from endothelial cells and tethered to the luminal endothelial cell membrane in an ultra-high-molecular-weight form via binding to the transmembrane protein P-selectin. Tethered vWF is rapidly processed into smaller fragments released into the plasma due to cleavage by the ADAMTS-13 metalloprotease. In individuals with a deficiency of ADAMTS-13 and TTP, the ultra-high-molecular-weight vWF remains attached to endothelial cells, leading to platelet adhesion, aggregation, and microvascular thrombosis. It is thought that these long vWF/platelet aggregates can detach and embolize, further facilitating microvascular thrombosis.

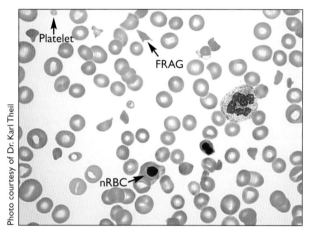

Photo courtesy of Dr. Karl Theil

Figure 14-13. Thrombotic thrombocytopenic purpura (TTP) is a microangiopathic hemolytic disorder that shows anemia and thrombocytopenia on the complete blood count. A peripheral blood smear typically shows few platelets (arrow) and prominent schistocytes, or erythrocyte fragments (FRAG). This smear also shows polychromasia with a leukoerythroblastic response including nucleated erythrocytes (nRBC). (Wright-stained blood smear, original magnification X100.)

immunohistochemistry, in contrast to the platelet-fibrin thrombi in DIC.

Plasma ADAMTS-13 activity in normal individuals typically ranges from 50% to 180%; patients with acute TTP usually have activity of <5%. Inhibitors of ADAMTS-13 are usually distinguished from enzyme deficiency by mixing studies or purification and testing of IgG from the patient's serum. Variably decreased activity can also be observed with liver disease, disseminated malignancy, inflammatory disorders, and during pregnancy. Infants may also have a lower activity than adults.

Traditional screening coagulation studies, such as PT and aPTT, are normal in patients with TTP. Assays for von Willebrand factor, such as vWF antigen, ristocetin cofactor, and factor VIII, are typically normal. However, vWF multimer analysis may show an increase in ultra-high-molecular-weight multimers during the acute disease presentation.

Thrombotic thrombocytopenic purpura can be diagnosed definitively in many patients based on the distinct clinical features and markedly decreased ADAMTS-13 activity. However, the ADAMTS-13 assays are not widely available in clinical laboratories, and many of these assays are not suitable for rapid testing. Additionally, many patients with clinical symptoms of TTP have only modestly decreased levels of ADAMTS-13, while other individuals with markedly decreased ADAMTS-13 levels may not have clinical symptoms of TTP.

Rapid diagnosis of TTP often rests on excluding other consumptive thrombocytopenic disorders. TTP and DIC are both thrombotic microangiopathies that exhibit schistocytes on the peripheral smear, but these disorders can be distinguished by the characteristic coagulation test abnormalities found in DIC, as indicated in chapter 13. Hemolytic-uremic syndrome is distinguishable by serology for *E. coli* 0157:H7, lack of abnormalities in ADAMTS-13, and more severe renal dysfunction. Patients with immune thrombocytopenic purpura typically lack hemolytic anemia and multiorgan failure and do not have schistocytes on the peripheral smear. Lack of association with transfusions or PL^{A1}/PL^{A2} genotype distinguish TTP from post-transfusion purpura. Distinction of TTP from drug-induced thrombocytopenia may be difficult in patients treated with multiple drugs, but these disorders usually lack microangiopathic thrombosis or organ failure. Heparin-induced thrombocytopenia is one drug-induced thrombocytopenia that is complicated by thrombosis, but HIT is preceded by heparin exposure 5 to 14 days prior, and is characterized by large vessel thrombosis and presence of anti-platelet factor 4 antibodies, as detailed in chapter 21. Paroxysmal nocturnal hemoglobinuria (PNH) may occasionally present with thrombocytopenia and microangiopathic hemolysis but is usually distinguishable by lack of renal dysfunction and flow cytometric analysis demonstrating decreased expression of the anchoring protein GpI and decreased expression of CD24, CD55, and CD59 in neutrophils and CD55 and CD59 on erythrocytes.

Hypertensive disorders of pregnancy with thrombocytopenia have many causes, including incidental thrombocytopenia of pregnancy, ITP, antiphospholipid antibody syndrome, and hypertensive disorders of pregnancy. Incidental thrombocytopenia of pregnancy is the most common cause of thrombocytopenia during pregnancy (up to 75% of cases) and is usually asymptomatic. The antiphospholipid antibody syndrome, lupus anticoagulant, and systemic lupus erythematosus are uncommon but are associated with significant pregnancy complications, including venous thrombosis and spontaneous abortion. Hypertensive disorders of pregnancy account for approximately 20% of pregnancy-associated thrombocytopenia. Preeclampsia and eclampsia are associated with graded degrees of hypertension and proteinuria, with headache, abdominal pain, and thrombocytopenia. Subsets of these patients have a syndrome called HELLP (hemolysis elevated liver enzymes and low platelets). There may be some overlap between patients with HELLP syndrome and gestational TTP or HUS.

Decreased Platelet Production

Thrombocytopenias due to decreased platelet production include both rare congenital and more common acquired causes; an algorithm is shown in Figure 14-11. Acquired causes of thrombocytopenia are generally due to bone marrow disorders associated with marrow failure, such as aplastic anemia or myelodysplasia.

Hereditary Thrombocytopenias with Decreased Platelet Production

Congenital thrombocytopenias due to decreased megakaryocytes with normal platelet size include thrombocytopenia with absent radii (TAR syndrome), X-linked amegakaryocytic thrombocytopenia (congenital amegakaryocytic thrombocytopenia [CAMT]), and familial thrombocytopenia-leukemia. Fanconi anemia is a congenital thrombocytopenia but is distinguished by erythroid hypoplasia and DNA instability. Other hereditary thrombocytopenias associated with large platelets have been covered previously (see "Thrombocytopenia with Increased Platelet Size").

Patients with **familial thrombocytopenia-leukemia** (Tel-AML1) have a relatively mild thrombocytopenia (80,000 to 100,000/μL) but often have a bleeding diathesis symptomatically out of proportion to the degree of thrombocytopenia, consistent with clinical platelet dysfunction. Platelet function testing usually reveals a storage pool disorder. Approximately half of the patients develop a malignancy, usually acute myeloid leukemia or solid tumors. Recently, mutations in the transcription factor CBFA2 have been identified in families with this disorder.

Patients with **congenital amegakaryocytic thrombocytopenia** (CAMT) have a genetic defect in the C-Mpl (1p34) gene, leading to a deficiency or dysfunction of the thrombopoietin receptor. They have a thrombocytopenia associated with marrow megakaryocytic hypoplasia that evolves into bone marrow aplasia.

Congenital thrombocytopenia with radio-ulnar synostosis (CTRUS) is due to a mutation in the HOXA11 gene (7p15-14). They have thrombocytopenia of normal platelet size with a general aplastic anemia. Other associated abnormalities include radio-ulnar synostosis and sensorineural hearing loss, together with other skeletal malformations. Bone marrow evaluation reveals reduced to absent megakaryocytes.

Thrombocytopenia with absent radii (TAR) is an autosomal recessive disorder characterized by severe thrombocytopenia of normal platelet size presenting within the first year of life. The thrombocytopenia progressively improves, with normal platelet count during adulthood. It is accompanied by bilateral radial aplasia and other skeletal abnormalities. Bone marrow evaluation shows reduced megakaryocytes. The genetic defect has not been characterized.

Others. Maturation of megakaryocytic and erythroid cell lines is dependent upon transcription factors GATA-1 and GATA-2, together with the cofactor FOG-1. GATA-1 is involved in both megakaryopoiesis and erythropoiesis, while GATA-2 stimulates megakaryopoiesis. Genetic defects in GATA-1 that prevent its interaction with FOG-1 have been associated with X-linked thrombocytopenia with dyserythropoiesis, a syndrome characterized by severe anemia, abnormal erythrocyte morphology, and RBC hemolysis. Other GATA-1 mutations that prevent its reaction with DNA are associated with X-linked thrombocytopenia with thalassemia (XLTT), which is characterized by thrombocytopenia, mild anemia, and imbalanced globin chain synthesis leading to a thalassemia minor phenotype.

Other congenital thrombocytopenia disorders are being identified as the megakaryocytic transcription factor pathways are being elucidated. Examples are the Paris-Trousseau type thrombocytopenia (TCPT) and Jacobsen syndrome, both caused by a deletion of 11q23.3, a chromosomal region that includes the genes for the transcription factors Ets-1 and Fli-1.

Acquired Thrombocytopenias with Decreased Platelet Production

This category of disorders is usually due to ineffective megakaryopoiesis. Rather than specific disorders, this finding is usually associated with many disorders of bone marrow failure, such as aplastic anemia, drug-induced aplasia, toxic chemicals, viral infection, radiation, myelophthisic disorders, paroxysmal nocturnal hemoglobinuria (PNH), and myelodysplasia. A few rare disorders, discussed below, affect platelet production only.

Acquired amegakaryocytic thrombocytopenic purpura (AATP) is the platelet analog to pure red cell aplasia. It is a rare disorder characterized by severe thrombocytopenia and lack of megakaryocytes in the bone marrow, with normal erythropoiesis and granulopoiesis. In contrast to immune thrombocytopenic purpura, no antiplatelet antibodies are detected. A mechanism has not been elucidated, but decreased megakaryopoiesis may be due to aberrant T cell suppression or an intrinsic colony forming unit-megakaryocyte (CFU-MK) defect. Some patients progress to aplastic anemia, while others develop myelodysplasia and acute leukemia.

Cyclic thrombocytopenia is characterized by fluctuations in platelet count, ranging from severe thrombocytopenia to marked thrombocytosis. It is observed more in women than men, and the platelet counts may fluctuate with menstrual cycles. The pathophysiology of the disorder is not known, but it is associated with cyclic failure of megakaryopoiesis.

Platelet Dysfunction Associated with Other Illnesses

Other significant disorders of platelet function with platelet counts in the normal range are usually acquired with the presence of another disease or drug therapy. These are, by far, more common than the disorders described above. Platelet dysfunction is often observed with chronic renal failure or liver disease in patients suffering from a variety of myeloproliferative and lymphoproliferative disorders (ie, polycythemia vera, myelofibrosis, paroxysmal nocturnal hemoglobinuria, AML, and hairy cell leukemia). Platelet dysfunction also may be associated with a variety of clinical scenarios, such as previous cardiopulmonary bypass, implantation of prosthetic materials such as vascular grafts and prosthetic heart valves, and ventricular assist devices. Platelet dysfunction in these disorders is usually difficult to characterize because nonspecific defects of platelet aggregation are usually observed.

Platelet Disorders Associated with Thrombosis or Platelet Activation

While many distinct platelet disorders associated with a bleeding diathesis have been characterized, distinct thrombotic disorders due to platelet dysfunction are not well established. Platelets play a widespread role in thrombotic disorders, including arterial thrombosis, stroke, cardiovascular disease, atherosclerosis, and thrombosis associated with cardiovascular devices and extracorporeal circulation. The involvement of platelets in thrombosis is likely a consequence of normal platelet activation mechanisms stimulated by other pathologic processes, from atherosclerosis to endothelial injury to implantation of cardiovascular devices. An example is heparin-induced thrombocytopenia (HIT), discussed previously and in chapter 21, in which normal platelet activation and aggregation mechanisms are pathologically stimulated by the presence of heparin/anti-platelet factor 4/ immunoglobulin complexes, resulting in clinical thrombosis. Specific genetic risk factors for platelet-mediated thrombosis do not necessarily need to be implicated, but these may play a significant role, as indicated below.

Platelet Activation

Platelet activation is associated with many other disease processes, from cardiovascular diseases to hypertension, diabetes, and vascular injury. Abnormal glycation is thought to contribute to abnormalities of platelet function in diabetic patients. Some patients with myeloproliferative disorders will show evidence of activated platelets or hyperplatelet function. Although most patients with essential thrombocythemia and polycythemia vera will display hypoaggregation, some will show an increased response to various agonists. As a caution, in vitro hypoplatelet or hyperplatelet function in patients with myeloproliferative disorders does not necessarily correlate with bleeding or thrombosis.

Activated platelets are known to bind to leukocytes. Platelet P-selectin, expressed on the outer membrane of activated platelets, can bind to leukocytes through P-selectin glycoprotein ligand-1 and through Mac-1 (CD11b/CD18). Platelet-neutrophil aggregates and platelet-monocyte aggregates can form during platelet activation and stimulate inflammation, which has been implicated as one mechanism of cardiovascular atherosclerosis. Activated platelets express the ligand for the CD40 receptor located on endothelial surfaces, which may be another mechanism for platelet/ endothelial interaction in cardiopulmonary bypass, as binding of activated platelets to CD40 may trigger an inflammatory response in endothelial cells.

Various techniques for the detection of in vivo platelet activation or increased platelet function are available. Spontaneous in vitro platelet aggregation or the finding of increased aggregation responses to low concentrations of platelet agonists suggests increased platelet function but is not specific for the cause. Circulating platelet aggregates may be detected either by flow cytometry or by collecting a whole blood sample into a fixative.

Platelet flow cytometry can be used to detect circulating activated platelet due to neoexpression of new surface markers (ie, P-selectin or CD63) or by detecting proteins bound to activated platelet surface glycoproteins (ie, fibrinogen, thrombospondin). Some caution must be exercised in the detection of increased platelet activation, because persistent in vivo activation can lead to degranulation of platelets, with the detection of hypoaggregation in the laboratory. Conversely, blood collection and processing alone can be associated with platelet activation and degranulation, so rigorous laboratory processing methods should be adopted prior to attempting these studies.

Genetic Polymorphisms Associated with Thrombosis and Cardiovascular Disease

Genetic polymorphisms of many platelet surface glycoproteins have been identified. Polymorphisms are stable DNA sequence variations that occur in greater than 1% of chromosomes in the general population. Genetic variability of these surface glycoproteins has been associated with some variability in platelet response to agonists, with hypoactivity and hyperactivity identified. Clinical studies aimed at determining the significance of genetic polymorphisms of platelet surface receptors as risk factors for thrombotic cardiovascular disease have provided intriguing results, but some conflicting data as well.

A polymorphism of glycoprotein IIIa (Leu33Pro) is commonly identified. The majority of individuals have leucine at amino acid 33 (PL^{A1} or HPA-1a), while the polymorphic allele results in proline at amino acid 33 (PL^{A2} or HPA-1b). This polymorphism has previously been associated with posttransfusion purpura and neonatal alloimmune thrombocytopenia. In 1996, a report associated the PL^{A2} allele with a higher risk of myocardial infarction. Since the initial report, many studies have investigated the PL^{A2} allele as a cardiovascular risk factor, with conflicting results. This is likely due to the choice of patient populations and the specific risk factors selected. Recent meta-analyses have concluded that the PL^{A2} allele is a modest risk factor for coronary thrombosis, but not stroke, with higher odds ratios for younger subjects and those undergoing revascularization procedures. The mechanism for the thrombotic tendency has not been elucidated, but some studies suggest that the Pro33 allele is associated with enhanced platelet adhesion to fibrinogen and increased outside-in signaling, which would favor stable thrombus formation.

Other polymorphisms of platelet surface glycoproteins have been associated with a thrombosis risk. The C807T polymorphism of glycoprotein Ia is linked to GPIa/IIa receptor density; the 807T allele is associated with higher receptor densities. The C807T polymorphism has been variably associated with a modestly increased risk for myocardial infarction and stroke, especially in younger patients.

The GPIb receptor is responsible for platelet adhesion to von Willebrand factor, and three polymorphic sites have been identified: the Kozak dimorphism (-5T/C), the Thr145Met dimorphism (Br or HPA-2), and the variable number of tandem repeats (VNTR) polymorphism. While many studies have examined the clinical associations of these polymorphisms, there is little consistent data indicating that these are associated with increased thrombotic risk.

Cardiovascular Devices

Both implanted and temporary cardiovascular devices have been widely employed for treatment of cardiovascular diseases. These devices range from small intracoronary stents to artificial heart valves, vascular grafts, and the total artificial heart. Temporary devices include hemodialysis, cardiopulmonary bypass (CPB), intra-aortic balloon counterpulsation (IABP), and extracorporeal membrane oxygenation (ECMO). With few exceptions, these devices are made of nonbiologic materials, such as polymers and metals, and they elicit a strong platelet-activating effect upon implantation. For this reason, anticoagulation and antiplatelet drug therapy is often employed in patients exposed to these devices.

Upon exposure of the blood to the material surfaces of the implanted cardiovascular devices, there is rapid adsorption and conformational alteration of plasma proteins, including fibrinogen and von Willebrand factor. Platelets rapidly adhere to the adsorbed vWF, and there is evidence that the conformationally-altered fibrinogen can adsorb nonactivated platelets by a calcium-dependent process. After adhesion to the protein layer, platelets activate and release their granule contents, as shown by increased plasma platelet activation markers: platelet factor 4, β-thromboglobulin, and thromboxane B2. In cardiopulmonary bypass, platelets can also be activated by the blood/gas interface of the bubble oxygenators. The adherent activated platelets form a phospholipid substrate for coagulation activation. Secondary to increased shear and complement activation, these activated platelets release submicron platelet microparticles into the plasma, which become systemic substrates for ongoing coagulation activation. Overt DIC is not uncommon with implantation of larger devices, such as ventricular assist devices. Platelet aggregation can occur to the layer of adherent platelets, but many small platelet aggregates likely embolize from the device surface, where they may lodge in the microvasculature and cause microvascular thrombosis, leading to tissue necrosis and further inflammation. Formation of larger platelet aggregates can lead to overt thromboembolism that may be clinically significant, especially with cardiac valve prostheses. Activated degranulated platelets can also dislodge singly from the material surface and circulate, but because they are already degranulated, they are less responsive to agonists. This may give rise to "exhausted platelets" or acquired storage pool disorder, where the circulating platelets are refractory to stimulation.

Flow cytometric analysis of circulating platelets has detected changes in the platelet surface glycoproteins during CPB or during acute implantation of larger cardiovascular devices. Glycoprotein Ib, the von Willebrand factor receptor, may be decreased during CPB, as is CD31. The fibrinogen receptor, GPIIb/IIIa, has also been shown to be variably decreased. P-selectin is rapidly expressed during CPB, a reflection of the rapid alpha granule release. The mechanism behind the changes in platelet surface glycoproteins during CPB is not firmly established, but mechanical forces and membrane fragmentation may be partly responsible. Plasminogen can bind to platelets, which, when converted to plasmin, can cleave platelet surface proteins.

Summary

The adequate function of platelets is crucial for normal hemostasis. Bleeding and thrombotic disorders can be associated with platelet dysfunction or abnormal platelet counts. The causes of platelet dysfunction, thrombocytosis, and thrombocytopenia are legion, and new etiologies are being elucidated continually. The laboratory evaluation of all of these disorders should start with measurement of the platelet count and evaluation of platelet morphology on a well-prepared blood smear. A careful and thorough patient history is essential, especially in exclusion of drugs or other substances that can interfere with platelet function. Appropriate platelet functional tests and other ancillary tests may be needed to positively diagnose many platelet disorders. Repeat testing is often necessary due to variability of platelet functional test results. It is hoped that the algorithmic approach discussed here will be helpful to the pathologist and clinician alike in facilitating the diagnosis of platelet disorders.

Suggested Reading

General Reviews

Bennett JS. Hereditary disorders of platelet function. In: Hoffman R, Benz E, Shattil S, Furie B, Cohen H. *Hematology: Basic Principles and Practice.* 4th ed. London: Churchill Livingstone; 2005: 2327-2345.

Freson K, Labarque V, Thys C, Wittevrongel C, Geet CV. What's new in using platelet research?: to unravel thrombopathies and other human disorders. *Eur J Pediatr.* 2007;166:1203-1210.

Kottke-Marchant K, Corcoran G. The laboratory diagnosis of platelet disorders: an algorithmic approach. *Arch Pathol Lab Med.* 2002;126:133-146.

Nurden AT. Qualitative disorders of platelets and megakaryocytes. *J Thromb Haemost.* 2005;3:1773-1782.

Nurden P, Nurden AT. Congenital disorders associated with platelet dysfunctions. *Thromb Haemost.* 2008;99:253-263.

White GC II. Congenital and acquired platelet disorders: current dilemmas and treatment strategies. *Semin Hematol.* 2006;43(suppl 1):S37-S41.

Platelet Dysfunction with Normal Platelet Count

Glycoprotein Disorders

Glanzmann Thrombasthenia

Clemetson KJ, Clemetson JM. Platelet adhesive protein defect disorders. In: Gresele P, Page C, Fuster V, Vermylen, eds. *Platelets in Thrombotic and Non-Thrombotic Disorders: Pathophysiology, Pharmacology and Therapeutics.* Cambridge, UK: Cambridge University Press; 2002: 639-654.

French DL, Seligsohn U. Platelet glycoprotein IIb/IIIa receptors and Glanzmann's thrombasthenia. *Arterioscler Thromb Vasc Biol.* 2000;20:607-610.

Glanzmann E. Hereditar hamorrhagische thrombasthenie: Ein Beitrag zur Pathologie der Blutplattchen. *J Kinderkranken.* 1918;88:113-117.

Linden MD, Frelinger AL 3rd, Barnard MR, Przyklenk K, Furman MI, Michelson AD. Application of flow cytometry to platelet disorders. *Semin Thromb Hemost.* 2004;30:501-511.

Peretz H, Rosenberg N, Usher S, et al. Glanzmann's thrombasthenia associated with deletion-insertion and alternative splicing in the glycoprotein IIb gene. *Blood.* 1995;85:414-420.

Vinciguerra C, Trzeciak MC, Philippe N, et al. Molecular study of Glanzmann thrombasthenia in 3 patients issue from 2 different families. *Thromb Haemost.* 1994;74:822-827.

Wilcox DA, White GC 2nd. Gene therapy for platelet disorders: studies with Glanzmann's thrombasthenia. *J Thromb Haemost.* 2003;1(11):2300-2311.

Platelet-Type von Willebrand Disease

Doggett TA, Girdhar G, Lawshe A, et al. Alterations in the intrinsic properties of the GPIbalpha-vWF tether bond define the kinetics of the platelet-type von Willebrand disease mutation, Gly233Val. *Blood.* 2003;102(1):152-160.

Miller JL. Platelet-type von Willebrand disease. *Thromb Haemost.* 1996;75:865-869.

Miller JL, Cunningham D, Lyle VA, Finch CN. Mutation in the gene encoding the a chain of platelet glycoprotein Ib in platelet type von Willebrand disease. *Proc Natl Acad Sci USA.* 1991;88:4761-4765.

Othman M, Notley C, Lavender FL, et al. Identification and functional characterisation of a novel 27bp deletion in the macroglycopeptide-coding region of the GPIb{alpha} gene resulting in platelet-type von Willebrand disease. *Blood.* 2005;105(11):4330-4336.

Russell SD, Roth GJ. Pseudo-von Willebrand disease: a mutation in the platelet glycoprotein Iba gene associated with a hyperactive surface receptor. *Blood.* 1993;81:1787-1791.

Sadler JE. New concepts in von Willebrand disease. *Annu Rev Med.* 2005;56:173-191.

Glycoprotein IV Disorders

Kashiwagi H, Honda S, Tomiyama Y, et al. A novel polymorphism in glycoprotein IV (replacement of proline 90 by serine) predominates in subjects with platelet GPIV deficiency. *Thromb Haemost.* 1993;69:481-484.

Yamamoto N, Ikeda H, Tandon NN, et al. A platelet membrane glycoprotein (GP) deficiency in healthy blood donors: Naka- platelets lack detectable GPIV (CD36). *Blood.* 1990;76:1698-1703.

Collagen Receptor Disorders

Arai M, Yamamoto N, Moroi M, Akamatsu N, Fukatake K, Tanoue K. Platelet with 10% of the normal amount of glycoprotein VI have an impaired response to collagen that results in a mild bleeding tendency. *Br J Haematol.* 1995;89;124-130.

Kahn ML. Platelet-collagen responses: molecular basis and therapeutic promise. *Semin Thromb Hemost.* 2004;30:419-426.

Nieuwenhuis HK, Sakariassen KS, Houdijk WPM, Nievelstein PFEM, Sixma JJ. Deficiency of platelet membrane glycoprotein Ia associated with a decreased platelet adhesion to subendothelium: a defect in platelet spreading. *Blood*. 1986;68:692-695.

ADP Receptor Abnormalities

Cattaneo M. The P2 receptors and congenital platelet function defects. *Semin Thromb Hemost*. 2005;31:168-173.

Cattaneo M, Lecchi A, Randi AA, McGregor JL, Mannucci PM. Identification of a new congenital defect of platelet function characterized by severe impairment of platelet responses to adenosine diphosphate. *Blood*. 1990;80:2787-2796.

Cattaneo M, Zighetti ML, Lombardi R, et al. Molecular bases of defective signal transduction in the platelet P2Y12 receptor of a patient with congenital bleeding. *Proc Nat Acad Sci USA*. 2003;100: 1978-1983.

Hollopeter G, Jantzen H-M, Vincent D, et al. Identification of the platelet ADP receptor targeted by antithrombotic drugs. *Nature*. 2001;409:202-207.

Nurden P, Savi P, Heilmann E, et al. An inherited bleeding disorder linked to a defective interaction between ADP and its receptor on platelets: its influence on glycoprotein IIb-IIIa complex function. *J Clin Invest*. 1995;95:1612-1622.

Oury C, Toth-Zsamboki E, Van Geet C, et al. A natural dominant negative P2X1 receptor due to deletion of a single amino acid residue. *J Biol Chem*. 2000;275: 22611-22614.

Storage Pool Disorders

Biddle DA, Neto TG, Nguyen ND. Platelet storage pool deficiency of α and β granules. *Arch Pathol Lab Med*. 2001;125:1125-1126.

Gunay-Aygun M, Huizing M, Gahl WA. Molecular defects that affect platelet dense granules. *Semin Thromb Hemost*. 2004;30:537-547.

Huizing M, Anikster Y, Gahl WA. Hermansky-Pudlak syndrome and Chediak-Higashi syndrome: disorders of vesicle formation and trafficking. *Thromb Haemost*. 2001;86:233-245.

Lorez HP, DaPrada M, Rendu F, Pletscher A. Mepacrine, a tool for investigating the 5-hydroxytrypatinineorganells of blood platelets by fluorescence microscopy. *J Clin Med*. 1977;89:200-206.

Weiss HJ, Lages B, Vicic W, Tsung JL, White JG. Heterogenous abnormalities of platelet dense granule ultrastructure in 20 patients with congenital storage pool deficiency. *Br J Haematol*. 1993;83:282-295.

White JG. Inherited abnormalities of the platelet membrane and secretory granules. *Hum Pathol*. 1987;18:123-139.

Signal Transduction Disorders

Gabbeta J, Yang X, Kowalska MA, Sun L, Dhanasekaran N, Rao AK. Platelet signal transduction defect with Galpha subunit dysfunction and diminished Galphaq in a patient with abnormal platelet responses. *Proc Natl Acad Sci USA*. 1997;94:8750-8755.

Mestel F, Oetliker O, Beck E, et al. Severe bleeding associated with defective thromboxane synthetase. *Lancet*. 1980;1:157.

Patel YM, Patel K, Rahman S, et al. Evidence for a role for Galphai1 in mediating weak agonist-induced platelet aggregation in human platelets: reduced Galphai1 expression and defective Gi signaling in the platelets of a patient with a chronic bleeding disorder. *Blood*. 2003;101:4828-4835.

Rao AK, Gabbeta J. Congenital disorders of platelet signal transduction. *Arterioscler Thromb Vasc Biol*. 2000;20:285-289.

Rao AK, Jalagadugula G, Sun L. Inherited defects in platelet signaling mechanisms. *Semin Thromb Hemost*. 2004;30:525-535.

Platelet Procoagulant Disorders

Dachary-Prigent J, Pasquet JM, Fressinaud E, Toti F, Freyssinet JM, Nurden AT. Aminophospholipid exposure, microvesiculation and abnormal protein tyrosine phosphorylation in the platelets of a patient with Scott syndrome: a study using physiologic agonists and local anaesthetics. *Br J Haematol*. 1997; 99:959-967.

Zwaal RFA, Schroit AJ. Pathophysiologic implications of membrane phospholipids asymmetry in blood cells. *Blood*. 1997;98:1121-1131.

Platelet Disorders with Thrombocytosis

Barbui T, Finazzi G. Thrombocytosis and thrombocythemia. In: Gresele P, Page C, Fuster V, Vermylen J, eds. *Platelets in Thrombotic and Non-Thrombotic Disorders: Pathophysiology, Pharmacology and Therapeutics*. Cambridge, UK: Cambridge University Press; 2002:623-638.

Hexner EO. JAK2 V617F: implications for thrombosis in myeloproliferative diseases. *Curr Opin Hematol*. 2007;14:450-454.

Lengfelder E, Merx K, Hehlmann R. Diagnosis and therapy of polycythemia vera. *Semin Thromb Hemost*. 2006;32:267-275.

Levine RL, Pardanani A, Tefferi A, Gilliland DG. Role of JAK2 in the pathogenesis and therapy of myeloproliferative disorders. *Nat Rev Cancer*. 2007;7:673-683.

Michiels JJ, Juvonen E. Proposal for revised diagnostic criteria of essential thrombocythemia and polycythemia vera by the Thrombocythemia Vera Study Group. *Semin Thromb Haemost*. 1997;23:339-347.

Reilly JT. Idiopathic myelofibrosis: pathogenesis to treatment. *Hematol Oncol*. 2006;24:56-63.

Thiele J, Kvasnicka HM. Clinicopathological criteria for differential diagnosis of thrombocythemias in various myeloproliferative disorders. *Semin Thromb Hemost*. 2006;32:219-230.

Vora AJ, Lilleyman JS. Secondary thrombocytosis. *Arch Dis Child*. 1993; 68:88-90.

Platelet Disorders with Thrombocytopenia

Thrombocytopenia with Increased Platelet Size

Macrothrombocytopenias with Neutrophilic Inclusions (MYH9 Disorders)

Balduini CL, Savoia A. Inherited thrombocytopenias: molecular mechanisms. *Semin Thromb Haemost*. 2004;30:513-523.

Kunishima S, Matsushita T, Kojima T, et al. Immunofluorescence analysis of neutrophil nonmuscle myosin heavy chain-A in MYH9 disorders: association of subcellular localization with MYH9 mutations. *Lab Invest*. 2003;83:115-122.

Seri M, Pecci A, Di Bari F, et al. MYH9-related disease: May-Hegglin anomaly, Sebastian syndrome, Fechtner syndrome, and Epstein syndrome are not distinct entities, but represent a variable expression of a single illness. *Medicine*. 2003;82:203-215.

Bernard-Soulier Syndrome

Bernard J, Soulier JP. Sur une nouvelle variete de dystrophie thrombocytair hemorragipar congenitale. *Sem Hop*. 1948:24:3217-3223.

Bunescu A, Lindahl T, Solum NO, et al. Partial expression of GP Ib measured by flow cytometry in two patients with Bernard-Soulier syndrome. *Thromb Res*. 1994;76:441-450.

Cines DB, Busssel JB, McMillan RB, Zehnder JL. Congenital and acquired thrombocytopenia. *Hematology Am Soc Hematol Educ Program*. 2004;390-406.

Drouin J, Carson NL, Laneuville O. Compound heterozygosity for a novel nine-nucleotide deletion and the Asn45Ser missense mutation in the glycoprotein IX gene in a patient with Bernard-Soulier syndrome. *Am J Hematol*. 2005;78(1):41-8.

Kenny D, Morateck PA, Gill JC, Montgomery RR. The critical interaction of glycoprotein (GP) Ibβ with GP IX. A genetic cause of Bernard-Soulier syndrome. *Blood*. 1999;93:2968-2975.

Kunishima S, Kamiya T, Saito H. Genetic abnormalities of Bernard-Soulier syndrome. *Int J Hematol*. 2002;76(4):319-327.

Li C, Pasquale DN, Roth GJ. Bernard-Soulier syndrome with severe bleeding: absent platelet glycoprotein Ib alpha due to a homozygous one-base deletion. *Thromb Haemost*. 1996;76:670-674.

Lopez JA, Andrews RK, Afshar-Kharghan V, Berndt MC. Bernard-Soulier syndrome. *Blood*. 1998; 91:4397-4418.

Noris P, Arbustini E, Spedini P, Belletti S, Balduini CL. A new variant of Bernard-Soulier syndrome characterized by dysfunctional glycoprotein (gp) Ib and severely reduced amounts of gpIX and gpV. *Br J Haematol*. 1998;103:1004-1013.

Tomer A, Scharf RE, McMillan R, Ruggeri ZM, Harker LA. Bernard-Soulier syndrome: quantitative characterization of megakaryocytes and platelet by flow cytometric and platelet kinetic measurements. *Eur J Haematol*. 1994;52:193-200.

Gray Platelet Syndrome

Falik-Zaccai TC, Anikster Y, Rivera CE, at al. A new genetic isolate of gray platelet syndrome (GPS): clinical, cellular and hematologic characteristics. *Mol Genet Metab*. 2001;74:303-313.

Gebrane-Younes J, Cramer EM, Orcel L, Caen JP. Gray platelet syndrome: dissociation between abnormal sorting in megakaryocyte α-granules and normal sorting in Weibel-Palade bodies of endothelial cells. *J Clin Invest*. 1993;92:3023-3028.

Jantunen E, Hanninen A, Naukkarienen A, Vornanen M, Lahtinen R. Gray platelet syndrome with splenomegaly and signs of extrameduallary hematopoiesis: a case report with review of the literature. *Am J Hematol*. 1994; 46:218-224.

Niewenhuis HK, Akkerman JW, Sixma JJ. Patients with a prolonged bleeding time and normal aggregation tests may have storage pool deficiency: studies on one hundred six patients. *Blood*. 1987;70:620-623.

White JG, Brunning RD. Neutrophils in the gray platelet syndrome. *Platelets*. 2004;15(5):333-340.

Others

Behrens WE. Mediterranean macrothrombocytopenia. *Blood*. 1975;46(2):199-208.

Evans DA, Metz J. Mediterranean macrothrombocytopenia revisited. *Med J Aust*. 1999;171(5):277-278.

Freson K, Devriendt K, Matthijs G, et al. Platelet characteristics in patients with X-linked macrothrombocytopenia because of a novel GATA1 mutation. *Blood*. 2001;98:85-92.

Krishnamurti L, Neglia JP, Nagarajan R, et al. Paris-Trousseau syndrome platelets in a child with Jacobsen's syndrome. *Am J Hematol*. 2001;66:295-299.

Mehaffey MG, Newton AL, Gandhi MJ, Crossley M, Drachman JG. X-linked thrombocytopenia caused by a novel mutation of GATA-1. *Blood*. 2001;98:2681-2688.

Okita JR, Frojmovic MM, Kristopet S, Wong T, Kunicki TJ. Montreal platelet syndrome: a defect in calcium-activated neutral proteinase (calpain). *Blood*. 1989;74:715-721.

Yu C, Niakan KK, Matsushita M, Stamatoyannopoulos G, Orkin SH, Raskind WH. X-linked thrombocytopenia with thalassemia from a mutation in the amino finger of GATA-1 affecting DNA binding rather than FOG-1 interation. *Blood*. 2002;100:2040-2045.

Thrombocytopenia with Decreased Platelet Size

Wiskott-Aldrich Syndrome (WAS) and X-Linked Thrombocytopenia (XLT)

Haddad E, Cramer E, Riviere C, et al. The thrombocytopenia of Wiskott Aldrich syndrome is not related to a defect in proplatelet formation. *Blood*. 1999;94:509-518.

Ochs HD, Thrasher AJ. The Wiskott-Aldrich syndrome. *J Allergy Clin Immunol*. 2006;117:725-738.

Semple JW, Siminovitch KA, Mody M, et al. Flow cytometric analysis of platelets from children with the Wiskott-Aldrich syndrome reveals defects in platelet development, activation and structure. *Br J Haematol*. 1997;97:747-754.

Snapper SB, Rosen FS. A family of WASPs. *N Engl J Med*. 2003;348:350-351.

Thrombocytopenia with Normal Platelet Size

Peripheral Platelet Destruction

Immune Destructive Thrombocytopenias
Immune Thrombocytopenic Purpura (ITP)

Cines DB, Bussel JB, McMillan RB, Zehnder JL. Congenital and Acquired Thrombocytopenia. *Hematology Am Soc Hematol Educ Program*. 2004; 390-406.

Cimo PL, Pisciotta AV, Desai RG, et al. Detection of drug-dependent antibodies by the 51Cr platelet lysis test: documentation of immune thrombocytopenia induced by diphenyl-hydantoin, diazepam, and sulfisoxazole. *Am J Hematol*. 1977;2:65-72.

Cines DB, Blanchette VS. Immune thrombocytopenic purpura. *N Engl J Med*. 2002;346;995-1008.

Coopamah MD, Garvey MB, Freedman J, Semple JW. Cellular immune mechanisms in autoimmune thrombocytopenic purpura: an update. *Transfus Med Rev*. 2003;17:69-80.

George JN, Woolf SH, Raskob GE, et al. Idiopathic thrombocytopenic purpura: a practice guideline developed by explicit methods for the American Society of Hematology. *Blood*. 1996;88:3-40.

Kiefel V, Santoso S, Weisheit M, Mueller-Eckhardt C. Monoclonal antibody-specific immobilization of platelet antigens (MAIPA): a new tool for the identification of platelet-reactive antibodies. *Blood*. 1987;70:1722-1726.

McMillan R. Antiplatelet antibodies in chronic adult immune thrombocytopenic purpura: assays and epitopes. *J Pediatr Hematol Oncol*. 2003;(suppl 1):57-61.

McFarland JG. Laboratory investigation of drug-induced immune thrombocytopenias. *Transfus Med Rev*. 1993;VII:275-287

Alloimmune Thrombocytopenias

Bussel JB, Zabusky MR, Berkowitz RL, McFarland JG. Fetal alloimmune thrombocytopenia. *N Engl J Med*. 1997;337:22-26.

Christie DJ, Pulkrabek S, Putnam JL, Slatkoff ML, Pischel KD. Posttransfusion purpura due to an alloantibody reactive with glycoprotein Ia/IIa (anti-HPA-5b). *Blood*. 1991;77:2785-2789.

Mueller-Eckhardt C, Lechner K, Heinrich D, et al. Posttransfusion thrombocytopenic purpura: immunological and clinical studies in two cases and review of the literature. *Blut*. 1980;40:249-257.

Murphy MF, Manley R, Roberts D. Neonatal alloimmune thrombocytopenia. *Haematologica*. 1999;84:110-114.

Drug-Induced Thrombocytopenias
Immune Drug-Induced Thrombocytopenias

Aster RH. Drug-induced thrombocytopenia. In: Michelson AD. *Platelets*. Amsterdam: Academic Press; 2002: 593-606.

George JN, Raskob GE, Rizvi Shah S, et al. Drug-induced thrombocytopenia: a systematic review of published case reports. *Ann Intern Med*. 1998;129:886-890.

Pedersen-Bjergarrd U, Andersen M, Hansen PB. Drug-specific characteristics of thrombocytopenia caused by non-cytotoxic drugs. *Eur J Clin Pharmacol*. 1998;54:701-706.

Nonimmune Destructive Thromboctyopenias
Thrombotic Thrombocytopenia Purpura

Bohm M, Vigh T, Scharrer I. Evaluation and clinical application of a new method for measuring activity on von Willebrand factor-cleaving metalloprotease (ADAMTS13). *Ann Hematol*. 2002;81:430-435.

Furlan M, Robles R, Solenthaler M, et al. Deficient activity of von Willebrand factor-cleaving protease in chronic relapsing thrombotic thrombocytopenic purpura. *Blood*. 1997;89:3097-3103.

George JN. Thrombotic thrombocytopenic purpura. *N Engl J Med*. 2006;354:1927-1935.

Kokame K, Matsumoto M, Fujimura Y, Miyata T. VWF73, a region from D1596 to R1668 of von Willebrand factor, provides a minimal substrate for ADAMTS-13. *Blood*. 2004;103:607-612.

Moake JL. Thrombotic microangiopathies. *N Engl J Med*. 2002;347:589-600.

Sadler JE, Moake JL, Miyata T, and Geroge JN. Recent advances in thrombotic thrombocytopenic purpura. *Hematology Am Soc Hematol Educ Program*. 2004; 407-423.

Soejima K, Nakagaki T. Interplay between ADAMTS13 and von Willebrand factor in inherited and acquired thrombotic microangiopathies. *Semin Hematol*. 2005;42(1):56-62.

Studt JD, Bohm M, Budde U, Girma JP, Varadi K, Lammle B. Measurement of von Willebrand factor-cleaving protease (ADAMTS-13) activity in plasma: a multicenter comparison of different assay methods. *J Thromb Haemost*. 2003;1:1882-1887.

Tsai H-M. Molecular mechanisms in thrombotic thrombocytopenic purpura. *Semin Thromb Haemost*. 2004;30:549-557.

Hypertensive Disorders of Pregnancy, Including Hemolysis, Increased LFTs, and Preeclampsia (HELLP)

Katz VL, Thorp JM Jr, Rozas L, Bowes WA Jr. The natural history of thrombocytopenia associated with preeclampsia. *Am J Obstet Gynecol*. 1990;163:1142-1143.

Mattar F, Sibai BM. Preeclampsia: clinical characteristics and pathogenesis. *Clin Liver Dis*. 2001;3:15-29.

Saphier CJ, Repke JT. Hemolysis, elevated liver enzymes and low platelets (HELLP) syndrome: a review of diagnosis and management. *Semin Perinatol*. 1998;22:118-133.

Decreased Platelet Production

Hereditary Thrombocytopenias with Decreased Platelet Production

Ballmaier M, Germeshausen M, Schulze H, et al. c-mpl mutations are the cause of congenital amegakaryocytic thrombocytopenia. *Blood*. 2001;97:139-146.

Greenhalgh KL, Howell RT, Bottani A, et al. Thrombocytopenia-absent radius syndrome: a clinical genetic study. *J Med Genet*. 2002;39:876-881.

Ihara K, Ishii E, Eguchi M, et al. Identification of mutations in the c-mpl gene in congenital amegakaryocytic thrombocytopenia. *Proc Natl Acad Sci USA*. 1999;96:3132-3136.

Song WJ, Sullivan MG, Legare RD, et al. Haploinsufficiency of CBFA2 causes familial thrombocytopenia with propensity to develop acute myelogenous leukaemia. *Nat Genet*. 1999;23:166-175.

Thompson AA, Woodruff K, Feig SA, Nguyen LT, Schanen NC. Congenital thrombocytopenia and radio-ulnar synostosis: a new familial syndrome. *Br J Haematol*. 2001;113:866-870.

Tijssen MR, di Suma F, van den Oudenrijn S, et al. Functional analysis of single amino-acid mutations in the thrombopoietin-receptor Mpl underlying congenital amegakaryocytic thrombocytopenia. *Br J Haematol*. 2008;141(6):808-813.

Acquired Thrombocytopenias with Decreased Platelet Production

Cohen T, Cooney DP. Cyclic thrombocytopenia: case report and review of literature. *Scand J Haematol*. 1974;12:9-17.

Nagasawa T, Hasegawa Y, Kamoshita M, et al. Megakaryopoiesis in patients with cyclic thrombocytopenia. *Br J Haematol*. 1995;91:185-190.

Platelet Disorders Associated with Thrombosis or Platelet Activation

Platelet Activation

Trip MD, Cats VM, van Capelle FJ, Vreeken J. Platelet hyperreactivity and prognosis in survivors of myocardial infarction. *N Engl J Med*. 1990;322:1549-1554.

Genetic Polymorphisms Associated with Thrombosis and Cardiovascular Disease

Afshar-Khargha V, Bray PF. Platelet polymorphisms. In: Michelson AD, ed. *Platelets*. Academic Press; 2002: 157-169.

Di Castelnuovo A, de Gaetano G, Donati MB, Iacoviello L. Platelet glycoprotein receptor IIa polymorphism PLA1/PLA2 and coronary risk: a meta-analysis. *Thromb Haemost*. 2001;85:626-633.

Kenny D, Muckian C, Fitzgerald DJ, Cannon CP, Shields DC. Platelet glycoprotein Ib alpha receptor polymorphisms and recurrent ischaemic events in acute coronary syndrome patients. *J Thromb Thrombolysis*. 2002;13:13-19.

Kottke-Marchant K. Genetic polymorphisms associated with venous and arterial thrombosis: an overview. *Arch Pathol Lab Med*. 2002;126:295-304.

Marian AJ, Brugada R, Kleiman NS. Platelet glycoprotein IIIa PlA polymorphism and myocardial infarction. *N Engl J Med.* 1996;335:1071-1072.

Moshfegh K, Wuillemin WA, Redondo M, et al. Association of two silent polymorphisms of platelet glycoprotein Ia/IIa receptor with risk of myocardial infarction: a case-control study. *Lancet.* 1999;353:351-354.

Yee DL, Bray PF. Clinical and functional consequences of platelet membrane glycoprotein polymorphisms. *Semin Thromb Hemost.* 2004;30:593-600.

Cardiovascular Devices

Clagett GP, Eberhart RC. Artificial devices in clinical practice. In: Colman RW, Hirsh J, Marder VJ, Salzman EW, eds. *Hemostasis and Thrombosis: Basic Principles and Clinical Practice.* 3rd ed. Philadelphia, Pa: JB Lippincott; 1994: 1486-1505.

Gemmell CH, Ramirez SM, Yeo EL, Sefton MV. Platelet activation in whole blood by artificial surfaces: identification of platelet-derived microparticles and activated platelet binding to leukocytes as material-induced activation events. *J Lab Clin Med.* 1995;125:276-287.

Kottke-Marchant K, Sapatnekar S. Hemostatic abnormalities in cardiopulmonary bypass: pathophysiologic and transfusion considerations. *Semin Cardiothoracic Vasc Anethesia.* 2001;5(3):187-206.

Ray MJ, Marsh NA, Hawson GA. Relationship of fibrinolysis and platelet function to bleeding after cardiopulmonary bypass. *Blood Coagul Fibrinolysis.* 1994;5:679-685.

CHAPTER 15

Abnormal Thrombin Time

Mark T. Cunningham, MD

Introduction

The thrombin time (TT) measures the time of fibrin clot formation after the addition of a standard concentration of thrombin to citrated plasma. Thrombin mediates fibrin clot formation by cleaving two molecules each of fibrinopeptide A and fibrinopeptide B from fibrinogen, producing fibrin monomer. Fibrin monomer then polymerizes in a linear and lateral fashion to form the fibrin clot. The technical aspects of the TT assay are discussed in detail in chapter 6.

This chapter will discuss the clinical indications for performing the TT, the differential diagnosis of an abnormal test result, and an algorithm for sequential test selection that can be used for evaluating an abnormal TT.

Clinical Indications

There are two clinical indications for measuring the TT. The first is to screen for bleeding disorders that characteristically produce a prolonged TT. These include afibrinogenemia, hypofibrinogenemia, dysfibrinogenemia, polyclonal and monoclonal autoantibodies against thrombin or fibrinogen/fibrin, acquired alloantibodies against topical bovine thrombin, or other conditions, such as disseminated intravascular coagulation. Although most of these disorders are very uncommon, they should be considered if there is an appropriate clinical history or if more common etiologies of bleeding are excluded.

The second clinical indication is to screen for anticoagulant drugs that interfere with other coag-

Case Study

A 17-year-old male with congenital heart disease was admitted for revision of a prosthetic cardiac valve. He previously had undergone three operations for correction of cardiac anomalies, with no history of bleeding problems. Topical bovine thrombin was used as a hemostatic agent during the two most recent operations. At the time of the present admission, the patient was not receiving heparin or a direct thrombin inhibitor drug. Preoperative laboratory testing yielded the following results:

Test	Value	Reference Range
PT (s)	12	11-13
aPTT (s)	31	27-33
Fibrinogen (mg/dL)	285	200-400
D-dimer (ng/mL FEU)	<500	<500
Platelet count (per µL)	340,000	150,000-400,000
TT (s)		
Bovine thrombin	>300	18-28
Human thrombin	21	18-28
TT 1:1 mix (s)		
Bovine thrombin	>300	18-28
Serum immunofixation electrophoresis	Negative	Negative

What is the most likely cause for the prolonged thrombin time?

Answer: Bovine thrombin inhibitor. The use of bovine thrombin during surgery can cause the development of alloantibodies against bovine thrombin. Bovine thrombin is used as a reagent in most TT assays. Due to species differences in the thrombin sequence, these inhibitors usually do not show cross-reactivity with human thrombin and thus are infrequently associated with bleeding complications. However if the antibody does cross-react with human thrombin, then bleeding complications can occur. The diagnosis is made in the appropriate clinical context (ie, history of topical bovine thrombin use) and by demonstrating a prolonged TT with bovine thrombin, a normal or prolonged TT with human thrombin, and a noncorrected TT 1:1 mixing study. The normal D-dimer level and normal serum immunofixation electrophoresis in this patient excludes disseminated intravascular coagulation and a monoclonal thrombin inhibitor.

Table 15-1. Differential Diagnosis of an Abnormal Thrombin Time (TT)

Disorder/Substance	TT	Mechanism of Effect
Amyloidosis, light chain type	Prolonged	Inhibition of fibrin monomer polymerization
Autoantibodies		
Monoclonal	Prolonged	Inhibition of thrombin; inhibition of fibrin monomer polymerization
Polyclonal	Prolonged	Inhibition of thrombin; inhibition of fibrin monomer polymerization
Drugs		
Bovine thrombin, topical	Prolonged	Development of alloantibodies against bovine thrombin; may cross-react with human thrombin
Direct thrombin inhibitors		
Argatroban	Prolonged	Inhibition of thrombin
Bivalirudin	Prolonged	Inhibition of thrombin
Lepirudin	Prolonged	Inhibition of thrombin
Melagatran	Prolonged	Inhibition of thrombin
Heparin	Prolonged	Inhibition of thrombin
Plasminogen activators		
Tissue plasminogen activator	Prolonged	Decreased fibrinogen substrate; inhibition of fibrin monomer polymerization due to production of fibrin degradation products
Urokinase	Prolonged	Decreased fibrinogen substrate; inhibition of fibrin monomer polymerization due to production of fibrin degradation products
Volume expanders		
Dextran	Shortened	Increased fibrin monomer polymerization
Hydroxyethyl starch	Shortened	Unknown
Fibrin degradation products	Prolonged	Inhibition of fibrin monomer polymerization
Fibrinogen abnormalities		
Afibrinogenemia	Prolonged	Absent fibrinogen substrate
Dysfibrinogenemia, acquired	Prolonged	Inhibition of fibrin monomer polymerization
Dysfibrinogenemia, inherited	Prolonged	Inhibition of fibrinopeptide A release; inhibition of fibrinopeptide B release; inhibition of fibrin monomer polymerization
	Shortened	Unknown
Hyperfibrinogenemia	Prolonged	Unknown
Hypofibrinogenemia	Prolonged	Decreased fibrinogen substrate
Radiocontrast agents	Prolonged	Inhibition of fibrinopeptide A release

ulation tests, such as the prothrombin time (PT), activated partial thromboplastin time (aPTT), and lupus anticoagulant assays. Drugs that inhibit these tests include heparins (unfractionated heparin, low-molecular-weight heparin) and direct thrombin inhibitors (argatroban, lepirudin, bivalirudin). The TT is very sensitive to these drugs; therefore, a normal TT would exclude them as a possible explanation for an abnormal PT, aPTT, or lupus anticoagulant assay.

Differential Diagnosis

The differential diagnosis of an abnormal TT and the effects of specific disorders and substances on the TT are shown in Table 15-1. These entities are discussed in more detail in the following.

Amyloidosis

Systemic immunoglobulin light chain amyloidosis is associated with a prolonged TT in 32% to 85% of cases. The mechanism appears to be a circulating inhibitor that interferes with the conversion of fibrinogen to fibrin. The identity and specific mechanism of the inhibitor has not been well characterized; however, in one patient it was shown to be a monoclonal lambda immunoglobulin that bound to fibrinogen and blocked fibrin monomer polymerization.

Based on multivariate analysis, the prolonged TT in primary amyloidosis is not associated with bleeding. It has been associated with hepatic amyloid infiltration, proteinuria, and hypoalbuminemia. Amyloidosis is also associated with a decrease in factor X, a prolongation of PT and aPTT, and an increased bleeding tendency in some patients. This is discussed in more detail in chapter 11.

Autoantibodies

Autoantibodies directed against thrombin are very rare; their descriptions are limited to case reports. The clinical presentation is either bleeding or thrombosis. Risk factors include autoimmune disease and plasma cell myeloma. The primary laboratory abnormality is a prolonged TT. These antibodies can be polyclonal or monoclonal, and have been shown to inhibit either thrombin proteolytic activity or the anion-binding exosite of thrombin. Inhibition of proteolytic activity is associated with bleeding, whereas inhibition of the anion-binding exosite is associated with recurrent arterial thrombosis. High levels of the immunoglobulin paraprotein in plasma cell dyscrasias may also prolong the TT due to interference in fibrin monomer polymerization.

Autoantibodies directed against fibrinogen/fibrin are also very rare. These antibodies are associated with bleeding, produce a prolonged TT, and mediate their effect by inhibiting fibrin monomer polymerization.

Drugs

Bovine Thrombin, Topical

Topical bovine thrombin has been used in surgical procedures to control localized intraoperative bleeding. This drug is infrequently associated with the development of alloantibodies against bovine thrombin. These antibodies cause TT prolongation by inhibiting the bovine thrombin used in most TT reagents. Human thrombin-based reagents do not typically show TT prolongation. In some cases, the antibodies cross-react with human thrombin and cause bleeding by interfering with physiologic thrombin function. A variety of thrombin functions can be inhibited, such as (1) serine protease activity against fibrinogen and (2) antithrombin binding. There may also be a development of a factor V inhibitor due to contamination of the topical bovine thrombin product with factor V.

An acquired inhibitor to bovine thrombin can be diagnosed by performing the TT, using both human and bovine thrombin reagents. In most cases, the TT is typically prolonged with the bovine thrombin, and mixing studies of the TT using a 1:1 mix of patient to normal plasma fail to correct the prolonged TT. The TT with human thrombin is not typically prolonged. The results of other clot-based assays, such as the PT and aPTT, are usually normal.

Direct Thrombin Inhibitors

Direct thrombin inhibitors (DTIs) are a class of anticoagulant drug used to prevent and treat venous thrombosis, and to manage percutaneous coronary interventions. Their clinical utilization and monitoring are discussed in detail in chapter 24. DTIs are particularly valuable in patients that cannot use heparin, such as those with a history of heparin-induced thrombocytopenia with thrombosis. Specific DTIs in current use include argatroban, melagatran, lepirudin, and bivalirudin. The mechanism involves binding to the catalytic site of thrombin; some of the larger molecules, such as lepirudin, also bind to thrombin's anion-binding exosite.

These drugs prolong the TT as well as the aPTT, with variable prolongation of the PT at therapeutic concentrations. They can interfere with a wide variety of coagulation assays, such as testing for a lupus anticoagulant, factor assays, some fibrinogen assays, and clottable protein C and S assays. Mixing studies of the TT using a 1:1 mix of patient to normal plasma show noncorrection of the prolonged TT with DTI drugs, and anti-Xa assays for heparin are negative. The DTI inhibitor effect cannot be neutralized in vitro, so further testing for factor assays or lupus anticoagulant is not recommended.

Heparin

Unfractionated heparin is a potent inhibitor of thrombin and subsequently prolongs all global coagulation tests, such as the TT, PT, and aPTT. Of these three tests, the TT is the most sensitive to heparin, particularly at therapeutic doses of 0.3 to 0.7 units/mL as measured by the anti-factor Xa activity assay.

The mechanism of unfractionated heparin-induced thrombin inhibition involves heparin binding to both antithrombin and thrombin, resulting in the formation of a covalent antithrombin-thrombin complex that inactivates thrombin's proteolytic activity. Heparin binding to antithrombin is mediated by a specific pentasaccharide region that induces an antithrombin conformational change; heparin binding to thrombin is mediated by anionic saccharide domains.

Low-molecular-weight heparins and fondaparinux (synthetic pentasaccharide) have very low to absent affinity for thrombin, as they are pre-

Table 15-2. Differential Diagnosis of Anticoagulant Contamination of Plasma Samples

Test	Unfractionated Heparin	Low-Molecular-Weight Heparin and Fondaparinux	Direct Thrombin Inhibitors
Prothrombin time (PT)	Normal	Normal	Variably prolonged
Activated partial thromboplastin time (aPTT)	Prolonged	Normal to slightly prolonged	Prolonged
Thrombin time (TT)	Prolonged	Usually normal	Prolonged
Thrombin time (TT) (1:1 mix)	Noncorrection	—	Noncorrection
Anti-factor Xa inhibition	Elevated	Elevated	Normal

dominantly anti-factor Xa inhibitors; consequently, these drugs do not prolong the TT at therapeutic doses. The differential diagnosis of anticoagulants and their effect on common clotting tests in plasma samples is presented in Table 15-2.

Plasminogen Activators

Plasminogen activators are a class of anticoagulant drug used to re-establish blood flow in thrombosed arteries (stroke, acute myocardial infarction) and veins. Specific plasminogen activators in current use include urokinase and tissue-type plasminogen activator. These agents mediate fibrinolysis by activating the zymogen plasminogen into the enzyme plasmin, which then cleaves fibrin at the amino-terminal side of lysine residues. Fibrin degradation products (FDPs) are the end result of fibrinolysis. Plasminogen activators produce a prolonged TT by decreasing fibrinogen levels and by producing FDPs that inhibit fibrin monomer polymerization.

Volume Expanders

Plasma volume expanders are used to establish and maintain intravascular volume and blood pressure. Two examples are hydroxyethyl starch and dextran. These agents produce a shortening of the TT. In the case of dextran, the mechanism involves potentiation of fibrin monomer polymerization. The mechanism of hydroxyethyl starch effect on the TT is unknown but may be similar to dextran.

Fibrin Degradation Products

Fibrin degradation products are produced by plasmin mediated cleavage of fibrin polymers, as discussed above (see "Plasminogen Activators"). A variety of clinical situations are associated with elevated plasma FDPs, most notably deep vein thrombosis, pulmonary embolism, disseminated intravascular coagulation, myocardial infarction, stroke, postoperative state, liver disease, and pregnancy. FDPs at high concentration can produce a prolonged TT by inhibiting fibrin monomer polymerization.

Fibrinogen Abnormalities

Dysfibrinogenemia

Dysfibrinogenemia is a blood clotting disorder caused by a structurally abnormal fibrinogen. Inherited dysfibrinogenemia is caused by mutations in the coding regions of the fibrinogen Aα, Bβ, or γ genes, and over 300 mutations have been described. It is rare, usually autosomal dominant, and can present with bleeding or thrombosis. Acquired dysfibrinogenemia is usually caused by disease of the liver or biliary tract, which leads to increased sialylation of carbohydrate side chains of the fibrinogen molecule.

The primary screening test for dysfibrinogenemia is the TT. The majority of cases show prolongation of the TT, and there is one case reported to cause shortening of the TT. There are three mechanisms by which the TT is affected: inhibition of fibrinopeptide A release, inhibition of fibrinopeptide B release, and inhibition of fibrin monomer polymerization.

The diagnosis of dysfibrinogenemia is confirmed by demonstrating a low fibrinogen activity to fibrinogen antigen ratio with an elevated TT and a prolonged reptilase time. The inherited form is diagnosed by showing a similar laboratory test abnormality in family members, or by analysis of the fibrinogen protein or fibrinogen gene in the patient. The acquired form is diagnosed by

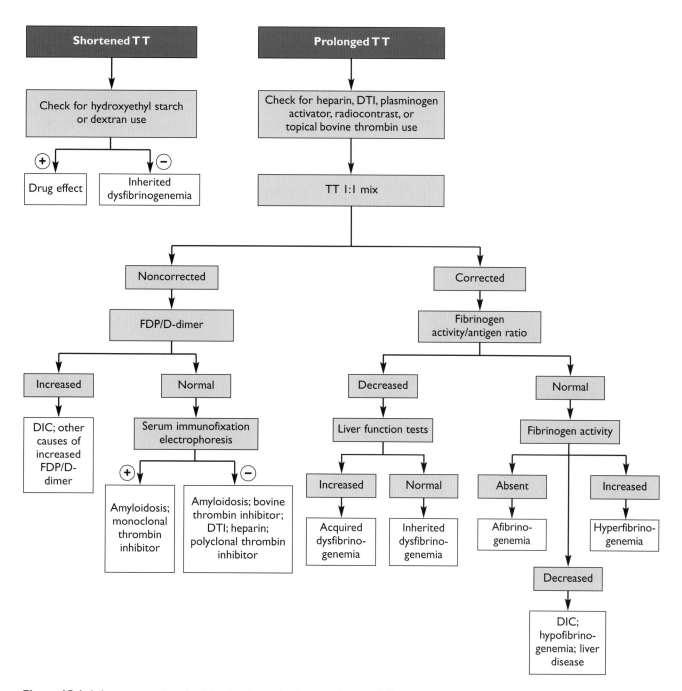

Figure 15-1. Laboratory testing algorithm for the evaluation of an abnormal thrombin time (TT). Initial evaluation of the TT is based upon whether or not the TT is normal, elevated, or decreased. The next step in the evaluation of a prolonged TT is to perform a TT mixing study using a 1:1 mix of patient to normal plasma. If the TT corrects into the normal range, this indicates that an inhibitor is not present. Subsequent testing would entail measurement of the fibrinogen clottable and immunologic assays, with determination of a fibrinogen activity/antigen ratio. A normal ratio usually indicates a quantitative fibrinogen abnormality due to decreased production (afibrinogenemia, hypofibrinogenemia, liver disease) or increased destruction (disseminated intravascular coagulation [DIC]). A decreased fibrinogen activity/antigen ratio suggests a dysfibrinogenemia. If the TT mixing study shows noncorrection, this suggests some type of inhibitor. Review of clinical records should help in the identification of heparin or direct thrombin inhibitor (DTI). If the fibrin degradation products (FDP)/D-dimer are elevated, this suggests DIC. In DIC, the TT mixing study does not always show an inhibitor effect, depending on the degree of elevation of the FDPs. If the FDP/D-dimer are normal, serum immunofixation electrophoresis should help to distinguish a monoclonal thrombin inhibitor from a polyclonal thrombin inhibitor, such as seen with topical bovine thrombin use. Acquired dysfibrinogenemia has been rarely reported to cause a noncorrected TT 1:1 mixing study.

demonstrating abnormal liver function tests and by ruling out dysfibrinogenemia in family members.

Hypofibrinogenemia

Hypofibrinogenemia is defined as a plasma fibrinogen concentration less than 150 mg/dL. Acquired hypofibrinogenemia is common and is caused by severe liver disease, disseminated intravascular coagulation, primary fibrinogenolysis, and high-volume perioperative fluid replacement. Impaired fibrinogen production by the liver is the basis for hypofibrinogenemia in liver disease. Increased fibrinogen consumption by thrombin and/or plasmin lowers fibrinogen levels in disseminated intravascular coagulation and primary fibrinogenolysis. High-volume fluid replacement, without sufficient fresh frozen plasma or cryoprecipitate transfusion, causes dilution of all plasma proteins including fibrinogen.

Inherited hypofibrinogenemia is very rare and is caused by mutations in the coding and noncoding regions of the fibrinogen Aα, Bβ, and γ genes. These mutations impair the synthesis or secretion of fibrinogen by hepatocytes.

The TT is sensitive to low plasma fibrinogen concentrations at levels below 100 mg/dL. The degree of TT prolongation is dependent on the type of TT reagent and the severity of the fibrinogen deficiency. In one study comparing five TT reagents, the TT prolongation ranged from 3 to 5 seconds and 6 to 14 seconds at fibrinogen levels approximating 100 mg/dL and 50 mg/dL, respectively.

Afibrinogenemia

Afibrinogenemia is a very rare, inherited, autosomal recessive fibrinogen deficiency with a prevalence of about 1 in 1,000,000. These patients are usually homozygous for a fibrinogen mutation or compound heterozygotes. The mutations responsible have been identified in all three fibrinogen genes, and approximately 30 mutations are known. The majority of the mutations are large or small deletions, but insertion mutations and nonsense mutations have also been described. The laboratory findings typically show unmeasurably long PT, aPTT, TT, and reptilase time, with fibrinogen <25 mg/dL and absent platelet aggregation.

Hyperfibrinogenemia

Hyperfibrinogenemia is defined as a plasma fibrinogen concentration greater than 450 mg/dL. Fibrinogen is an acute phase reactant; therefore, acute inflammation is a common cause of hyperfibrinogenemia. The effect of hyperfibrinogenemia on the TT appears to be dependent on the TT method. For example, one method showed minimal effect on the TT even up to fibrinogen concentrations of 1350 mg/dL, whereas another method showed significant prolongation of the TT at fibrinogen concentrations greater than 500 mg/dL.

Radiocontrast Agents

A variety of radiocontrast agents can prolong the TT in a concentration-dependent manner. These include diatrizoate, ioxaglate, ioxithalamate, iopamidol, and metrizamide. The mechanism involves inhibition of fibrinopeptide A release from fibrinogen. Thrombin's protease activity is not affected.

Algorithm

A laboratory testing algorithm for the evaluation of an abnormal TT is shown in Figure 15-1. Testing is directed toward evaluating the differential diagnosis just presented, using a scheme that proceeds from the most simple testing to the more complex. This algorithm is just one of many possible strategies that could be devised.

Suggested Reading

Amyloidosis

Gambia G, Montani N, Anesi E, et al. Clotting alterations in primary systemic amyloidosis. *Haematologica*. 2000;85:289-292.

Gastineau DA, Gertz MA, Daniels TM, Kyle RA, Bowie EJ. Inhibitor of thrombin time in systemic amyloidosis: a common coagulation abnormality. *Blood*. 1991;77:2637-2640.

Mumford AD, O'Donnell J, Gillmore JD, Manning RA, Hawkins PN, Laffan M. Bleeding symptoms and coagulation abnormalities in 337 patients with AL-amyloidosis. *Br J Haematol*. 2000;110:454-460.

Autoantibodies

Arnaud E, Lafay M, Gaussem P, et al. An autoantibody directed against human thrombin anion-binding exosite in a patient with arterial thrombosis: effects on platelets, endothelial cells, and protein C activation. *Blood*. 1994;84:1843-1850.

Bajaj SP, Rapaport SJ, Barelay S, Herbst KD. Acquired hypoprothrombinemia due to nonneutralizing antibodies to prothrombin: mechanism and management. *Blood*. 1985;65:1538-1543.

Colwell NS, Tollefson DM, Blinder M. Identification of a monoclonal thrombin inhibitor associated with multiple myeloma and a severe bleeding disorder. *Br J Haematol.* 1997;97:219-226.

Costa JM, Fiessinger JN, Capron L, Aiach M. Partial characterization of an autoantibody recognizing the secondary binding site(s) of thrombin in a patient with recurrent spontaneous arterial thrombosis. *Thromb Haemost.* 1992;67:193-199.

Galanakis DK, Ginzler EM, Fikrig SM. Monoclonal IgG anticoagulants delaying fibrin aggregation in two patients with systemic lupus erythematosus (SLE). *Blood.* 1978;52:1037-1046.

La Spada AR, Skalhegg BS, Henderson R, Schmer G, Pierce R, Chandler W. Fatal hemorrhage in a patient with an acquired inhibitor of human thrombin. *N Engl J Med.* 1995;333:494-497.

Scully MF, Ellis V, Kakkar VV, Savidge GF, Williams YF, Sterndale H. An acquired coagulation inhibitor to factor II. *Br J Haematol.* 1982;50:655-664.

Sie P, Bezeaud A, Dupouy D, et al. An acquired antithrombin autoantibody directed toward the catalytic center of the enzyme. *J Clin Invest.* 1991;88:290-296.

Wisloff F, Michaelsen TE, Keirulf P, Godal HC. The molecular localization of the ability of certain monoclonal immunoglobulins to interfere with fibrin polymerization. *Thromb Res.* 1985;40:473-482.

Drugs

Bovine Thrombin, Topical

Banninger H, Hardegger T, Tobler A, et al. Fibrin glue in surgery: frequent development of inhibitors of bovine thrombin and human factor V. *Br J Haematol.* 1993;85:528-532.

Flaherty MJ, Henderson R, Wener MH. Iatrogenic immunization with bovine thrombin: a mechanism for prolonged thrombin times after surgery. *Ann Intern Med.* 1989;111:631-634.

Lawson JH, Pennell BJ, Olson JD, Mann KG. Isolation and characterization of an acquired antithrombin antibody. *Blood.* 1990;76:2249-2257.

Direct Thrombin Inhibitors

Shammas NW. Bivalirudin: pharmacology and clinical applications. *Cardiovasc Drug Rev.* 2005;23:345-360.

Walenga JM, Hoppensteadt D, Koza M, Wallock M, Pifarre R, Fareed J. Laboratory assays for the evaluation of recombinant hirudin. *Haemostasis.* 1991;21(suppl 1):49-63.

Warkentin TE, Greinacher A, Craven S, Dewar S, Sheppard JA, Ofosu FA. Differences in the clinically effective molar concentrations of four direct thrombin inhibitors explains their variable prothrombin time prolongation. *Thromb Haemost.* 2005;94:958-964.

Heparin

Favaloro EJ, Bonar R, Sioufi J, et al; Royal College Pathologists of Australasia Quality Assurance Program in Haematology. An international survey of current practice in the laboratory assessment of anticoagulant therapy with heparin. *Pathology.* 2005;37:234-238.

Penner JA. Experience with a thrombin clotting time assay for measuring heparin activity. *Am J Clin Pathol.* 1974;61:645-653.

Plasminogen Activators

Kane KK. Fibrinolysis: a review. *Ann Clin Lab Sci.* 1984;14:443-449.

Volume Expanders

Lindblad B, Bergqvist D, Hallbook T, Lindhagen A, Hedner U. Postoperative haemostatic changes in patients given thromboembolic prophylaxis with dextran 70-alone or in combination with dihydroergotamine. *Acta Chir Scand.* 1984;150:525-529.

Strauss RG, Stump DC, Henriksen RA, Saunders R. Effects of hydroxyethyl starch on fibrinogen, fibrin clot formation, and fibrinolysis. *Transfusion.* 1985;25:230-234.

Fibrin Degradation Products

Arnesen H. Studies on the thrombin clotting time, II: the influence of fibrin degradation products. *Scand J Haematol.* 1973;10:291-297.

Belitser VA, Lugovskoy EV, Musjalkovskaja AA, Gogolinskaja GK. Quantitation of the inhibitory effect of fibrinogen and its degradation products on fibrin polymerization. *Thromb Res.* 1982;27:261-269.

Furlan M, Rupp C, Beck EA. Inhibition of fibrin polymerization by fragment D is affected by calcium, Gly-Pro-Arg and Gly-His-Arg. *Biochim Biophys Acta.* 1983;742:25-32.

Fibrinogen Abnormalities

Carr ME, Gabriel DA. Hyperfibrinogenemia as a cause of prolonged thrombin clotting time. *South Med J.* 1986;79:563-570.

Cunningham MT, Brandt JT, Laposata M, Olson JD. Laboratory diagnosis of dysfibrinogenemia. *Arch Pathol Lab Med.* 2002;126:499-505.

Flanders MM, Crist R, Rodgers GM. Comparison of five thrombin time reagents. *Clin Chem.* 2003;49:169-172.

Moen JL, Lord ST. Afibrinogenemias and dysfibrinogenemias. In: Colman RW, Marder VJ, Clowes AW, George JN, Goldhaber SZ, eds. *Hemostasis and Thrombosis: Basic Principles and Clinical Practice.* 5th ed. Philadelphia, Pa: Lippincott, Williams and Wilkins; 2006: 939-952.

Neerman-Arbez M. The molecular basis of inherited afibrinogenaemia. *Thromb Haemost.* 2001;86:154-163.

Van Cott EM, Smith EY, Galanakis DK. Elevated fibrino-gen in an acute phase reaction prolongs the reptilase time but typically not the thrombin time. *Am J Clin Pathol.* 2002;118:263-268.

Radiocontrast Agents

Belleville J, Baguet J, Paul J, Clendinnen G, Eloy R. In vitro study of the inhibition of coagulation induced by different radiocontrast molecules. *Thromb Res.* 1985;38:149-162.

CHAPTER 16

Von Willebrand Disease

Elizabeth M. Van Cott, MD

Introduction

Von Willebrand disease is the most common inherited bleeding disorder, arising from quantitative or qualitative deficiencies of von Willebrand factor (vWF). The estimated incidence ranges from 1.3 per 100 to 1 per 33,000. The incidence may be underestimated because many cases are mild and therefore remain undiagnosed. Variable expression and reduced penetrance, commonly observed in von Willebrand disease type 1, add to the difficulty in establishing the true incidence.

Von Willebrand Factor

Hemostasis at sites of injured endothelium is achieved by the formation of both a platelet plug and a fibrin clot; von Willebrand factor plays a role in both of these events. First, vWF mediates platelet adhesion to injured endothelium, which is the first step in the formation of the platelet plug. Von Willebrand factor mediates platelet adhesion by binding to glycoprotein Ib (GPIb) on the platelet surface and by attaching to the subendothelium exposed upon endothelial injury. Second, vWF has an indirect role in the formation of the fibrin clot by functioning as a protective carrier protein for factor VIII. Factor VIII participates in the coagulation cascade, which culminates in the formation of a fibrin clot. Factor VIII levels are often decreased in von Willebrand disease, because the half-life of factor VIII is shortened when it is not bound to vWF in the plasma. Thus, when the quantity or quality of vWF is abnormal, hemostasis is impaired, which can lead to a bleeding disorder.

Von Willebrand factor is a protein synthesized in endothelial cells and megakaryocytes. A 741-amino acid propeptide antigen (VW AgII) is cleaved from vWF to result in a mature 2050-amino acid subunit (Figure 16-1). The factor VIII binding region is in the N-terminal portion of the mature subunit, while the binding sites for GPIb,

heparin, and collagen span the A1 to A3 domains. The cleavage site for the ADAMTS-13 (a disintegrin and metalloprotease with thrombospondin-1-like domains) is also in this region. There is an arginine-glycine-aspartic acid (RGD) peptide near the C-terminal end of the subunit that mediates platelet aggregation by binding to the GPIIb/IIIa receptor on platelets. The protein subunits are heavily glycosylated and are arranged in cross-linked aggregates called "multimers," which range in size from 600,000 to 20,000,000 daltons. The smallest multimer is a dimer, and the largest multimers contain up to 100 subunits. Von Willebrand factor is secreted from endothelial cells into the plasma and into the subendothelium and is also present in platelets. The plasma forms of vWF do not readily bind to platelets but undergo a conformational change that opens up the GPIb binding site, induced by shear stress or binding to extracellular matrix components. This conformational change can also be induced by ristocetin.

Clinical Features

Bleeding is variable in patients with von Willebrand disease, even among members of the same kindred. Bleeding symptoms in mildly to moderately affected individuals may improve with age. The most common bleeding symptoms are epistaxis, easy bruising, and bleeding after dental extractions. Other common bleeding symptoms include menorrhagia, bleeding after trauma or surgery, gingival bleeding, and postpartum bleeding. Gastrointestinal bleeding may also occur. Unlike in hemophilia, hemarthrosis is uncommon, occurring in only the most severely affected patients. Bleeding in patients with von Willebrand disease is rarely fatal.

The three types of von Willebrand disease are described below. Clinically, type 1 von Willebrand disease is a mild to moderate bleeding disorder. Type 2 can be severe but is most often mild to

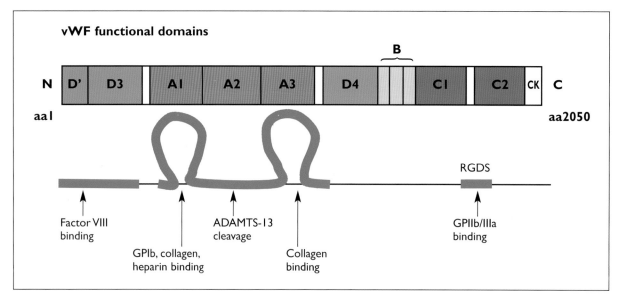

Figure 16-1. Domain structure and function of von Willebrand Factor (vWF). The factor VIII binding region is in the N-terminal portion of the mature subunit, while the binding sites for glycoprotein Ib (GPIb), heparin, and collagen span the A1 to A3 domains. The cleavage site for the ADAMTS-13 metalloprotease is also in this region (in the A2 domain). A disulfide loop is present in both the A1 and the A3 domains. There is an arginine-glycine-aspartic acid (RGD) peptide near the C-terminal end of the subunit that mediates platelet aggregation by binding to the GPIIb/IIIa receptor on platelets. A cystine knot region (CK) in the C-terminal end is necessary for von Willebrand dimer formation.

moderate in severity. Type 3 von Willebrand disease is a severe bleeding disorder.

Classification

Many variants of von Willebrand disease have been described, but the classification scheme has been simplified into three types. Type 1 is the most common form, accounting for more than 70% to 80% of cases. Type 1 von Willebrand disease is characterized by a partial quantitative deficiency of vWF. Although the quantity of vWF is reduced, the functional quality of the individual vWF molecules that are synthesized is normal.

Type 2 von Willebrand disease is characterized by qualitative deficiencies with functional abnormalities of vWF. Often the quantity of vWF is also reduced. Type 2 von Willebrand disease accounts for approximately 20% to 30% of cases.

Type 3 von Willebrand disease is rare, with an estimated incidence of 0.5 to 5.3 per million. Type 3 is characterized by a quantitative deficiency of vWF that is so severe that vWF is typically undetectable.

Type 2 von Willebrand disease is further subdivided into four categories. Type 2A accounts for 10% to 15% of von Willebrand disease cases; type 2B accounts for less than 5% of von Willebrand disease cases; and types 2M and 2N are rare. In a study of the French population, the distribution was reported as 30% type 2A, 28% type 2B, 34% type 2N, and 8% type 2M or unclassified.

Types 2A and 2B are characterized by a loss of high-molecular-weight multimers of vWF. The multimers with the highest molecular weight have more hemostatic function than the lower-molecular-weight multimers. Therefore, in these disorders, the overall function relative to the quantity of vWF molecules is reduced. In type 2A, two mechanisms for the loss of high-molecular-weight multimers have been described. Some mutations cause defective assembly and secretion of the multimers. Other mutations lead to normal secretion of multimers, but the multimers are proteolyzed at an increased rate.

Type 2B von Willebrand disease mutations lead to increased binding of vWF to GPIb, the platelet vWF receptor. Platelets coated with vWF are cleared from the bloodstream at an increased rate, leading to loss of high-molecular-weight multimers as well as thrombocytopenia.

Platelet-type or pseudo-von Willebrand disease is a rare disorder that is similar to type 2B von Willebrand disease, except that the mutation is located in the platelet GPIb gene instead of in the *VWF* gene. The mutation in platelet-type von Willebrand disease leads to increased binding of vWF to GPIb, resulting in the same findings described above for type 2B von Willebrand disease. This disorder is described further in chapter 14.

Type 2M *VWF* mutations cause decreased function despite the presence of normal-sized multimers (subtle multimer abnormalities are present in some variants). The *VWF* abnormality is qualitative. Type 2M mutations have been found in the GPIb binding domain of vWF, impairing the ability of vWF to bind to platelet GPIb.

Type 2N *VWF* mutations (von Willebrand disease Normandy) are located within the factor VIII binding site of vWF. Factor VIII and vWF are normally bound together in the circulation. In patients with type 2N von Willebrand disease, the factor VIII binding ability of vWF is impaired, and the half-life of factor VIII is consequently shortened. Thus, vWF is normal in quantity and has normal platelet adhesion function, but factor VIII levels are decreased. As a result, patients with type 2N von Willebrand disease are frequently misdiagnosed as having hemophilia A.

Inheritance

Type 1 deficiencies are inherited in an autosomal dominant manner. Type 2A, 2B, and 2M deficiencies also generally exhibit autosomal dominant inheritance. Type 2N is an autosomal recessive disorder, although it has also been inherited in a compound heterozygous state, with a type 2N mutation on one allele and a type 1 mutation on the other allele. Type 3 is usually an autosomal recessive disorder, with neither parent affected. However, in some type 3 cases, parents or other heterozygous relatives do have mild bleeding symptoms.

Laboratory Testing

Laboratory testing for von Willebrand disease is summarized in Table 16-1. The diagnosis of von Willebrand disease relies extensively on laboratory testing (which often requires repeat testing) and assessment of bleeding symptoms and inheritance. Before discussing the tests for vWF, two important diagnostic points must be made. First, both vWF and factor VIII become elevated during acute phase reactions. An elevation of a low or borderline value for vWF into the normal range during an acute phase reaction often masks the diagnosis of von Willebrand disease. Therefore, measurement of another acute phase reactant, such as fibrinogen, is helpful in assessing whether or not a patient is in an acute phase reaction at the time of testing. Von Willebrand factor levels typically increase during pregnancy and may mask a mild von Willebrand disease, especially in the third trimester of pregnancy. Second, vWF levels are also influenced by ABO blood group. In a large study of normal persons, the mean vWF level was 75% in blood type O, 106% in type A, 117% in type B, and 123% in type AB individuals. The overall mean vWF level was 100%. The mechanism underlying this association is not certain, but ABO carbohydrate antigens are present on circulating vWF and may affect the vWF level. In addition, blood type appears to influence proteolysis of vWF by the ADAMTS-13 metalloprotease, with vWF from blood type O individuals being more susceptible to proteolysis than vWF from the other blood types (the order of susceptibility decreases from O > B > A > AB). Because of the variability with blood group type, some laboratories stratify their von Willebrand test reference ranges by blood group.

Three assays, which measure ristocetin cofactor, vWF antigen, and factor VIII, constitute the typical von Willebrand disease test panel. Measurement of the prothrombin time (PT) and activated partial thromboplastin time (aPTT) are typically included to help exclude other coagulation disorders. Platelet function testing (for example, using the PFA-100 platelet function analyzer) can also be included to help distinguish von Willebrand disease from platelet dysfunction. Ristocetin-induced platelet aggregation may also be used to distinguish between some of the type 2 subtypes and platelet-type von Willebrand disease.

The ristocetin cofactor assay assesses vWF function by measuring ristocetin-mediated binding of vWF to platelet GPIb. Ristocetin was once used as an antibiotic, but ristocetin-induced platelet aggregation led to the discontinuation of its clinical use. The ristocetin cofactor assay is performed by mixing the patient's plasma with ristocetin and commercially available formalin-fixed normal platelets and measuring the amount of platelet agglutination in an aggregometer (Figure 16-2). The speed of platelet agglutination is compared to a calibration curve using normal plasma. One disadvantage of the ristocetin cofactor assay is that the coefficient of variation (CV) is high, which can result in variable results among different laboratories. Proficiency testing by the College of American Pathologists has found interlaboratory CVs of 23% to 31%.

The second assay in the panel is the vWF antigen assay, which measures the quantity of vWF regardless of the quality of vWF function. Current methods include enzyme-linked immunosorbent

Table 16-1. Laboratory Abnormalities in von Willebrand Disease

Laboratory Test	Type of von Willebrand Disease							
	I	2A	2B	2M	2N	3	Platelet Type	Acquired
PT	Normal	Normal	Normal	Normal	Normal	Normal	Normal	Normal
aPTT	Normal or elevated	Normal or elevated	Normal or elevated	Normal or elevated	Elevated	Markedly elevated	Normal or elevated	Usually elevated
Factor VIII	Low	Low	Low or normal	Low or normal	Very low	<10%	Low or normal	Low or very low
von Willebrand factor antigen	Low	Low	Low or normal	Low	Normal[a]	Undetectable	Low or normal	Low or very low
Ristocetin cofactor	Low	Very low	Very low	Very low	Normal[a]	Undetectable	Very low	Low or very low
Ristocetin-induced platelet aggregation	Low or normal	Low	Increased	Low or normal	Normal[a]	Very low	Increased	Low or very low
Multimer analysis	Normal or all sizes decreased	Absent large and intermediate sized multimers	Absent large multimers	Normal (type 2M Vicenza also has ultralarge multimers)	Normal[a]	All sizes absent	Absent large multimers	Type 1 or type 2 pattern (case reports of type 3 pattern)
Mean ristocetin cofactor:vWF antigen ratio	1	0.3	0.6	<1	1[b]	Both undetectable	<1	≤1
Thrombocytopenia	No	No[c]	Yes	No	No	No	Yes	No, not unless it results from the underlying disease

[a] Type 2N patients who have co-inherited type 1 von Willebrand disease may show decreased von Willebrand factor (vWF) antigen, ristocetin cofactor, ristocetin-induced platelet aggregation, and a decrease of all multimer sizes.

[b] The more important ratio in type 2N is the mean factor VIII to vWF antigen ratio, which is 0.28. Ratios are from Meyer (1997).

[c] Rare type 2A variants do show thrombocytopenia.

assays (ELISAs) or automated latex immuno-assays.

The third assay in the panel is a factor VIII assay. The most common method is an aPTT-based clotting assay that measures factor VIII activity. As mentioned above, vWF serves as the carrier protein for factor VIII, and levels of factor VIII are often decreased when vWF is decreased. The performance of this assay is discussed in chapter 6.

If the results of the above panel are suggestive of type 2 or type 3 von Willebrand disease (discussed below), a vWF multimer analysis can be performed to confirm the diagnosis (Figure 16-3). Multimer analysis involves separation of multimers by size using agarose gel electrophoresis of

the patient's plasma, followed by multimer detection using a radiolabeled or enzyme-linked anti-vWF antibody.

If the results of the multimer analysis are suggestive of type 2B von Willebrand disease, a ristocetin-induced platelet aggregation test is performed to confirm or refute a diagnosis of type 2B. This test is similar in principle to the ristocetin cofactor assay. It is performed by mixing the patient's platelets and plasma with ristocetin and measuring platelet aggregation in an aggregometer. This assay is less sensitive than the ristocetin cofactor assay for diagnosing von Willebrand disease, but it is useful for confirming a diagnosis of type 2B, as discussed below. There is no consensus

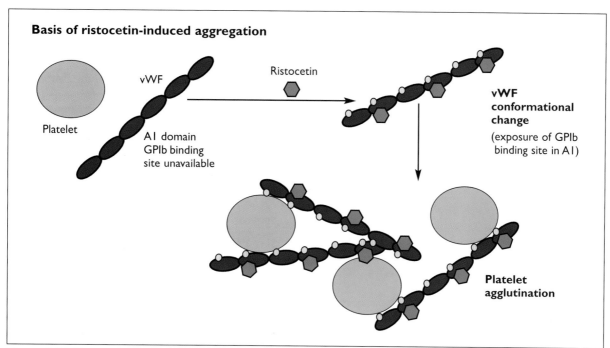

Figure 16-2. Ristocetin cofactor assay. Ristocetin induces a conformational change in von Willebrand factor (vWF) that allows binding to the glycoprotein Ib (GPIb) receptor on platelet surfaces. The vWF in the patient's plasma, in the presence of ristocetin, results in the aggregation of fixed reagent platelets.

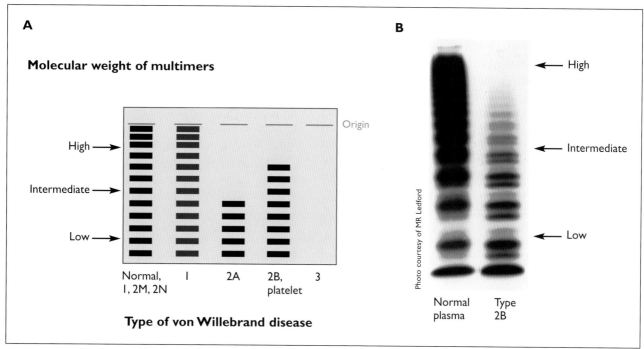

Figure 16-3. Von Willebrand factor (vWF) multimer analysis. A. Multimer analysis involves separation of multimers by size using agarose gel electrophoresis of the patient's plasma, followed by multimer detection using a radiolabeled or enzyme-linked anti-vWF antibody. Multimer analysis is typically normal in types 1, 2N, or 2M von Willebrand disease. Although all multimer sizes are decreased in quantity in patients with type 1, this typically appears normal upon multimer analysis because many patients with type 1 are only mildly affected. In severe type 1 disease, all multimers may appear decreased (gray shading in the figure). High-molecular-weight multimers are absent in types 2A and 2B. Intermediate-sized multimers are also absent in most patients with type 2A. A severe decrease or absence of all multimers is found in type 3 von Willebrand disease. B. Autoradiograph of sodium dodecyl sulfate agarose (1.4%) electrophoresis using I^{125}-labeled anti-vWF antibody. The absence of high-molecular-weight multimers is evident in the patient with type 2B von Willebrand disease.

regarding what concentration of ristocetin should be used in this assay, and a recent survey by the North American Specialized Coagulation Laboratory Association revealed that various concentrations are used. One example is to use 1200, 900, 600, and 300 µg/mL ristocetin. Normal individuals typically show greater than 65% aggregation in response to high concentrations (eg, 1000 to 1500 µg/mL) and minimal or no aggregation with low concentrations (eg, 200 to 500 µg/mL). At the low ristocetin concentrations, specimens from patients with type 2B or platelet-type von Willebrand disease show more aggregation than do those of normal controls. Specimens from patients with other types of von Willebrand disease show either normal or reduced aggregation in this assay, depending on the severity of the decrease in their vWF activity.

More recently, in addition to the ristocetin cofactor activity assay, some alternative activity assays have become available. Collagen binding assays involve an ELISA method that uses collagen as the antigen and test the ability of vWF to bind collagen. Some collagen binding assays have performed well in comparison to the ristocetin cofactor assay in some studies, depending on the method. Another commercially available vWF activity test is an ELISA that uses a monoclonal antibody against an epitope on vWF that interacts with platelet GPIb. It has not performed as well as the ristocetin cofactor activity assay in some studies. More recently, a latex turbidimetric immunoassay using a monoclonal anti-vWF antibody directed against the platelet binding site on vWF (the platelet GPIb receptor) has been developed to measure activity, as has a turbidimetric ristocetin cofactor assay that can be performed on certain automated coagulation analyzers.

DNA testing is offered by a limited number of reference laboratories, but it is not routinely needed. For example, DNA sequencing of exon 28 of *VWF* for type 2A, 2B, and 2M, as well as DNA sequencing of exons 18, 19, 20, 23, and 24 of *VWF* for type 2N von Willebrand disease, can be performed.

The bleeding time is often prolonged in von Willebrand disease, and in the past it was used as a screening test; however, it is neither a necessary nor a reliable test for diagnosis. More recently, the PFA-100 platelet function analyzer has been developed. Studies have shown that the PFA-100 results are abnormal in many cases of moderate or severe von Willebrand disease, but that it is not as sensitive in detecting mild cases. The PFA-100 meas-

ures the time in seconds that it takes to occlude a channel that is coated with collagen and adenosine diphosphate (ADP) or collagen and epinephrine, which activate platelets, as the patient's whole-blood sample flows through the channel.

The typical laboratory findings for each of the three types of von Willebrand disease are described below. An evidence-based guideline for von Willebrand disease diagnosis has been recently published by the National Heart, Lung, and Blood Institute (NHLBI; www.nhlbi.nih.gov/guidelines/vwd).

Laboratory Findings in Type 1 von Willebrand Disease

In type 1 von Willebrand disease, the laboratory results for ristocetin cofactor, vWF antigen, and factor VIII are all decreased to approximately the same degree. Levels <30 IU/dL are designated as the level for a definitive diagnosis of von Willebrand disease. All multimer sizes are decreased in quantity (the multimer bands on the multimer analysis gel might appear lighter in intensity), but this typically appears normal upon multimer analysis because many patients with type 1 von Willebrand disease are only mildly affected. Levels of vWF can vary in patients with type 1 von Willebrand disease and can become intermittently normal, so repeat testing is often required. Because of the wide variation in vWF levels in normal individuals, a firm diagnosis of type 1 von Willebrand disease may often be problematic, especially if the von Willebrand assays are only minimally decreased. The recent NHLBI guidelines suggest that mild reductions in vWF levels may not necessarily indicate von Willebrand disease, and that such patients can be said to have "low von Willebrand factor" rather than von Willebrand disease.

Laboratory Findings in Type 2 von Willebrand Disease

Patients with type 2A or type 2B von Willebrand disease typically show a decrease of ristocetin cofactor out of proportion to the decrease in vWF antigen and factor VIII. The mean (ristocetin cofactor activity to vWF antigen) ratio was 0.3 in a series of patients with type 2A and was 0.6 in patients with type 2B. Multimer analysis shows the absence of high-molecular-weight multimers in both type 2A and type 2B. Individuals with type 2A also demonstrate loss of intermediate-sized multimers.

Ristocetin-induced platelet aggregation is performed to further distinguish between type 2A and type 2B von Willebrand disease. Unlike in the ristocetin cofactor assay, the patient's own platelets are used in this assay. Platelets from patients with type 2B become abnormally coated with vWF in vivo due to the mutation in vWF that increases its affinity for platelet GPIb. As a result, the patient's platelets show increased aggregation in this assay. Platelet-type von Willebrand disease also shows increased aggregation in this assay due to a mutation in platelet GPIb that increases its affinity for vWF. In contrast, other types of von Willebrand disease (except type 2N) show decreased ristocetin-induced platelet aggregation due to decreased vWF quantity and/or function. Performing platelet aggregation with added cryoprecipitate or the addition of desialyated (asialo) vWF to the patient's platelet-rich plasma in EDTA, without ristocetin, can induce platelet agglutination in specimens from patients with platelet-type von Willebrand disease, but not in those from patients with type 2B von Willebrand disease.

Performing a platelet count is also suggested because occasionally the multimer analysis and/or the vWF activity-to-antigen ratio may not reliably identify cases of type 2B von Willebrand disease. Type 2B is the only variant associated with thrombocytopenia. It has also been suggested to perform the platelet count before and after a patient's first trial with desmopressin acetate (DDAVP) because a decrease in platelet count would be suggestive of type 2B.

In type 2M von Willebrand disease, ristocetin cofactor is typically decreased more than the vWF antigen and factor VIII, similar to types 2A and 2B. However, unlike types 2A and 2B, multimer analysis is normal in type 2M. Type 2M Vicenza is a variant of von Willebrand disease that has ultra-large multimers in addition to the normal sizes of multimers. Type 2M Vicenza has recently been proposed to be reclassified as type 1 by the International Society on Thrombosis and Haemostasis Subcommittee on von Willebrand Factor.

Patients with type 2N (Normandy) von Willebrand disease have normal levels of ristocetin cofactor and vWF antigen, with decreased factor VIII. Factor VIII levels are generally above 5%. The mean (factor VIII to vWF antigen) ratio in a series of 51 unrelated patients with type 2N von Willebrand disease was 0.28. Multimer analysis is normal. Patients who have inherited a type 2N mutation on one allele and a type 1 mutation on the other allele will have similarly decreased levels of ristocetin cofactor and vWF antigen levels, with a disproportionately decreased factor VIII. Patients with type 2N von Willebrand disease may be misdiagnosed as having mild hemophilia A. The two disorders may be distinguished by the inheritance pattern. Type 2N von Willebrand disease is inherited as an autosomal recessive disorder, or as a compound heterozygous entity with type 1 von Willebrand disease. In contrast, hemophilia A is an X-linked recessive disorder. An assay that measures the ability of vWF to bind factor VIII is available in some reference laboratories. Tests for the most common DNA mutations that cause type 2N are also available.

Laboratory Findings in Type 3 von Willebrand Disease

Ristocetin cofactor, vWF antigen, and factor VIII are severely decreased and often undetectable in patients with type 3 von Willebrand disease. Factor VIII levels are usually detectable but are less than 10% of normal. Multimer analysis shows a marked decrease or absence of all multimers.

The Molecular Basis of von Willebrand Disease

A question regarding von Willebrand disease is whether type 3 disease arises from homozygous type 1 mutations: Is type 1 the same condition as heterozygous type 3, or is the mechanism of type 3 different from that of type 1? Emerging molecular studies are providing information regarding this issue. Type 1 von Willebrand disease has recently been shown to arise most often from missense mutations, but small deletions/insertions, splice site, promoter, and nonsense mutations also occur. Only a minority of the mutations are predicted to cause a null (nonexpressed) allele. In contrast, null alleles are thought to be the typical mutation seen in individuals with type 3 and include deletions, insertions, frameshift and splicing mutations, nonsense mutations, and, uncommonly, missense mutations. In addition, although type 3 heterozygotes bleed more than normal, they experience significantly less bleeding than do individuals with type 1 von Willebrand disease. Type 1 and type 3 mutations are scattered throughout the VWF gene (Figure 16-4).

In contrast, most type 2 mutations are single-base substitution missense mutations located at sites associated with specific vWF functions. Type

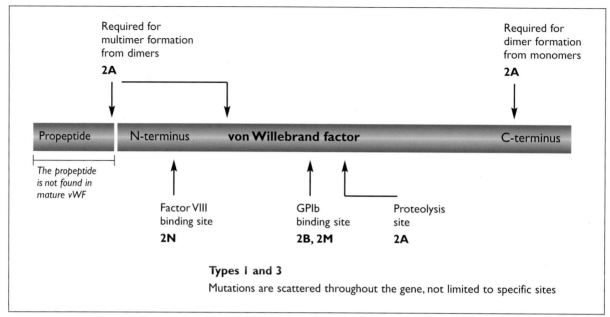

Figure 16-4. The molecular basis of von Willebrand disease. The von Willebrand factor gene is located on chromosome 12. DNA mutations that cause type 2A von Willebrand disease are mostly clustered in a region of the gene that encodes for an important site for proteolysis in the natural degradation of von Willebrand factor (vWF). These mutations lead to increased proteolysis and sometimes impaired secretion of vWF. Other, less common type 2A mutations impair dimerization or multimerization of vWF. Type 2B mutations are clustered in a region of the gene that encodes the binding site for platelet glycoprotein Ib (GPIb), leading to increased binding of vWF to GPIb. Type 2M mutations are clustered in the platelet GPIb binding region of the gene, but these mutations lead to decreased binding of vWF to GPIb. Type 2N mutations are located in the portion of the gene that encodes the factor VIII binding site, leading to decreased binding of vWF to factor VIII. Types 1 and 3 von Willebrand disease mutations do not appear to be clustered in any particular functional domain.

2A mutations are mostly clustered in or near domain A2 in exon 28, a region of the *VWF* gene that encodes for an important site for proteolysis in the natural degradation of vWF. These mutations lead to increased proteolysis of vWF and sometimes also impaired secretion of vWF. Other type 2A mutations cause defective multimer assembly and are located in domain D2 in the propeptide, the cystine knot domain, or the D3 domain. Type 2B mutations are clustered in or near domain A1 in exon 28 in a region of the gene that encodes for the binding site for platelet GPIb. These type 2B mutations lead to increased binding of vWF to GPIb. Type 2M mutations are also clustered in the platelet GPIb binding region of the *VWF* gene, but these mutations lead to decreased binding of vWF to GPIb. Type 2N mutations are located in the portion of the gene that encodes the factor VIII binding site, leading to decreased binding of vWF to factor VIII.

Acquired von Willebrand Disease

Acquired von Willebrand disease is a rare condition that can resemble type 2 or, less commonly, type 1 von Willebrand disease. In contrast to the familial forms, the acquired form is a bleeding diathesis with an onset late in life in a patient with no prior personal or family history of bleeding. It can occur in association with an underlying disorder, particularly certain hematologic disorders such as lymphoma, myeloma, myeloproliferative disorders, leukemia, Waldenstrom macroglobulinemia, monoclonal gammopathy of undetermined significance, hairy cell leukemia, and thalassemia. Other underlying conditions include autoimmune disorders, hypothyroidism, mitral valve prolapse, ventricular septal defects, Wilms tumor, angiodysplasia, gastrointestinal telangiectasia, and the use of valproic acid.

Several different mechanisms appear to be involved in the development of acquired von Willebrand disease. Some cases arise from the development of an autoantibody directed against vWF. Other mechanisms include adsorption of vWF onto malignant cells, decreased production of vWF (as in hypothyroidism), or increased destruction of vWF by proteolysis or high shear stress (as with cardiac valve abnormalities). Laboratory testing for acquired von Willebrand disease can include measurement of the von Willebrand propeptide antigen and ristocetin cofactor mixing studies. The von Willebrand propeptide antigen is a peptide that is cleaved from vWF to produce the mature vWF subunit. In acquired von Willebrand disease, the levels of propeptide antigen are normal compared with decreased levels of vWF. In contrast, congenital forms of von Willebrand disease usually have decreased propeptide antigen. Ristocetin cofactor mixing studies, in which the patient's plasma is mixed with normal plasma and the ristocetin cofactor assay is performed, can distinguish acquired von Willebrand disease with circulating antibodies from other types of acquired and congenital von Willebrand disease because the mixing does not correct the abnormal test result. However, the mixing study does correct the test result to normal in the types of acquired von Willebrand disease that do not involve an antibody, or that do involve an antibody that leads to clearance of vWF but does not directly inhibit von Willebrand activity.

Treatment can be difficult because the half-lives of endogenous and transfused vWF and factor VIII are shortened because of the above-mentioned mechanisms.

Treatment

Desmopressin acetate (DDAVP), an analogue of vasopressin, can be used to elevate plasma levels of endogenous factor VIII and vWF by 2- to 3-fold in approximately 80% of patients with type 1 von Willebrand disease. A minority of patients do not respond to DDAVP. Therefore, shortly after a diagnosis of type 1 von Willebrand disease is made, a DDAVP trial is performed to determine if the patient responds to DDAVP. This is determined by performing vWF assays, including von Willebrand antigen, ristocetin cofactor, and factor VIII, before and after DDAVP administration. The results determine whether or not DDAVP is the appropriate treatment for future bleeding episodes. DDAVP is available as an intravenous product and for intranasal delivery. The usual dose is 0.3 µg/kg for the intravenous form or 300 µg for the intranasal form. The intravenous route has produced more consistent elevations in vWF than has intranasal administration. The rise in vWF occurs within 30 minutes of intravenous DDAVP infusion and lasts for 6 to 12 hours. Patients should be monitored for tachyphylaxis and, although rare, hyponatremia from water retention.

Patients with type 2A von Willebrand disease demonstrate variable responses to DDAVP, and the vWF released into the circulation is dysfunctional. Many physicians consider DDAVP to be contraindicated in patients with type 2B because it may lead to or worsen thrombocytopenia; however, the assumption that DDAVP is contraindicated in patients with type 2B has been challenged by the report of its successful use in a small series of type 2B cases. Patients with type 2M may respond to DDAVP, but often it is not successful. In patients with type 2N, DDAVP may not effectively maintain factor VIII levels because the vWF has a decreased ability to bind factor VIII. In patients with type 3, levels of vWF are too severely decreased for DDAVP to have much effect, but it can be considered a second-line therapy for some cases.

Epsilon aminocaproic acid (Amicar), a fibrinolytic inhibitor, is often used for dental procedures in patients with von Willebrand disease. Menorrhagia has been successfully controlled with oral estrogens.

Certain factor VIII concentrates contain substantial amounts of vWF in addition to factor VIII and are therefore useful in patients for whom DDAVP is ineffective or unsuitable. Two such brands available in the United States are Humate-P (Aventis Behring) and Alphanate (Grifols). In addition, a vWF concentrate essentially devoid of factor VIII is available in Europe. A recombinant form of vWF is under development.

Humate-P and Alphanate have been cleared by the Food and Drug Administration for use in von Willebrand disease. The Humate-P dosage protocol involves maintaining trough vWF levels greater than 50% for major bleeding episodes. For minor bleeding, a single dose of 40 to 50 units/kg can be administered, with a second dose 12 hours later if needed. The Alphanate protocol for adults is similar: maintain levels >40% to 50% for 1 to 3 days after a minor procedure and for ≥3 to 7 days after a major procedure. Other, similar dosage protocols have been published, including maintaining trough vWF levels greater than 50% for 5 to 10 days after major surgery and greater than 30% for 2 to 4 days after minor surgery. Factor VIII levels should be monitored once daily during treatment with concentrates (twice daily on the first day), both to ensure adequate levels and to prevent overdosing, because high levels may increase the risk for thrombosis. The use of vWF-containing concentrates in patients with type 3 von Willebrand disease with gene deletions has been complicated by the development of an inhibitor, that is, an antibody against vWF.

Treatment guidelines have not been established for platelet-type von Willebrand disease, in part because the molecular defect lies in platelet GPIb rather than in vWF. The use of platelet transfusions with relatively low-dose vWF replacement has been reported to be successful.

Suggested Reading

Reviews

Ewenstein BM. von Willebrand disease. *Ann Rev Med.* 1997;48:525-542.

Mannucci PM. Treatment of von Willebrand disease. *New Engl J Med.* 2004;351:683-694.

Nichols W, Ginsberg D. von Willebrand disease. *Medicine.* 1997;76:1-20.

Nichols WL, Hultin MB, James AH, et al. *Diagnosis, Evaluation, and Management of von Willebrand Disease.* Washington, DC: US Department of Health and Human Services. National Institute of Health. National Heart, Lung, Blood Institute (NHLBI); 2007. NIH publication number 08-5832.

Nichols WL, Hultin MB, James AH, et al. von Willebrand disease (VWD): evidence-based diagnosis and management guidelines, the National Heart, Lung, and Blood Institute (NHLBI) Expert Panel report (USA). *Haemophilia.* 2008;14(2):171-232.

Rick ME. Laboratory diagnosis of von Willebrand disease. *Clin Lab Med.* 1994;14:781-794.

Sadler JE, Blinder M. von Willebrand disease: diagnosis, classification and treatment. In: Colman RW, Hirsh J, Marder VJ, Clowes AW, George JN, eds. *Hemostasis and Thrombosis: Basic Principles and Clinical Practice.* 4th ed. Philadelphia, Pa: Lippincott, Williams and Wilkins; 2001: 825-837.

Sadler JE, Budde U, Eikenboom JC, et al. Update on the pathophysiology and classification of von Willebrand disease: a report of the Subcommittee on von Willebrand Factor. *J Thromb Haemost.* 2006; 4:2103-2114.

Sadler JE, Mannucci PM, Berntorp E, at al. Impact, diagnosis and treatment of von Willebrand disease. *Thromb Haemost.* 2000;84:160-174.

Mutation Databases and Genetic Analyses

Baronciani L, Cozzi G, Canciani MT, et al. Molecular characterization of a multiethnic group of 21 patients with type 3 von Willebrand disease. *Thromb Haemost.* 2000;84:536-540.

Ginsberg D, Sadler JE. von Willebrand disease: a database of point mutations, insertions, and deletions. *Thromb Haemost.* 1993;69:177-184.

ISTH SSC VWF Information Homepage. Up-to-date databases of point mutations, insertions, deletions, and polymorphisms found in the gene for human von Willebrand factor. UK: University of Sheffield. Available at: http://www.vwf.group.shef.ac.uk/. Accessed July 15, 2008.

Meyer D, Fressinaud E, Gaucher C, et al. Gene defects in 150 unrelated French cases with type 2 von Willebrand disease: from the patient to the gene. *Thromb Haemost.* 1997;79:451-456.

Peerlink K, Eikenboom JC, Ploos Van Amsterl HK, et al. A patient with von Willebrand disease characterized by a compound heterozygosity for a substitution of Arg854 by Gln in the putative factor-VIII-binding domain of von Willebrand factor (vWF) on one allele and very low levels of mRNA from the second vWF allele. *Br J Haematol.* 1992;80:358-363.

Genetic Analyses and Type 1 versus Type 3 von Willebrand Disease

Castaman G, Rodeghiero F, Tosetto A, et al. Hemorrhagic symptoms and bleeding risk in obligatory carriers of type 3 von Willebrand disease: an international, multicenter study. *J Thromb Haemost.* 2006;4:2164-2169.

Goodeve A, Eikenboom J, Castaman G, et al. Phenotype and genotype of a cohort of families historically diagnosed with type 1 von Willebrand disease in the European study, Molecular and Clinical Markers for the Diagnosis and Management of Type 1 von Willebrand Disease (MCMDM-1VWD). *Blood.* 2007; 109:112-121.

James PD, Notley C, Hegadom C, et al. The mutational spectrum of type 1 von Willebrand disease: results from a Canadian cohort study. *Blood.* 2007;109:145-154.

Meyer D, Fressinaud E, Gaucher C, et al. Gene defects in 150 unrelated French cases with type 2 von Willebrand disease: from the patient to the gene. INSERM Network on Molecular Abnormalities in von Willebrand Disease. *Thromb Haemost.* 1997;78(1):451-456.

Montgomery RR. When it comes to von Willebrand disease, does 1 + 1 = 3? *J Thromb Haemost.* 2006;4:2162-2163.

Blood Type Effect on von Willebrand Factor

Bowen DJ. An influence of ABO blood group on the rate of proteolysis of von Willebrand factor by ADAMTS13. *J Thromb Haemost.* 2003;1:33-40.

Gill JC, Endres-Brooks J, Bauer PJ, Marks WJ, Montgomery RR. The effect of ABO blood group on the diagnosis of von Willebrand disease. *Blood*. 1987;69:1691-1695.

O'Donnell J, Boulton FE, Manning RA, Laffan MA. Amount of H antigen expressed on circulating von Willebrand factor is modified by ABO blood group genotype and is a major determinant of plasma von Willebrand factor antigen levels. *Arterioscler Thromb Vasc Biol*. 2002;22:335-341.

Von Willebrand Assay Methods

Hayes TE, Brandt JT, Chandler WL, et al. External peer review quality assurance testing in von Willebrand disease: the recent experience of the United States College of American Pathologists proficiency testing program. *Semin Thromb Hemost*. 2006;32:499-504.

Ledford-Kraemer M. Testing for VWF: the laboratory's armamentarium. *The Clotting Times*. 2006;5:8-12.

Moffat KA, Ledford-Kraemer MR, Nichols WL, Hayward CPM. Variability in clinical laboratory practice in testing for disorders of platelet function: results of two surveys of the North American Specialized Coagulation Laboratory Association. *Thromb Haemost*. 2005;93:549-553.

Collagen Binding and ELISA Activity Assays

Favaloro EJ. Detection of von Willebrand disorder and identification of qualitative von Willebrand factor defects: direct comparison of commercial ELISA-based von Willebrand factor activity options. *Am J Clin Pathol*. 2000;114:608-618.

Federici AB, Canciani MT, Forza I, Cozzi G. Ristocetin cofactor and collagen binding activities normalized to antigen levels for a rapid diagnosis of type 2 von Willebrand disease. *Thromb Haemost*. 2000;84:1127-1128.

Paczuski R. Determination of von Willebrand factor activity with collagen-binding assay and diagnosis of von Willebrand disease: effect of collagen source and coating conditions. *J Lab Clin Med*. 2002;140:250-254

Von Willebrand Factor Activity Assays by Latex Immunoassay or Automated Turbidimetric Method

Favaloro EJ, Bonar R, Meiring M et al. 2B or not 2B? Disparate discrimination of functional vWF discordance using different assay panels or methodologies may lead to success or failure in the early identification of type2B vWD. *Thromb Haemost*. 2007;98:346-358.

Salem RO, Van Cott EM. A new automated screening assay for the diagnosis of von Willebrand's disease. *Am J Clin Pathol*. 2007;127:730-735.

PFA-100 Platelet Function Analyzer

Hayward CPM, Harrison P, Cattaneo M, et al. Platelet function analyzer (PFA)-100 closure time in the evaluation of platelet disorders and platelet function. *J Thromb Haemost*. 2006;4:312-319.

Schlammadinger A, Kerenyi A, Muszbek L, Boda Z. Comparison of the O'Brien filter test and the PFA-100 platelet function analyzer in the laboratory diagnosis of von Willebrand disease. *Thromb Haemost*. 2000;84:88-92.

Acquired von Willebrand Disease

Federici AB, Rand JH, Bucciarelli P, et al. Acquired von Willebrand syndrome: data from an international registry. *Thromb Haemost*. 2000;84:345-349.

Jakway JL. Acquired von Willebrand disease. *Hematol Oncol Clin North Am*. 1992;6:1409-1419.

van Genderen PJ, Boertjes RC, van Mourik JA. Quantitative analysis of von Willebrand factor and its propeptide in plasma in acquired von Willebrand syndrome. *Thromb Haemost*. 1998;80(3):495-498.

Veyradier A, Jenkins CS, Fressinaud E, Meyer D. Acquired von Willebrand syndrome: from pathophysiology to management. *Thromb Haemost*. 2000;84:175-182.

Treatment

Bichet DG, Mohammad R, Longergan M, et al. Hemodynamic and coagulation responses to 1-desamino [8-D-arginine] vasopressin in patients with congenital nephrogenic diabetes insipidus. *N Engl J Med*. 1988;318:881-887.

Casonato A, Pontara E, Dannhaeuser D, et al. Re-evaluation of the therapeutic efficacy of DDAVP in type IIB von Willebrand disease. *Blood Coagul Fibrinolysis*. 1994;5:959-964.

Menache D, Aronson DL. New treatments of von Willebrand disease: plasma derived von Willebrand factor concentrates. *Thromb Haemost*. 1997;78:566-570.

Schwarz HP, Turecek PL, Pichler L, et al. Recombinant von Willebrand factor. *Thromb Haemost*. 1997;78:571-576.

Platelet-type von Willebrand Disease

Franchini M, Montagnana M, Lippi G. Clinical, laboratory and therapeutic aspects of platelet-type von Willebrand disease. *Int J Lab Hematol*. 2008;30(2):91-94.

Miller JL. Platelet-type von Willebrand disease. *Thromb Haemost*. 1996;75:865-869.

Miller JL, Ruggeri ZM. Von Willebrand factor binds to platelets and induces aggregation in platelet-type but not type IIB von Willebrand disease. *J Clin Invest*. 1983;72:1532-1542.

Section 4

Algorithmic Approach to Thrombophilic Disorders

Charles Eby, MD, and John D. Olson, MD, PhD, Editors

Chapter 17. Arterial and Venous Thrombosis in Adults

Chapter 18. Unique Issues of Thrombophilia in Women and Children

Chapter 19. Laboratory Diagnosis of Inherited Thrombophilia

Chapter 20. Fibrinolytic Thrombotic Disorders

Chapter 21. Heparin-Induced Thrombocytopenia

Chapter 22. Antiphospholipid Antibodies

Arterial and Venous Thrombosis in Adults

John D. Olson, MD, PhD
Charles Eby, MD

Introduction

The outcome of the growth in the study of thrombosis and vascular biology in the prior three decades has been the discovery of a number of pathways involved in the regulation of hemostasis. Abnormalities in some components of these pathways have been elucidated and, after evaluation of patients, it has been discovered that affected families and populations are at increased risk of thrombosis, and the term "thrombophilia" is now used to describe this condition. An outcome of this increase in information has been a concomitant growth in the number of plasma or molecular components that can be measured in the clinical laboratory. With tests being readily available, both laboratories and clinicians struggle with issues surrounding this testing, such as whom to test, when to test, what tests should be performed, what are the best test methods, and so forth.

This problem is exemplified by studies of McKenzie et al (2000) and Somma et al (2006). These studies, performed 6 years apart and in different parts of the United States, show difficulties in testing the "right" patients at the "right" time, avoiding testing patients receiving anticoagulants, and other issues.

The XXXVI Consensus Conference of the College of American Pathologists (CAP), "Diagnostic Issues in Thrombophilia," held November 9-11, 2001, in Atlanta, Georgia, was convened to provide recommendations concerning these important clinical issues. Thirty-one participants, 27 of whom were in attendance at the conference, contributed to this project. Members of the Biochemical and Molecular Genetics Resource Committee and the Coagulation Resource Committee of the CAP invited other pathologists and clinicians to participate in the conference. As a result, 16 manuscripts dealing with specific thrombophilic risk factors and 2 manuscripts dealing with the application of the factors in thrombotic conditions were prepared by the participants.

On the basis of their review of the literature and their experience, the authors suggested specific recommendations. At the conference, the conclusions of the manuscripts were presented, and the recommendations were discussed and voted on by the participants in attendance. Both the inherited and the acquired thrombophilias were discussed. Recommendations regarding patient management were not addressed at the conference; however, in the manuscripts, some of the authors included their own observations regarding the impact of testing on clinical decisions: testing for thrombophilia in case of venous thromboembolism may be valuable for (1) knowing etiology and communicating etiologic factors to patients, (2) influencing treatment decisions (duration of treatment or prevention), (3) influencing future approaches to the patient (threshold/intensity of preventive measures in future high-risk situations), and (4) influencing future approaches to affected family members (continuous primary prophylaxis, threshold/intensity of preventive measures in future high-risk situations).

Conference recommendations were approved and included in the final publication only if more

The content of this chapter has been extracted in large part from the work of Olson (College of American Pathologists Consensus Conference XXXVI: Diagnostic Issues in Thrombophilia) and Van Cott et al (Laboratory Evaluation of Hypercoagulability With Venous or Arterial Thrombosis) from the College of American Pathologists Consensus Conference XXXVI, "Diagnostic Issues in Thrombophilia," the results of which were published in the November 2002 issue of *Archives of Pathology & Laboratory Medicine*. The text has been reorganized and updated, but the content is predominantly the product of the consensus conference. The reader is referred to this citation for additional information and references. Modified with permission from *Archives of Pathology & Laboratory Medicine*. Copyright 2002. College of American Pathologists.

Table 17-1. Criteria Used for the Level of Evidence Assigned to Recommendations

Level 1
❏ The recommendation is based on one or more well-designed prospective studies or two or more well-designed retrospective studies

Level 2
❏ The recommendation is based on retrospective studies or multiple anecdotal studies that reach consensus

Level 3
❏ The recommendation is based on isolated anecdotal studies and/or the consensus of expert practitioners

than 70% of the participants present agreed with the content and wording of the recommendation. Assessing the level of evidence for testing for thrombophilia is challenging because the purpose of this testing can be manifold. The conference assigned a "Level of Evidence" to each recommendation (Table 17-1). After the conference, the manuscripts were finalized, submitted to the *Archives of Pathology & Laboratory Medicine,* and subjected to the journal's peer review process.

As might be expected, many of the recommendations will be controversial in the opinion of the reader. Other experts will find some items with which they agree and some with which they strongly disagree. The same was true of the conference participants. Nevertheless, the recommendations, as presented, were agreed upon by a 70% majority and should provide useful guidance for clinicians and laboratorians.

General Considerations

The specific defects that lead to thrombophilia are addressed in the latter part of this chapter and in chapters 9, 18, 19, and 20. A few issues have general application or merit additional emphasis.

Consent and Counseling

Thrombophilia is a complex condition in which a positive test result has a low probability of affecting the patient's health, yet may lead to changes in therapeutic and counseling approaches. For example, the use of anticoagulant prophylaxis at times of increased risk for thrombosis may, despite some

risk, be beneficial. In addition, there can be financial consequences, such as reduced insurability or increased cost of insurance, which may also have an impact. The concepts of thrombophilia are often unclear to clinicians and can certainly be confusing to patients. For these reasons, it is recommended that informed consent be obtained before testing for thrombophilia. This recommendation is particularly important when planning to test the asymptomatic relatives of an affected patient. In some jurisdictions, regulations require that written informed consent be obtained for any "genetic" testing, and in those locales, written informed consent must be obtained. However, in general, oral consent for thrombophilia testing with appropriate notation in the medical record should be sufficient. In addition, patients who are found to have thrombophilia should also be counseled regarding such issues as the risks of thrombosis to themselves and their family members, the importance of early recognition of the signs and symptoms of venous thromboembolism (VTE) or arterial thrombosis that would require immediate medical attention, and the risks and benefits of antithrombotic prophylaxis in situations when the risk of thrombosis is increased, such as in perioperative or puerperal periods.

Assay Calibration, Reference Range, and Assay Reliability

Most of the assays of the function and antigen of the thrombophilic factors suffer from imprecision and inaccuracy. These problems are exemplified by the results of proficiency testing. The coefficient of variation of the results, for example, of the assay for free protein S antigen was 16% and for protein S activity was 17.2% (College of American Pathologists. 2007. Coagulation Education Survey CED-B. Participant Summary Report). In addition, the analytes that are involved in thrombophilia are often affected by a variety of acquired conditions and medications. For this reason, it is important to control all possible variables. When available, World Health Organization (WHO) standards or standards that can be linked to the WHO standard should be used to calibrate functional and antigenic assays. For many of these factors, the effects of age and gender have been demonstrated to affect the result and should always be taken into consideration when interpreting results. Whenever possible, reference ranges that are age and gender specific should be developed.

All of these issues (assay variability, biologic variability, effect of medication, effect of underlying medical conditions, and others) frequently lead to a test result that is below the reference range in individuals who do not have inherited disease. Therefore, before concluding that a patient has an inherited thrombophilia, functional and antigenic assays should be repeated after excluding acquired etiologies of the defect. General recommendations regarding thrombophilia testing are found in Table 17-2.

Combined Deficiency

There have been repeated reports of the increased risk of thrombosis as a result of the additive (or more than additive) effect of multiple thrombophilic defects occurring in the same patient. Because of the remarkable increase in risk seen in some of the combined deficiencies, patients with a known inherited thrombophilia deserve to be tested for the presence of other risk factors for thrombosis.

Testing for Thrombophilic Risk in Patients with Venous or Arterial Thrombosis

Thrombosis on the venous and arterial sides of the circulation affect a substantial portion of the population every year. In the laboratory evaluation for hypercoagulability in affected patients, test selection differs depending on location and type of thrombosis. The remainder of this chapter addresses testing strategies for evaluating thrombophilic risk in patients with venous and arterial thrombosis, with emphasis on myocardial infarction, thrombotic stroke, and cerebral venous thrombosis. The supporting evidence for test selection is analyzed, and consensus recommendations for test selection are presented.

Thrombophilic Risk Factor Test Selection in Provoked and Spontaneous Venous Thromboembolism

The decision to perform a "thrombophilia workup" requires a thorough evaluation of the circumstances surrounding the venous thromboembolic event, knowledge of the patient's personal and family medical history, physical examination,

Table 17-2. Thrombophilia Testing: General Recommendations

- Patients and especially asymptomatic family members require informed consent before testing for thrombophilia. Molecular testing, such as for factor V Leiden or prothrombin G20210A, requires written consent in some jurisdictions, but, in general, oral consent for testing with appropriate notation in the medical record should be sufficient. *Level 3*
- Individuals testing positive for a thrombophilia require counseling regarding:
  - the risks of thrombosis to them and their family members,
  - the importance of early recognition of the signs and symptoms of venous thromboembolism that would require immediate medical attention, and
  - the risks and benefits of antithrombotic prophylaxis in situations when their risk of thrombosis is increased, such as perioperative or peripartum. *Level 3*
- Laboratory testing for other inherited and acquired thrombophilias should be considered even after the identification of a known thrombophilia, because more than one thrombophilia could coexist, compounding the risk for thrombosis in many cases. *Level 1*
- When available, World Health Organization (WHO) standards, or standards that can be linked to the WHO standards, should be used to calibrate functional and antigenic assays. *Level 1*
- The effects of age and gender should always be taken into consideration when interpreting the results of antigenic and functional assays, and, whenever possible, reference ranges that are age and gender specific should be developed. *Level 1*
- Before concluding that a patient has an inherited thrombophilia, diagnostic assays for function or antigen should be repeated after excluding acquired etiologies of the defect. *Level 3*

Case Study 1

A 38-year-old man made an appointment with his internist and reported 10 days of persistent pain and progressive swelling in his left calf. He felt well otherwise and had no other health problems. The patient's left calf circumference was 2 cm greater than his right calf, and flexion of his left ankle caused an accentuation of the pain in his left calf. The remainder of the physical exam and routine laboratory tests (complete blood cell count, initial chemistries, and urinalysis) were normal. Because of a high clinical suspicion of deep vein thrombosis (DVT), a compression ultrasound was performed, which confirmed a proximal DVT extending into the distal femoral vein. Outpatient anticoagulation with weight-adjusted low-molecular-weight heparin was started and continued until a therapeutic international normalized ratio (2 to 3) was obtained with oral warfarin. The patient's left calf symptoms quickly resolved, and he experienced no hemorrhagic or thrombotic complications during 6 months of therapeutic oral anticoagulation. Family history was notable for his mother, who had taken warfarin for treatment of a postpartum DVT. Should this patient be tested for inherited thrombophilia risk factors?

Case Study 2

A 49-year-old woman twisted her right knee severely while snow skiing. Magnetic resonance imaging confirmed a torn anterior cruciate ligament, and arthroscopic repair was performed promptly. Before and after surgery, she continued to take hormone replacement therapy for menopausal symptoms. One week after surgery, she noted pain and swelling in her right calf, and a compression ultrasound confirmed a DVT in the popliteal vein. Hormone replacement therapy was stopped, and outpatient therapeutic anticoagulation with weight-adjusted low-molecular-weight heparin was started, continued for 3 months, and then stopped. Should this patient be tested for inherited thrombophilia risk factors?

and initial hematology and chemistry tests. In the first case study presented in this chapter, the patient suffered an unprovoked DVT at a young age, had no signs or symptoms suspicious for an occult malignancy (which may present with a spontaneous VTE), and had a positive family history of VTE. Laboratory testing for inherited and acquired types of thrombophilia would be appropriate for this patient.

In the second case study, the patient had several acquired and reversible VTE risk factors when she developed a symptomatic DVT (hormone replacement therapy, immobilization, and both accidental and surgical trauma), and it is very unlikely that inherited thrombophilia risk factors played an important role in her thrombosis or that she is at high risk for recurrent VTE. Therefore, the duration of anticoagulation therapy can be shortened, and no thrombophilia testing should be performed.

Most episodes of VTE occur in patients older than 50 years who have one or more identifiable acquired risk factors (Table 17-3). Expensive workups for inherited hypercoagulable risk factors are not indicated in these situations. Once the decision to perform testing for *inherited* thrombophilia is made, the following important questions should be addressed:

1. What tests should be performed?
2. When should testing be done?

And for each test that is performed, the following questions should be addressed:

1. What are the potential causes of inaccurate results? (See other chapters on test methodology.)
2. How will the results affect the patient and the patient's relatives?

Some hypercoagulable risk factors are *acquired* or a combination of a genetic predisposition and specific clinical setting. Laboratory testing related to these risk factors will be briefly reviewed, and the reader is referred to other chapters that address them in detail.

What Tests Should Be Performed?

Although the prevalences of Factor V Leiden, prothrombin gene mutation, and elevated factor VIII activity and hyperhomocysteinemia are considerably higher than the prevalences of deficiencies of protein C, protein S, and antithrombin (Table 17-4), it is appropriate to test for all recognized inherited risk factors in selected patients with VTE to identify both the common and uncommon risk factors as well as the co-inheritance of two or more

Table 17-3. Thrombophilia Risk Factors

Acquired	Inherited
Major surgery	Factor V Leiden
Trauma, accidental and surgical	Prothrombin gene mutation G20210A
Solid and hematopoietic malignancies	Antithrombin deficiency
Cancer therapies	Protein C deficiency
Central venous catheters	Protein S deficiency
Limb immobilization	Hyperhomocysteinemia*
Hip, knee replacement	Factor VIII activity*
Prolonged cast or splinting	Dysfibrinogenemia
Stroke	Controversial
Bedridden due to acute illness	Factor IX and XI
Cardiopulmonary disease	Plasminogen deficiency
Infectious disease	Hypofibrinolysis
Inflammatory disease	
Malignancy	
Long-distance air travel	
Antiphospholipid antibody syndrome	
Heparin-induced thrombocytopenia	
Paroxysmal nocturnal hemoglobinuria	
Advancing age, especially >50 y	
Pregnancy	
Oral contraceptive use	
Hormone replacement therapy	

* Hereditary and environmental contributions

Table 17-4. Prevalence of Inherited Thrombophilia Risk Factors

Risk Factor	General Populations	First Venous Thromboembolism Patients
Factor V Leiden	3% – 7%	10% – 20%
Prothrombin gene mutation[1]	1% – 3%	5%
Antithrombin	~0.05%	1% – 3%
Protein C	0.33% – 0.5%	1% – 3%
Protein S	~0.2%	1% – 3%
Homocysteine >95th percentile[2]	5%	10%
Factor VIII activity >150%[3]	10%	25%

1. Applies to people of European ancestry (Caucasians). Prevalences are much lower among other indigenous ethnic and racial populations.
2. den Heijer M, Koster T, Blom HJ, et al. Hyperhomocysteinemia as a risk factor for deep-vein thrombosis. *N Engl J Med.* 1996;334(12):759-762.

3. O'Donnell J, Tuddenham EG, Manning R, Kemball-Cook G, Johnson D, Laffan M. High prevalence of elevated factor VIII levels in patients referred for thrombophilia screening: role of increased synthesis and relationship to the acute phase protein. *Thromb Haemost.* 1997;77:825-828.

inherited risk factors in some individuals (Figure 17-1). A consensus is lacking among hemostasis experts regarding the prothrombotic potential of abnormal results for unmodified activated protein C resistance, dysfibrinogenemia (see chapter 20), protein C pathway, heparin cofactor II, and endogenous thrombin generation tests; and the

clinical significance of in vitro hypofibrinolysis results for plasminogen antigen/activity, thrombin-activatable fibrinolysis inhibitor, and global in vitro fibrinolysis time tests. Therefore, these aforementioned tests are not included in the standard thrombophilia laboratory evaluation.

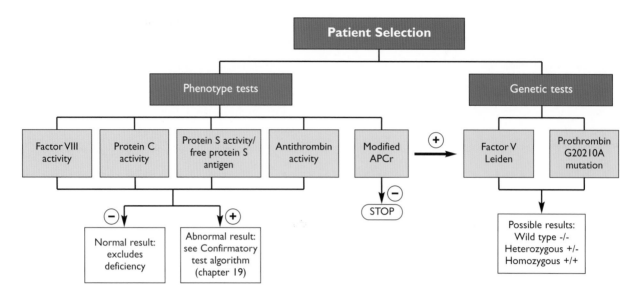

Figure 17-1. Inherited thrombophilia testing algorithm: a stepwise approach to evaluating patients for inherited thrombophilia. (See the text of chapters 18 and 19 for details.) Abbreviation: APCr, activated protein C resistance.

All patients with acute VTEs will have an elevated D-dimer unless there is a delay of several weeks from thrombus formation to evaluation and diagnosis or anticoagulation therapy has been started before D-dimer testing is performed. However, a negative D-dimer in a patient suspected of having an acute VTE can be useful in excluding a VTE (see chapter 20).

When Should Thrombophilia Testing Be Performed?

The motivation to provide an explanation for a spontaneous DVT or pulmonary embolism is very strong at the time of diagnosis. However, not all thrombophilia tests should be performed immediately because of the potential for inaccurate results. Evidence-based guidelines for initiation of anticoagulation therapy are based on the clinical circumstances at the time of the thromboembolic event, not the results of thrombophilia tests. In selected patients (like the man in the first case study) with idiopathic VTEs, recommendations pertaining to the duration of anticoagulation therapy may be influenced by thrombophilia test results. Ideally, comprehensive thrombophilia testing would be done months after the VTE and at least 30 days after stopping warfarin. However, some patients may experience a recurrent VTE while waiting to be tested. In a prospective study by Palareti and colleagues (2006) of the negative and positive predictive values of an elevated D-dimer measured 30 days after completion of at least 3 months of oral anticoagulation therapy, 5 of

619 patients (0.8%) had a recurrent DVT or pulmonary embolism during the waiting period. An alternative to simply discontinuing warfarin is to initiate prophylactic or therapeutic doses of a low-molecular-weight heparin until thrombophilia testing is performed and results are available for patient counseling.

Recommendations for Testing of Individual Analytes

Protein C, Protein S, and Antithrombin Deficiency

There is convincing evidence, derived from prospective studies, that deficiencies of protein C, protein S, and antithrombin are etiologic factors for thrombosis. However, for these deficiencies, the best available evidence is based on retrospective cohort analysis of family studies. These studies indicate that the risk in deficient individuals is increased approximately 10-fold in comparison with nondeficient individuals. The retrospective and prospective cohort studies in asymptomatic family members with these deficiencies, identified through deficient symptomatic probands, showed comparable results with regard to the annual incidence rates. In the literature, the estimated annual incidence of a first idiopathic venous thrombosis in asymptomatic individuals with deficiencies of antithrombin, protein C, and protein S ranges for all from about 0.4% to 3.0%. Finding a thrombophilic condition in a patient with thrombosis cannot be interpreted as having identified "the cause of" the event, but rather "a contributing or associated factor" to the event.

There is limited but consistent evidence that protein C, protein S, and antithrombin deficiencies increase the recurrence rate of venous thrombosis and thus may influence decisions about the duration of anticoagulation therapy. On the basis of retrospective cohort studies and a review of family studies, the annual incidence of recurrent venous thrombosis in deficient individuals has been estimated to be 5% to 17%, and decision analysis indicates that prolongation of anticoagulation may be worthwhile, especially in younger patients.

If therapeutic anticoagulation therapy is discontinued in a patient with an inherited protein C, protein S, or antithrombin deficiency, it might seem logical to lower the threshold for starting prophylaxis, to give higher doses of prophylactic agents, or to prolong prophylaxis in patients with an underlying thrombophilic condition. There have been no reports in the literature on the results of these practices. It is important to note, however, that in any patient with a prior venous thromboembolic event, a vigorous approach to prophylaxis in the case of a high-risk situation is warranted.

There is conflicting evidence about these deficiencies in regard to counseling asymptomatic affected family members. While routinely screening women for protein C, protein S, and antithrombin deficiencies (with a combined frequency of 1 in 500 to 1 in 300 in the general population) prior to starting oral contraceptives is clearly not cost effective, this might not be the case for women who are a first-degree family member of a person who experienced a spontaneous venous thromboembolic event and has an inherited thrombophilic risk factor. In a retrospective study, the combined annual incidence of VTE in antithrombin-, protein C- or protein S-deficient women has been estimated to be about 4.3%. A similar observation is true for pregnancy, in which the risk per pregnancy is estimated at 5% for antithrombin-, protein C-, or protein S-deficient women. For the usual high-risk situations, the risk also seems clearly increased among individuals who have deficiencies of protein C, protein S, or antithrombin. Hormonal therapy and pregnancy are addressed in greater detail in chapter 18.

Because there are no alternative, more aggressive approaches to prophylaxis that have been documented to be safe and effective, testing for protein C, protein S, and antithrombin deficiencies could have consequences for counseling, but not for evidence-based decisions of different prophylactic approaches.

Factor V Leiden and Prothrombin G20210A Mutations

There is convincing evidence that the factor V Leiden and prothrombin G20210A mutations are etiologic factors for thrombotic risk. The evidence can be labeled as level 2 because, by the nature of their design, almost all studies have been retrospective. The increased risk of thrombosis has been documented to be 3- to 7-fold for factor V Leiden, with prothrombin G20210A a relatively weaker risk factor. It is important to keep in mind that there is wide variation in the distribution of these genetic defects by race, being more common in people of European descent, less common in the African American population, and rare in Asians and Native Americans. Both retrospective and prospective cohort studies have estimated the annual incidence of a first episode of venous thrombosis in heterozygous carriers of factor V Leiden to be about 0.3 to 0.5%. Again, it should be realized that finding a thrombophilic condition in a patient with thrombosis cannot be interpreted as having identified the cause of the event, but rather a contributing or associated factor to the event.

Whether carriers of these mutations have additional increased risk of recurrent venous thrombosis is still controversial, with widely ranging estimates reported in the literature. For factor V Leiden and prothrombin G20210A, the reports on the increased risk of recurrent events are contradicting, and the risk is probably not dramatically increased. Currently, the determination of these abnormalities has no impact on duration of treatment. However, homozygous patients and patients with combined defects have consistently been shown to be correlated with an increased risk of recurrent VTE.

Although it might seem logical to lower the threshold for starting prophylaxis, to give higher doses of prophylactic agents, or to prolong prophylaxis in patients who are heterozygous for factor V Leiden and prothrombin G20210A, there have been no reports in the literature on the results of these practices. It should be realized, however, that in any patient with a prior venous thromboembolic event, a vigorous approach to prophylaxis in case of a high-risk situation is warranted.

Although in the general population (with a frequency of 1 in 14 for factor V Leiden and 1 in 28 for prothrombin G20210A), screening of women who begin taking oral contraceptives is clearly not cost effective, this might not be the case for women who have a first-degree family member that is a carrier of such a thrombophilic defect and has experienced a VTE. In these women, the risk of inheriting a defect is at least 1 in 2, and those who do are at greater risk of VTE if also using oral contraceptives. The estimated annual VTE incidence in carriers of the factor V Leiden mutation has been estimated to be about 0.48% to 2.0%.

For pregnancy, the risk per pregnancy is estimated at 2% to 3% for factor V Leiden or prothrombin G20210A carriers. Therefore, testing women of child-bearing age who have a family member with a history of VTE and who are positive for factor V Leiden or prothrombin G20210A might alter the counseling and prophylactic approach for these individuals. There is no clear evidence that the risk for thrombosis in these carriers is increased for the other high-risk situations. Hormonal therapy and pregnancy are addressed in greater detail in chapter 18.

Hyperhomocysteinemia and High Concentrations of Factor VIII

For hyperhomocysteinemia, the best evidence available is based on case-control studies, with a moderate (2- to 3-fold) increase in VTE risk. Thus, testing could add to the knowledge of etiology. Currently, there are no data to suggest a role in other aspects of treatment, although elevated homocysteine levels can be lowered with vitamins B_{12} and B_6 and folate. Studies to date have failed to demonstrate that the lowering of homocysteine will lead to a reduction in initial and recurrent arterial or venous thrombotic events.

Persistent elevation in factor VIII has also been associated with a moderate (2- to 3-fold) increased risk of VTE. Elevations of factor VIII are seen as the result of acute phase reaction; because of this, the timing of testing and the need for repeat testing to confirm persistent elevations are important considerations.

The most commonly used method to measure factor VIII activity is an activated partial thromboplastin time (aPTT)-based clotting assay that compares patient values to normal pooled plasma. Several dilutions in factor VIII-deficient plasma are measured. These methods were developed to detect deficiency of factor VIII and have wider coefficients of variation at higher levels of the pro-

tein. In 2004, the College of American Pathologists proficiency testing survey for factor VIII activity reported data from more than 500 laboratories. The all method mean for an elevated specimen was 287%, with a range from 163% to 537% with a coefficient of variation of 20% (Coagulation Survey CG2-B. Participant Summary Report). From these data, it is clear that if a cutoff of >150% (a level used in some reports) is used to define an increase in thrombotic risk, many samples would be interpreted as normal by some laboratories and as associated with an increased thrombotic risk by others. Thus, measuring factor VIII can be of value but must be done carefully, keeping in mind these limitations. Factor VIII is discussed in more detail in chapter 19.

Acquired Thrombophilic Risk Factors

Antiphospholipid Antibodies

Case-control studies have demonstrated a substantial (5-fold) increase in risk of thrombosis in individuals with antiphospholipid antibodies. Thus, these tests may be helpful in understanding the etiology of arterial and venous thrombosis. In addition, there is convincing evidence that the duration of treatment should be prolonged in patients with thrombosis and antiphospholipid antibodies, based on consistent high recurrence rates in the absence of treatment and documented efficacy of oral anticoagulant treatment in this group of patients. Thus, testing for antiphospholipid antibodies can influence treatment decisions. Because antiphospholipid antibodies are not an inherited disorder, testing asymptomatic relatives is not relevant. Additional information on antiphospholipid antibodies can be found in chapter 22.

Heparin-Induced Thrombocytopenia

Heparin-induced thrombocytopenia should be considered in any patient who experiences venous thrombosis while exposed to heparin or within 30 days of heparin exposure. Heparin-induced thrombocytopenia should be considered if the patient also has a drop in platelet count of more than 50% from baseline while receiving heparin therapy. Additional information on a diagnostic approach to heparin-induced thrombocytopenia can be found in chapter 21. Conclusions and recommendations regarding testing in VTE can be found in Table 17-5.

Table 17-5. Venous Thromboembolism: Conclusion and Recommendations

Conclusion
1. The identification of protein C, protein S, or antithrombin deficiency or combined thrombophilias may be used to influence decisions about treatment duration or prophylaxis. *Level 2*

Recommendations
1. Testing for factor V Leiden (an activated protein C resistance assay with factor V-deficient plasma can be used as an initial test), functional protein C, functional protein S or free protein S antigen, functional antithrombin, and prothrombin G20210A mutation is appropriate in patients with venous thromboembolism (VTE), particularly for idiopathic VTE, younger patients, and/or those with a family history of thrombosis. *Level 2*
2. Anticardiolipin antibody and lupus anticoagulant assays are appropriate for patients with VTE, particularly if it is idiopathic or associated with autoimmune disease or in the absence of a family history of venous thrombosis. *Level 2*
3. Testing for thrombophilia (tests noted in recommendations 1 and 2) is appropriate for patients with the following characteristics:
   – a history of recurrent VTE;
   – VTE before the age of 50 years;
   – unprovoked VTE at any age; however, testing for protein C, protein S, and antithrombin deficiency may be of lower diagnostic yield in patients with a first lifetime VTE after age 50 years;
   – VTE at unusual sites (eg, cerebral, mesenteric, portal, hepatic);
   – VTE patients with a positive family history of VTE; and
   – VTE secondary to pregnancy, oral contraceptives, or hormone replacement therapy. *Level 2*
4. Although there is an association between homocysteine and venous thrombosis, the implications for testing are controversial. *Level 2*
5. Heparin-induced thrombocytopenia should be considered for any patient who experiences venous thrombosis while exposed to heparin, or within 30 days of heparin exposure, with a decrease in platelet count to less than 50% of baseline.
6. Prior to pregnancy or oral contraceptive use, it may be worthwhile to test asymptomatic female first-degree relatives of a proband with a defined inherited thrombophilia (for that identified defect). This testing is especially important for families with known antithrombin deficiency. *Level 2*

Controversial Recommendations
1. Testing for thrombophilia is controversial in patients with the following characteristics:
   – a first, provoked VTE in older patients (age >50 years); in general, testing for thrombophilia is not recommended for VTE associated with active cancer or an intravascular device in adults;
   – a first VTE related to selective estrogen receptor modulators (SERMs) or tamoxifen. *Level 2*
2. After counseling, testing for thrombophilia is appropriate in asymptomatic first-degree relatives of a proband with a known inherited thrombophilia. Such testing may be particularly useful in families with deficiencies of protein C, protein S, or antithrombin. *Level 2*

Thrombophilic Risk Factor Test Selection in Arterial Thrombosis

Arterial thrombosis, such as myocardial infarction (MI) or peripheral arterial disease, most often occurs in association with atherosclerosis (arterial disease affecting the cerebrovascular system is dis-cussed in the neurologic section of this chapter). Laboratory markers of arterial thrombosis risk differ from laboratory markers of venous hypercoagulability. One major difference is that elevated cholesterol is an established risk factor for arterial thrombosis associated with atherosclerosis. In addition, a number of studies have confirmed that lowering low-density lipoprotein (LDL) cholesterol reduces the risk. It is recommended that a fasting lipid profile (total cholesterol, LDL cholesterol, high-density lipoprotein [HDL] cholesterol, and triglyceride) be measured every 5 years in all adults age 20 or older. More frequent testing may be indicated in high-risk patients, those with abnormal lipid profiles, and those undergoing lipid-modifying treatment.

Antiphospholipid Antibodies (Lupus Anticoagulant and Anticardiolipin Antibodies)

Antiphospholipid antibodies are associated with an increased risk of venous or arterial thrombosis. The risk of any thrombosis is higher among individuals who have had a history of previous thrombosis or who have anticardiolipin antibodies greater than 40 IgG phospholipid units (GPL).

Antiphospholipid antibodies are associated with an increased risk of MI. Healthy men with elevated LDL-cholesterol at baseline have a 2-fold increased risk of MI or cardiac death if they also have elevated anticardiolipin antibody titers. Among patients undergoing coronary artery bypass graft surgery, there is an increased risk of graft occlusion at 1-year follow-up in those patients with preoperative anticardiolipin antibodies.

The antigen recognized by lupus anticoagulants is often prothrombin bound to phospholipid, and the antigen recognized by anticardiolipin antibodies is beta 2-glycoprotein I (β_2GPI) bound to phospholipid. Therefore, newer immunoassays have been developed that use prothrombin (and less frequently other vitamin K-dependent coagulation factors) or β_2GPI as the antigen. More information regarding these assays can be found in chapter 22. Among healthy men with high LDL-cholesterol at baseline, a 2.5-fold increased risk of MI or cardiac death has been demonstrated among those with elevated antiprothrombin antibody titers. There was no significant association noted with anti-β_2GPI antibodies in this group of patients.

Anticardiolipin antibodies are much more common in patients with peripheral arterial thrombosis than in healthy blood donor controls, but the antibodies have not been associated with an increased risk for restenosis 6 months after angioplasty. However, individuals with lupus anticoagulants or anticardiolipin antibodies have been shown to develop more postoperative progression of lower extremity arterial occlusive disease than did those without anticardiolipin antibodies.

In summary, antiphospholipid antibodies are associated with arterial as well as venous thrombosis. Although MI is not the most common type of arterial thrombosis observed among individuals with antiphospholipid antibodies, several prospective studies suggest that the risk of MI or recurrent MI appears higher with elevated anticardiolipin antibody titers. The association with ischemic stroke is discussed in the neurologic section of this chapter.

C-Reactive Protein

C-reactive protein (CRP) is a sensitive marker of inflammation. Prospective studies in healthy populations indicate that elevated CRP is associated with future risk for MI in men, combined future risk for MI and other atherosclerotic disease in women, and peripheral arterial disease in men and women. At the time of acute myocardial infarction, elevated CRP has predicted overall mortality, cardiac death, reinfarction, and/or other cardiac complications in a number of prospective studies. Similar results have been observed for unstable angina in most, but not all, prospective studies.

The use of aspirin has been shown to have its greatest benefit in patients with the highest CRP, raising speculation that part of the benefit of aspirin in preventing heart disease is its anti-inflammatory properties, in addition to its antiplatelet effect. 3-hydroxy-s-methyl-glutaryl coenzyme-A reductase inhibitors, referred to as statin agents, decrease CRP, leading to speculation that part of the benefit of statin therapy relates to an anti-inflammatory effect. Clinical trials to determine if lowering CRP levels in patients without dyslipidemia reduces the risk of future cardiovascular events are ongoing.

When assessing baseline CRP in apparently healthy individuals, the newer, highly sensitive methods should be used, and, ideally, individuals should be tested in the absence of an acute phase reaction (eg, infection, inflammation, trauma) that will elevate CRP.

Lipoprotein (a)

Lipoprotein (a) (Lp(a)) is an LDL-cholesterol particle with an additional protein covalently attached called apolipoprotein (a) (apo(a)). Apolipoprotein (a) has structural similarities to plasminogen but without any enzymatic activity. Indications are that elevated Lp(a) is associated with future risk for MI or related atherosclerotic disease among healthy or general populations and among patients with coronary artery disease, but not all prospective studies have found an association. Whether Lp(a) is predictive of future peripheral arterial disease in apparently healthy individuals at baseline remains to be demonstrated because, currently, there are conflicting results in the literature.

Lp(a) testing can be complicated by the fact that Lp(a) can be elevated by acute phase reactions. In addition, assays that detect a repeat domain in Lp(a) called kringle 4 can overestimate Lp(a) if the patient has large apo(a) isoforms and can underestimate Lp(a) if the patient has small apo(a) isoforms, because the number of kringle 4 repeats is higher in the large isoforms. Smaller isoforms are associated with a higher risk than larger isoforms. Some retrospective studies may be hindered if any observed Lp(a) elevations are a result of thrombosis-induced acute phase reactions.

Further study is needed to determine if Lp(a)-reducing treatment reduces subsequent arterial thrombosis risk. Preliminary evidence indicates that reducing Lp(a) with estrogen therapy in postmenopausal women is associated with decreased risk for MI, but not cardiac death. An earlier study provided preliminary evidence that reducing Lp(a) with LDL apheresis may reduce restenosis after angioplasty.

Homocysteine

Homocysteine, in addition to being implicated as a risk factor for venous thrombosis, has also been associated with arterial thrombosis and atherosclerosis. In some patients with atherosclerosis and elevated homocysteine, lowering homocysteine with folic acid, vitamin B_{12}, and vitamin B_6 slows the progression of carotid artery atherosclerosis. Recent randomized trials in patients with cardiovascular disease and elevated homocysteine have shown reduction of homocysteine levels but not in subsequent ischemic events for subjects randomized to B vitamin supplementation.

Homocysteine testing is complicated by the fact that homocysteine can increase by day 3 after an MI and may remain elevated 6 months or longer. Because pre-MI homocysteine levels were unknown, it is not clear if homocysteine becomes temporarily decreased during the acute phase of an MI or if homocysteine becomes elevated as a consequence of the MI. Homocysteine levels are similar on day 1 and at 6 months after an MI.

Many prospective studies in healthy or general populations show an association between homocysteine and subsequent risk for MI or other coronary artery disease complications, such as cardiac death, but many other such studies have not found an association. For example, a prospective study involving 229 cases of fatal coronary artery disease among 21,520 healthy men age 35 to 64 years at baseline found that homocysteine levels were significantly higher in the men with coronary artery disease than in the control population, and that the upper quartile of homocysteine levels (≥ 15.17 µmol/L) was associated with a 3-fold increased risk for ischemic heart disease. In this study, the risk started to increase with homocysteine values above 10.25 µmol/L. In contrast, homocysteine was not a risk factor in another study of 232 cases of MI among 15,792 middle-aged men and women. Similar negative results were found in a study involving 213 such cases among 2290 men age 45 to 59 years at baseline and in a study involving 240 such cases among 12,866 men age 35 to 57 years at baseline.

Prospective studies involving patients with coronary artery disease at baseline have generally found an association between homocysteine and subsequent risk of MI and/or cardiovascular or all-cause mortality. Similar results have been found among patients with end-stage renal disease or systemic lupus erythematosus. Studies with negative results have generally had a very short follow-up period.

Homocysteine has been useful in predicting the development of peripheral arterial disease in initially healthy individuals.

Although elevated homocysteine can be an acquired condition, it also may arise from a genetic predisposition. A common mutation in the methylene tetrahydrofolate reductase (MTHFR) gene is designated *MTHFR C677T*, which indicates a C to T substitution at nucleotide 677. When *MTHFR C677T* is present in the homozygous form, it can predispose an individual to elevations in homocysteine. However, results have been conflicting among the many retrospective studies involving this mutation and the risk of arterial thrombotic disease.

In summary, some, but not all, of the prospective studies in healthy or general populations have found an association between homocysteine and future risk of MI, cardiac death, and/or all-cause mortality. On the other hand, most of the prospective studies in various patient populations, for example, patients with existing atherosclerosis, have found such an association. However, because lowering homocysteine levels by treatment with B vitamins (folate and vitamins B_{12} and B_6) has not been shown to reduce thrombotic risk, measuring homocysteine in patients with existing atherosclerosis would seem of little value at this time. It is not yet clear if screening healthy individuals for homocysteine has any benefit. There is not yet a clear benefit to testing for the *MTHFR C677T* mutation instead of, or in addition to, homocysteine itself in the laboratory investigation of coronary artery or peripheral vascular disease.

Prothrombin G20210A Mutation

Most studies investigating the association between the prothrombin G20210A mutation and MI or other arterial thrombotic disease were not prospective. In the prospective, nested, case-control Physician's Health Study involving 404 cases of MI among 14,916 men, there was no association between prothrombin G20210A and MI (relative risk = 0.8). A number of case-control (not prospective) studies also found no association between prothrombin G20210A and myocardial infarction. Arruda et al (1998) reported a case-control (not prospective) study that did find prothrombin G20210A more frequently among cases of MI than in controls.

When particular subgroups are analyzed, several additional case-control studies did find an association between prothrombin G20210A and MI. The presence of prothrombin G20210A (or factor V Leiden, *MTHFR C677T*, or apolipoprotein E4), when combined with hypertension, high cholesterol, or diabetes, was associated with a 9-fold increased risk. When any one of these four laboratory markers was combined with smoking, the risk was increased 18-fold.

No association was found between prothrombin G20210A and history of MI in 2210 consecutive angiography patients (mean age was approximately 60 years). There was also no association with coronary artery disease—except that in certain subgroups, individuals with the prothrombin G20210A mutation had more coronary artery disease than did wild-type individuals in the subgroup. These subgroups include individuals with elevated Lp(a) and abnormal apolipoprotein A1 to apolipoprotein B ratios, or individuals not receiving aspirin.

In summary, prothrombin G20210A may be more prevalent among patients with MI who do not have other evidence of atherosclerosis and/or who have certain other risk factors (smoking, hypertension, obesity, high cholesterol, or diabetes), when compared with control groups.

Factor V Leiden
(Activated Protein C Resistance)

Most studies investigating the association between the factor V Leiden mutation and MI were not prospective. In the prospective, nested, case-control Physician's Health Study involving 374 cases of MI among 14,916 healthy men, factor V Leiden was not significantly associated with an increased risk for MI. In another study involving 289 MI or angina cases among 5201 men and women over the age of 65 years, with no history of MI at baseline, followed for 3.4 years, factor V Leiden was not a risk factor for future MI or angina. Subjects had no history of MI at baseline.

However, as mentioned before, the presence of factor V Leiden (or prothrombin G20210A, *MTHFR C677T*, or apolipoprotein E4), when combined with hypertension, high cholesterol, or diabetes, was associated with a 9-fold increased risk. When any one of these four laboratory markers was combined with smoking, the risk was increased 18-fold.

In summary, factor V Leiden may be more prevalent among patients with MI who do not have atherosclerosis and/or among young patients with certain other risk factors (smoking, hypertension, obesity, high cholesterol, or diabetes), when compared with control groups.

Deficiencies of Protein C, Protein S, and Antithrombin

Hereditary deficiencies of protein C, protein S, and antithrombin are not believed to be etiologic factors in atherosclerosis and are rare among patients with arterial thrombosis, even in young patients without atherosclerosis. There is no clear evidence that the incidence of arterial thrombosis is increased among patients with protein C, protein S, or antithrombin deficiency. The ratio of venous thrombotic events to arterial thrombotic events among symptomatic patients with protein C, protein S, or antithrombin deficiency is approximately 24 to 1. The prevalence of arterial thrombosis with hereditary antithrombin deficiency was 2% in a literature review. The rare occurrence of homozygous antithrombin deficiency due to heparin-binding mutations has been described to have an increased incidence of arterial thrombosis in addition to severe venous thrombosis.

In summary, hereditary antithrombin, protein C, and protein S deficiencies are uncommon in general populations and in persons with arterial thrombosis. In most cases, it is difficult to know with certainty if individuals with these deficiencies have any increased risk of arterial thrombosis, because of their low prevalence.

Coagulation Factors

Elevated fibrinogen, which is an acute phase reactant, as is C-reactive protein (CRP), is associated with an increased risk for MI in prospective studies. Elevated factor VIII and von Willebrand factor, which are also acute phase reactants, are associated with an increased risk for MI in prospective studies, but the association is often eliminated after adjusting for other risk factors. High levels of factor VII and/or certain factor VII polymorphisms have been associated with an increased risk of MI, but several other studies have not confirmed an association. Factor VII levels may be affected by dietary lipids, and factor VII levels correlate with triglyceride and cholesterol levels. Finally, a factor XIII polymorphism (Val34Leu) is possibly protective against MI.

Heparin-Induced Thrombocytopenia

Heparin-induced thrombocytopenia should be considered in any patient who experiences arterial thrombosis while exposed to heparin or within 30 days of heparin exposure. Heparin-induced thrombocytopenia should be considered if the patient also has a reduction in platelet count of more than 50% from baseline. A detailed discussion of heparin-induced thrombocytopenia is included in chapter 21. Conclusions and recommendations regarding thrombophilia testing in arterial thrombosis can be found in Table 17-6.

Thrombophilic Risk Factor Test Selection in Neurovascular Thrombosis

Laboratory testing for thrombophilic risk factors in patients with thrombosis in the cerebrovascular system—in particular, arterial ischemic stroke, transient ischemic attack (TIA), or cerebral venous thrombosis—are discussed in this section. Laboratory testing for the risk of arterial ischemic stroke or cerebral venous thrombosis shares similarities with laboratory markers of other arterial and venous hypercoagulability, respectively, but there are also several differences. The risk of stroke with elevated cholesterol, decreased HDL, or other lipid abnormalities is not as well understood

as it is for myocardial infarction. CRP may predict future risk of stroke in men and women, similar to the apparent association between CRP and risk of MI. Additional laboratory tests pertinent to neurologic thrombosis are discussed below.

Antiphospholipid Antibodies (Lupus Anticoagulant and Anticardiolipin Antibodies)

Antiphospholipid antibodies are associated with an increased risk of venous or arterial thrombosis. Most studies involving these antibodies have been retrospective. In a prospective study that followed 360 individuals with antiphospholipid antibodies (lupus anticoagulants and/or anticardiolipin antibodies) for a median of 3.9 years, 4.7% of patients developed venous thrombosis and 4.7% developed arterial thrombosis during the follow-up period. All but one of these arterial thrombotic events were strokes or TIAs. The risk for any thrombosis was higher among individuals who had a history of previous thrombosis or who had anticardiolipin antibodies greater than 40 GPL (IgG phospholipid units).

Other studies have not found an association between baseline IgG anticardiolipin antibodies and risk of future stroke. Strokes appear to be the most common type of arterial thrombosis associated with these antibodies. Among patients with stroke, some prospective studies suggest that the risk of future thrombotic events and death is higher with elevated baseline anticardiolipin antibody titers.

Lipoprotein (a)

Most studies investigating the relationship between Lp(a) and stroke have not been prospective. Several studies have not found an association between Lp(a) and either TIA or stroke. In another prospective study involving healthy elderly individuals, Lp(a) was higher in those who subsequently developed stroke (with or without dementia), but the trend did not reach significance ($P = .06$). One prospective study involving 151 patients with ischemic stroke or TIA found no relation between Lp(a) at baseline and subsequent risk of recurrent stroke, MI, vascular death, or death.

Lp(a) testing can be complicated by the fact that Lp(a) can become elevated by acute phase reactions; however, there appears to be no significant change in Lp(a) during the first 4 weeks after a stroke. Retrospective studies may be hindered if any observed Lp(a) elevations are a result of thrombosis-induced acute phase reactions, as might be seen with stroke. Lastly, as discussed in

Table 17-6. Arterial Thrombosis: Conclusions and Recommendations

Conclusions

1. C-reactive protein appears to predict future risk for coronary artery disease in healthy individuals as well as in various patient populations, such as MI or unstable angina patients. *Level 1* However, there is no consensus on which persons should be tested or how the test result should affect patient management (eg, statin, aspirin therapies).

2. Most, but not all, prospective studies find an association between lipoprotein (a) (Lp(a)) and myocardial infarction or related atherosclerotic disease. However, there is no clear consensus regarding who should be tested. Lipoprotein (a) can be reduced by estrogen (in women) or niacin therapy, but confirmation that such treatments will reduce future risk is awaiting further study. *Level 1, with some conflicting results*

Recommendations

1. Antiphospholipid antibody (lupus anticoagulant and anticardiolipin antibody) assays can be considered for patients with arterial thrombosis, particularly in a young person or a person with no documented atherosclerosis. *Level 1, with relatively few prospective studies to date*

2. Consider measuring homocysteine for patients with documented atherosclerotic arterial occlusive disease. *Level 1* Homocysteine concentration can be reduced by therapy with vitamins B_6 and B_{12} and folate, but studies to date have not confirmed that such treatments will reduce the risk of future cardiovascular events.

3. Routine testing for factor V Leiden and prothrombin G20210A is not recommended in patients with arterial thrombotic disease that is associated with atherosclerosis. However, these tests can be considered in certain unusual situations, such as patients with unexplained arterial thrombosis without atherosclerosis or young patients who smoke. *Level 2*

4. Routine testing for protein C, protein S, and antithrombin is not recommended for patients with arterial thrombotic disease that is associated with atherosclerosis. However, these assays can be considered in certain unusual situations, such as young patients with unexplained arterial thrombosis without atherosclerosis. *Level 3*

5. Heparin-induced thrombocytopenia should be considered in any patient who experiences arterial thrombosis while exposed to heparin or within 30 days of heparin exposure, with a decrease in platelet count to less than 50% of baseline.

Although the following recommendation does not relate specifically to thrombophilia, it is included because of its well-documented relationship to arterial vascular disease:

6. A fasting lipid profile (total cholesterol, LDL cholesterol, HDL cholesterol, and triglycerides) should be measured every 5 years in all adults age 20 years or older. More frequent testing may be indicated in certain individuals. *Level 1*

the arterial section, assays that detect a repeat domain in Lp(a) called kringle 4 can overestimate Lp(a) if the patient has large apo(a) isoforms and can underestimate Lp(a) if the patient has small apo(a) isoforms.

In summary, results of prospective studies investigating the relationship between Lp(a) and stroke or TIA have been conflicting. Further study is needed to determine if Lp(a)-reducing treatment reduces subsequent risk.

Homocysteine

Homocysteine, in addition to being implicated as a risk factor for venous thrombosis, coronary artery disease, and MI, has also been associated with stroke.

As with MI, homocysteine levels might be different in the acute setting of stroke compared with the convalescent recovery phase after a stroke. When compared with immediately after a stroke, median homocysteine levels rise 3 to 4 µM approximately 1.5 years later—a change not observed in control subjects. Because prestroke homocysteine levels are not known, it is not clear if homocysteine becomes temporarily decreased during the acute phase of a stroke or if homocysteine subsequently becomes elevated as a consequence of the stroke.

Most observational studies in healthy or general populations (including men, women, middle-aged, and/or elderly persons) show an association between homocysteine and subsequent risk for stroke; however, data are conflicting. Results of retrospective studies on the *MTHFR C677T* mutation and stroke have been conflicting in children. Most studies involving the association of MTHFR and stroke in adults have been negative.

In summary, most of the prospective observational studies in healthy or general populations have found an association between homocysteine and future risk of stroke. However, because lowering homocysteine levels by treatment with B vitamins (folate and vitamins B_{12} and B_6) has not been shown to reduce thrombotic risk, measuring homocysteine in patients with existing atherosclerosis would seem of little value at this time. There are no prospective data linking *MTHFR C677T* to stroke.

Prothrombin G20210A Mutation

Most studies investigating the association between the prothrombin G20210A mutation and ischemic arterial stroke or cerebral venous thrombosis were not prospective. In the prospective, nested, case-control Physician's Health Study involving 259 stroke cases among 14,916 men, there was no association between prothrombin G20210A and stroke. A number of case-control (no prospective follow-up) studies also found no association between prothrombin G20210A and stroke or TIA, including studies involving young adults (age <45 to 50 years).

Several case-control, nonprospective studies involving children and infants (not neonates) suggest that prothrombin G20210A may be a risk factor for thrombosis in younger populations, but other studies were negative.

In summary, prothrombin G20210A may be more prevalent among pediatric patients who have experienced a stroke as compared with controls, but the data are conflicting and prospective studies are lacking. Case-control, nonprospective studies also suggest that prothrombin G20210A is more prevalent among patients with cerebral venous thrombosis than among controls.

Factor V Leiden
(Activated Protein C Resistance)

Most studies investigating the association between the factor V Leiden mutation and ischemic arterial stroke or cerebral venous thrombosis were not prospective. In the prospective, nested, case-control Physician's Health Study involving 209 stroke cases among 14,916 healthy men, factor V Leiden was not significantly associated with an increased risk for stroke, and other studies have shown similar findings.

A number of case-control studies (no prospective follow-up) also found no association between factor V Leiden and mild or moderate stroke, stroke or TIA, or TIA/minor stroke in older persons as well as in young adults.

In several nonprospective studies, factor V Leiden was more common in patients with cerebral venous thrombosis than in controls.

In summary, factor V Leiden may be more prevalent among pediatric patients who have experienced a stroke as compared with controls (see chapter 18). Prospective studies are lacking in this age group. Case-control, nonprospective studies also suggest that factor V Leiden is more prevalent among patients with cerebral venous thrombosis than in controls.

Deficiencies of Protein C, Protein S, and Antithrombin

A prospective study found that baseline protein C levels had a borderline significant association with the risk of stroke over a 6- to 9-year follow-up, but that baseline antithrombin levels had no significant association with future risk of stroke. In this study, it is not known which, if any, of the patients had hereditary protein C or antithrombin deficiency. Also, in this prospective study, baseline protein C values were significantly associated with cerebral infarctions identified by magnetic resonance imaging that was performed at a 6-year follow-up examination (baseline magnetic resonance imaging was not performed for comparison). In another study, during 160 patient-years of prospective follow-up, no cases of stroke were observed among 36 patients with hereditary protein C deficiency, 36 patients with hereditary protein S deficiency, and 9 patients with hereditary antithrombin deficiency.

In general, other studies involving protein C, protein S, or antithrombin in stroke have not been prospective. Among 26,800 cardiovascular inpatients in Japan, the age at onset of stroke was significantly younger in patients with hereditary protein C deficiency as compared with patients with normal protein C. In a retrospective study of 150 families, 0% of antithrombin-deficient, 1.6% of protein C-deficient, and 4.8% of protein S-deficient

Table 17-7. Neurovascular Thrombosis: Conclusion and Recommendations

Conclusion

1. It is difficult to make a recommendation regarding Lp(a) and stroke because results of the few level 1 studies that have been completed are conflicting.

Recommendations

1. Testing for antiphospholipid antibodies (anticardiolipin antibodies and lupus anticoagulants) should be considered in patients with unexplained stroke, particularly in a young person or a patient with autoimmune disease. *Level 1, with few prospective studies to date and some conflicting results* Testing should also be considered for patients with cerebral venous thrombosis. *Level 3*

2. Consider measuring homocysteine in patients with documented stroke or existing cerebrovascular disease. *Level 1* Homocysteine concentration can be reduced by therapy with vitamins B_6 and B_{12} and folate, but studies to date have not confirmed that such treatments will reduce the risk of future cerebrovascular events.

3. Routine testing for factor V Leiden and prothrombin G20210A is not recommended in adult patients with arterial stroke. However, these tests can be considered in certain unusual situations, such as pediatric patients with stroke. These assays may also be useful for patients with cerebral venous thrombosis. *Level 2*

4. Routine testing for protein C, protein S, and antithrombin deficiency is not recommended for adult patients with stroke. These assays can be considered in certain unusual situations, such as young patients with stroke *(Level 2)* or patients who also have a personal or family history of venous thrombosis or paradoxical emboli *(Level 3)*. These assays are appropriate for patients with cerebral venous thrombosis. *Level 3*

5. Heparin-induced thrombocytopenia should be considered in any patient who experiences ischemic stroke or cerebral venous sinus thrombosis while exposed to heparin or within 30 days of heparin exposure, with a decrease in platelet count to less than 50% of baseline.

subjects experienced stroke as compared to 0.6% of relatives without a deficiency. (P value not given except to state that the overall rate of arterial thrombosis [MI or stroke] was not significantly different between deficient subjects and nondeficient subjects).

Testing for hereditary deficiencies of protein C, protein S, or antithrombin is complicated by the fact that acquired deficiencies are much more common than hereditary deficiencies, as described in chapter 19. Acquired, transient decreases that disappear upon later retesting are commonly encountered in stroke patients.

In summary, hereditary antithrombin, protein C, and protein S deficiencies are uncommon in general populations and in persons with stroke. There is some evidence to suggest an association with stroke. However, it is difficult to know with certainty if individuals with these deficiencies have any increased risk of stroke, because of their low prevalence and the complexities of distinguishing acquired from hereditary deficiencies.

Heparin-Induced Thrombocytopenia

Heparin-induced thrombocytopenia should be considered in any patient who experiences ischemic stroke or cerebral venous thrombosis while exposed to heparin or within 30 days of heparin exposure. Heparin-induced thrombocytopenia should be considered if the patient also has a drop in platelet count of more than 50% from baseline. Conclusions and recommendations regarding thrombophilia testing in neurovascular thrombosis can be found in Table 17-7.

Suggested Reading

General

McKenzie SB, Clare CN, Smith LA, Lee Sang JE. Laboratory test utilization in the diagnosis of hypercoagulability. *Clin Lab Sci.* 2000;13:215-221.

Olson JD. College of American Pathologists Consensus Conference XXXVI. Diagnostic Issues in Thrombophilia: introduction and general considerations. *Arch Pathol Lab Med.* 2002;126:1277-1280.

Somma J, Sussman II, Rand JH. An evaluation of thrombophilia screening in an urban tertiary care medical center: a "real world" experience. *Am J Clin Pathol.* 2006;126:120-127.

Van Cott EM, Laposata M, Prins MH. Laboratory evaluation of hypercoagulability with venous or arterial thrombosis. *Arch Pathol Lab Med.* 2002;126:1281-1295.

Venous Thromboembolism

De Stefano V, Leone G, Mastrangelo S, et al. Clinical manifestations and management of inherited thrombophilia: retrospective analysis and follow-up after diagnosis of 238 patients with congenital deficiency of antithrombin III, protein C, protein S. *Thromb Haemost.* 1994;72:352-358.

den Heijer M, Koster T, Blom HJ, et al. Hyperhomocysteinemia as a risk factor for deep-vein thrombosis. *N Engl J Med.* 1996;334(12):759-762.

Falcon CR, Cattaneo M, Panzeri D, Martinelli I, Mannucci PM. High prevalence of hyperhomocyst(e)inemia in patients with juvenile venous thrombosis. *Arterioscler Thromb.* 1994;14:1080-1083.

Finazzi G, Brancaccio V, Moia M, et al. Natural history and risk factors for thrombosis in 360 patients with antiphospholipid antibodies: a four-year prospective study from the Italian Registry. *Am J Med.* 1996; 100:530-536.

Kearon C, Kahn SR, Agnelli G, et al. Antithrombotic therapy for venous thromboembolic disease: American College of Chest Physicians Evidence-Based Clinical Practice Guidelines (8th Edition). *Chest.* 2008; 133(suppl 6):454S-545S.

Koster T, Rosendaal FR, Briët E, et al. Protein C deficiency in a controlled series of unselected outpatients: an infrequent but clear risk factor for venous thrombosis (Leiden Thrombophilia Study). *Blood.* 1995; 85:2756-2761.

Kuipers S, Schreijer AJ, Cannegieter SC, Buller HR, Rosendaal FR, Middeldorp S. Travel and venous thrombosis: a systematic review. *J Intern Med.* 2007;262(6):615-634

Kyrle PA, Minar E, Hirschl M, et al. High plasma levels of factor VIII and the risk of recurrent venous thromboembolism. *N Engl J Med.* 2000;343:457-462.

Middeldorp S, Henkens CMA, Koopman MMW, et al. The incidence of venous thromboembolism in family members of patients with factor V Leiden mutation and venous thrombosis. *Ann Int Med.* 1998;128:15-20.

O'Donnell J, Tuddenham EG, Manning R, Kemball-Cook G, Johnson D, Laffan M. High prevalence of elevated factor VIII levels in patients referred for thrombophilia screening: role of increased synthesis and relationship to the acute phase protein. *Thromb Haemost.* 1997;77:825-828.

Pabinger I, Brucker S, Kyrle PA, et al. Hereditary deficiency of antithrombin III, protein C and protein S: prevalence in patients with a history of venous thrombosis and criteria for rational patient screening. *Blood Coagul Fibrinolysis.* 1992;3:547-553.

Palareti G, Cosmi B. Predicting the risk of recurrence of venous thromboembolism. *Curr Opin Hematol.* 2004;11(3):192-197.

Palareti G, Cosmi B, Legnani C, et al; PROLONG Investigators. D-dimer testing to determine the duration of anticoagulation therapy. *N Engl J Med.* 2006;355(17):1780-1789. Erratum in: *N Engl J Med.* 2006;355(26):2797.

Ridker PM, Miletich JP, Stampfer MJ, Goldhaber SZ, Lindpainter K, Hennekens CH. Factor V Leiden and risks of recurrent idiopathic venous thromboembolism. *Circulation.* 1995;92:2800-2802.

Tait RC, Walker ID, Perry DJ, et al. Prevalence of antithrombin deficiency in the healthy population. *Br J Haematol.* 1994;87:106-112.

Tait RC, Walker ID, Reitsma PH, et al. Prevalence of protein C deficiency in the healthy population. *Thromb Haemost.* 1995;73:87-93.

Arterial Thrombosis

Arruda VR, Siquiera LH, Chiaparini LC, et al. Prevalence of prothrombin gene variant 20210 G-->A among patients with myocardial infarction. *Cardiovasc Res.* 1998;37:42-45.

Bonaa KH, Njolstad I, Ueland PM, et al. Homocysteine lowering and cardiovascular events after acute myocardial infarction. *N Engl J Med.* 2006;354:1578-1588.

Bostom AG, Cupples LA, Jenner JL, et al. Elevated plasma lipoprotein (a) and coronary heart disease in men aged 55 years and older: a prospective study. *JAMA.* 1996;276:544-548.

Bostom AG, Silbershatz H, Rosenberg IH, et al. Nonfasting plasma total homocysteine levels and all-cause and cardiovascular disease mortality in elderly Framingham men and women. *Arch Intern Med.* 1999;159:1077-1080.

Brattstrom L, Wilcken DEL, Ohrvik J, Brudin L. Common methylenetetrahydrofolate reductase gene mutation leads to hyperhomocysteinemia but not to vascular disease: the result of a meta-analysis. *Circulation.* 1998;98:2520-2526.

Cushman M, Rosendaal FR, Psaty BM, et al. Factor V Leiden is not a risk factor for arterial vascular disease in the elderly: results from the Cardiovascular Health Study. *Thromb Haemost.* 1998;79:912-915.

Dacosta A, Tardy-Poncet B, Isaaz K, et al. Prevalence of factor V Leiden (APCR) and other inherited thrombophilias in young patients with myocardial infarction and normal coronary arteries. *Heart*. 1998;80:338-340.

Danesh J, Whincup P, Walker M, et al. Low grade inflammation and coronary heart disease: prospective study and updated meta-analyses. *BMJ*. 2000;321:199-204.

den Heijer M, Williams HPJ, Blom HJ, et al. Homocysteine lowering by B vitamins and the secondary prevention of deep vein thrombosis and pulmonary embolism: a randomized, placebo-controlled, double-blind trial. *Blood*. 2007;109:139-144.

Egerton WB, Silberberg JMB, Crooks R, Ray C, Xie L, Dudman N. Serial measures of plasma homocyst(e)ine after acute myocardial infarction. *Am J Cardiol*. 1996;77:759-761.

Folsom AR, Wu KK, Rosamond WD, Sharrett AR, Chambless LE. Prospective study of hemostatic factors and incidence of coronary heart disease: the Atherosclerosis Risk in Communities (ARIC) Study. *Circulation*. 1997;96:1102-1108.

Knekt P, Alfthan G, Aromaa A, et al. Homocysteine and major coronary events: a prospective population study amongst women. *J Intern Med*. 2001;249:461-465.

Kronenberg F, Kronenberg MF, Kiechl S, et al. Role of lipoprotein (a) and apolipoprotein (a) phenotype in atherogenesis: prospective results from the Bruneck Study. *Circulation*. 1999;100:1154-1160.

Ridker PM, Hennekens CH, Miletich JP. G20210A mutation in prothrombin gene and risk of myocardial infarction, stroke, and venous thromboembolism in a large cohort of US men. *Circulation*. 1999;99:999-1004.

Ridker PM, Manson JE, Buring JE, Shih J, Matias M, Hennekens CH. Homocysteine and risk of cardiovascular disease among postmenopausal women. *JAMA*. 1999;281:1817-1821.

Sakata T, Kario K, Katayama Y, Matsuyama T, Kato H, Miyata T. Studies on congenital protein C deficiency in Japanese: prevalence, genetic analysis, and relevance to the onset of arterial occlusive diseases. *Semin Thromb Hemost*. 2000;26:11-16.

Schnyder G, Roffi M, Pin R, et al. Decreased rate of coronary restenosis after lowering of plasma homocysteine levels. *N Engl J Med*. 2001;345(22):1593-1600.

Seed M, Ayres KL, Humphries SE, Miller GJ. Lipoprotein (a) as a predictor of myocardial infarction in middle-aged men. *Am J Med*. 2001;110:22-27.

Vaarala O, Manttari M, Manninen V, et al. Anti-cardiolipin antibodies and risk of myocardial infarction in a prospective cohort of middle-aged men. *Circulation*. 1995;91:23-27.

Neurovascular Thrombosis

Ahmed E, Stegmayr B, Trifunovic J, Weinehall L, Hallmans G, Lefvert AK. Anticardiolipin antibodies are not an independent risk factor for stroke: an incident case-referent study nested within the MONICA and Vasterbotten Cohort Project. *Stroke*. 2000;31:1289-1293.

Alfthan G, Pekkanen J, Jauhiainen M, et al. Relation of serum homocysteine and lipoprotein (a) concentrations to atherosclerotic disease in a prospective Finnish population based study. *Atherosclerosis*. 1994;106:9-19.

Gaustadnes M, Rudiger N, Moller J, et al. Thrombophilic predisposition in stroke and venous thromboembolism in Danish patients. *Blood Coagul Fibrinolysis*. 1999;10:251-259.

Ginsburg KS, Liang MH, Newcomer L, et al. Anticardiolipin antibodies and the risk for ischemic stroke and venous thrombosis. *Ann Intern Med*. 1992;117:997-1002.

Green D. Thrombophilia and stroke. *Top Stroke Rehabil*. 2003;10(3):21-33.

Knuiman MW, Folsom AR, Chambless LE, Liao D, Wu KK. Association of hemostatic variables with MRI-detected cerebral abnormalities: the atherosclerosis risk in communities study. *Neuroepidemiology*. 2001;20:96-104.

Levine SR. Hypercoagulable states and stroke: a selective review. *CNS Spectr*. 2005;10(7):567-578.

Martinelli I, Sacchi E, Landi G, Taioli E, Duca F, Mannucci PM. High risk of cerebral-vein thrombosis in carriers of a prothrombin-gene mutation and in users of oral contraceptives. *N Engl J Med*. 1998;338:1793-1797.

Reuner KH, Ruf A, Grau A, et al. Prothrombin gene G20210-->A transition is a risk factor for cerebral venous thrombosis. *Stroke*. 1998;29:1765-1769.

Ridker PM, Stampfer MJ, Hennekens CH. Plasma concentration of lipoprotein (a) and the risk of future stroke. *JAMA*. 1995;273:1269-1273.

Unique Issues of Thrombophilia in Women and Children

Charissa A. Bailey, MD

John D. Olson, MD, PhD

Introduction

Hemostasis is a complex mechanism that involves a precarious balance among coagulation, fibrinolysis, and thrombosis. There are several conditions unique to women that are responsible for shifting this balance. Combination oral contraceptive use, oral hormone replacement therapy, pregnancy, and the puerperium are all conditions in women that predispose to thrombosis and that warrant special consideration in the setting of thrombophilia. In addition, thrombophilic risk factors in the mother can present peril to the developing fetus. Infants and children also present some unique considerations regarding the occurrence of thrombosis, predisposing factors, and the prophylaxis and management of such events.

This chapter addresses the role of laboratory testing for both acquired and inherited thrombophilic risk factors in women and children.

Pregnancy and Hormonal Therapy

Oral Contraceptives and Risk of Venous Thromboembolism

The use of oral estrogens in women has been well recognized as a risk factor for venous thrombogenesis. Multiple studies have documented a 3- to 6-fold increase in risk of venous thromboembolism (VTE) with oral contraceptive use in otherwise healthy premenopausal women. This risk appears to remain elevated despite the use of newer, low-dose hormone contraceptives. Several pathogenic mechanisms have been proposed to explain this shift in the hemostatic balance. An increase in multiple prothrombotic factors, including factors VII, VIII, and X, as well as an increase in fibrinogen and prothrombin has been detected in oral contraceptive users. Oral contraceptive use is also a known cause of acquired activated protein C (APC) resistance that can partially be explained by an observed decrease in protein S.

The type of oral contraceptive also appears to affect the degree of thrombosis risk. Third-generation oral contraceptives containing the progestins desogestrel and gestodene have consistently been shown to incur a higher risk of VTE than second-generation contraceptives containing levonorgestrel and norgestrel. The article published by Brenner et al (2002) cited 16 studies that compared third-generation oral contraceptives to second-generation contraceptives. Only three of these studies found no difference between the two groups, whereas 13 studies demonstrated a 1.4- to 4-fold increase in risk. This difference can possibly be explained by an observed increase in acquired APC resistance that has been more pronounced in third-generation than second-generation oral contraceptive users.

The presence of an inherited thrombophilic defect in women using oral contraceptives appears to have a supra-additive effect on VTE risk. Factor V Leiden heterozygotes who are taking oral contraceptives have a 35-fold increase in risk for VTE as compared with nonusers with wild-type factor V, whereas factor V Leiden homozygotes have an even greater increase in risk. The presence of the prothrombin G20210A mutation among contraceptive users causes a similar increase in the risk of VTE, including cerebral

The content of this chapter has been extracted in large part from the work of Brenner et al (Diagnostic Studies for Thrombophilia in Women on Hormonal Therapy and During Pregnancy, and in Children) from the College of American Pathologists Consensus Conference XXXVI, "Diagnostic Issues in Thrombophilia," the results of which were published in the November 2002 issue of *Archives of Pathology & Laboratory Medicine*. The text has been reorganized and updated, but the content is predominantly the product of the consensus conference. The reader is referred to this citation for additional information and references. Modified with permission from *Archives of Pathology & Laboratory Medicine*. Copyright 2002. College of American Pathologists.

Case Study

A 32-year-old Hispanic woman presented to the emergency department with right lower extremity swelling and pain, which had been worsening over the past 3 days. She was 6 days status post spontaneous vaginal delivery of a baby boy. Her past medical history included a right lower-extremity deep vein thrombosis (DVT) 7 years ago after surgery for a ruptured appendix. In addition, she previously had three second-trimester spontaneous abortions. Her D-dimer was 2750 ng/mL.

A Doppler ultrasound examination demonstrated an obstructive thrombus extending into the right common femoral vein. Baseline laboratory values included:

Test	Value	Reference Range
WBC (per µL)	9000	4000-10,000
Hematocrit (%)	42	36-44
Platelet count (per µL)	90,000	150,000-450,000
PT (INR)	1.1	0.8-1.2
aPTT (s)	78	25-36

On further testing, the mixing study (see chapters 6 and 10) failed to correct the aPTT, indicating the presence of an inhibitor. The dilute Russell viper venom time demonstrated that the inhibition was phospholipid dependent, confirming the presence of a lupus anticoagulant (see chapters 9, 19, and 22). The findings of recurrent fetal loss (RFL), DVT, thrombocytopenia, and lupus anticoagulant are consistent with the antiphospholipid syndrome. Management of this patient to prevent thrombosis and to improve the probability of successful future pregnancies is discussed here and in chapter 17.

vein thrombosis. Likewise, in the same population, deficiencies in antithrombin, protein S, and protein C have also been associated with an increased risk of VTE.

The utility of screening for thrombophilic defects among oral contraceptive users has not been established. The highest risk for users of oral contraceptives has been observed in carriers of factor V Leiden. Despite this, the overall incidence of fatal VTE in factor V Leiden heterozygotes taking oral contraceptives remains low, at approximately 5 in 100,000, compared to an overall incidence of fatal VTE in wild-type oral contraceptive users of 0.7 in 100,000. The cost effectiveness of screening all oral contraceptive users for factor V

Leiden remains questionable. The actual cost of an individual test for factor V Leiden varies; cost-benefit analysis has suggested that screening would be cost effective if the cost per test were less than US $9.00 (in the year 1996). In the future, tests such as APC-resistance assays and a protein C global assay might be suitable for screening populations with an especially high factor V Leiden carrier rate. There is significant evidence to support the practice of testing oral contraceptive users with a prior history of VTE and users who have first-degree relatives with VTE or a known thrombophilic risk factor.

Hormone Replacement Therapy and Venous Thromboembolism

Data from randomized trials and observational data have consistently demonstrated an increase in the risk of venous thromboembolism among postmenopausal women using oral hormone replacement therapy (HRT). Oral HRT typically includes an estrogen with or without a progestogen, such as medroxyprogesterone, but at lower levels of ethinylestradiol than used in oral contraceptive formulations. The risk conferred by HRT in this population appears to be slightly less than that observed in users of oral contraceptives; the thrombosis risk is highest in the first year of therapy and decreases with time. However, the absolute risk increase actually may be higher because HRT users tend to be of an older age, which poses an additional independent risk factor for thrombosis. However, as in oral contraceptive users, the presence of factor V Leiden in users of HRT further increases their risk of VTE. The combination of HRT use and the presence of factor V Leiden appear to have an additive effect on VTE risk. More studies are needed to examine the role of other thrombophilic defects and their influence on thrombosis in this population. Recent case-controlled trials have shown that, in contrast to orally administered estrogen, transdermal estrogen may confer less risk of thrombosis in women with other thrombophilic risk factors. On the basis of these preliminary studies, the route of administration may be important, but these findings await further confirmation. Recommendations regarding oral contraceptives and hormonal therapy are presented in Table 18-1.

Women with breast cancers that express hormone receptors are often treated with selective estrogen receptor modulators (SERMs), such as tamoxifen or raloxifene. Treatments with these medications are associated with a 1.5- to 2.3-fold increase in the incidence of VTE, without significant increases in myocardial infarction or stroke. The thrombophilic effect is seen only while the patients are taking the medication and is not associated with increases in VTE after the medications are discontinued. Similar rates have been reported in women with high risk for breast cancer who are taking SERMs for prevention. It is interesting to note that, in this latter group, the presence of factor V Leiden or prothrombin G20210A did not increase the apparent risk.

Pregnancy-Related Venous Thromboembolism

A normal pregnancy involves a dramatic disruption of hemostasis, resulting in a progressive prothrombotic state that reaches its peak in the early puerperium. This situation may result from an increase in several clotting factors such as factor VIII and fibrinogen, increased von Willebrand factor, an acquired APC resistance, a decrease in protein S, and decreased fibrinolysis. During gestation the risk for VTE is increased 3- to 4-fold, with an estimated incidence of 1.72 per 1000 deliveries and 1.1 maternal deaths per 100,000 deliveries. Although uncommon, VTE is a significant source of morbidity and mortality in pregnancy, and pulmonary embolism is the most common cause of maternal death.

The risk of perinatal thrombosis may be associated with increases in maternal age and obesity; peripartum thrombosis has been shown to be more common with cesarean delivery as compared with vaginal delivery and even more prevalent with emergency cesarean deliveries. The presence of a congenital thrombophilic risk factor

Table 18-1. Recommendation Regarding Hormone Therapy

> ❑ Testing for thrombophilia is recommended in women who experience VTE as cerebral venous thrombosis during oral contraceptive therapy or HRT. *Level 1**

* Criteria Used for the Level of Evidence Assigned to Recommendations

Level 1: The recommendation is based on one or more well-designed prospective studies or two or more well-designed retrospective studies

Level 2: The recommendation is based on retrospective studies or multiple anecdotal studies that reach consensus

Level 3: The recommendation is based on isolated anecdotal studies and/or the consensus of expert practitioners

further increases the risk of thromboembolic events during pregnancy. In the majority of women who present with a pregnancy-related VTE, the presence of another thrombophilic risk factor can be detected. The most common mutation detected in pregnant white women with VTE is factor V Leiden, found in 30% to 60%. The absolute risk of gestational VTE in factor V Leiden heterozygotes is 1 in 500. The prothrombin G20210A mutation and the presence of antiphospholipid antibodies each account for 10% to 20% of cases. The absolute VTE risk during pregnancy for prothrombin mutation carriers is approximately 1 in 200. The less common risk factors, such as protein C, protein S, and antithrombin deficiencies, together account for another 10% to 20% of cases. The absolute risk for women with more than one mutation is significantly greater than would be expected from merely adding the individual risks; women who are carriers for both prothrombin G20210A and factor V Leiden have an absolute risk of 1 in 20 for gestational VTE.

Thrombophilia and Recurrent Fetal Loss

Recurrent fetal loss (RFL), traditionally defined as three or more spontaneous abortions, is a common and yet often unexplained problem of childless couples. During their reproductive years, 1% to 2% of women experience three or more pregnancy losses, and 5% of women experience two or more losses. Although many factors have been implicated, such as chromosomal disruptions, endocrine or autoimmune disease, infectious agents, and anatomic causes, many women who experience such losses are never given a diagnosis.

There is a well-established association between certain acquired thrombophilic risk factors, such as the antiphospholipid syndrome, and RFL. Several different antiphospholipid antibodies have recently been detected in women with unexplained early fetal loss as a retrospective independent risk factor, including antiphosphatidylethanolamine, anti-annexin V IgG antibodies, and the lupus anticoagulant. These four markers were found to be associated with a significant risk of fetal loss in subsequent pregnancies despite low-dose aspirin treatment.

More recently, multiple observations have suggested an association between RFL and inherited thrombophilia. Brenner et al (2002) reviewed various articles exploring this link. One study found that among women with protein C, protein S, or antithrombin deficiency, 22% of pregnancies were spontaneously aborted in comparison with only 11% of pregnancies in the control group. A similar study followed pregnant women with dysfibrinogenemia associated with thrombosis. In this population, 39% of pregnancies ended in abortion and 9% resulted in intrauterine fetal death. In addition, factor V Leiden and the prothrombin G20210A mutations are associated with an approximately 3-fold increase in the risk of late fetal loss. Out of 67 women with late fetal loss, 11 (16%) had either the factor V Leiden or the prothrombin mutation versus 13 (6%) of the 232 control subjects. The relative risk of late fetal loss in carriers of the factor V Leiden and prothrombin mutations was 3.2 and 3.3, respectively. Factor V Leiden has been detected in a significant proportion of white women with RFL, with a 2.3-fold increase in the prevalence of factor V Leiden in women with RFL. Because of the high prevalence of this mutation in women with RFL, several studies suggest evaluation for its presence may be warranted among white women with unexplained RFL. It should be emphasized, however, that some other reports have not demonstrated this strong association between the factor V Leiden mutation and RFL.

In populations in which homozygosity for factor V Leiden is prevalent, such as populations of Northern European descent, a significant association between this condition and RFL can also be demonstrated. The risk for RFL is twice as high in homozygotes as in heterozygotes with the factor V Leiden mutation. Siblings of thrombophilic women with the factor V Leiden mutation are also at an increased risk for RFL. Women with thrombophilia appear to be at a greater risk for fetal loss in the second and third trimesters than in the first trimester and experience an increased percentage of losses at later stages of gestation. However, there is also an association between factor V Leiden mutation and recurrent first-trimester pregnancy loss.

Activated protein C resistance (APCr) in the absence of the factor V Leiden mutation has also been associated with pregnancy loss, possibly as a result of other mutations that confer APCr. A progressive acquired resistance to APC is observed throughout a healthy pregnancy as part of the shift toward hypercoagulability, which includes increases in factor VIII, fibrinogen, and factor V, and decreases in protein S levels. This transient physiologic development of APCr, documented even in women with normal factor V genotypes,

has an additive effect on APCr in women with a factor V Leiden mutation.

Both acquired and genetic conditions can cause hyperhomocysteinemia. The homozygous MTHFR T/T genotype, resulting from a 677C→T mutation in the 5,10-methylenetetrahydrofolate reductase (MTHFR) gene, results in higher total plasma homocysteine level. Despite its association with hyperhomocysteinemia, the role of this mutation in RFL has not been established. Although contradictory studies have been published, the majority of recent studies have found that the MTHFR T/T genotype is not a significant risk factor for early RFL. In contrast, multiple studies have documented a significant associated between elevated fasting and post–methionine-loading homocysteine levels and early fetal loss, implicating other effects of hyperhomocysteinemia. Although still being actively studied, there is animal and population evidence that has associated MTHFR mutations with neural tube defects. Such effects may also contribute to fetal loss in a manner unrelated to thrombosis and may provide an alternative justification for testing for MTHFR mutations in women with RFL.

As one would expect, women with combined thrombophilic defects are at even higher risk for fetal loss. It appears that multiple factors are necessary for the development of thromboembolic complications in pregnancy. Hereditary thrombophilic defects may be detected in up to 20% of women with normal pregnancies, in contrast to detection in up to 50% of women with recurrent venous thromboembolism. These findings provide support for the utilization of thrombophilic testing only in the setting of previous VTE or RFL.

Thrombophilia and Late Gestational Vascular Complications

An impairment of uteroplacental perfusion is suspected to play a role in the pathogenesis of several pregnancy-related complications, including placental abruption, intrauterine fetal death, fetal growth retardation, and preeclampsia. The exact etiology of preeclampsia has not been elucidated; however, it is likely that placental insufficiency and hypoxia trigger the release of numerous chemical factors that cause endothelial damage, blood coagulation, and various metabolic changes. In light of this mechanism, a relationship between thrombophilic defects and preeclampsia has been suspected. Several recent reports have found an association between APC resistance due to the factor V Leiden mutation and early-onset severe preeclampsia. A study by Dizon-Townson et al (1996) demonstrated that 8.9% of women with severe preeclampsia were factor V Leiden carriers as compared with only 4.2% of women in a control group. Similarly, in a study by Nagy et al (1998), the factor V Leiden mutation was more than twice as prevalent in women with preeclampsia than in women without preeclampsia. An association also has been found between APCr and placental abruption; one study by Wiener-Megnagi et al (1998) found that APCr was more than four times as prevalent among patients with placental abruption than among control patients.

An association between hyperhomocysteinemia and vascular complications has been implicated. Hyperhomocysteinemia was detected in 26% of women presenting with placental abruption, 11% of women presenting with intrauterine fetal death, and 38% of women delivering babies with intrauterine growth retardation. In comparison, hyperhomocysteinemia was found in only 2% to 3% in the general control population. Likewise, hyperhomocysteinemia was documented in 31% of women with previous placental infarcts or abruption as compared with 9% of control subjects. In a large study by Vollset et al (2000), 14,492 gestations were evaluated in 5883 women with hyperhomocysteinemia. An increased risk for preeclampsia, stillbirth, early labor, and placental abruption was confirmed.

Treatment of Pregnant Women with Thrombophilias

The association between thrombosis and both gestational VTE and vascular gestational complications suggests that early intervention with antithrombotic medications could significantly improve maternal and fetal outcomes. Studies suggest that low-molecular-weight heparin may be a safe and effective therapy for pregnant women with inherited thrombophilias.

In a collaborative study by Gris et al (1999), the use of low-molecular-weight heparin was associated with successful pregnancies in 89% of women with a history of RFL as compared with only a 20% success in historical controls with no anticoagulation. The same study found that use of low-molecular-weight heparin was associated with successful outcomes in 28 of 28 gestations in women with history of previous preeclampsia. In women with inherited thrombophilia, 46 (75%) of 61 pregnancies treated with low-molecular-weight heparin

Table 18-2. Conclusions and Recommendations Regarding Pregnancy

Conclusions
- The risk of VTE during gestation increases 3- to 4-fold. *Level 1*
- Thrombophilia can be identified in the majority of women with gestational VTE. *Level 1*
- Thrombophilia is associated with unexplained pregnancy loss (especially second and third trimester). *Level 2*
- Other gestational vascular complications (preeclampsia, intrauterine growth retardation, placental abruption) are associated with thrombophilia. *Level 3*
- Combined thrombophilic conditions increase the risk for gestational complications. *Level 2*
- Patients with prior VTE during pregnancy who have a thrombophilic state are at high risk for recurrence during subsequent pregnancy, may receive antithrombotic prophylaxis during gestation, and should receive antithrombotic prophylaxis in the postpartum period. *Level 2*
- Prevention of pregnancy loss in women with thrombophilia by antithrombotic therapy is currently being evaluated in prospective randomized trials.

Recommendations
- Women with VTE during pregnancy or in the postpartum period should be evaluated for thrombophilia. *Level 1*
- Women with pregnancy loss that is either recurrent or late in the pregnancy (second and third trimester) should be evaluated for thrombophilia. *Level 1*
- Whether women with other gestational vascular complications should be evaluated for thrombophilia is controversial. *Level 3*
- Testing results for APC resistance and protein S obtained during pregnancy or the postpartum period should be interpreted with caution in view of physiologic changes.

resulted in a live birth as compared with a success rate of only 20% in these same 50 women in prior gestations without antithrombotic therapy. Although low-molecular-weight heparin appears to be successful at preventing pregnancy-associated thrombophilic complications, the optimal dosage of low-molecular-weight heparin has yet to be established by randomized trials. A recent study by Brenner et al (2005) has shown that low-molecular-weight heparin at doses of 40 mg/day or 80 mg/day has low risk of complication and can reduce abortion rates in women with RFL. Conclusions and recommendations regarding treatment of thrombophilia in pregnancy are presented in Table 18-2. Vitamin K antagonists like warfarin are not used in this setting because of the teratogenic complications.

Thrombophilia in Neonates, Infants, and Children

Description of the Problem

Pediatricians frequently remind other practitioners that children are not "little adults." This is particularly true when considering thrombosis and thrombophilia in infants and children. Although thrombosis is not a common problem in the pediatric population, both venous and arterial thrombi have been diagnosed in infancy and childhood more frequently in recent years. This is probably secondary to improvements in diagnostic capabilities and improvements in supportive care, resulting in prolonged survival in premature infants and children with serious medical conditions that predispose to thrombosis. In addition, an increase in the use of central venous catheters has significantly contributed to an increased incidence of VTE in children. The number of studies addressing prophylaxis and treatment of VTE specifically in children has been limited; as a result, pediatric

guidelines have been derived from those based on adults. Recently, more data have been accumulating that address some of the differences in etiology, pathophysiology, and complications of thromboembolism in children.

Neonates, particularly those who are premature, are at risk for thrombosis, in part because the liver is not fully developed and naturally occurring inhibitors of coagulation are present in reduced concentration as compared with adults. As discussed below, venous thrombosis in both neonates and children does not occur in the same anatomic locations as in adults and has a high association with thrombophilic risk factors. Testing for thrombophilic risk factors in the child described in the case study would be indicated, as would be the testing of the child's siblings. In addition, as pointed out previously in this chapter, the development of late gestational vascular complications of pregnancy (ie, preeclampsia) is also associated with risk factors for thrombophilia.

An annual incidence of symptomatic thromboembolism has been reported by Andrew (1995) to be 0.07 per 10,000 children in the general population, whereas an incidence of 5.3 per 10,000 has been reported in children in an inpatient setting, and an incidence of 2.4 per 1000 of newborns admitted to intensive care units. Children appear to be at a lower risk for primary VTE than adults throughout the majority of childhood, possibly secondary to age-associated changes in regulation of hemostasis. There are ages, however, that are exceptions to this principle. In childhood, thromboembolic events tend to occur in an almost bimodal distribution. The first peak occurs during the first year of life, possibly because of the lower concentrations of antithrombin, heparin cofactor II, protein S, and protein C found in neonates, in addition to a decreased capacity for fibrinolysis. The second peak occurs during adolescence and puberty, again in association with reduced fibrinolysis.

In infancy and childhood, secondary factors play a significant role in the etiology of VTE. Numerous environmental effects and generalized disease states cause an increase in thrombin production and a corresponding hypercoagulability (Table 18-3). Such acquired conditions include birth asphyxia, neonatal infections, fetal diabetes, the use of central intravascular lines, trauma or surgery, dehydration, malignant diseases, renal diseases, autoimmune diseases, and the use of oral contraceptives in adolescent girls. Studies have shown that the use of central venous lines and

Case Study

After an uncomplicated pregnancy, a 34-year-old Hispanic woman, the mother of two healthy children, developed signs of preeclampsia in the 33rd week of her pregnancy. She delivered a boy prematurely, with Apgar scores of 6 and 9. In the neonatal intensive care unit, the child was doing well for the first 48 hours, after which he developed a seizure, somnolence, and irritability. The child had a fever (39.5°C) and a normocytic anemia for age (hemoglobin, 10 g/dL). A magnetic resonance imaging (MRI) scan of the head demonstrated a superior sagittal sinus thrombosis. He was treated with low-molecular-weight heparin followed by warfarin therapy. His symptoms resolved, and at 6 months of age he had no identifiable sequelae.

intra-arterial catheters are the most important causes of thrombosis in children.

Various inherited thrombophilic defects, such as the factor V Leiden mutation and the prothrombin G20210A mutation, as well as antithrombin, protein C, and protein S deficiencies, have been recognized in children as well as adults as predisposing to thrombosis. More recently, metabolic diseases, such as hyperhomocysteinemia and increased concentrations of lipoprotein (a), have also been shown to increase the risk of venous and arterial thrombosis. As in adults, the combination of genetic prothrombotic risk factors and acquired environmental or clinical conditions greatly increases the risk of thrombosis in children.

Although rare, the homozygous deficiencies of protein C, protein S, and antithrombin all manifest in childhood, most frequently in the neonatal period. They often present as purpura fulminans. The diagnosis of the inherited deficiencies of these factors is difficult in neonates because the plasma levels of these factors, especially protein C and protein S, are significantly lower in the normal neonate than in an adult and rise as the liver function matures.

The anatomic distribution of VTEs in children tends to vary significantly from that observed in adults. The most common sites of thrombus formation in neonates are the renal veins, vena cava, and other vessels where occlusion produces a thromboembolic infarct. The high rate of catheter-related thrombosis in pediatric patients results in an increased frequency of thrombosis in the upper venous system, in contrast to adults. Central venous lines can lead to thrombogenesis near the catheter implantation site, especially when prothrombotic risk factors are involved. Other sites of childhood thromboembolism include cerebral vein thrombosis and portal and mesenteric vein thrombosis. Arterial vascular occlusions, which are commonly catheter related, result in thrombosis of the aorta, femoral artery, subclavian artery, and other vessels.

Purpura fulminans, a rare condition occurring in children, can be a life-threatening complication of thrombophilia, resulting from small vessel thrombosis with subsequent tissue infarction. This condition usually presents with the sudden onset of confluent and symmetric tender ecchymoses that enlarge rapidly, often with the development of hemorrhagic bullae and central necrosis. Hemorrhagic necrosis of the adrenal glands (Waterhouse-Friderichsen syndrome) and renal cortical necrosis may also occur. Although usually

Table 18-3. Acquired Risk Factors for Pediatric Thromboembolism

Perinatal Diseases	Birth asphyxia Respiratory distress syndrome Infants of diabetic mothers Neonatal infections Necrotizing enterocolitis Dehydration Congenital nephrotic syndrome Polycythemia
Medical Interventions	Central lines Surgery Renal transplantation Immobilization Plaster casts Extracorporeal membrane oxygenation
Acute Diseases	Trauma Sepsis Dehydration Acute rheumatic diseases Nephrotic syndrome Acute lymphoblastic leukemia
Chronic Diseases	Malignancies Renal diseases Cardiac malformations Chronic rheumatic diseases
Drugs	L-asparaginase Prednisone Coagulation factor concentrates Heparins Antifibrinolytic agents Oral contraceptives

associated with disseminated intravascular coagulation or bacterial septicemia, purpura fulminans in neonates is very suggestive of a thrombophilic defect. Congenital deficiencies of protein C or protein S and the presence of a homozygous or a heterozygous factor V Leiden mutation are all known causes of purpura fulminans.

Testing for Thrombophilia in Children

Children who are carriers of inherited thrombophilias are at increased risk of thrombosis or recurrence after a prior thrombosis, and they may benefit from antithrombotic prophylaxis during high-risk situations. As a result, screening for genetic risk factors in children and infants who present with arterial and venous thrombosis is indicated, especially in the setting of spontaneous or recurrent thrombosis. Among children with recurrent vascular occlusion after a first episode of spontaneous VTE, there is a subgroup of pediatric patients suffering from combined prothrombotic risk factors who are at an even higher risk for recurrent thrombosis. In selected patient populations, evaluation for multiple risk factors is justified. In addition, evaluation of asymptomatic siblings and first-degree relatives of patients presenting with early thrombosis is also recommended.

In such a setting, suitable assays for APC resistance, protein C activity, free and total protein S antigen, antithrombin activity, fibrinogen concentration, plasminogen activity, factor VIII, lipoprotein (a) (Lp(a)), and fasting homocysteine levels should be considered. It is important to use age-appropriate reference ranges for many of these assays, especially protein C and protein S, because the levels of many of these proteins are different in children as compared with adults. In addition, DNA-based assays should be used where appropriate, such as testing for factor V Leiden mutation if indicated by APC-resistance assay, and possibly the *MTHFR C677T* genotype if an elevation of homocysteine is detected. A DNA-based assay is necessary for evaluation for the prothrombin G20210A mutation. Lp(a) assays should be interpreted cautiously in light of the lack of standardization and considerable variability of current commercially available tests. A recent study demonstrated that individual assays for the measurement of apolipoprotein (a) were biased toward higher or lower Lp(a) values, so that at a given Lp(a) threshold, some assays overestimate and some underestimate the thrombotic risk. For this reason and because Lp(a) serum levels are determined mainly by a genetically determined size polymorphism of its main protein component, some have recommended inclusion of the analysis of apolipoprotein (a) phenotypes.

Molecular (DNA or RNA) assays have several advantages, including unambiguous results and greater ease of standardization among laboratories. It is most important to note that molecular assays are not influenced by the presence of an acute thrombus, by anticoagulation, or by thrombolytic therapy. Therefore, molecular assays can be performed immediately at the onset of a thrombotic event. In contrast, protein C, protein S, and antithrombin plasma samples ideally should be obtained 3 to 6 months or more after the thrombotic episode to reduce the interference of transient thrombus-related abnormalities. If the patient is taking oral anticoagulants, the results of plasma-based assays can be affected; therefore, samples

Table 18-4. Conclusions and Recommendations Regarding Pediatrics

Conclusions

❏ Thrombophilia is commonly found in children (particularly in infants) with VTE or stroke. *Level 1*

❏ The association of multiple thrombophilias greatly increases the risk of thrombosis and/or recurrence of thrombosis in infants and children, as it does in adults. *Level 2*

❏ The distribution of prothrombotic risk factors varies with respect to the ethnic background and the number of patients/controls investigated. *Level 3*

Recommendations

❏ Testing for thrombophilia in children with venous or arterial thrombosis is recommended. The etiology and prevalence of thrombophilia differ when comparing children and adults. *Level 1*

❏ Age-specific reference ranges should be used to interpret the results of thrombophilia testing in the pediatric and neonatal age groups.

❏ Routine evaluation for thrombophilia for asymptomatic children of adult probands with inherited thrombophilia may be delayed until puberty. *Level 3*

❏ Evaluation for thrombophilia for the siblings of probands with early symptomatic thromboembolism is recommended. *Level 3*

should be freshly drawn at least 30 days after withdrawal of therapy.

In addition to inherited defects, all children with thrombophilia should be tested for antiphospholipid or anticardiolipin antibodies and for lupus anticoagulants. In symptomatic children with a strong family history of thrombosis but negative results for the more prevalent thrombophilias mentioned above, testing for rare thrombophilic defects, such as dysfibrinogenemia, hypoplasminogenemia, dysplasminogenemia, heparin cofactor II deficiency, and increased levels of histidine-rich glycoprotein, may be considered.

In the setting of disseminated intravascular coagulation, possibly complicated by purpura fulminans, a functional fibrinogen assay, functional antithrombin assay, and D-dimer assay should be obtained in addition to an activated partial thromboplastin time (aPTT), prothrombin time (PT), and platelet count. In the absence of laboratory evidence of accelerated fibrinolysis, purpura fulminans is unlikely. Testing for protein C, protein S, and the factor V Leiden mutation should be considered in neonates presenting with purpura fulminans.

For all plasma-based assays, an abnormality should only be documented if the result is outside of the age-dependent normal reference range. In order to make a diagnosis of hereditary thrombophilia, confirmation of a suspected defect is necessary by obtaining a second sample 3 to 6 months after the first sample (in the absence of oral anticoagulation). Identification of a causative gene mutation is also helpful for confirmation, but this may only be possible for factor V Leiden, prothrombin G20210A, and *MTHFR* mutations. Routine laboratory testing for genetic mutations of protein C, protein S, antithrombin, and plasminogen is not widely available. The conclusions and recommendations regarding thrombophilia testing in infants and children are presented in Table 18-4. In a recent review, Kenet and Nowak-Göttl (2006) have recommended that, in the case of perinatal thrombosis, thrombophilia assessment should be performed in the affected child, the siblings, and the mother.

Suggested Reading

General

Brenner BR, Nowak-Göttl U, Kosch A, Manco-Johnson M, Laposata M. Diagnostic studies for thrombophilia in women on hormonal therapy and during pregnancy, and in children. *Arch Pathol Lab Med.* 2002; 126:1296-1303.

Lindhoff-Last E, Luxembourg B. Evidence-based indications for thrombophilia screening. *Vasa.* 2008;37(1): 19-30.

Martinelli I. Thromboembolism in women. *Semin Thromb Hemost.* 2006;32:709-715.

Oral Contraceptive and Hormone Replacement Therapy

Abramson N, Costantino JP, Garber JE, Berliner N, Wickerham DL, Wolmark N. Effect of factor V Leiden and prothrombin G20210A mutations on thromboembolic risk in the national surgical adjuvant breast and bowel project breast cancer prevention trial. *J Natl Cancer Inst.* 2006;98:904-910.

Canonico M, Oger E, Conard J, et al; EStrogen and THromboEmbolism Risk (ESTHER) Study Group. Obesity and the risk of venous thromboembolism among postmenopausal women: differential impact of hormone therapy by route of estrogen administration. *J Thromb Haemost.* 2006;4:1259-1265.

Vandenbroucke JP, van der Meer FJ, Helmerhorst FM, Rosendaal FR. Factor V Leiden: should we screen oral contraceptive users and pregnant women? *BMJ.* 1996;313:1127-1130.

Pregnancy and Venous Thromboembolism

James AH, Jamison MG, Brancazio LR, Myers ER. Venous thromboembolism during pregnancy and the postpartum period: incidence, risk factors, and mortality. *Am J Obstet Gynecol.* 2006;194:1311-1315.

Thrombophilia and Recurrent Fetal Loss

Brenner B, Bar J, Ellis M, Yarom I, Yohai D, Samueloff A; Live-Enox Investigators. Effects of enoxaparin on late pregnancy complications and neonatal outcome in women with recurrent pregnancy loss and thrombophilia: results from the Live-Enox study. *Fertil Steril.* 2005;84(3):770-773.

de Vries JIP, Dekker GA, Huijgens PC, Jakobs C, Blomberg BM, van Geijn HP. Hyperhomocysteinaemia and protein S deficiency in complicated pregnancies. *Br J Obstet Gynaecol.* 1997;104:1248-1254.

Gris JC, Quéré I, Monpeyroux F, et al. Case-control study of the frequency of thrombophilic disorders in couples with late foetal loss and no thrombotic antecedent: the Nimes Obstetricians and Haematologists Study5 (NOHA5). *Thromb Haemost.* 1999;81:891-899.

Nelen WL, Blom HJ, Steegers EA, den Heijer M, Eskes TK. Hyperhomocysteinemia and recurrent early pregnancy loss: a meta-analysis. *Fertil Steril.* 2000; 74:1196-1199.

Northrup H, Volcik K. Spina bifida and other neural tube defects [review]. *Curr Probl Pediatr.* 2000;30:313-332.

Thrombophilia and Late Gestational Vascular Complications

Cadroy Y, Grandjean H, Pichon J, Desprats R, Berrebi A, Boneu B. Evaluation of six markers of haemostatic system in normal pregnancy and pregnancy complicated by hypertension or pre-eclampsia. *Br J Obstet Gynaecol.* 1993;100:416-420.

Dizon-Townson DS, Nelson LM, Easton K, Ward K. The factor V Leiden mutation may predispose women to severe preeclampsia. *Am J Obstet Gynecol.* 1996; 175:902-905.

Goddijn-Wessel TA, Wouters MG, van de Molen EF, et al. Hyperhomocysteinemia: a risk factor for placental abruption or infarction. *Eur J Obstet Gynecol Reprod Biol.* 1996;66:23-29.

Nagy B, Toth T, Rigo J Jr, Karadi I, Romics L, Papp Z. Detection of factor V Leiden mutation in severe pre-eclamptic Hungarian women. *Clin Genet.* 1998;53: 478-481.

Vollset SE, Refsum H, Irgens LM, et al. Plasma total homocysteine, pregnancy complications, and adverse pregnancy outcomes: the Hordaland Homocysteine Study. *Am J Clin Nutr.* 2000;71:962-968.

Wiener-Megnagi Z, Ben-Shlomo I, Goldberg Y, Shalev E. Resistance to activated protein C and the Leiden mutation: high prevalence in patients with abruptio placentae. *Am J Obstet Gynecol.* 1998;179:1565-1567.

Thrombophilia in Neonates and Children

Andrew M. Developmental hemostasis: relevance to thromboembolic complications in pediatric patients. *Thromb Haemost.* 1995;74:415-425.

Kenet G, Nowak-Göttl U. Fetal and neonatal thrombophilia. *Obstet Gynecol Clin North Am.* 2006 Sep;33(3):457-466.

Marcovina SM, Albers JJ, Scanu AM, et al. Use of a reference material proposed by the International Federation of Clinical Chemistry and Laboratory Medicine to evaluate analytical methods for the determination of plasma lipoprotein(a). *Clin Chem.* 2000;46:1956-1967.

Sébire G, Tabarki B, Saunders DE, et al. Cerebral venous sinus thrombosis in children: risk factors, presentation diagnosis and outcome. *Brain.* 2005;128:477-489.

Thornburg CD, Dixon N, Paulyson-Nunez K, Ortel T. Thrombophilia screening in asymptomatic children. *Thromb Res.* 2008;121(5):597-604.

Tormene D, Gavasso S, Rossetto V, Simioni P. Thrombosis and thrombophilia in children: a systematic review. *Semin Thromb Hemost.* 2006;32:724-728.

CHAPTER 19

Laboratory Diagnosis of Inherited Thrombophilia

Charles Eby, MD

Introduction

Laboratory evaluation of patients for thrombophilia risk factors constitutes a major proportion of the testing performed in specialized coagulation laboratories. In this chapter, an algorithmic testing approach for various causes of inherited thrombophilia will be reviewed. Equally important is the laboratory evaluation of selected patients for acquired thrombophilia. Lupus anticoagulant and antiphospholipid antibody testing is reviewed in chapter 22.

Algorithmic Approach to Laboratory Diagnosis of Thrombophilic Risk Factors

If thrombophilia testing is done during the acute phase of a venous thromboembolism (VTE) and normal results are obtained, they are unlikely to be false-negatives, and the risk factors can be dismissed. However, if abnormal results are obtained, the false-positive rates are very high, due to concurrent anticoagulation therapy or acute inflammatory responses, and repeat testing when the patient is in stable health and not anticoagulated is mandatory to prevent an erroneous diagnosis that may be used to justify an inappropriate recommendation of indefinite anticoagulation therapy. An abnormal result should always be repeated to confirm a diagnosis. If the repeat value is still abnormal, then the repeat testing would support a consistent abnormality, either inherited or acquired, depending upon the analyte.

Clinical factors and preanalytical and analytical variables should always be taken into consideration in the interpretation of test results. As seen in Figure 19-1, if an abnormal prothrombin time (PT) or activated partial thromboplastin time (aPTT) result is obtained, sample collection variables

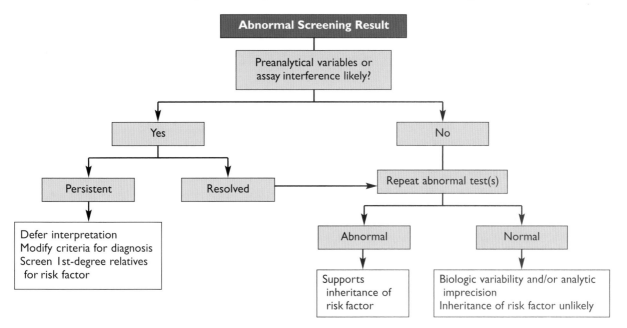

Figure 19-1. Inherited risk factor confirmatory test algorithm. This algorithm can be generally applied to measurement of any of the tests in the thrombophilia laboratory eval- uation. Careful consideration should be made of a patient's medical condition and sample collection and processing conditions, which can affect each assay.

Case Study

A 45-year-old male accountant, previously healthy, developed shortness of breath and chest pain while at work. He was evaluated in the emergency department and found to have a pulmonary embolus. He had no previous thrombotic events and had no family history of thrombosis. He was hospitalized for anticoagulation treatment and further evaluation. Upon admission, he was started on intravenous unfractionated heparin therapy and simultaneously was administered warfarin at 5 mg/day. Two days after admission, a physician ordered a testing panel for thrombophilic risk factors. The results are as follows:

Test	Value	Reference Range
INR	1.8	0.8-1.2
aPTT (s)	48.1	24.6-32.8
Fibrinogen (mg/dL)	450	200-400
Factor VIII	292%	49%-134%
Protein C activity	68%	76%-147%
Protein S activity	41%	59%-131%
Antithrombin activity	70%	84%-138%
APCr ratio	2.3	2.0-2.7
Homocysteine (µmol/L)	12.0	6.4-13.7
Prothrombin G20210A	Wild type	(normal)

In this patient, there are many abnormal laboratory values. However, the elevated aPTT can be explained by the heparin therapy. The decreased antithrombin activity may be associated with the heparin therapy, because antithrombin levels usually decrease during heparin therapy, but antithrombin levels are also decreased by an acute thrombus due to consumption. The elevated, but subtherapeutic, INR is probably due to early warfarin therapy, as are the decreased levels of protein C and protein S. The protein S is relatively more decreased than the protein C, but this result can be explained by the combined effect of warfarin and the elevated factor VIII; high levels of factor VIII may spuriously lower protein S in some clot-based assays. The elevated factor VIII is accompanied by an elevated fibrinogen. These last two abnormalities are probably due to an acute phase reaction and would likely be accompanied by an elevated C-reactive protein. The APCr ratio, prothrombin genotyping, and homocysteine levels were normal. Thus, despite many apparent laboratory value abnormalities, all that can be surmised is that the patient does not have elevated homocysteine and does not have the thrombophilic mutations factor V Leiden, based on a normal APCr ratio, which is a sensitive screening test, or prothrombin G20210A. The remainder of the laboratory results can be explained by the effect of the acute thrombus, anticoagulant therapy, or acute phase response. This case highlights why it is *not* advisable to perform thrombophilia testing in the setting of an acute thrombus or during anticoagulant therapy, because these settings may result in many misleading test abnormalities.

should be considered, such as a short-draw tube or high hematocrit, which can affect the results of clot-based tests. In addition, therapeutic anticoagulants, such as heparin, warfarin, or direct thrombin inhibitors, can affect the results of clot-based tests and mixing studies. If the interference is persistent, such as is seen with suppression of clottable protein S values in patients with factor V Leiden, an alternative test method, unaffected by the interfering substance, or testing of first-degree relatives may be helpful to establish a diagnosis.

Figures 19-2, 19-3, and 19-5 are examples of algorithms to follow when evaluating patients for possible protein C, protein S, and antithrombin deficiencies. Reduced protein C, protein S, and antithrombin activities may occur as a result of consumption of coagulation regulators during thrombus formation, and both protein C and protein S activities decline during oral anticoagulation with a vitamin K antagonist such as warfarin.

Protein C

Protein C levels are affected by the patient's age, hepatic synthesis, vitamin K status, acute phase response, and consumptive coagulopathies. It is recommended to screen patients for protein C deficiency using a functional assay, such as a chromogenic or clot-based assay. As indicated in Figure 19-2, if the initial protein C assay result is normal, further testing is not warranted, and the patient is unlikely to have protein C deficiency.

To prevent erroneous diagnoses of inherited protein C deficiency in patients with acquired protein C deficiency, the following steps should be followed, as illustrated in Figure 19-2.

1. Screen with a prothrombin time/international normalized ratio (PT/INR) assay and aPTT. If prolonged, consider acquired causes, such as liver disease, disseminated intravascular coagulation, or vitamin K deficiency/warfarin therapy. Additional tests that may help clarify the diagno-

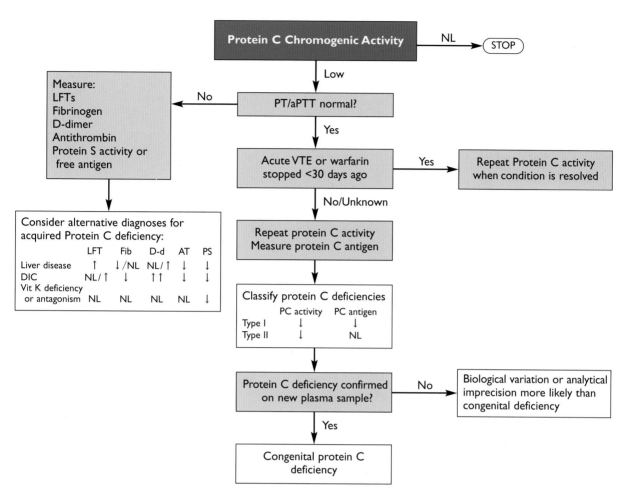

Figure 19-2. Protein C testing algorithm. This algorithm details the laboratory diagnosis of protein C deficiency and shows steps to distinguish acquired protein C deficiency from congenital deficiency. Knowledge of hepatic function, measurement of other coagulation tests, and history of anti-coagulant therapy or acute thrombosis is very helpful in diagnostic interpretation. Abbreviations: AT, antithrombin; D-d, D-dimer; Fib, fibrinogen; LFT, liver function tests; NL, normal; PC, protein C; PS, protein S; VTE, venous thromboembolism.

sis include liver function tests (especially albumin and total protein), fibrinogen, D-dimer, antithrombin, and protein S. Combined "deficiency" of protein C and protein S only is most likely due to vitamin K deficiency or oral anticoagulant therapy. Low levels of several proteins, including fibrinogen, protein C, protein S, and antithrombin, with abnormal liver function tests suggests decreased protein C synthesis. A consumptive coagulopathy is likely with marked elevations of D-dimer in the setting of normal liver function tests and decreased clotting factor levels.

2. Do not test for protein C activity during an acute venous thromboembolic event or other medical, surgical, or trauma-related acute illness. An acute thrombotic event is often accompanied by decreased levels of one or more natural anticoagulants, such as protein C, protein S, or antithrombin.

3. Wait at least 30 days after discontinuation of vitamin K antagonist (warfarin) to measure protein C activity.

4. If low protein C activity is obtained, measurement of at least two vitamin K-dependent coagulation factor activities may be helpful to rule out mild, nonspecific, acquired hyposynthesis.

5. Confirm protein C deficiency by repeating the test on a new plasma sample and measuring protein C antigen. This is especially important when the initial protein C activity is between two and three standard deviations below the mean of the reference range, a region where false-positive results are likely, reflecting poor specificity. If a repeat test sample does not show decreased protein C activity, the initial abnormal result was more likely due to biological variation or analytical imprecision than a congenital deficiency.

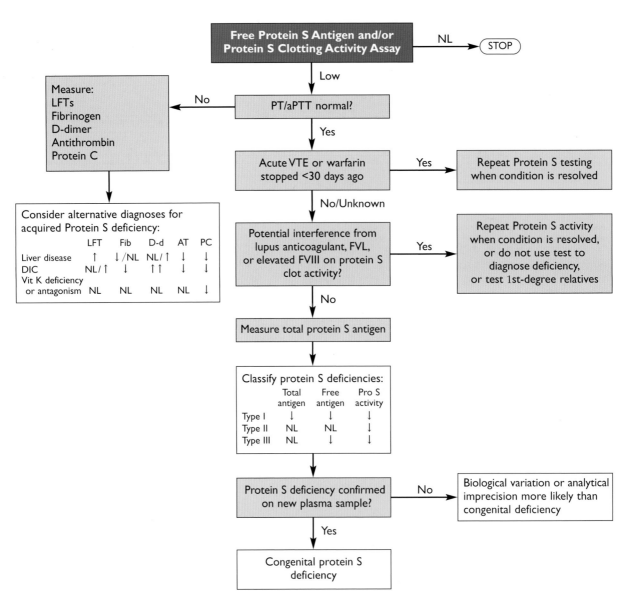

Figure 19-3. Protein S testing algorithm. Diagnosis of protein S deficiency must take into account hepatic function, consumptive coagulopathy, and vitamin K status, as well as presence of acute thrombosis. Use of three assays, protein S clotting activity, total protein S, and free protein S, is necessary to distinguish protein S deficiency types I, II, and III. Abbreviations: AT, antithrombin; D-d, D-dimer; Fib, fibrinogen; LFT, liver function tests; FVL, factor V Leiden; FVIII, factor VIII; NL, normal; PC, protein C; PS, protein S; VTE, venous thromboembolism.

Testing of protein C antigen may be helpful to subclassify type I versus type II protein C deficiency, but this is rarely necessary for clinical diagnosis or treatment.

6. Screening first-degree relatives may confirm the diagnosis by demonstrating an autosomal dominant inheritance pattern of protein C deficiency.

Protein S

Because protein S is a vitamin K-dependent protein produced in the liver, like protein C, some of

the considerations for protein S testing are the same as for protein C. See Figure 19-3 for a diagnostic algorithm for protein S. The clot-based protein S activity assays can be affected by factor V Leiden, elevated factor VIII, lupus anticoagulants, and the presence of heparin or direct thrombin inhibitors (Figure 19-4). Factor V Leiden and elevated factor VIII can diminish the ability of activated protein C (APC)/protein S to degrade factor Va and VIIIa and falsely depress protein S functional activity. Conversely, the presence of a high-titer lupus anticoagulant, heparin concentration

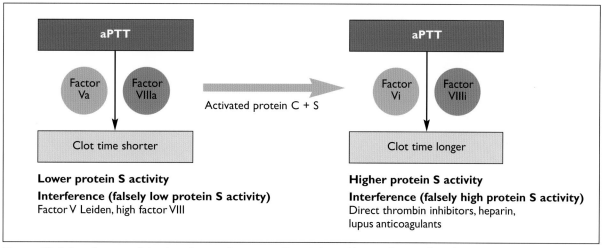

Figure 19-4. Interference of clot-based assays for protein S. Testing for protein S function using a clot-based assay is com-

plicated by interference from factor VIII, factor V Leiden, lupus anticoagulants, heparin, and direct thrombin inhibitors.

greater than 1 U/mL, or a direct thrombin inhibitor can prolong the clotting assay, falsely elevating the protein S functional result. If the patient is receiving an interfering anticoagulant, repeating the protein S activity test when the patient is no longer receiving the anticoagulant is recommended. However, the presence of factor V Leiden, elevated factor VIII, or a lupus anticoagulant usually cannot be altered, and diagnosis of protein S deficiency should be made based on antigenic assays or testing of first-degree relatives. Measurement of total and free protein S antigen, together with protein S activity, can be used to subclassify protein S deficiency as type I, II, or III. Some caution is advised in the diagnosis of type III protein S deficiency: normal total protein S, but decreased free and functional protein S. Protein S is carried by C4b binding protein (C4bBP), and an elevation of C4bBP in response to inflammatory cytokines may mimic type III protein S deficiency by increasing bound protein S and decreasing free and functional protein S levels.

Antithrombin

Initial screening for antithrombin deficiency should be performed using a chromogenic, heparin cofactor-based antithrombin activity assay. An algorithm for antithrombin testing can be seen in Figure 19-5. A normal antithrombin result excludes a congenital or acquired deficiency. Antithrombin is produced in the liver and is also one of the first proteins to decline during disseminated intravascular coagulation; therefore, assays for liver function, fibrinogen, D-dimer, protein C, and protein S would be recommended if the

antithrombin level is decreased and the PT and/or aPTT are abnormal. Concurrent medical situations, such as acute thromboembolism, heparin therapy, L-asparaginase therapy, nephrotic syndrome, or cardiopulmonary bypass, can also lower antithrombin levels. Antithrombin activity is temporarily reduced by 10% to 20% during therapeutic unfractionated heparin treatment, but not during low-molecular-weight heparin therapy. Acquired antithrombin deficiency may also arise from hepatic hyposynthesis during treatment with L-asparaginase chemotherapy for acute lymphoblastic leukemia, shortened half-life due to increased renal excretion in nephrotic syndrome, and hemodilution associated with cardiopulmonary bypass. If the antithrombin level is normal on the repeated test sample, the initial abnormal result was more likely due to assay imprecision or biological variability than to congenital deficiency. Elevation of antithrombin activity during oral anticoagulation therapy has been reported, but compelling evidence that this could lead to misclassifying a partial antithrombin-deficient patient as "normal" is lacking.

If acquired causes of antithrombin deficiency can be excluded, the functional antithrombin assay should be repeated on a new sample by heparin cofactor-dependent and non-heparin-dependent methods, together with determination of antithrombin antigen concentration. Type I antithrombin deficiency is diagnosed when there is a decrease in both activity assays and antithrombin antigen. Type II antithrombin deficiency is seen with decrease in antithrombin activity with normal antigen. Diagnosis of the type II heparin-

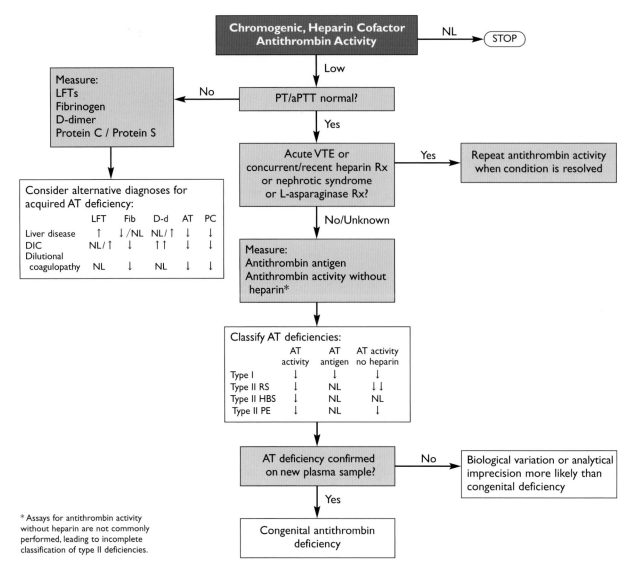

Figure 19-5. Antithrombin testing algorithm. The algorithm presented in this figure distinguishes acquired antithrombin deficiency from congenital deficiency. In addition to hepatic disease and disseminated intravascular coagulation, assessment of low antithrombin levels should take in account heparin therapy, nephrotic syndrome, and L-asparaginase therapy. Types I and II antithrombin deficiency can be distinguished on the basis of a combination of functional and anti- genic antithrombin assays. While type II antithrombin defi- ciency subtypes can be distinguished by comparing heparin- dependent and heparin-independent functional assays, the lat- ter is not commonly performed in clinical laboratories. Abbreviations: AT, antithrombin; D-d, D-dimer; Fib, fibrinogen; HBS, heparin-binding site; LFT, liver function tests; NL, normal; PE, pleotropic; RS, reactive site; VTE, venous thromboembolism.

binding site deficiency can be diagnosed by nor- mal antithrombin activity in a non-heparin- dependent assay, but by decreased antithrombin activity in a heparin-dependent assay.

Factor VIII

Factor VIII activity typically increases in response to exercise, pregnancy, trauma, and acute inflam- matory states, including VTE. Therefore, an ele- vated factor VIII activity, obtained at presentation with a VTE or during initiation of anticoagulation

therapy may be an acquired anomaly rather than evidence of an inherited hypercoagulable risk fac- tor. However, when factor VIII activity or antigen is measured several months after a VTE, elevated levels (usually defined as greater than 150% or greater than the 90th percentile) typically occur in about 25% of patients. In this setting, elevated fac- tor VIII levels do not correlate with other markers of inflammation (C-reactive protein or erythrocyte sedimentation rate), remain elevated when repeat- ed over months to years, and, in selected families,

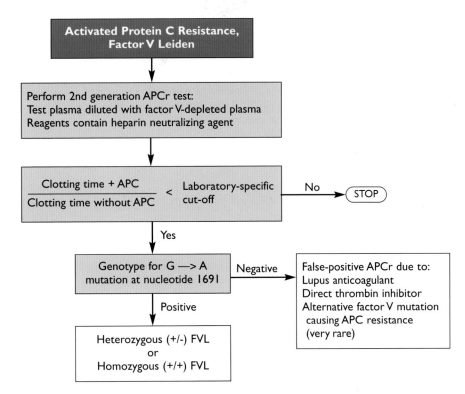

Figure 19-6. Activated protein C resistance/factor V Leiden testing algorithm. Use of a functional, second-generation, activated protein C resistance (APCr) test is recommended as the first step in diagnosing factor V Leiden. If an abnormally low APCr ratio is obtained, factor V Leiden genotyping should be done to distinguish heterozygotes from homozygotes. Occasionally, a discrepancy will be seen: low APCr ratio and normal factor V Leiden genotyping, usually because of the presence of an inhibitor (false-positive APCr), or a factor V mutation at a different location than factor V Leiden that causes resistance to cleavage by activated protein C (false-negative genetic test). Abbreviations: APC, activated protein C; APCr, activated protein C resistance; FVL, factor V Leiden.

elevated factor VIII patterns appear to be inherited. In addition, factor VIII activities that are greater than the 90th percentile were associated with a relative risk of 6.7 for recurrent VTE. Unfortunately, the precision and accuracy of one-stage factor VIII activity assays is poor, making it difficult to apply the findings from clinical research studies to routine VTE risk factor assessment.

Activated Protein C Resistance and Factor V Leiden

The activated protein C resistance (APCr) functional assay is usually recommended as the initial screening assay for the presence of the factor V Leiden mutation. The various APCr assays are variations of routine clotting methods (aPTT, PT, and dilute Russell viper venom time) and can be performed in most coagulation laboratories. The APCr is recommended as the initial screening test because of its wide availability, lower cost, and ability to detect any factor V mutation that increas-

es resistance to cleavage by activated protein C. Modifications of the original APCr method include dilution of test plasma with factor V-depleted plasma, addition of heparin-neutralizing substances, and alternative activation strategies, making "second-generation" APCr tests more robust. However, a "resistant" result with a decreased APCr ratio should still be confirmed by genotyping for factor V Leiden to determine whether a patient is heterozygous or homozygous for the mutation (Figure 19-6). Genetic tests (factor V Leiden) are unaffected by concurrent medications or health issues and can be performed at any time.

Occasionally, an abnormally decreased APCr ratio will be observed with negative factor V Leiden genotyping. Possible causes for this apparent discrepancy can usually be discovered by close inspection of the assay results. If the clotting time without APC is elevated even after dilution with factor V-depleted plasma, as can be seen with a lupus anticoagulant or direct thrombin inhibitor, then the test result may not be proportionately prolonged after addition of APC, and a decreased ratio may result. To detect a lupus anticoagulant, aPTT mixing studies or dilute Russell viper venom time testing may be helpful, whereas a prolonged thrombin time or clinical history usually can detect direct thrombin inhibitor therapy. If a lupus anticoagulant and direct thrombin inhibitor therapy can be excluded as the cause of the decreased APCr ratio, then factor V sequencing may be helpful to detect factor V 306 mutations (factor V Cambridge or Hong Kong) or the HR2 haplotype.

Prothrombin G20210A Mutation

It is recommended to perform prothrombin G20210A genotyping for thrombophilia risk assessment rather than the measurement of factor II levels. This test can be performed at any time because the genotyping test is not affected by pre-analytical variables or disease state.

Homocysteine

Genetic predispositions and acquired conditions can contribute to hyperhomocysteinemia. Homozygosity for *MTHFR C677T*, which is a polymorphism in the gene for 5,10-methylenetetrahydrofolate reductase (MTHFR), occurs in about 10% to 12% of Caucasians and is associated with about a 25% increase in mean total plasma homocysteine (tHcy) concentrations. A meta-analysis of epidemiologic studies evaluating homocysteine metabolism and risk of VTE concluded that homozygosity for *MTHFR 677TT* was not associated with an increased risk for VTE in the United States, but there was a weak increased risk associated with the mutation in studies performed in other countries. The different outcomes are probably due to higher levels of folate fortification of foods in the United States. However, a recent, large, Dutch case-control study (MEGA trial, N = 4375 patients with a first VTE and age-matched controls [Bezemer ID et al, 2007]) did not find an association between *MTHFR 677TT* and risk of venous thrombosis. Genetic testing for *MTHFR C677T* is offered by many molecular diagnostic laboratories as part of thrombophilia test panels. Some laboratories also screen for another *MTHFR* single-nucleotide polymorphism, *A1298C*. Homozygosity (CC) for *1298* reduces enzyme activity by 50% to 60%, but there are no persuasive data to support an association between *1298CC* and hyperhomocysteinemia or increased risk of VTE.

Elevated fasting tHcy concentration has been consistently associated with a mild increased risk of venous and arterial thromboembolic complications. A subset of patients with normal fasting tHcy concentrations will have abnormally high levels when measured hours after a "loading dose" of methionine. However, clinicians and clinical pathologists generally do not favor routine measurement of both fasting and post-methionine-loading tHcy concentrations. Figure 19-7 is an example of an algorithm for laboratory evaluation of homocysteine metabolism.

A causal relationship between elevated tHcy and vascular disease has not been confirmed in secondary prevention studies of venous and arterial thromboembolic events in patients with hyperhomocysteinemia who were randomized to receive placebo or B vitamin supplementation. Subjects who took B vitamin supplements lowered their tHcy concentrations but did not have fewer recurrent vascular events as compared with subjects who took a placebo. Total plasma homocysteine concentration is not affected by acute thrombosis or anticoagulant therapy, but a marked elevation may be a sign of concurrent folate or vitamin B_{12} nutritional deficiency. This finding would prompt a review of the complete blood count and peripheral blood smear, an alertness for findings on history and physical examination suggestive of neurologic complications of B_{12} deficiency, and measurement of serum concentrations of these essential vitamins. Elevated tHcy can also be seen with renal insufficiency, so checking blood urea nitrogen or creatinine levels may be helpful in assessing homocysteine elevation. False-positive tHcy elevation due to leakage of homocysteine from red blood cells is avoidable if plasma is promptly removed or if whole blood is stored at 4°C until processed. Total plasma homocysteine can also be elevated in a hemolyzed sample.

Thrombophilia Testing, Duration of Anticoagulation Therapy, and Screening First-Degree Relatives

Determining the optimal duration of anticoagulation therapy for patients who have experienced an unprovoked VTE is complicated by the following factors: (1) gaps in our knowledge about the risk of recurrent VTE for patients with uncommon inherited risk factors such as protein C and protein S deficiencies, (2) different interpretations of results

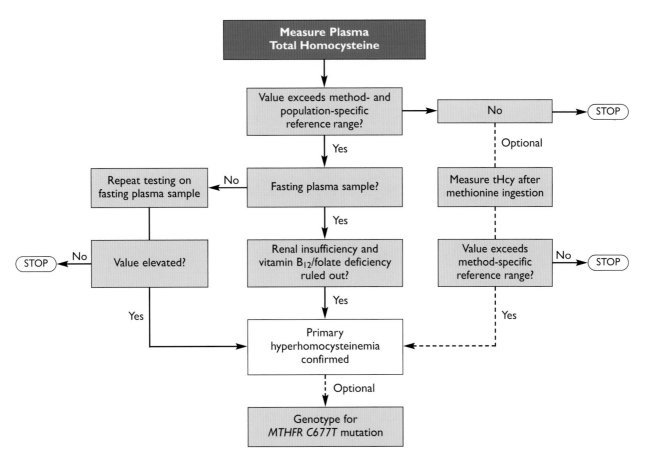

Figure 19-7. Homocysteine testing algorithm. In this algorithm, homocysteine should initially be assessed by measurement of total plasma homocysteine (tHCy). If the tHCy is elevated, it should be determined whether the patient was fasting or not at the time of sample collection. Some institutions perform methionine-loading tests and repeat the tHCy measurement after methionine ingestion. Some diseases and metabolic conditions, such as renal failure and vitamin B_{12}/folate deficiency, can contribute to homocysteine elevation. Genotyping for the *MTHFR C677T* mutation may be done, but it might only add limited clinical information.

from prospective studies comparing extended therapy with either standard (target INR 2 to 3) or low-intensity (target INR 1.5 to 2.0) oral anticoagulation, (3) recent intriguing results supporting a termination of anticoagulation in patients who do not have an elevated D-dimer 1 month after completing at least 3 months of warfarin therapy, (4) assessing a patient's risk of bleeding complications, and (5) importance of patient education and input into the decision process. The eighth edition of the American College of Chest Physicians Antithrombotic and Thrombolytic Therapy Evidence-Based Guidelines provides recommendations on the duration of anticoagulation therapy for VTEs that combine clinical circumstances, location of VTE, estimate of patients' bleeding risk, and patient preferences to arrive at individualized treatment decisions (Table 19-1). The eighth edition does not consider positive tests for inherited thrombophilia risk factors to be major determinants of recurrent VTE risk or duration of anticoagulation therapy, which is a departure from previous editions. Several longitudinal studies of patients with spontaneous VTE indicate that recurrence rates are high when anticoagulation therapy is stopped, and the increased risk of recurrent thromboses is relatively small for most thrombophilia risk factors, with the exception of phospholipid antibodies and combined factor V Leiden and prothrombin 20210 mutations. Since patient-specific risks and benefits of anticoagulation therapy vary considerably and can change over time, clinicians and clinical pathologists are likely to identify many patients for whom thrombophilia testing will provide information that is useful for patient education and management decisions.

A consensus is also lacking regarding testing of asymptomatic first-degree relatives of VTE patients who have also inherited one or more thrombophilia risk factors. Retrospective studies

Table 19-1. Guidelines for Duration of Anticoagulation Therapy for Venous Thrombosis

Clinical and Laboratory Findings	Duration of Anticoagulation
First VTE, reversible risk factor(s)	3 months VKA
First unprovoked distal DVT	3 months VKA
First unprovoked proximal DVT or PE	Minimum 3 months VKA; recommend long-term treatment if compatible with assessment of bleeding risk and patient preference
VTE and cancer	Therapeutic LMWH for 3-6 months; continued anticoagulation therapy until cancer is in remission
Two or more unprovoked VTEs	Long-term VKA

Abbreviations: DVT, deep vein thrombosis; LMWH, low-molecular-weight heparin; PE, pulmonary embolism; VKA, vitamin K antagonist anticoagulant; VTE, venous thromboembolism.

Based on: Antithrombotic and Thrombolytic Therapy, American College of Chest Physicians Evidence-Based Clinical Practice Guidelines (8th Edition), 2008.

of patients who have had a VTE and are factor V Leiden positive confirm a 3- to 4-fold increased risk of venous thrombosis among asymptomatic first-degree relatives who are factor V Leiden positive as compared with relatives who are factor V Leiden negative. However, the absolute risk for factor V Leiden carriers is low, approximately 0.45% to 0.53% per year, and many would not be prevented by avoidance of oral contraceptives or prophylactic anticoagulant therapy during pregnancy, surgery, or trauma, because about half of the VTEs are spontaneous. Accurate estimates of the thrombotic risk for asymptomatic relatives of VTE probands who are protein C, protein S, or antithrombin deficient, or heterozygous for the prothrombin gene mutation, are not available. It is important to recognize that an informed asymptomatic carrier of an inherited thrombophilia risk factor may choose to accept the low absolute risk for VTE associated with certain health care choices, such as oral contraceptive use, hormone replacement therapy, and pregnancy.

In conclusion, the decision to test patients for inherited thrombophilia risk factors is a multistep process. First, only patients who experience a spontaneous VTE with no, or perhaps weak, concurrent acquired risk factors should be candidates for testing. Second, testing should be comprehensive, including the five established inherited risk factors (ie, protein C, protein S, antithrombin,

APCr screen for factor V Leiden, and prothrombin mutation); risk factors that involve both inherited and acquired mechanisms (ie, factor VIII activity and tHcy concentration [unless additional secondary prevention of VTE studies clearly show no benefit to lowering tHcy]); and antiphospholipid antibodies (see chapter 22). Testing should be performed when temporary, false-positive results are least likely to be obtained, for instance after patients have not taken a vitamin K antagonist such as warfarin for at least 30 days. Because functional and immunologic assays are vulnerable to both preanalytical and analytical sources of errors, it is necessary to repeat and verify any abnormal result in order to prevent erroneous diagnoses of inherited thrombophilia. Additional tests for uncommon (dysfibrinogenemia) or controversial (plasminogen, undiluted APCr, factor IX, factor XI, fibrinolytic activity) risk factors may be advocated by some hemostasis specialists in certain clinical settings but should not be considered part of the standard evaluation. Positive thrombophilia test results may lead to important clinical recommendations for patients as well as their first-degree relatives, who may be encouraged to be screened. It is essential that laboratories performing thrombophilia testing do their utmost to provide clinicians with accurate test results and appropriate guidance regarding patient selection and test interpretation.

Suggested Reading

General

Bauer KA, Rosendaal FR, Heit JA. Hypercoagulability: Too many tests, too much conflicting data. *Hematology Am Soc Hematol Educ Program*. 2002;353-368.

Jennings I, Cooper P. Screening for thrombophilia: a laboratory perspective. *Br J Biomed Sci*. 2003;60:39-51.

Machin SJ. Pros and cons of thrombophilia testing: cons. *J Thromb Haemost*. 2003;1:412-413.

Martinelli I. Pros and cons of thrombophilia testing: pros. *J Thromb Haemost*. 2003;1:410-411.

Merriman L, Greaves M. Testing for thrombophilia: an evidence-based approach. *Postgrad Med J*. 2006; 82:699-704.

Simioni P. Who should be tested for thrombophilia? *Curr Opin Hematol*. 2006;13:337-343.

Tripodi A, Mannucci PM. Laboratory investigation of thrombophilia. *Clin Chem*. 2001;47:1597-1606.

Inherited Thrombophilia Testing Guidelines

Goodwin AJ, Rosendaal FR, Kottke-Marchant K, Bovill EG. A review of the technical, diagnostic, and epidemiologic considerations for protein S assays. *Arch Pathol Lab Med*. 2002;126:1349-1366.

Kottke-Marchant K, Comp P. Laboratory issues in diagnosing abnormalities of protein C, thrombomodulin, and endothelial cell protein C receptor. *Arch Pathol Lab Med*. 2002;126:1337-1348.

Kottke-Marchant K, Duncan A. Antithrombin deficiency: issues in laboratory diagnosis. *Arch Pathol Lab Med*. 2002;126:1326-1336.

McGlenne RC, Key NS. Clinical and laboratory management of the prothrombin G20210A mutation. *Arch Pathol Lab Med*. 2002;126:1319-1325.

Press RD, Bauer KA, Kujovich L, Heit JA. Clinical utility of factor V Leiden (R506Q) testing for the diagnosis and management of thromboembolic disorders. *Arch Pathol Lab Med*. 2002;126:1304-1318.

Factor VIII

Koster T, Vandenbroucke JP, Rosendaal FR, Blann AD, Briët E. Role of clotting factor VIII in effect of von Willebrand factor on occurrence of deep-vein thrombosis. *Lancet*. 1995;345:152-155.

Kraaijenhagen RA, in't Anker PS, Koopman MM, et al. High plasma concentration of factor VIIIc is a major risk factor for venous thromboembolism. *Thromb Haemost*. 2000;83:5-9.

Kyrle PA, Minar E, Hirschl M, et al. High plasma levels of factor VIII and the risk of recurrent venous thromboembolism. *N Engl J Med*. 2000;343:457-462.

O'Donnell J, Mumford AD, Manning RA, Laffan M. Elevation of FVIII:C in venous thromboembolism is persistent and independent of the acute phase response. *Thromb Haemost*. 2000;83:10-13.

Homocysteine

Bezemer ID, Doggen CJ, Vos HL, Rosendaal FR. No association between common MTHFR 677 C–>T polymorphism and venous thrombosis. *Arch Intern Med*. 2007;167:497-505.

Bonaa KH, Njolstad I, Ueland PM, et al. Homocysteine lowering and cardiovascular events after acute myocardial infarction. *N Engl J Med*. 2006;354:1578-1588.

den Heijer M, Williams HPJ, Blom HJ, et al. Homocysteine lowering by B vitamins and the secondary prevention of deep vein thrombosis and pulmonary embolism: a randomized, placebo-controlled, double-blind trial. *Blood*. 2007;109:139-144.

Domagala TB, Adamek L, Nizankowska M, et al. Mutations C677T and A1298C of the 5,10 methylenetetrahydrofolate reductase gene and fasting plasma homocysteine levels are not associated with the increased risk of venous thromboembolic disease. *Blood Coagul Fibrinolysis*. 2002;13:423-431.

Lonn E, Yusuf S, Arnold MJ, et al. Homocysteine lowering with folic acid and B vitamins in vascular disease. *N Engl J Med*. 2006;354:1567-1577.

Thrombophilia Testing, Duration of Anticoagulation Therapy

Dalen JE. Should patients with venous thromboembolism be screened for thrombophilia? *Am J Med*. 2008;121:458-463.

Kearon C, Kahn SR, Agnelli G, Goldhaber S, Raskob GE, Comerota AJ. Antithrombotic therapy for venous thromboembolic disease: American College of Chest Physicians Evidence-Based Clinical Practice Guidelines (8th Edition). *Chest*. 2008;133(suppl):454S-545S.

Palareti G, Cosmi B, Legnani C, et al. D-dimer testing to determine the duration of anticoagulation therapy. *N Engl J Med*. 2006;355:1780-1789.

Fibrinolytic Thrombotic Disorders

Timothy E. Hayes, DVM, MD

Introduction

After a clot is formed, the healing process begins. During this phase, blood vessel patency is restored through the slow enzymatic digestion of the intravascular clot by plasmin. As with the clotting factor cascade of secondary hemostasis, fibrinolysis has its own set of activators and inhibitors that influence the rate of clot lysis. Digestion of the fibrin strand leads to the liberation of fibrin degradation products, including D-dimer. See chapter 3 for an in-depth discussion of the physiology of fibrinolysis.

Hereditary and acquired abnormalities in fibrinolysis appear to be much less common as compared with defects in primary and secondary hemostasis. Hyperfibrinolysis may result in rapid clot dissolution before the healing process has concluded, resulting in hemorrhage (see chapter 13). Theoretically, hypofibrinolysis may result in accumulation of clot and manifest as thrombosis.

Laboratory tests, such as the D-dimer, are increasingly playing a role in evaluating patients for the presence of venous thromboembolism (VTE). Understanding the limitations of these assays and how they fit appropriately into clinical decision making are critical to good patient outcomes.

Thrombosis and Defects of Fibrinolysis

Decreased fibrinolysis may be seen with impaired formation of plasmin (decreased plasminogen, decreased tissue plasminogen activator [tPA], increased plasminogen activator inhibitor-1 [PAI-1]) or impaired plasmin function (increased alpha 2-antiplasmin, dysfibrinogenemia, increased thrombin activatable fibrinolysis inhibitor, and antifibrinolytic medications).

Plasminogen Deficiency

Plasminogen is the zymogen of plasmin, the central enzyme of the fibrinolytic system. Both plasminogen and tPA have kringle-domain lysine binding sites, which facilitate their binding to lysine residues within fibrin strands during the clotting process. In this process, fibrin serves as a catalyst for the conversion of plasminogen to plasmin by tPA and also positions the newly formed plasmin enzyme in close approximation to its fibrin substrate.

Both quantitative (hypoplasminogenemia) and qualitative (dysplasminogenemia) forms of hereditary plasminogen deficiency have been identified. These defects are most commonly inherited in an autosomal dominant pattern. Although mouse models of homozygous plasminogen deficiency have demonstrated a decreased lysis of intravascular clots, a cause-and-effect relationship has not been convincingly proven in humans. There are two lines of evidence suggesting that heterozygous plasminogen deficiency is not a significant thrombotic risk factor. First, most family studies have failed to demonstrate a relationship between heterozygous plasminogen deficiency and thrombosis in family members of the propositus. Second, studies have shown that the incidence of plasminogen deficiency is higher in the general population than it is in a population of thrombophilic patients. Therefore, the value of testing thrombophilic patients for plasminogen deficiency remains uncertain and is not routinely recommended.

Tissue Plasminogen Activator Deficiency

Tissue plasminogen activator is typically present in the circulation at very low basal levels. Certain stimuli will release tPA into the circulation, and a portion of this tPA is bound to its inhibitor, PAI-1, which also fluctuates in concentration. All of these factors complicate the documentation of a true tPA deficiency. Although tPA deficiency has been described in a few thrombophilic patients, a cause-and-effect relationship has not been proven. Routine testing of thrombophilic patients for tPA deficiency is not recommended.

Plasminogen Activator Inhibitor Excess

Plasminogen activator inhibitor type 1 is the serine protease inhibitor (SERPIN) most responsible for inactivating tPA and thereby dampening plasmin formation and subsequent fibrinolysis. Several studies have described a relationship between elevated levels of PAI-1 and both venous and arterial thrombosis. The "metabolic syndrome" (also known as syndrome X, the insulin resistance syndrome, the deadly quartet, and the obesity dyslipidemia syndrome) includes a constellation of findings, such as obesity (especially abdominal obesity), insulin resistance, hyperinsulinemia, hyperglycemia, dyslipidemia (elevated triglycerides, low levels of high-density lipoproteins), hypertension, vascular endothelial cell dysfunction, and vascular inflammation. This syndrome is typically associated with type 2 diabetes mellitus and atherosclerotic cardiovascular disease. The metabolic syndrome has been documented as a proinflammatory and prothrombotic state and is associated with elevated levels of PAI-1, C-reactive protein, and inflammatory cytokines, such as interleukin-6.

In addition to the metabolic syndrome, PAI-1 is elevated in other clinical settings that are associated with thrombosis, such as cancer, surgery, pregnancy, and sepsis. Because PAI-1 is an acute phase reactant, and its synthesis is induced by insulin and high plasma lipid levels, confirming a cause-and-effect relationship between elevated PAI-1 and thrombosis in these clinical settings has been difficult.

Polymorphisms in the PAI-1 promoter region have been described, including the 4G/5G insertion/deletion and the A-844G polymorphism. Higher levels of PAI-1 have been found in the 4G/4G genotype compared to the 5G/5G genotype. Most studies have shown no increased risk of venous thrombosis with an isolated 4G/4G genotype. Some studies have shown an increased risk of venous thrombosis when the 4G/4G genotype is inherited with another thrombotic risk factor, such as factor V Leiden.

The value of routinely testing thrombophilic patients for elevated levels of PAI-1, or for the PAI-1 polymorphisms, remains uncertain and is not recommended.

Alpha 2-Antiplasmin Excess

Alpha 2-antiplasmin, a serine protease inhibitor, contains lysine residues that are able to bind to the kringle-domains of plasminogen and plasmin. This binding may serve to prevent plasminogen from being incorporated into an enlarging fibrin clot and also increases the ability of the serine protease inhibitor to inactivate plasmin. Although an excess of alpha 2-antiplasmin may theoretically contribute to impaired fibrinolysis and thrombosis, it has not been reported. Testing thrombophilic patients for elevated alpha 2-antiplasmin is not recommended.

Euglobulin Clot Lysis Time Assay

The euglobulin clot lysis (ECL) time is a global assay used to assess the overall state of the fibrinolytic system. Citrated plasma is diluted with an acidic low-ionic-strength solution, which precipitates the euglobulin fraction. This precipitate includes fibrinogen, tPA, and plasminogen, but not the fibrinolytic inhibitors (PAI-1). The precipitate is resuspended and clotted by adding thrombin. The time for the clot to lyse is then measured. A shortened ECL time may be seen in hyperfibrinolytic states, whereas a prolonged ECL time is associated with impaired fibrinolysis. The ECL is not reliably prolonged in the setting of hypofibrinolysis and is therefore not routinely considered as part of a thromboembolic evaluation.

Thrombin-Activatable Fibrinolysis Inhibitor

Plasmin normally is incorporated into a clot via the interaction between its lysine binding sites and the lysine moieties within the fibrin polymer. This binding facilitates the interaction between enzyme and substrate. Thrombin-activatable fibrinolysis inhibitor (TAFI) is a procarboxypeptidase that is activated by thrombin when the latter is bound to thrombomodulin on the endothelial cell surface. The activated form of TAFI is then able to cleave arginine and lysine moieties from the carboxy terminus of fibrin degradation products being liberated during early fibrinolysis. After removal of these amino acids from fibrin degradation products, plasmin is less able to bind to its substrate, resulting in impaired fibrinolysis.

Excess TAFI has been found in some, but not all, patients with venous thromboemboli, but a cause-and-effect relationship has not been proven. Some recent reports of genetic variations in TAFI, such as 505A/G, and their association with a thrombophilic risk await confirmation. The value of measuring TAFI in thrombophilic patients remains uncertain and is not routinely recommended.

Dysfibrinogenemia

Inherited abnormal fibrinogens (dysfibrinogens) are rare, and most cases are asymptomatic. However, approximately 25% of such cases are associated with increased bleeding, and another 25% are associated with increased risk of thrombosis. A 1995 report from the International Society on Thrombosis and Haemostasis Scientific and Standardization Committee, Subcommittee on Fibrinogen, found a significant relationship between some forms of congenital dysfibrinogenemia and venous thrombosis. The best-characterized prothrombotic dysfibrinogen involves a substitution of cysteine for arginine at position 554 in the Aα fibrinogen chain (fibrinogen Chapel Hill). The mechanism of thrombosis is unknown. In theory, thrombosis may result from either increased fibrin formation or impaired fibrinolysis.

The commonly available assays for detecting dysfibrinogenemia (thrombin time, reptilase time, clot-based fibrinogen activity, and fibrinogen antigen) lack both sensitivity and specificity. In addition, congenital dysfibrinogenemias appear to be rare, and the potential for thrombotic risk is incompletely defined. Therefore, routine testing for dysfibrinogenemia in a thrombophilic patient is not recommended, but it may be considered in a patient with a thrombotic history and persistently decreased fibrinogen levels without signs and symptoms of disseminated intravascular coagulation (DIC) or liver disease.

Factor XII (Hageman Factor) Deficiency

Severe factor XII deficiency is inherited in an autosomal recessive pattern. Because of factor XII's position in the initial contact activation stage of the intrinsic clotting pathway, a severe deficiency is typically associated with a markedly prolonged activated partial thromboplastin time (aPTT). Despite this abnormal clotting time, factor XII deficiency is not associated with a bleeding risk, even in the homozygous state. Instead, there have been reports of increased risk of thrombosis with partial and complete deficiencies of factor XII. A possible mechanism is reduced factor XIIa-kallikrein–high-molecular-weight kininogen-dependent cell surface activation of urokinase

plasminogen activator leading to diminished in vivo fibrinolysis. Several single-nucleotide polymorphisms of factor XII, including 46C—>T polymorphism, have been reportedly associated with a thrombophilic risk, but this is controversial. However, a convincing cause-and-effect relationship between factor XII deficiency and thrombosis has yet to be proven, and testing thrombophilic patients for factor XII deficiency is not recommended.

Lipoprotein (a) Excess

As noted above, fibrin acts as a catalyst for the activation of plasminogen to plasmin by tPA by providing lysine binding sites to these molecules. An alternative catalyst is endothelial cell-bound annexin II, which also binds tPA and plasminogen through lysine binding sites.

Lipoprotein (a) (Lp(a)) is a modified form of low-density lipoprotein cholesterol, in which apolipoprotein (a) is covalently bound to apolipoprotein B-100. Apolipoprotein (a) contains kringle-domains similar to those of plasminogen. As such, Lp(a) competes with plasminogen for the catalytic sites on fibrin and annexin II. High concentrations of Lp(a) may result in decreased activation of plasminogen and reduced rate of fibrinolysis and may theoretically result in increased risk of thrombosis.

Testing for Lp(a) is complicated by the variable detection of the different apolipoprotein (a) isoforms (smaller isoforms are thought to be more thrombogenic) and by the fact that Lp(a) is an acute phase reactant. Elevated levels of Lp(a) have been inconsistently linked to future risk for myocardial infarction and peripheral arterial disease in otherwise healthy individuals. Further study is required before routine testing of thrombophilic patients for elevated Lp(a) can be recommended.

Disseminated Intravascular Coagulation

Disseminated intravascular coagulation is a complex entity in which there is activation of coagulation, excessive thrombin generation, consumption of clotting factors and platelets, intravascular fibrin formation, and a compensatory increase in fibrinolysis. Depending on the relative activity of these opposing processes, there may be severe bleeding, thrombosis, or both. Thrombosis is not uncommon in the setting of chronic (compensated) DIC, where there is often a low-grade stimulus, such as occurs with certain malignancies and inflammatory conditions. Microthrombi may occlude small blood vessels and contribute to multiorgan failure. See chapter 13 for additional discussion of DIC.

Antifibrinolytic Medications

Plasminogen binds to fibrin through kringle-domain lysine binding sites. Tranexamic acid and aminocaproic acid are lysine analogues. They impair fibrinolysis by interfering with the binding of plasmin to fibrin. Aprotinin reversibly binds to and inhibits several serine proteases. It is thought that aprotinin inhibits fibrinolysis via its inhibition of kallikrein-mediated activation of fibrinolysis and also through direct inhibition of plasmin. These medications may be used to decrease bleeding due to hyperfibrinolysis, which may occur secondary to thrombolytic therapy and cardiopulmonary bypass and, very rarely, as primary hyperfibrinolysis in association with certain cancers. These medications may also be used to retard fibrinolysis at mucocutaneous sites, including the oral cavity and genitourinary tract, when excessive spontaneous or posttraumatic bleeding

Case Study

A 52-year-old man was admitted to the emergency department with a swollen right leg of 1 day's duration. The patient had been previously well, but 2 days before admission, he drove from California to Cleveland after taking his daughter back to college. On examination, the leg was erythematous and painful, with limited range of motion of the knee joint. Most of the swelling was associated with the knee. A Wells score showed a moderate pretest probability of +2 due to the leg swelling and history of immobilization from the long car trip (Table 20-1). A complete blood count showed an elevated white blood cell count of 14,000/µL, and review of a peripheral smear showed a granulocytic left shift. The result of a high-sensitivity D-dimer (degradation product of fibrin) enzyme-linked immunosorbent assay (ELISA) was 200 ng/mL fibrinogen equivalent units (FEU; normal range = <500 ng/mL). A Doppler ultrasound imaging of the leg did not show venous thrombosis. An aspirate of the right knee joint showed a leukocytosis with negatively birefringent crystals, consistent with gout. In this case, the normal D-dimer assay was helpful to rule out a deep vein thrombosis.

occurs. One potential side effect of these medications is thrombosis. It is especially important to exclude DIC as the underlying cause for bleeding before administering these drugs. As noted above, DIC is a disorder in which fibrin strands are being deposited in the intravascular space and lysed by the fibrinolytic system. This compensatory fibrinolysis may be critical for maintaining ongoing organ perfusion. In this clinical setting, administration of an antifibrinolytic drug may result in unopposed fibrin formation and the potential for catastrophic thrombosis. Likewise, antifibrinolytic drugs are contraindicated in patients with active kidney or bladder hemorrhage, because stabilization of fibrin clots may cause ureter or bladder outlet obstruction.

The Role of D-Dimer in Excluding Venous Thromboembolism

Venous thromboembolism is a relatively common disorder, affecting 1 in 1000 Americans per year. VTE is primarily comprised of two main clinical entities: deep vein thrombosis (DVT) and pulmonary embolism (PE). Typically, DVT is further divided into distal (calf) and proximal (thigh) vein thrombosis of the lower extremity. The reason for this differentiation is because PE is most commonly due to emboli that originate from proximal vein thrombosis. About 50% of patients with untreated proximal DVT may progress to a PE in the ensuing days or weeks. Pulmonary embolism is of marked clinical significance. Untreated PE carries a 30% mortality rate, and more than 60,000 Americans die each year from acute PE. Timely diagnosis is critical because effective treatment may reduce the mortality rate to less than 8%.

A diagnosis of VTE cannot be established without objective testing. Only 10% to 25% of patients who are initially suspected of having a lower-extremity DVT actually are proven to have the disorder. This is highlighted in the case study. Therefore, the need for inexpensive noninvasive testing for screening purposes is of considerable importance in this clinical setting. D-dimer is a plasmin-mediated degradation product of fibrin that is composed of two cross-linked fibrin D-domains (see chapter 8 for assays for D-dimer). The use of D-dimer for evaluating patients with suspected VTE has been studied extensively. In general, nearly all patients with an acute VTE have an elevated D-dimer. Unfortunately, the D-dimer may be elevated in a multitude of clinical settings, including DIC, venous thrombotic disorders (DVT, PE), arterial thrombotic disorders (myocardial infarction, stroke), recent surgery, hemorrhage, trauma, liver disease, cancer, pregnancy (normal pregnancy, preeclampsia, eclampsia), increasing age, sepsis, and the systemic inflammatory response syndrome. Therefore, an *elevated* D-dimer level is not specific for VTE and cannot be used to *confirm* such a diagnosis. Instead, much of the literature has focused on the role of a *negative* D-dimer concentration, below a clinically validated cutoff value, in *excluding* VTE—in other words, the negative predictive value of the D-dimer.

Because of the high risks associated with missing a diagnosis of VTE, the D-dimer must have a high negative predictive value to be of clinical use. Unlike sensitivity and specificity metrics, the predictive value (positive or negative) of an assay is dependent upon the prevalence of the disease being assessed. The screening tests for VTE are of most clinical value when they are combined with a measure of pretest probability of VTE. The Wells clinical score (Table 20-1) has been one of the most widely used and studied scoring systems for VTE. Patients with a low pretest probability score (low prevalence of DVT) have an 87% to 100% (median = 96%) negative predictive value for DVT. The combination of low pretest probability with a negative D-dimer increases the negative predictive value to 96% to 100% (median = 99%). As the pretest probability (prevalence of disease) increases, the negative predictive value of the D-dimer diminishes. For example, a *negative* D-dimer result does *not* exclude PE in more than 15% of patients with a high pretest probability clinical assessment. In this setting, additional testing (spiral chest computed tomographic angiography, ventilation-perfusion scan, pulmonary angiography, or lower-extremity imaging to confirm a DVT) is required to either confirm or exclude the presence of PE.

D-Dimer Testing

There are several challenges surrounding the widespread adoption of the D-dimer for excluding VTE, including variation in sensitivity among assay types, selecting the appropriate cutoff point, and understanding the units of measure.

Variations in the Types of D-Dimer Assays

D-dimer assays have been divided into seven broad categories, including enzyme-linked immunosorbent assay (ELISA), quantitative rapid ELISA, semiquantitative rapid ELISA, qualitative rapid ELISA, quantitative latex, semiquantitative

Table 20-1. Pretest Probability (PTP) Calculations for Deep Vein Thrombosis (DVT) and Pulmonary Embolism (PE)

Clinical Model for DVT	
Clinical Parameter	**Score**
Active cancer	+1
Paralysis, paresis, recent casting of leg	+1
Bedridden (>3 days) or major OR (12 weeks)	+1
Entire leg swollen	+1
Calf swelling (>3 cm) compared to other leg	+1
Pitting edema greater in symptomatic leg	+1
Collateral nonvaricose superficial veins	+1
Localized tenderness along deep venous system	+1
Previously documented DVT	+1
Alternative diagnosis as or more likely than DVT	-2

Score: Low PTP, <0; Moderate PTP, 1 or 2; High PTP, >2
Wells et al (1997 and 2003).

Clinical Model for PE	
Clinical Parameter	**Score**
Clinical signs and symptoms of DVT	**+3**
Heart rate greater than 100/min	+1.5
Hemoptysis (coughing up blood)	+1
Active cancer	+1
Bedridden (≥3 days) or major OR (4 weeks)	+1.5
Previous objectively diagnosed DVT or PE	+1.5
PE most likely diagnosis	+3

Score: Low PTP, <2; Moderate PTP, 2 to 6; High PTP, >6
Wells et al (2000).

latex, and whole blood methods. For the purposes of excluding VTE, Stein et al (2004) stratified the various D-dimer assays into "high sensitivity" (including ELISA, quantitative rapid ELISA, and qualitative rapid ELISA), "moderate sensitivity" (including semiquantitative rapid ELISA, quantitative latex, and whole blood methods), and "low sensitivity" (semiquantitative latex). The semiquantitative latex assays are not appropriate for excluding VTE because they lack sufficient negative predictive value. Wells et al (2006) utilized this stratification to demonstrate that the diagnosis of DVT could be safely excluded in patients with both a low clinical probability for DVT and a *negative* (normal) D-dimer, using either high-sensitivity or moderate-sensitivity assays. This approach may safely exclude the diagnosis of DVT in up to 40% of patients with suspected DVT. By using a high-sensitivity D-dimer assay, DVT can be excluded not only in the low-probability patients, but also in those with moderate probability.

It is imperative that both the laboratory and the ordering clinician clearly understand which D-dimer methodology is being utilized in order to avoid an adverse clinical outcome (eg, prevent a diagnosis of VTE from being erroneously excluded in a patient with a proximal vein thrombosis and/or a PE).

Selecting the Appropriate Cutoff and Units for Excluding Venous Thromboembolism

As noted in chapter 8, D-dimers may be reported in D-dimer units or in fibrinogen equivalent units (FEU). The test result in FEU is twice that of D-dimer units (ie, a D-dimer result of 500 ng/mL FEU = 250 ng/mL D-dimer units). Most manufacturers calibrate their D-dimer assay in FEU, but some do report results in D-dimer units, and the clinical D-dimer literature contains many articles that do not specify the units used in the study. One of the most commonly selected cutoffs between normal and abnormal is 500 ng/mL FEU. It is imperative that the testing laboratory makes clear to the ordering clinician what the appropriate cutoff is for excluding venous thromboembolism. For example, a D-dimer result of 480 ng/mL may be below the cutoff for a particular assay if the result is in FEU, whereas if this result is in D-dimer units, a diagnosis of VTE cannot be excluded, because this is actually above the cutoff (480 ng/mL D-dimer units = 960 ng/mL FEU).

Ideally, the appropriate cutoff for each D-dimer assay should be determined and validated through prospective management trials. Results of these types of studies may not be available for each of the various D-dimer assays. Only a limited number of quantitative D-dimer assays have been studied with sufficient rigor to be approved by the US Food and Drug Administration for use in the exclusion of DVT or PE. Comparison studies and meta-analyses may assist in selecting the appropriate cutoff for an individual assay.

Summary

Although the fibrinolytic system has an important role in the remodeling and degradation of fibrin clots, abnormalities in fibrinolytic factors—either decreased profibrinolytic factors or decreased fibrinolytic inhibitors—have not been established as significant thrombophilic risk factors. Testing for fibrinolytic parameters is not typically a part of routine thrombophilia evaluations, but it may be helpful in patients with thrombophilia in whom an established cause cannot be documented. On the other hand, the D-dimer assay has been established as a helpful negative predictive marker in evaluating patients for DVT or PE, but only if quantitative assays and well-established cutoffs are used together with an assessment of clinical probability of thrombosis.

Suggested Reading

Hypofibrinolysis and Thrombosis

Ay C, Tengler T, Vormittag R, et al. Venous thromboembolism: a manifestation of the metabolic syndrome. *Haematologica*. 2007;92(3):374-380.

Brandt JT. Plasminogen and tissue-type plasminogen activator deficiency as risk factors for thromboembolic disease. *Arch Pathol Lab Med*. 2002;126:1376-1381.

Francis CW. Plasminogen activator inhibitor-1 levels and polymorphisms associated with venous thrombosis. *Arch Pathol Lab Med*. 2002;126:1401-1404.

Francis CW, Marder VJ. Clinical disorders of fibrinolysis. In: Colman RW, Clowes AW, Goldhaber SZ, Marder VJ, George JN, eds. *Hemostasis and Thrombosis: Basic Principles and Clinical Practice*. 5th ed. Philadelphia, Pa: Lippincott Williams and Wilkins; 2006: 1035-1044.

Girolami A, Randi ML, Gavasso S, Lombardi AM, Spiezia F. The occasional venous thromboses seen in patients with severe (homozygous) FXII deficiency are probably due to associated risk factors: a study of prevalence in 21 patients and review of the literature. *J Thromb Thrombolysis*. 2004;17(2):139-143.

Hayes T. Dysfibrinogenemia and thrombosis. *Arch Pathol Lab Med*. 2002;126:1387-1390.

Leurs J, Hendriks D. Carboxypeptidase U (TAFIa): a metallocarboxypeptidase with a distinct role in haemostasis and a possible risk factor for thrombotic disease. *Thromb Haemost*. 2005;94(3):471-487.

Lippi G, Targher G, Franchini M. Lipoprotein(a), thrombophilia and venous thrombosis. *Acta Haematol*. 2007;117(4):246-247.

Martini CH, Brandts A, de Bruijne EL, et al. The effect of genetic variants in the thrombin activatable fibrinolysis inhibitor (TAFI) gene on TAFI-antigen levels, clot lysis time and the risk of venous thrombosis. *Br J Haematol*. 2006;134(1):92-94.

Tirado I, Soria JM, Mateo J, et al. Association after linkage analysis indicates that homozygosity for the 46C—>T polymorphism in the F12 gene is a genetic risk factor for venous thrombosis. *Thromb Haemost.* 2004;91(5):899-904.

D-Dimer and the Diagnosis of Venous Thromboembolism

Goodacre S, Sampson FC, Sutton AJ, Mason S, Morris F. Variation in the diagnostic performance of D-dimer for suspected deep vein thrombosis. *QJM.* 2005;98:513-527.

Stein PD, Hull RD, Kalpesh CP, et al. D-dimer for the exclusion of acute venous thrombosis and pulmonary embolism: a systematic review. *Ann Intern Med.* 2004;140:589-602.

Stein PD, Woodward PK, Weg JG, et al. Diagnostic pathways in acute pulmonary embolism: recommendations of the PIOPED II investigators. *Am J Med.* 2006;119:1048-1055.

Stevens SM, Elliot CG, Woller SC, et al. The use of a fixed high sensitivity to evaluate five D-dimer assays' ability to rule out deep venous thrombosis: a novel approach. *Br J Haematol.* 2005;131:341-347.

Tamariz LJ, Eng J, Segal JB, et al. Usefulness of clinical prediction rules for the diagnosis of venous thromboembolism: a systematic review. *Am J Med.* 2004; 117:676.

Wells PS, Anderson DR, Bormanis J, et al. Value of assessment of pretest probability of deep-vein thrombosis in clinical management. *Lancet.* 1997; 350:1795-1798.

Wells PS, Anderson DR, Rodger M, et al. Derivation of a simple clinical model to categorize patients probability of pulmonary embolism: increasing the models utility with the SimpliRED D-dimer. *Thromb Haemost.* 2000;83:416-420.

Wells PS, Anderson DR, Rodger M, et al. Evaluation of D-dimer in the diagnosis of suspected deep-vein thrombosis. *N Engl J Med.* 2003;349:1227-1235.

Wells PS, Owen C, Doucette S, Fergusson D, Tran H. Does this patient have deep venous thrombosis? *JAMA.* 2006;295:199-207.

CHAPTER 21

Heparin-Induced Thrombocytopenia

Jerry B. Lefkowitz, MD

Introduction

Heparin is a glycosaminoglycan that is typically isolated from animal tissues and used pharmaceutically as an anticoagulant. The classic story of the discovery of heparin as told to generations of medical students has been that a second-year medical student working at Johns Hopkins University first isolated the material in 1916. This young man purified the anticoagulant from the canine liver; the Greek name for liver is *epar*—hence, the name "heparin." Although this story is historically accurate, the student did not isolate what we now call heparin; there are those who suggest that others should also be given credit for the discovery.

In the late 1930s, workers in North America and Scandinavia pioneered the use of heparin as a treatment for thrombosis. In the 1970s, the first well-documented report appeared in the medical literature associating therapeutic heparin use with thrombocytopenia and thrombosis. Since that report, the association of thrombocytopenia and paradoxical thrombosis with the pharmaceutical use of the anticoagulant heparin has been a diagnostic conundrum for both clinicians and laboratorians alike.

Clinical Presentation

Heparin-induced thrombocytopenia (HIT) is an immune-mediated thrombocytopenia associated with heparin therapy (also called type II) that is clinically important. A nonimmune heparin-associated thrombocytopenia (HAT; also called type I) has been described to occur soon after heparin administration and begins with a mild decrease in platelet count, which is thought to result from passive coating and elimination of platelets exposed to pharmaceutical heparin.

As the case study suggests, the rather nonspecific hallmark of HIT is a decline in the platelet count after initiation of heparin therapy. HIT develops anywhere from 5 to 10 days after initiation of treatment. The thrombocytopenia is associated with a platelet count drop of more than 50% during heparin therapy or development of absolute thrombocytopenia (<150,000/µL), with a nadir variously reported between 60,000/µL and 150,000/µL. The antibodies in HIT can persist for approximately 3 months. If a patient has had

Case Study

A 66-year-old man was admitted to the hospital to have an elective knee joint replacement. Upon admission, the patient had a complete blood count demonstrating a normal white blood cell count, a normal hematocrit, and a normal platelet count of 250,000/µL. Immediately postoperatively, the patient was started on prophylactic injections of unfractionated heparin (UFH) twice a day to prevent venous thrombosis. On postoperative day 5, a drop in the patient's platelet count to 150,000/µL was noted. On postoperative day 6, the hospital staff noted some dark, dusky-colored patches of skin near the heparin injection sites. On that same day, the patient's platelet count had fallen to 75,000/µL, while the white blood cell count had increased and the hematocrit had stayed within the reference range. Using available clinical observations and laboratory test data, the attending physician suspected a diagnosis of heparin-induced thrombocytopenia (HIT).

The attending physician immediately discontinued all heparin administration and started the patient on the direct thrombin inhibitor argatroban. For laboratory confirmation, the attending physician ordered an enzyme immunoassay (EIA) test for HIT. Later that day, the nursing staff noticed that the patient's left foot was cold and pulseless. Doppler studies suggested no blood flow to the foot, and angiographic studies demonstrated no arterial blood flow to the extremity. The patient was referred for immediate vascular surgery evaluation; it was thought that the patient's foot might be salvageable. Several days after this incident, the EIA test for HIT returned with a positive result.

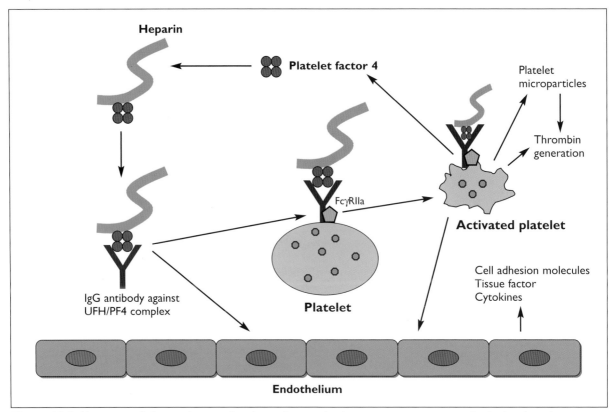

Figure 21-1. Depiction of the mechanism thought to be responsible for heparin-induced thrombocytopenia. Platelet factor 4 (PF4) is a low-molecular-weight tetrameric peptide that binds heparin. The unfractionated heparin (UFH)/PF4 complex forms an immunogenic neoantigen that can elicit an antibody response. The IgG anti-UFH/PF4 antibody can bind to the FcγRIIa receptor on the platelet membrane and cause platelet activation. With platelet activation, the ensuing platelet aggregation can cause both the thrombocytopenia and thrombosis seen with heparin-induced thrombocytopenia. Platelet aggregation can release thrombogenic platelet microparticles into the circulation, and the activated platelet surface can act as a substrate for thrombin generation. The IgG anti-UFH/PF4 antibody can bind to and cause endothelium to release a variety of activation molecules, including adhesive moieties, selectins, and cytokines.

exposure to heparin and then is re-exposed within 100 days of the first exposure, it is possible to have a rapid-onset form of HIT, with platelet counts falling within 24 hours. Usually, platelet counts start to return to normal within a few days after heparin therapy is discontinued. Finally, a delayed form of HIT has been reported, with thrombocytopenia commencing after termination of heparin therapy, but this presentation is uncommon.

The paradox with HIT is that the clinically relevant symptoms are thrombotic, unlike with other thrombocytopenias, which are frequently associated with an increased risk of hemorrhage. The spectrum of clinical presentations includes venous, arterial, and microvascular thromboses. The most common thrombotic symptoms are venous, which can include deep venous thrombosis, pulmonary embolism, venous-associated limb gangrene, adrenal hemorrhage, and cerebral venous sinus thrombosis. Arterial thromboses can include lower-limb circulation thrombosis, stroke, myocardial infarction, and other sites. Skin necrosis and erythematous plaques can occur at the site of heparin injections. Finally, acute reactions, such as fevers, chills, flushing, and transient amnesia, have also been reported. However, despite the often profound thrombocytopenia, bleeding complications are unusual in HIT.

Commonly, HIT is seen with intravenous administration of heparin; however, it is important to be cognizant that any exposure, even the small amount of heparin used to maintain the patency of intravenous catheters, can cause HIT. Unfractionated heparin (UFH) is the form of the drug most often associated with HIT. Low-molecular-weight heparin also can cause HIT, although not nearly as frequently as UFH. Fondaparinux (Arixtra), a synthetic glycosaminoglycan resembling heparin, is believed not to cause HIT and has been used to treat the condition, but definitive

Table 21-1. Pretest Probability of Heparin-Induced Thrombocytopenia (HIT) Using the Four Ts

Points	2	1	0
Thrombocytopenia	Platelet count drop >50% during heparin therapy	30%-50% fall in platelet count or nadir <150,000/μL	<30% fall in platelet count or nadir <10,000/μL
Timing of platelet count fall	Onset between days 5-10 or <1 day if heparin exposure in last 100 days	Onset of thrombocytopenia after day 10	Platelet count drop before day 5 without heparin exposure in last 100 days
Thrombosis or skin lesions at heparin injection site	New thrombosis, skin necrosis, post heparin bolus acute systemic reaction	Progressive or recurrent thrombosis, erythematous skin lesions, thrombosis not yet proven	None
O**T**her causes for thrombocytopenia	No other causes present	Possible cause is present	Definite other cause present

Points are tallied from each individual row listed above to create the pretest probability score. The maximum score for each row is 2 points. 6-8 points = high probability of HIT; 4-5 points = intermediate probability of HIT; <4 points = low probability of HIT. From: Warkentin TE, Heddle NM. Laboratory diagnosis of immune heparin-induced thrombocytopenia. *Curr Hematol Rep.* 2003;2:148-157. Copyright 2003 Current Medicine Group LLC. Adapted with permission.

studies to confirm the safety of fondaparinux in patients with HIT have not been completed.

The incidence of HIT also varies with the type of patient. Surgical patients have a higher incidence of HIT than do medical patients. About 5% of patients undergoing an orthopedic surgical procedure will develop clinical HIT. Almost half of cardiac surgery patients on cardiopulmonary bypass develop HIT antibodies, but only about 5% acquire clinical HIT. Although heparin is used in children and infants, rarely do these groups develop clinically significant HIT. In patients who develop HIT, approximately 25% to 50% will develop thrombosis.

Pathophysiology

Figure 21-1 illustrates the mechanism that is thought to be responsible for the pathology observed in HIT. Platelet factor 4 (PF4) is a low-molecular-weight tetrameric peptide that binds heparin and is found in the platelet alpha granule. The UFH/PF4 complex forms an immunogenic neoantigen that can elicit an antibody response. Although IgM and IgA antibodies can form against the UFH/PF4 antigen, the IgG response is thought to be the only clinically important antibody. The IgG anti-UFH/PF4 antibody can bind to the FcγRIIa receptor on the platelet membrane and cause platelet activation. With platelet activation, the ensuing platelet aggregation can cause both the thrombocytopenia and thrombosis seen with HIT. Platelet aggregation can release thrombogenic platelet microparticles into the circulation, and the activated platelet surface can act as a sub-

strate for thrombin generation. The IgG anti-UFH/PF4 antibody can bind to and cause endothelium to release a variety of activation molecules, including adhesive moieties, selectins, and cytokines. Monocytes also can be activated to release tissue factor by the IgG anti-UFH/PF4.

Clinical Laboratory Diagnosis and Algorithm

Thrombocytopenia in HIT is usually isolated and not associated with other complete blood count abnormalities. Examination of the peripheral smear confirms the low platelet count, usually without platelet clumping. If extensive in vivo platelet activation is occurring, some of the circulating platelets may appear degranulated on the peripheral smear. If the sample is drawn into heparin anticoagulant, instead of EDTA, platelet clumping may be observed on the blood smear. The presence of schistocytes would suggest etiologies other than HIT, such as thrombotic thrombocytopenic purpura or disseminated intravascular coagulation.

The diagnosis of HIT relies on clinicopathologic findings; the laboratory is only an adjunct to aid in diagnosis. Because of the inability of any single laboratory test to provide an unequivocal identification, experts in this field recommend estimating pretest probability using the patient's clinical symptoms, platelet count, and drug history. Table 21-1 summarizes this suggested scoring system.

Figure 21-2 illustrates an algorithmic approach using a combination of pretest probability and lab-

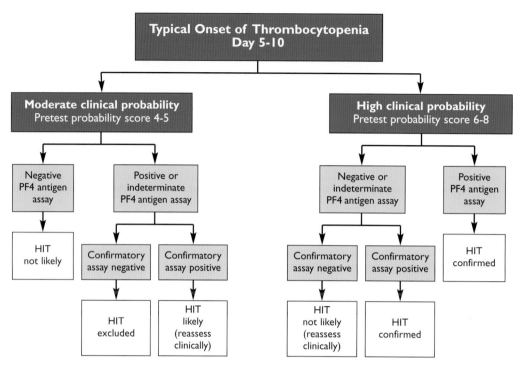

Figure 21-2. Algorithm for diagnosis of heparin-induced thrombocytopenia (HIT) using platelet factor 4 (PF4) antigen assay and confirmatory assay (serotonin release assay and heparin-induced platelet aggregation). Most patients with an onset thrombocytopenia typical for HIT (presenting on days 5 to 10) have a moderate to high pretest probability.

oratory testing in a patient developing thrombocytopenia 5 to 10 days after the start of initial heparin therapy. For rapid-onset HIT, where the platelet count drops within 4 days of heparin re-exposure in a patient with a history of heparin therapy in the preceding 3 months, the algorithm in Figure 21-3 is suggested.

The routinely available laboratory testing uses either an antibody detection method or functional platelet activation method. The most commonly available enzyme immunoassay (EIA) method is performed in microtiter plates detecting anti-PF4/heparin antibodies. The EIA methods take several hours to perform and are not typically available as stat tests. Two kits that are commonly available in the United States are manufactured by GTI Diagnostics (Waukesha, Wisconsin) or Diagnostica Stago (Parsippany, New Jersey). The GTI kit uses a heparin-like material, polyvinylsulfonate complexed to PF4, whereas the Diagnostica Stago kit uses heparin complexed to PF4.

With the GTI kit, a second step can be used to try to confirm that the antibody reaction is heparin specific. The assay is repeated after a positive result, except that this time excess heparin in a final concentration of 100 U/mL is added to the reaction. In the presence of anti-PF4/heparin anti-bodies, the excess heparin interferes with the heparin-specific antibody binding. The instructions with the GTI kit provide a formula for calculating inhibition by excess heparin. If the addition of excess heparin inhibits the reaction by 50% or more, then the reaction is confirmatory. If inhibition is less than 50%, then the result is considered equivocal or indeterminate and not positive for heparin-specific antibodies. An indeterminate test result could be due to non–heparin-dependent anti-PF4 antibodies.

The EIA methods tend to have a high negative predictive value and sensitivity greater than 95%. The specificity of these assays is typically much less—generally about 80%. Because the EIA specificity tends to be lower, a positive test needs to be interpreted in terms of the clinical presentation of the patient. The lower specificity of the EIA tests is thought to be due to the fact that these assays detect all species of anti-PF4/heparin antibodies: IgG, IgM, and IgA. Because it is thought that only the IgG anti-PF4/heparin antibodies are important in causing HIT pathology, the specificity of the EIA could be improved by detecting only the IgG anti-PF4/heparin antibodies. An assay to detect only IgG anti-PF4/heparin antibodies has been approved for use in the US, but the clinical utility for this kit remains to be established.

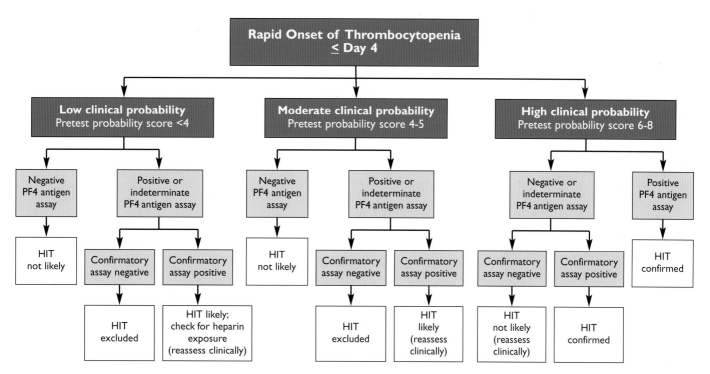

Figure 21-3. Algorithm for diagnosis of rapid-onset heparin-induced thrombocytopenia (HIT) using platelet factor 4 (PF4) antigen assay and confirmatory functional assays. The algorithm is based on pretest probability for HIT.

The confirmatory assays are platelet activation assays. One such assay, the serotonin release assay (SRA), is considered the "gold standard." This assay uses washed donor platelets loaded with radioactive ^{14}C-labeled serotonin, heat-inactivated serum from the patient, and heparin. The heat inactivation step is thought to denature residual thrombin, which is a potent platelet activator. The patient's serum and low-concentration heparin (typically 0.1 U/mL) are mixed with ^{14}C-serotonin-loaded normal donor platelets, and platelet activation is detected by release of radioactivity in the supernatant. The reaction is repeated with the patient's serum and excess heparin, typically at 100 U/mL. The excess heparin inhibits the ability of the anti-PF4/heparin antibodies to activate the platelets by disrupting the antibody/antigen complexes. The SRA test is more sensitive and specific than the UFH/PF4 EIA methods; specificity has been reported to be up to 100%, with sensitivity at 88%. The positive predictive value of the SRA has been reported as high as 100%.

Like all laboratory tests, the SRA does have limitations. The test is technically demanding, uses radioactivity, and is time consuming; hence, it is usually only available at reference laboratories, with an expected test turnaround time of several days. In addition, up to 5% of samples tested will give an equivocal or indeterminate result. For the SRA, an equivocal result is seen when all concentrations of added heparin cause radioactivity release. Resubmitting another specimen from the patient for SRA testing often yields a clear answer. However, consistently indeterminate SRA results are seen with some patients; for those, the EIA methods then become the preferred laboratory testing modality.

The heparin-induced platelet aggregation (HIPA) assay is another functional assay that uses normal donor, citrate-anticoagulated, platelet-rich plasma (PRP) and the patient's plasma or serum. This assay is performed frequently, but it is not recommended by experts in this area. The HIPA assay is performed in a platelet aggregometer. An aliquot of normal donor platelets is first tested with a platelet agonist, such as adenosine diphosphate (ADP), to ensure that the platelets are functional. Next, an aliquot of donor platelets is mixed with the patient's serum or plasma and then observed for 4 to 5 minutes to exclude any nonspecific aggregation response. Then heparin, at a concentration of between 0.1 and 1.0 U/mL, is added and aggregation is monitored for approximately 15 minutes. An aggregation response of 25% or greater above baseline is interpreted as an initial positive response. To help determine if the aggregation response is heparin specific, a second aliquot, consisting of donor platelets plus plasma or serum from the patient, is mixed with heparin to a final concentration of 100 U/mL. If this high-concentration heparin inhibits platelet aggregation, it suggests that the test is detecting anti-PF4/heparin antibodies.

The HIPA assay using PRP is confounded by numerous variables that result in false-positive and false-negative test results. Problems with the HIPA assay that make it difficult to perform in the laboratory include (1) some donor's platelets are unresponsive, (2) potential donors should be screened with known-positive plasma, and (3) optimally, platelets from three donors should be pooled. There may be some donor reactivity variation due to Fc receptor phenotype. Some advocate using platelets from donors with blood type O to avoid ABO antibody response. Occasionally, a positive aggregation response is seen without added heparin in patients who have alloantibodies with cross-reactivity to the donor platelets. Conversely, aggregation can be falsely negative in patients receiving intravenous glycoprotein IIb/IIIa (GPIIb/IIIa) antagonists, due to inhibition of donor platelets.

A variety of other assays for detection of HIT have been reported. Many employ some variation of the HIPA or platelet activation assays. HIPA using washed platelets is suggested to be more sensitive and specific than the comparable assay using PRP. This assay is more technically demanding than the HIPA using PRP, but it is also subject to some of the same problems detailed for HIPA with PRP.

A lumiaggregometry-based method using the Chrono-log Platelet Aggregometer (Havertown, Pennsylvania) has been reported. Lumiaggregometry measures both platelet aggregation by conventional spectrophotometry and also secretion of adenosine triphosphate (ATP) using an ATP-dependent luminescent enzyme system. This assay uses the release of ATP from platelet dense granules as the endpoint of platelet activation by anti-PF4/heparin antibodies.

Flow cytometry procedures have been developed to study an assortment of platelet activation markers. The flow cytometry assays use donor platelets from one to several donors. As with other functional assays, the serum from a patient who is suspected of having HIT is mixed with donor platelets along with low-concentration heparin and fluorescein-tagged antibodies against a platelet activation marker. The sample is then analyzed in a flow cytometer to look for markers of platelet activation. To demonstrate that the platelet activation is truly due to anti-PF4/heparin antibodies, a second sample is run, looking for inhibition of platelet activation with the addition of excess heparin. Platelet markers that have been used in these assays include P-selectin and annexin V, but others have also been used. A variation of this assay is to look for platelet-derived microparticles as a marker of platelet activation. These microparticles can be quantified by size using flow light scatter and specifically by tagging their glycoprotein Ib (GPIb) receptor with fluorescent antibody. The quantity of microparticles versus platelet numbers is the endpoint for this assay.

A rapid-turnaround-time assay for anti-PF4/heparin antibodies is currently marketed in the United States by Akers Biosciences (Thorofare, New Jersey). The test principle uses a particle immunofiltration assay with dyed microparticles coated with PF4. A positive test result is the presence of no blue color in the test window of the device, with red color present in the control window. There is limited experience with this assay at the present time.

A newer assay method on the horizon, but not currently available in the United States, uses red-dyed polymer particles coated with human PF4 to detect anti-PF4/heparin antibody complexes (PaGIA, Diamed, Switzerland). The patient's serum is incubated with the polymer particles and applied to the top of a gel column, which is then centrifuged. If the antibody is present, the immune complexes bind to the polymer particles and are trapped at the top of the gel. Otherwise, the polymer particles are forced to the bottom of the gel.

Treatment

The cardinal rule is that if HIT is suspected, heparin therapy should be discontinued immediately, and the patient should be anticoagulated with a direct thrombin inhibitor. These drugs are described in chapter 24. Low-molecular-weight heparins are contraindicated as a heparin replacement if HIT is suspected, because of high rates of antibody cross-reactivity. Fondaparinux, a synthetic pentasaccharide anti-Xa inhibitor, has not been associated with risk of HIT, but definitive studies for use of fondaparinux as an alternative to heparin in HIT have not been performed. Oral anticoagulation with warfarin should not be started, because of risk of venous limb gangrene, until substantial platelet recovery has occurred, with a platelet count of at least 150,000/μL. Warfarin should be initiated concurrently with a direct thrombin inhibitor until a therapeutic international normalized ratio (INR) is achieved.

The direct thrombin inhibitors approved for use in treating HIT in the United States include argatroban, lepirudin, and bivalirudin. They are discussed in detail in chapter 24. Argatroban is approved by the Food and Drug Administration (FDA) for use as a prophylaxis or treatment in HIT, including for patients undergoing percutaneous cardiac treatments. Lepirudin is only FDA approved for HIT associated with thrombosis, whereas bivalirudin is FDA approved for patients at risk of HIT who are undergoing percutaneous cardiac treatments. None of these agents has an antidote, and their use can be associated with an increased bleeding risk.

The anticoagulant effect of these agents can be monitored by the activated partial thromboplastin time (aPTT). The effects of argatroban and bivalirudin can also be followed by the activated whole-blood clotting time. Bivalirudin has the shortest half-life, 36 minutes, whereas the half-life of argatroban varies between 39 and 51 minutes. The longest half-life agent is lepirudin, at 1.7 hours. Argatroban is eliminated by the liver; the other two drugs are predominantly cleared by the kidneys, although bivalirudin also has an enzymatic elimination component. The choice of any one of these agents depends upon the patient's condition and the indications for use.

Cardiac surgery can present a special case for treatment of HIT because most protocols depend on heparin anticoagulation during the surgery. Many workers in this field suggest minimization of heparin exposure and perhaps limiting exposure to only during the surgery. Before and after surgery, some alternative anticoagulation should be considered. Results from experience with alternative anticoagulants suggest that these agents, especially bivalirudin, may be used during all phases of cardiac surgeries, although this is not currently a routine option.

Suggested Reading

Introduction

Girolami B, Girolami A. Heparin-induced thrombocytopenia: a review. *Semin Thromb Hemost.* 2006; 32(8):803-809.

Warkentin TE. Drug induced thrombocytopenia: from purpura to thrombosis. *N Engl J Med.* 2007;356:891-893.

Wintrobe MM, ed. *Blood, Pure and Eloquent: A Story of Discovery, of People, and of Ideas.* New York, NY: McGraw-Hill; 1980: 601-657.

Clinical Presentation

Spinler SA. New concepts in heparin-induced thrombocytopenia: diagnosis and management. *J Thromb Thrombolysis.* 2006;21:17-21.

Warkentin TE. Clinical picture of heparin-induced thrombocytopenia. In: Warkentin TE, Greinacher A, eds. *Heparin-Induced Thrombocytopenia.* 3rd ed. New York, NY: Marcel Dekker; 2004.

Pathophysiology

Poncz M. Mechanistic basis of heparin-induced thrombocytopenia. *Semin Thorac Cardiovasc Surg.* 2005; 17(1):73-79.

Visentin GP, Bacsi SG, Aster RH. Molecular immunopathogenesis of heparin-induced thrombocytopenia. In: Warkentin TE, Greinacher A, eds. *Heparin-Induced Thrombocytopenia.* 3rd ed. New York, NY: Marcel Dekker; 2004.

Clinical Laboratory Diagnosis and Algorithm

Greinacher A, Juhl D, Strobel U, et al. Heparin induced thrombocytopenia: a prospective study on the incidence, platelet activating capacity and clinical significance of antiplatelet factor 4/heparin antibodies of the IgG, IgM, and IgA classes. *J Thromb Haemost.* 2007;5:1666-1673.

Lee DP, Warkentin TE, Denomme GA, Hayward CPM, Kelton JG. A diagnostic test for heparin-induced thrombocytopenia: detection of platelet microparticles using flow cytometry. *Br J Haematol.* 1996;95:724-731.

Poley S, Mempel W. Laboratory diagnosis of heparin-induced thrombocytopenia: advantages of a functional flow cytometric test in comparison to the heparin-induced platelet-activation test. *Eur J Haematol.* 2001;66:253-262.

Pouplard C, Amiral Borg J-Y, Laporte-Simitsidis S, Delahousse B, Gruel Y. Decision analysis for use of platelet aggregation test, carbon 14-serotonin release assay, and heparin-platelet factor 4 enzyme-linked immunosorbent assay of diagnosis of heparin induced thrombocytopenia. *Am J Clin Pathol.* 1999;111:700-706.

Pouplard C, Gueret P, Foucassier M, et al. Prospective evaluation of the "4Ts" score and particle gel immunoassay specific to heparin/PF4 for the diagnosis of heparin-induced thrombocytopenia. *J Thromb Haemost.* 2007;5:1373-1379.

Stewart MW, Etches WS, Boshkov LK, Gordon PA. Heparin-induced thrombocytopenia: an improved method of detection based on lumi-aggregometry. *Br J Haematol.* 1995;91:173-177.

Tomer A, Masalunga C, Abshire TC. Determination of heparin-induced thrombocytopenia: a rapid flow cytometric assay for direct demonstration of antibody-mediated platelet activation. *Am J Hematol.* 1999;61:53-61.

Warkentin TE. Platelet count monitoring and laboratory testing for heparin-induced thrombocytopenia: recommendations of the College of American Pathologists. *Arch Pathol Lab Med.* 2002;126:1415-1423.

Warkentin TE, Greinacher A. Laboratory testing for heparin-induced thrombocytopenia. In: Warkentin TE, Greinacher A, eds. *Heparin-Induced Thrombocytopenia.* 3rd ed. New York, NY: Marcel Dekker; 2004.

Warkentin TE, Heddle NM. Laboratory diagnosis of immune heparin-induced thrombocytopenia. *Curr Hematol Rep.* 2003;2:148-157.

Warkentin TE, Sheppard JA. Testing for heparin-induced thrombocytopenia antibodies. *Transfus Med Rev.* 2006;20(4):259-272.

Treatment

Koster A, Dyke CM, Aldea G, et al. Bivalirudin during cardiopulmonary bypass in patients with previous or acute heparin-induced thrombocytopenia and heparin antibodies: results of the CHOOSE-ON trial. *Ann Thorac Surg.* 2007;83:572-577.

Levy JH, Hursting MJ. Heparin-induced thrombocytopenia, a prothrombotic disease. *Hematol Oncol Clin North Am.* 2007;21:65-88.

McRae SJ, Ginsberg JS. New anticoagulants for venous thromboembolic disease. *Curr Opin Cardiol.* 2005; 20:502-508.

Papadopoulos S, Flynn JD, Lewis DA. Fondaparinux as a treatment option for heparin-induced thrombocytopenia. *Pharmacotherapy.* 2007;27(6):921-926.

Warkentin TE, Greinacher A. Heparin-induced thrombocytopenia and cardiac surgery. *Ann Thorac Surg.* 2003;76:638-648.

Warkentin TE, Greinacher A, Koster A, Lincoff AM. Treatment and prevention of heparin-induced thrombocytopenia: American College of Chest Physicians Evidence-Based Clinical Practice Guidelines (8th Edition). *Chest.* 2008;133(suppl 6): 340S-380S.

CHAPTER 22

Antiphospholipid Antibodies

Elizabeth M. Van Cott, MD
Charles Eby, MD

Introduction

Antiphospholipid antibodies are acquired autoantibodies directed against phospholipid-protein complexes. These antibodies are associated with an increased risk for venous and arterial thrombosis as well as miscarriage. The two main types of antiphospholipid antibodies are lupus anticoagulants and anticardiolipin antibodies. The antiphospholipid antibody syndrome is diagnosed in patients with a history of thrombosis or miscarriage plus the presence of objective laboratory criteria that persist for more than 12 weeks.

Lupus anticoagulants are acquired autoantibodies that recognize epitopes on selected proteins when bound to phospholipid surfaces. They can be identified by their behavior, in vitro, of prolonging a variety of phospholipid-dependent clotting tests. Although originally identified in patients with systemic lupus erythematosus, leading to the name lupus anticoagulant, this heterogeneous group of antibodies rarely causes excessive bleeding, and although lupus anticoagulants are common in patients with systemic lupus erythematosus and other autoimmune disorders, most patients with a lupus anticoagulant do not have an autoimmune condition.

Anticardiolipin antibodies recognize a complex of cardiolipin (a phospholipid normally found in mitochondria) bound to a protein called beta 2-glycoprotein I (β_2GPI). The in vivo function of β_2GPI is not certain, but it is known to bind to anionic phospholipid membranes. Recent evidence shows that it inhibits von Willebrand factor. Lupus anticoagulants are more heterogeneous. Laboratory investigations have identified several target proteins (prothrombin, β_2GPI, protein C, protein S, and annexin V) and have confirmed that the serum from a patient who is identified as having lupus anticoagulants typically contains autoantibodies that recognize more than one epitope.

The pathologic mechanisms involved in antiphospholipid antibody-associated thrombotic complications remain obscure. Many different mechanisms have been proposed, but one mechanism implicates antiphospholipid antibodies in the disruption of annexin V (annexin A5) binding to phospholipid membranes. Annexin V is a protein that prevents the formation of coagulation complexes on phospholipid surfaces; decreased binding caused by antiphospholipid antibodies could lead to increased coagulation complex formation and thrombosis. In addition, there is evidence that anti-β_2GPI antibodies neutralize β_2GPI inhibition of von Willebrand factor.

Laboratory Testing for Antiphospholipid Antibodies

Patients suspected of having antiphospholipid antibody syndrome should be tested for both lupus anticoagulants and anticardiolipin antibodies because patients with this syndrome can have lupus anticoagulants alone, anticardiolipin antibodies alone, or both. The case study on the following page provides an example of the laboratory evaluation for phospholipid antibodies.

Lupus Anticoagulant Testing Considerations and International Society on Thrombosis and Haemostasis (ISTH) Criteria

In the hemostasis laboratory, the presence of a lupus anticoagulant (LA) is detected indirectly by:

❏ observing a prolongation of a phospholipid-dependent clotting test designed to be sensitive to LA;
❏ demonstrating an inhibitor effect and ruling out a coagulopathy by showing incomplete correction of the prolonged clotting time in a 50:50 mix of patient and normal pooled plasma;
❏ demonstrating phospholipid dependence, typically shown by shortening of the clotting time with the addition of more phosopholipid; and

Case Study

A 23-year-old woman with a history of Hashimoto thyroiditis was admitted with a fever and rash on her face, chest, and back, as well as a 2-week history of generalized stiffness in her joints. Routine admission coagulation tests revealed the following:

Test	Value	Reference Range
PT (s)	12.5	11.2-13.2
aPTT (s)	53.2	22.1-34.1
aPTT after heparinase* (s) (*degrades heparin if present)	52.8	
Fibrinogen (mg/dL)	273	150-400

The prolonged aPTT was unexplained and unexpected. The aPTT prolongation is not due to heparin, because the aPTT remained prolonged after treating the specimen with heparinase. Therefore, the clinicians asked the laboratory to identify the etiology of the aPTT prolongation. Further studies revealed the following: aPTT 1:1 mix 37.0 seconds at time 0, 37.9 seconds at 30 minutes, and 38.7 seconds at 60 minutes. Because these clotting times were prolonged (not corrected), lupus anticoagulant (LA) assays were performed, yielding the following results:

Test	Value	Reference Range
PTT-LA (s) (lupus-sensitive aPTT)	70.2 = positive screen	29-45
PTT-LA 1:1 mix (s)	58.1 (not corrected) = positive inhibitor mix	
Hexagonal phase STACLOT LA® positive (96.5 s with buffer, 69.8 s with hexagonal phase phospholipids = Δ 27.0, where Δ >8 is positive)	= confirms phospholipid dependency of inhibitor activity	
dRVVT screen (s)	55	<41
dRVTT mix (s)	48	<40
dRVVT confirm (s)	35	
Ratio dRVVT screen:confirm	1.57	<1.22

Test	Value	Reference Range
IgG anticardiolipin antibody (IgG phospholipid units)	48.1	<15
IgM anticardiolipin antibody (IgM phospholipid units)	27.9	<15

To complete the evaluation, aPTT factor assays were performed to ensure that an underlying factor deficiency (or inhibitor) was not present, given the prolonged aPTT. The results were as follows:

	Assay Dilution				
	1:5	1:10	1:20	1:40	1:80
Factor VIII (%)	–	86	105	126	142
Factor IX (%)	36	43	47	51	56
Factor XI (%)	16	36	51	54	59
Factor XII (%)	32	50	70	74	80

The factor assay results increase with increasing dilution, which is characteristic of LA interference in these aPTT-based assays. This does not always occur with LA, and it is also possible for the inhibitory interference to be seen in some of the factor assays but not others. The fact that the factor levels are normal or near normal at the highest dilution indicates that there is no laboratory evidence for the presence of a specific factor inhibitor. These results also demonstrate the importance of routinely performing factor assays at two or more dilutions. For example, if only the 1:5 or 1:10 dilution had been performed, the factor levels would have been misdiagnosed as being deficient.

In conclusion, the aPTT prolongation is due to a lupus anticoagulant. As a result of the patient's clinical evaluation during her hospitalization, she was diagnosed with systemic lupus erythematosus. Her antiphospholipid antibodies persisted beyond 12 weeks, and therefore she would qualify for a diagnosis of antiphospholipid antibody syndrome if she developed thrombosis or miscarriage within 5 years of the positive tests. After several years of follow-up, she has not developed thrombosis or miscarriages and therefore does not have antiphospholipid antibody syndrome.

❏ evaluating for the possibility of a co-existing specific factor inhibitor, particularly against factor VIII, or an anticoagulant drug such as heparin or direct thrombin inhibitor.

Because of both LA heterogeneity and variation in methods, reagents, and instrumentation, no single clotting test provides adequate sensitivity for detection of lupus anticoagulants. Therefore, in 1995 the International Society on Thrombosis and Haemostasis Scientific Subcommittee on Lupus Anticoagulants/Phospholipid-Dependent Antibodies recommended that at least two sensitive screening tests for LA that assess different components of the coagulation pathway—intrinsic (activated partial thromboplastin time, kaolin clotting time), extrinsic (dilute prothrombin time), and common pathway (dilute Russell viper venom time)—be employed, and if positive, mixing and confirmation steps be performed using the same test method. Multiple methods and reagents have been proposed for LA testing, and laboratories use a variety of combinations of in-house and commercial assays. As a result, proficiency testing results for LA have been poor for borderline- and weak-positive samples, whereas better agreement has been reported for negative or strongly positive LA plasmas. Presently, a lupus anticoagulant standard does not exist, although β_2GPI monoclonal antibodies have been used successfully for LA proficiency testing. The strength or potency of a positive LA plasma has not been shown to correlate with the risk of thrombotic complications, and therefore, LA test results are usually reported as positive or negative. In order to fulfill the consensus laboratory criteria for antiphospholipid antibody syndrome, persistence of a lupus anticoagulant must be confirmed by repeating the test at an interval of greater than or equal to 12 weeks. LA test results that are initially weakly positive are often transient, reinforcing the importance of confirmatory testing.

Preanalytical Variables

There are numerous preanalytical variables that can affect LA test results. Residual platelets in the test plasma (and control plasma used for mixing steps), especially if samples are frozen and thawed at a later time for testing, can produce platelet microparticles that absorb and neutralize LA antibodies, thus producing a false-negative screening or mixing test. To avoid this interference, quality control measures should be undertaken to ensure that residual platelet counts are less than 10,000/μL. Filtration of test plasma through a 0.22-μm filter before freezing or after thawing will also remove residual platelets and microparticles. The normal pooled plasma used in mixing studies should not be lyophilized.

Patients with inherited or acquired coagulopathies, or who are taking an oral anticoagulant (warfarin), may have false-positive LA screen results, because warfarin may cause an elevation of a dilute prothrombin time (dPT), activated partial thromboplastin time (aPTT), and dilute Russell viper venom time (dRVVT). Ideally, repeating the screening test on a 50:50 mix of test and control plasma will produce a substantial correction with factor deficiencies and warfarin therapy, but false-positive mixing studies can occur with warfarin. In general, although lupus anticoagulants are associated with a positive 50:50 mix-

ing study, false-negative mixing studies may occur with some weak lupus anticoagulants.

Plasma that contains heparin may produce false-positive LA screening tests and complicate the interpretation of mix-and-confirm steps. Performing a thrombin time, and, if prolonged, performing a thrombin time after treating the specimen with protamine or heparinase, or a reptilase time, which is insensitive to heparin, will identify heparin-contaminated samples. Options include neutralizing or degrading heparin in the test plasma before performing LA testing, using commercial LA reagents that contain a heparin-neutralizing material, or obtaining a new sample that is free of heparin. Even if the LA reagent contains a heparin neutralizer, caution should be used because heparin concentrations often exceed the heparin-neutralizing capability of the reagent (typical neutralization up to 1 U/mL anti-Xa activity). Direct thrombin inhibitors (DTIs), such as argatroban, bivalirudin, and lepirudin, may cause false-positive LA results by blocking the active site on thrombin molecules. A thrombin time is a sensitive screen for DTIs as well, but it will remain prolonged after a heparin neutralization step if a DTI is present. Because DTIs cannot be neutralized or absorbed, LA testing should not be performed when the presence of a DTI is suspected, and a new specimen should be obtained after the DTI therapy has been discontinued.

Elevated factor VIII or baseline short clotting times might cause false-negative LA tests by preventing what would otherwise be a prolonged clotting time.

Testing for a Lupus Anticoagulant

Lupus anticoagulants prolong various phospholipid-dependent clotting times because lupus anticoagulants are antiphospholipid antibodies, and phospholipid is essential for several steps in the coagulation cascade. If LA is present, it binds to phospholipid-protein complexes in the test tube, thereby interfering with the coagulation cascade and prolonging the clotting times. A testing algorithm for laboratory diagnosis of the presence of LA is shown in Figure 22-1 and described below.

The most commonly used screening test for LA is an aPTT-based method. Laboratories can make their own sensitive aPTT reagent, for example, using the Bell and Alton extract, but most use a commercial LA-sensitive aPTT product. The diversity of reagent (quantity and quality of phospholipid) and instrument combinations is respon-

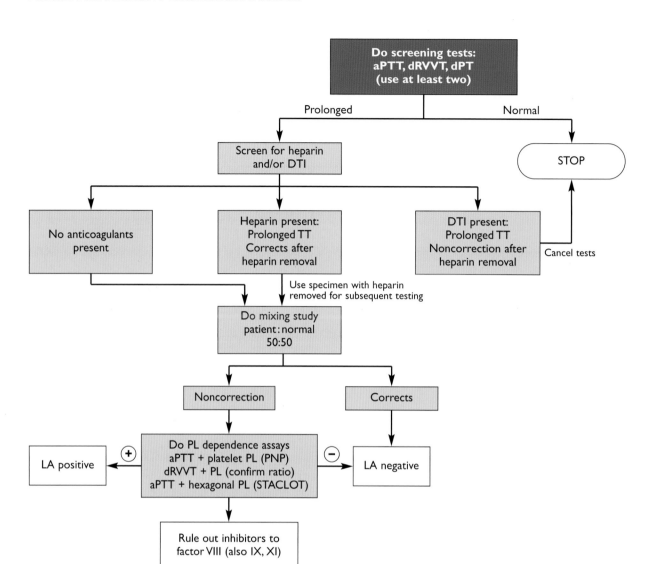

Figure 22-1. There are many possible laboratory algorithms for diagnosis of the presence of a lupus anticoagulant (LA). Testing usually starts with at least two phospholipid-dependent assays, such as the activated partial thromboplastin time (aPTT), the dilute Russell viper venom time (dRVVT), or the dilute prothrombin time (dPT). If the screening test result is prolonged, a heparin or direct thrombin inhibitor (DTI) effect should be excluded. If heparin is present, the sample can be tested further after heparin removal.

A mixing study, using a 50:50 mix of the patient's plasma with normal pooled plasma, will not show correction with a lupus anticoagulant. Demonstration of phospholipid (PL) dependence can be performed by repeating the abnormal screening assay with the addition of phospholipid and demonstrating a normalization of the clotting time. A factor inhibitor should also be excluded by performing factor assays. Abbreviations: PNP, platelet neutralization procedure; TT, thrombin time.

sible for substantial variability in aPTT sensitivity for LA.

When the screening aPTT is prolonged, and the presence of an anticoagulant such as heparin or DTI is ruled out, a 50:50 mix with normal pooled plasma would be the next step. Often, mixing studies are performed before and after incubation of the mixture of patient's plasma with normal pooled plasma at 37°C to distinguish an LA from a specific factor inhibitor. Lupus anticoagulants typically show an inhibitory effect immediately upon mixing the patient's plasma with normal plasma, whereas factor V and factor VIII inhibitors show an inhibitory effect only after prolonged incubation. The performance and interpretation of mixing study assays has been covered in detail in chapter 6. If the aPTT is not adequately corrected (there is no consensus on the criteria for evaluating 50:50 mix results), a confirmation step is performed by repeating the aPTT with addition of extra phospholipid or by performing the platelet neutralization procedure (PNP).

The PNP is performed by repeating the aPTT with a platelet lysate as a source of phospholipids; if a lupus anticoagulant is present, the aPTT should become shorter as a result of binding by the excess membrane phospholipids. Platelet lysate preparations are available commercially or can be prepared from outdated platelets from the blood bank. To prepare the lysate, platelets are centrifuged at 197g to remove red blood cells and washed three times with Tris-buffered saline. The platelet count is adjusted to 200,000 to 300,000 platelets/µL, and aliquots are frozen at -20°C. To perform the PNP, a platelet lysate aliquot is thawed and two aPTTs are performed in parallel: (1) lysate aPTT performed with 0.1 mL aPTT reagent + 0.1 mL test plasma + 0.1 mL frozen platelet suspension/excess phospholipid; and (2) saline aPTT performed with 0.1 mL aPTT reagent + 0.1 mL test plasma + 0.1 mL normal saline. The mixtures are incubated at 37°C for 5 minutes, 0.1 mL 25 mM CaCl$_2$ is added, and clotting times are determined. The PNP difference (delta) is calculated as follows:

$$\Delta = \text{saline aPTT} - \text{lysate aPTT}$$

Each laboratory should determine an upper limit for the delta based on a local reference population for each lot of platelet lysate; if a patient's delta exceeds the cutoff, this is indicative of phospholipid dependence.

The hexagonal phase phospholipid neutralization (STACLOT LA; Diagnostica Stago, France) is a popular and sensitive commercial aPTT-based LA test that combines screen, mix, and confirm steps. The source of extra phospholipid is a hexagonal phase phospholipid. The screen step is an aPTT performed with the LA-sensitive phospholipid activator provided with the kit, and a positive cutoff is determined from a reference population. The mix-and-confirm step is reserved for plasmas with a positive screen result. Skipping the screen step and performing mix-and-confirm STACLOT LA testing on all plasmas submitted for LA testing appears to improve sensitivity. The mix-and-confirm step requires preparation of two tubes containing a 50:50 mix of test plasma and normal pooled plasma (provided in the kit, with polybrene to neutralize heparin). Buffer is added to tube 1, and a phospholipid (phosphatidylethanolamine) extract from soybeans that retains a hexagonal structure and avidly binds antiphospholipid antibodies is added to tube 2. After incubation, a lupus-sensitive aPTT reagent is added to each tube, followed by CaCl$_2$, and clotting times

are determined. If a lupus anticoagulant is present, a shorter aPTT should be obtained with the addition of hexagonal phase phospholipid to tube 2, and the difference (delta), calculated as Δ = tube 1 aPTT – tube 2 aPTT, will be positive. The manufacturer recommends a cutoff of greater than or equal to 8 seconds as diagnostic for the presence of LA when testing is performed on a manual ST4 (STAGO) instrument; however, laboratories should establish their own lot-specific reference ranges for the assay. When the test is used on other platforms, an instrument-specific cutoff should be determined. False-positive STACLOT LA results have been reported when a moderate or strong factor VIII inhibitor is present, which could delay recognition of a serious acquired bleeding disorder.

The dilute Russell viper venom time is one of the most commonly performed LA screening tests. Russell viper venom activates factor X to initiate the common coagulation pathway. "Dilute" indicates that a dilute (minimal) amount of phospholipid is present, and phospholipid is required for clotting, in particular to form the prothrombinase complex that converts prothrombin into thrombin. If LA is present, it should bind to the phospholipid, thereby prolonging the dRVVT clotting time. The major advantage of the dRVVT method is that it is unaffected by alterations in the extrinsic or intrinsic coagulation pathways (inhibitors, coagulopathies, elevated factor VIII activity). However, common pathway factor deficiencies or inhibitors, excess heparin, and direct thrombin inhibitor anticoagulants can produce false-positive results.

The variable sensitivity of dRVVT methods used to screen for LA is due primarily to different sources and concentrations of Russell viper venom and phospholipid reagents. Commercial dRVVT kits provide reagents for both LA screen and confirm steps and typically contain a heparin-neutralizing agent. Typically, the dRVVT assay employs three phases: (1) dRVVT screen, (2) dRVVT mix, and (3) dRVVT confirm. The cutoff for a positive dRVVT screen is greater than twice the standard deviation (SD) of the dRVVT mean from a reference population (>mean+2SD). If the screen is positive, the dRVVT mix is performed using a 50:50 mixture of patient plasma and normal pooled plasma (NPP). A dRVVT mix ratio is calculated as follows:

$$\text{dRVVT mix ratio} = \frac{\text{dRVVT (50:50 mix patient:NPP)}}{\text{dRVVT (NPP)}}$$

A typical cutoff for a positive mix ratio is greater than twice the SD of the mean of the ratio of a reference population. If the screen and mix are both positive, the dRVVT confirm step is performed using added phospholipid. The dRVVT confirm ratio is calculated as follows:

$$\text{dRVVT confirm ratio} = \frac{\text{dRVVT (patient screen)}}{\text{dRVVT (patient + phospholipid)}}$$

The cutoff for the confirm ratio is greater than twice the SD of the mean of the confirm ratio of a reference population. Laboratories should compare commercial dRVVT kits and select the most sensitive one.

The kaolin clotting time (KCT) was first described as a screen for LA in 1978 by Exner et al. It is considered to be one of the most sensitive LA screen methods and continues to be popular in European hemostasis laboratories. As with aPTT-based tests, the test is based on the principle that activation of factor X and prothrombin on the surface of residual phospholipid in platelet-poor plasma is markedly sensitive to interference from LA antibodies. The activator (kaolin) is added to different ratios of platelet-poor test plasma and normal pooled plasma without an exogenous source of phospholipid, and clotting times are monitored. The delta KCT is a variation of the original method. Two preparations of test plasma are used: diluted 1:4 with normal pooled plasma and undiluted. In step 1, 0.2 mL of 1:4 or undiluted test plasma is combined with 0.1 mL 2% kaolin and incubated for 3 minutes. In step 2, 0.2 mL of 25 mM $CaCl_2$ is added and clotting time is measured. Then, the following calculation is made:

$$\Delta \text{KCT} = \text{KCT (undiluted)} - \text{KCT(1:4 test:NPP)}$$

A typical cutoff value for LA is a delta greater than 14 seconds. The KCT method is an extremely sensitive screening test for most lupus anticoagulants, is inexpensive, and can be automated. However, the KCT is also exquisitely sensitive to residual platelets, and platelet-poor plasma must be filtered (0.2-μm filter) before testing, or before freezing for later testing, to prevent false-negative results.

Another "in-house" or commercially available LA screening technique is the dilute prothrombin time (dPT), also known as the tissue thromboplastin inhibition or thromboplastin dilution test. Routine prothrombin time (PT) reagents contain a large amount of phospholipid, such that PT clotting times are typically normal in the presence of LA. Performing a PT with a diluted commercial thromboplastin (range 1:50 to 1:1000) enhances sensitivity to the presence of a lupus anticoagulant by diluting the amount of phospholipid. In other words, LA can prolong the clotting time of a dPT reagent, whereas undiluted PT reagents usually contain too much phospholipid for LA to interfere. However, the sensitivity of dPT LA screen is dependent on the type (recombinant human versus rabbit or bovine tissue-derived) and brand of thromboplastin, the coagulation instrumentation, and most important, the dilution factor that is used. Because there is no accepted standard method, the reported sensitivity of the dPT LA screen is variable.

Exclusion of Other Abnormalities (Factor Inhibitors)

Certain inhibitors can cause false-positive LA tests. Conversely, lupus anticoagulants can cause false-positive Bethesda assays. If LA tests are positive, efforts should be made to ensure that a different type of inhibitor is not present. This should include excluding the possibility of heparin or a direct thrombin inhibitor (such as argatroban, bivalirudin, or lepirudin), as described above. If the routine aPTT and PT are normal, the laboratory can be reasonably sure that a specific inhibitor (such as a factor VIII inhibitor) is not present. If the routine aPTT is prolonged and no anticoagulants are present, a factor VIII assay should be considered to ensure that a factor VIII inhibitor is not present. Assays for factors IX and XI can also be considered if the factor VIII activity is normal. If the PT is prolonged and no anticoagulants are present, assays for factors II, V, VII, X, and fibrinogen can be considered.

Two or more dilutions should be used in the factor assays. If a lupus anticoagulant, heparin, or a direct thrombin inhibitor is present, the measured factor level will be lower at the lower dilution (eg, 1:10) than at a higher dilution (eg, 1:20). If a specific factor inhibitor is present, the activity of the targeted factor is usually less than 10% with minimal increase upon dilution, and more normal results are typically obtained for other factors (although the other factor assays may show a mild increase upon dilution; see the example factor assays in the case study presented in this chapter).

If a markedly low factor VIII is found in a patient who also tests positive for LA, a chromogenic factor VIII assay might be useful to distinguish between a true decrease in factor VIII due to a factor VIII inhibitor versus an unusually strong interference in the factor VIII assay due to the LA. Lupus anticoagulants do not interfere with chromogenic factor VIII assays; therefore, if the low factor VIII activity (by the routine aPTT-based assay) is due to LA interference, the chromogenic factor VIII result will be normal. If, on the other hand, the low factor VIII activity in the routine aPTT-based assay is due to a true decrease resulting from a factor VIII inhibitor, the chromogenic factor VIII result will also be low. Rare patients may demonstrate both LA and factor VIII inhibitor activity.

Indeterminate Results: When Samples Do Not Meet All International Society on Thrombosis and Haemostasis Criteria

Occasionally, lupus anticoagulant results meet some, but not all, International Society on Thrombosis and Haemostasis criteria for positive LA. When one LA confirmatory test is positive but another is negative (for example, positive hexagonal phase with negative dRVVT), the results are probably positive for LA, because no single test is 100% sensitive for LA. However, possible causes of false-positive results in the positive assay should be considered, and it is recommended that the test be repeated 12 or more weeks later with the same method to determine if the LA is persistent. When the mixing step is positive but the confirm step is negative, the results should be considered negative for LA, and other causes for prolonged mixing studies should be investigated. Lastly, the mixing step may be negative (corrects to normal) while the phospholipid confirmatory step is positive. There is no consensus regarding how to interpret this scenario, but the results are probably positive for a weak LA, if reasons for false-positive confirmatory results can be excluded. It is possible that a mixing study can be falsely negative because of dilution of a weak LA.

Anticardiolipin Antibody Testing and Other Antiphospholipid Antibody Immunoassays

Testing for antiphospholipid antibodies, such as anticardiolipin antibodies, is performed by immunoassays (not measurement of clotting times), in particular, an enzyme-linked immunosorbent assay (ELISA). Several anticardiolipin antibody ELISA kits are commercially available. Results among the different assays are variable. Typically, the kits provide a 96-well plate that is coated with cardiolipin. The patient's serum is added, and if anticardiolipin antibodies are present, they bind to the cardiolipin on the plate. Washing steps are followed by steps that detect the presence of the antibody on the plate. Because the antigen target of anticardiolipin antibodies is beta 2-glycoprotein I (β_2GPI) bound to cardiolipin, commercial ELISA kits for anti-β_2GPI antibodies are also available. Results among different anti-β_2GPI assays are variable. Anti-β_2GPI antibodies are suspected to be more specific than anticardiolipin antibody assays, but if an anti-β_2GPI assay is used instead of an anticardiolipin antibody assay, some cases of antiphospholipid antibody syndrome could be missed.

Because the most common antigen target of LA is prothrombin bound to phospholipid, anti-prothrombin antibody immunoassays are offered by some specialized laboratories. Other types of immunoassays are also available, including assays for antiphosphatidylserine, antiphosphatidylethanolamine, and antiphosphatidylinositol antibodies. The clinical relevance of these antibodies is not well understood, and they are not currently considered diagnostic of antiphospholipid antibody syndrome. Therefore, at present, their utility is largely limited to research.

Since cardiolipin is the antigen used for syphilis screening tests (such as the Venereal Disease Research Laboratory [VDRL] test and the rapid plasma reagin [RPR] test), false-positive syphilis tests may occur in patients with anticardiolipin antibodies. Conversely, true syphilis infections can cause false-positive anticardiolipin antibody test results.

Clinical Features and Diagnosis

Antiphospholipid antibodies are relatively common among patients with venous thrombosis. Among 200 consecutive patients with thrombosis, 12% were found to have antiphospholipid antibodies. However, approximately 1% to 5% of healthy control individuals have positive test results for an anticardiolipin antibody/lupus anticoagulant. Not all individuals with antiphospholipid antibodies have clinical symptoms, and transient antiphospholipid antibodies are common, especially associated with infections and inflammatory disorders. However, the presence of an antiphospholipid antibody is considered a risk factor for thrombosis. In a prospective study involving 360 patients with a lupus anticoagulant

Table 22-1. Revised Sapporo Criteria for Diagnosis of the Antiphospholipid Antibody Syndrome

Clinical Criteria*	Laboratory Criteria**
Vascular thrombosis • Venous • Arterial • Small vessel Pregnancy complications • Fetal loss at or beyond gestational week 10 • 3 or more fetal loss events before gestational week 10 • Premature delivery due to eclampsia, preeclampsia, or placental insufficiency	Lupus anticoagulant IgG or IgM anticardiolipin antibody (>40 U or greater than 99th percentile) Medium or high-titer IgG or IgM beta 2-glycoprotein I antibodies (greater than 99th percentile)
*The thrombosis or pregnancy complication must occur within 5 years of the positive test result	**At least one of the above tests must be positive on two separate occasions at least 12 weeks apart

Source: Miyakis S et al. International consensus statement on an update of the classification criteria for definite antiphopholipid syndrome (APS). *J Thromb Haemost.* 2006;4:295-306.

and/or IgG anticardiolipin antibody, the incidence of thrombosis per year was 1% among individuals with no history of thrombosis, 4% among those with systemic lupus erythematosus, 5.5% among those with a history of thrombosis, and 6% among those with high-titer (>40 U) IgG anticardiolipin antibodies.

According to an international consensus statement (revised Sapporo criteria [Table 22-1]), the diagnosis of antiphospholipid antibody syndrome requires a positive antiphospholipid antibody test on two separate occasions at least 12 weeks apart, in the setting of thrombosis or a pregnancy complication occurring within 5 years of the positive test. The positive test can be a lupus anticoagulant and/or medium- or high-titer (eg, >40 U or greater than the 99th percentile) IgG or IgM anticardiolipin antibody. In the newest version of these criteria, medium- or high-titer (eg, greater than the 99th percentile) IgG or IgM beta 2-glycoprotein I antibodies may also be used to diagnose the syndrome, although the authors did not reach consensus on this issue. Current criteria do not include IgA anticardiolipin or IgA anti-β_2GPI antibodies as sufficient for diagnosis. Thrombosis can be venous, arterial, or small vessel thrombosis. Pregnancy complications are defined as fetal loss at or beyond the tenth week of gestation, three or more fetal loss events before the tenth week of gestation, or prematurity due to eclampsia, preeclampsia, or placental insufficiency.

One of the reasons that the diagnostic criteria require demonstration of persistence of the antibody is that antiphospholipid antibodies can be transient, for example, in association with a medication (eg, hydralazine or phenytoin) or infection. The antibody disappears when the medication is discontinued or the infection resolves. Transient antiphospholipid antibodies associated with infection may not be associated with clinical symptoms of antiphospholipid antibody syndrome. In addition, with infection, anticardiolipin antibodies can be seen that bind to cardiolipin without β_2GPI.

Bleeding is not usually a complication associated with antiphospholipid antibodies. However, bleeding can occur in two situations: thrombocytopenia due to antiplatelet antibodies or acquired hypoprothrombinemia. Thrombocytopenia is common with antiphospholipid antibodies, but it is usually mild. In rare instances, more severe thrombocytopenia may develop, which might lead to bleeding. Hypoprothrombinemia is also a relatively rare complication associated with LA. The antigen targeted by LA commonly is prothrombin bound to phospholipid, but the LA usually does not affect the prothrombin (factor II) activity. Rarely, the LA leads to accelerated clearance of factor II, resulting in hypoprothrombinemia. Depending on the severity of the decrease, a bleeding risk may be present. If a moderate PT prolongation is present and unexplained in a patient with LA, hypoprothrombinemia should be considered. If all PT factor assays are found to be normal (factors II, V, VII, X, and fibrinogen), then a second possible explanation for the PT prolongation would be LA interference in the PT assay itself. Depending on the amount of phospholipid in the PT reagent, LA may occasionally slightly prolong the PT by interfering in the assay.

Monitoring Anticoagulation in Patients with Lupus Anticoagulants

Lupus anticoagulants commonly prolong the aPTT. When patients with LA are treated with heparin, monitoring heparin with the aPTT can be difficult because it is not known how much of the prolongation is due to heparin and how much is due to the LA. Even if the aPTT is normal at baseline, it is possible that the LA will prolong the aPTT beyond what is expected when heparin is added. One approach to such patients is to treat them with low-molecular-weight heparin, which is not monitored with the aPTT and has a more predictable dose-response relationship. If low-molecular-weight heparin is contraindicated and heparin must be used, a chromogenic heparin assay (anti-factor Xa assay) can be performed. If heparin assays are performed, it is suggested that aPTT also be measured on the same specimen to help assess which aPTT values are within the therapeutic range for the patient, because aPTT tests are usually much more readily available.

It is less common that LA may prolong the PT, depending on the PT reagent. When patients with LA are treated with warfarin, monitoring can occasionally be difficult because the PT/INR (prothrombin time/international normalized ratio) might be prolonged by the lupus anticoagulant, even if the PT is normal at baseline. One approach is to measure a chromogenic factor X level when the INR has reached the therapeutic range. A clot-based factor X activity can probably suffice if a chromogenic factor X assay is not available, but the LA, in theory, could interfere with the PT-based factor X assay. The laboratory should establish the range of factor X values that correspond to an INR of 2 to 3 among patients on warfarin. An approximation is that an INR of 2 to 3 corresponds roughly to chromogenic factor X values of 20% to 40% and PT-based factor X values of 5% to 15%.

Suggested Reading

Reviews

Levine JS, Branch W, Rauch J. The antiphospholipid syndrome. *New Engl J Med*. 2002;346:752-763.

Triplett DA. Antiphospholipid antibodies. College of American Pathologists Consensus Conference XXXVI: Diagnostic Issues in Thrombophilia. *Arch Pathol Lab Med*. 2002;126:1424-1429.

Lupus Anticoagulant Tests

Bell WN, Alton HG. A brain extract as a substitute for platelet suspensions in the thromboplastin generation test. *Nature*. 1954;174(4436):880-881.

Brandt JT, Triplett DA, Alving B, Scharrer I. Criteria for the diagnosis of lupus anticoagulants, an update: on behalf of the Subcommittee on Lupus Anticoagulants/Antiphospholipid Antibodies of the ISTH. *Thromb Haemost*. 1995;74(6):1597-1603.

deLaat B, Derksen RH, Urbanus RT, Roest M, de Groot PG. Beta 2-glycoprotein I-dependent lupus anticoagulant highly correlates with thrombosis in the antiphospholipid syndrome. *Blood*. 2005;104;3598-3602.

Exner T. Comparison of two simple tests for the lupus anticoagulant. *Am J Clin Pathol*. 1985;83(2):215-218.

Exner T, Rickard KA, Kroneberg H. A sensitive test demonstrating lupus anticoagulant and its behavioural patterns. *Br J Haematol*. 1978;40:143-151.

Gibson J, Starling E, Date L, Rickard KA, Kronenberg H. Simplified screening procedure for detecting lupus inhibitor. *J Clin Pathol*. 1988;41:226-231.

Mannucci PM, Canciani MT, Mari D, Meucci P. The varied sensitivity of partial thromboplastin and prothrombin time reagents in the demonstration of the lupus-like anticoagulant. *Scand J Haematol*. 1979;22:423-432.

Pengo V, Ruffatti A, Iliceto S. The diagnosis of the antiphospholipid syndrome. *Pathophysiol Haemost Thromb*. 2006;35(1-2):175-180.

Rand JH, Wu XX, Quinn AS, et al. Human monoclonal antiphospholipid antibodies disrupt the annexin A5 anticoagulant crystal shield on phospholipids bilayers: evidence from atomic force microscopy and functional assay. *Am J Pathol*. 2003;163:1193-1200.

Schleider MA, Nachman RL, Jaffe EA, Coleman M. A clinical study of the lupus anticoagulant. *Blood*. 1976;48(4):499-509.

Thiagarajan, P, Pengo V, Shapiro SS. The use of the dilute Russell viper venom time for the diagnosis of lupus anticoagulants. *Blood*. 1986;68:869-874.

Triplett DA, Barna LK, Unger GA. A hexagonal (II) phase phospholipid neutralization assay for lupus anticoagulant identification. *Thromb Haemost*. 1993:70(5):787-793.

Triplett DA, Brandt JT, Kaczor D, Schaeffer J. Laboratory diagnosis of lupus inhibitors: a comparison of the tissue thromboplastin inhibition procedure with a new platelet neutralization procedure. *Am J Clin Pathol*. 1983;79(6):678-682.

Anticardiolipin, Beta 2-Glycoprotein I, and Prothrombin Antibodies

Galli M, Luciani D, Bertolini G, Barbui T. Anti-beta 2-glycoprotein I, antiprothrombin antibodies, and the risk of thrombosis in the antiphospholipid syndrome. *Blood*. 2003;102:2717-2723.

Hulstein JJ, Lenting PJ, de Laat B, Derksen RH, Fijnheer R, de Groot PG. β2-glycoprotein I inhibits von Willebrand factor-dependent platelet adhesion and aggregation. *Blood*. 2007;110:1483-1491.

Nash MJ, Camilleri RS, Kunka S, Mackie IJ, Machin SJ, Cohen H. The anticardiolipin assay is required for sensitive screening for antiphospholipid antibodies. *J Thromb Haemost*. 2004;2:1077-1081.

Elevated Factor VIII and Short Baseline Clotting Times Causing False-Negative Lupus Anticoagulant Tests

Guermazi S, Gorgi Y, Ayed K, Dellagi K. Interference de l'exces de facteur VIII sur la detection des lupus anticoagulants par le temps de cephaline active. *Pathol Biol*. 1997;45:28-33.

Jacobsen EM, Wisloff F. False negative screening tests for lupus anticoagulants: an unrecognized problem? *Thromb Res*. 1996;82:445-451.

Lupus Anticoagulant False-Negative Mixing Studies and Potential Interference from Anticoagulants

Thom J, Ivey L, Eikelboom J. Normal plasma mixing studies in the laboratory diagnosis of lupus anticoagulant. *J Thromb Haemost*. 2003;1:2689-2691.

Tripodi A, Biasiolo A, Chantarangkul V, Pengo V. Lupus anticoagulant (LA) testing: performance of clinical laboratories assessed by a national survey using lyophilized affinity-purified immunoglobulin with LA activity. *Clin Chem*. 2003;49:1608-1614.

Zhang L, Whitis JG, Embry MB, Hollensead SC. A simplified algorithm for the laboratory detection of lupus anticoagulants: utilization of two automated integrated tests. *Am J Clin Pathol*. 2005;124:894-901.

Factor VIII Inhibitor Interference in Lupus Anticoagulant Assays

Blanco AN, Cardozo MA, Candela M, Santarelli MT, Bianco RP, Lazzari MA. Anti-factor VIII inhibitors and lupus anticoagulants in haemophilia A patients. *Thromb Haemost*. 1997;77:656-659.

Chandler WL, Ferrell C, Lee J, Tun T, Kha H. Comparison of three methods for measuring factor VIII levels in plasma. *Am J Clin Pathol*. 2003;120:34-39.

de Maistre E, Wahl D, Perret-Guillaume C, et al. A chromogenic factor VIII assay allows reliable measurement of factor VIII levels in the presence of strong lupus anticoagulants. *Thromb Haemost*. 1998;79:237-238.

Goudemand J, Caron C, De Prost D, et al. Evaluation of sensitivity and specificity of a standardized procedure using different reagents for the detection of lupus anticoagulants. *Thromb Haemost*. 1997;77(2):336-342.

Clinical Features and Diagnosis

Bertolaccini ML, Khamashta MA, Hughes GR. Diagnosis of antiphospholipid syndrome. *Nat Clin Pract Rheumatol*. 2005 Nov;1(1):40-46.

Doig RG, O'Malley CJ, Dauer R, et al. An evaluation of 200 consecutive patients with spontaneous or recurrent thrombosis for primary hypercoagulable states. *Am J Clin Pathol*. 1994;102(6):797-801.

Finazzi G, Brancaccio V, Moia M, et al. Natural history and risk factors for thrombosis in 360 patients with antiphospholipid antibodies: a four year prospective study from the Italian registry. *Am J Med*. 1996;100:530-536.

Miyakis S, Lockshin MD, Atsumi T, et al. International consensus statement on an update of the classification criteria for definite antiphospholipid syndrome (APS). *J Thromb Haemost*. 2006;4:295-306.

Monitoring Anticoagulation with Lupus Anticoagulants

Arpino PA, Demirjian Z, Van Cott, EM. Use of the chromogenic factor X assay to predict the international normalized ratio in patients transitioning from argatroban to warfarin. *Pharmacotherapy*. 2005;25:157-164.

Moll S, Ortel TL. Monitoring warfarin therapy in patients with lupus anticoagulants. *Ann Intern Med*. 1997;127:177-185.

Section 5

Antiplatelet and Anticoagulant Drugs

John T. Brandt, MD, Editor

Chapter 23. Antithrombotic Agents

Chapter 24. Anticoagulant Agents

Chapter 25. Antiplatelet Agents

Antithrombotic Agents

John T. Brandt, MD

Introduction

Antithrombotic drugs are therapeutic agents that are commonly used in both the inpatient and outpatient settings. The overall therapeutic goal for antithrombotic drugs is to prevent the formation or extension of a thrombus, such as formation of a clot in a venous access catheter, an extracorporeal blood circuit, or a coronary artery after placement of a drug-eluting stent. When used appropriately, antithrombotic agents may be lifesaving, as in patients with venous thromboembolism, with pulmonary embolism, or during chronic anticoagulation in patients with atrial fibrillation. Because antithrombotic agents inhibit thrombus formation, they have the potential to increase the risk of

bleeding complications, which can be quite severe or life threatening. Therefore, it is imperative that the antithrombotic drug and level of hemostatic inhibition be matched to the clinical setting. The therapeutic window for each antithrombotic drug has largely been set by controlled clinical trials in which the balance of efficacy and bleeding has been assessed; it is important to remember that therapeutic windows developed for specific indications may differ for the same drug for different indications.

Antithrombotic agents can be classified into those that primarily affect the formation of a fibrin clot (anticoagulants) and those that inhibit platelet function (antiplatelet drugs). These agents can be categorized further on the basis of route of administration—parenteral (intravenous or subcuta-

Table 23-1. Anticoagulant Drugs

Drug	Mechanism of Action	Route of Administration
Unfractionated heparin	Indirect, irreversible inhibition of thrombin and factor Xa through antithrombin	Parenteral
Low-molecular-weight heparin	Indirect, irreversible inhibition of factor Xa and thrombin through antithrombin (Xa > thrombin)	Parenteral
Fondaparinux	Indirect, irreversible inhibition of factor Xa through antithrombin	Parenteral
Lepirudin	Direct, irreversible inhibition of thrombin	Parenteral
Desirudin	Direct, irreversible inhibition of thrombin	Parenteral
Bivalirudin	Direct, reversible inhibition of thrombin	Parenteral
Argatroban	Direct, reversible inhibition of thrombin	Parenteral
Warfarin	Decreased synthesis of γ-carboxylated vitamin K-dependent proteins	Oral
Acenocoumarol	Decreased synthesis of γ-carboxylated vitamin K-dependent proteins	Oral
Phenprocoumon	Decreased synthesis of γ-carboxylated vitamin K-dependent proteins	Oral
Fluindione	Decreased synthesis of γ-carboxylated vitamin K-dependent proteins	Oral

Table 23-2. Antiplatelet Drugs

Drug	Mechanism of Action	Route of Administration
Abciximab	Inhibition of glycoprotein IIb/IIIa	Parenteral
Eptifibatide	Inhibition of glycoprotein IIb/IIIa	Parenteral
Tirofiban	Inhibition of glycoprotein IIb/IIIa	Parenteral
Aspirin	Inhibition of cyclo-oxygenase	Oral
Clopidogrel	Inhibition of platelet P2Y$_{12}$ receptor	Oral
Ticlopidine	Inhibition of platelet P2Y$_{12}$ receptor	Oral
Dipyridamole	Inhibition of nucleoside transporter of cell membranes Inhibition of phosphodiesterase	Oral
Cilostazol	Inhibition of phosphodiesterase type-3	Oral

neous) versus oral—and mechanism of action (Tables 23-1 and 23-2). Each of these agents is discussed in more detail in chapters 24 and 25. Additional features, such as potency, half-life, and nontarget toxicity, contribute to the profile that guides clinical utilization of each of these agents.

Target Therapeutic Range and Monitoring Considerations

A number of factors contribute to the identification of the appropriate level of anticoagulation for a given indication. The intensity of anticoagulation required is often dictated by the clinical setting. For example, lower levels of heparin are required for prevention of thrombus formation in surgical patients than are required for treatment of a patient with established deep vein thrombosis. Unfortunately, animal models are not very predictive of the concentration of an antithrombotic agent necessary to prevent thrombus formation in humans for any clinical setting. Therefore, the relationship between inhibition of thrombus formation and risk of bleeding must be established for each indication by clinical trials. Ultimately, decisions regarding the therapeutic range are driven by the observed clinical events, although biomarkers such as prothrombin fragment 1.2, D-dimer, thrombin-antithrombin complex, and release of neutrophil products during dialysis or membrane oxygenation have been used to help justify a given dose.

For some agents there is a very predictable dose-response relationship between the amount of drug administered and the level of hemostatic inhibition achieved. In this setting, a drug may be administered by an established protocol, with an expectation that the patient will achieve therapeutic benefit from the drug. For other drugs, there is marked interindividual variation in the bioavailability or metabolism of the drug, or response to the drug. The variability in response may depend in part on the clinical setting. For example, the response to heparin is quite different in patients with an increase in acute phase reaction proteins than it is in an otherwise healthy individual presenting for elective surgery. The individual response to the drug usually needs to be monitored when significant interindividual variability is expected or known to occur. In other settings, the interindividual variation may be low, but the consequences of inadequate or excessive anticoagulation may be unacceptable. In these settings, the response should be monitored to assure that a therapeutic response has been achieved.

Other key factors to consider when choosing the appropriate antithrombotic drug include the rate of onset of anticoagulant activity, reversibility, and half-life. In some settings, such as percutaneous coronary intervention, a rapid-acting agent with a short half-life is desirable. In other settings, such as chronic therapy, a longer half-life may be acceptable, or even desirable, especially if it is accompanied by less variation in the level of antithrombotic activity between doses. Few of the available agents have antidotes, such as protamine sulfate for heparin, that can rapidly reverse the antithrombotic activity. In the case of antiplatelet agents, the effects of aspirin and clopidogrel last for the life span of the affected platelets, which can be an issue if serious bleeding occurs or if an urgent surgical procedure, such as cardiopulmonary bypass, is required.

Common Clinical Settings Where Antithrombotic Agents Are Used

Acute Venous Thromboembolism

Acute deep venous thromboembolism is a serious clinical disorder that can be life threatening. Clinical trials have demonstrated unequivocally that anticoagulants are more effective than antiplatelet agents in patients with acute venous thromboembolism. Thus, treatment with an immediate-acting anticoagulant is indicated once the diagnosis is established in order to prevent pulmonary embolism or local extension of the thrombus and to decrease the chance of post-phlebitic symptoms. Because treatment needs to extend for 3 to 6 months or longer, therapy with an immediate-acting parenteral agent is usually converted to an oral anticoagulant such as warfarin. Typically, therapy with warfarin (or other vitamin K antagonist) is started at the same time as the immediate-acting agent because it takes several days of warfarin therapy to achieve a therapeutic effect.

In some patients, it may be necessary to continue anticoagulant therapy for an indefinite period of time after an episode of venous thromboembolism. Patients at high risk of recurrent thrombosis, whether based on the diagnosis of congenital thrombophilia or perhaps the antiphospholipid syndrome, may benefit from extended therapy. The concentration of D-dimer after a 3- to 6-month course of therapy has been explored as a possible indicator of the need for continued anticoagulation. In many patients, the potential benefit of continued anticoagulation must be balanced against the risk of serious bleeding.

Prophylaxis for Venous Thromboembolism

Patients who undergo major orthopedic surgery, such as total hip or knee replacement, or major abdominal surgery are at increased risk of developing deep vein thrombosis with its potential sequelae. These patients now commonly receive prophylactic antithrombotic therapy. Anticoagulants have been found to be more effective than antiplatelet agents for prophylaxis of deep vein thrombosis. Either parenteral or oral agents may be used for elective surgery, but recent evidence suggests that low-molecular-weight heparin or fondaparinux may provide greater efficacy than unfractionated heparin, without an increased bleeding risk. Hospitalized patients, particularly those in the intensive care unit or who are otherwise immobilized, are also at increased risk of developing deep vein thrombosis. Prophylactic therapy with an anticoagulant should be considered in such patients.

Prophylaxis in Patients with Atrial Fibrillation

Clinical studies have demonstrated that patients with chronic atrial fibrillation and a moderate-to-high risk of stroke benefit from long-term anticoagulant therapy by prevention of atrial thromboemboli. Antiplatelet therapy with either aspirin or aspirin plus clopidogrel has been explored as an alternative to chronic warfarin therapy, but neither aspirin nor aspirin plus clopidogrel is as effective as warfarin. Chronic warfarin therapy requires diligent monitoring of the pharmacodynamic response as measured by the international normalized ratio (INR). When warfarin therapy is well controlled, the benefit-risk relationship strongly favors treatment. However, the quality of warfarin therapy in the routine outpatient office setting is not as good as that obtained in well-managed clinical trials. Patients with atrial fibrillation, as a group, are undertreated, perhaps because of the difficulty of managing chronic warfarin therapy.

Acute Coronary Syndromes

Modern management of patients with an acute coronary syndrome, which includes unstable angina, non-ST elevation myocardial infarction, and ST elevation myocardial infarction, is dependent on aggressive antithrombotic therapy. Patients usually receive a combination of anticoagulant and antiplatelet agents during an acute episode. Aspirin is given as soon as possible after the onset of symptoms. Heparin, low-molecular-weight heparin, or a direct thrombin inhibitor is often given, at least through the completion of the diagnostic process, with higher-intensity anticoagulation during percutaneous coronary intervention. In certain high-risk patients, a potent platelet glycoprotein IIb/IIIa inhibitor may also be administered before and/or during percutaneous coronary intervention. Thienopyridine (eg, clopidogrel) therapy is often added, and dual antiplatelet therapy is continued for 6 to 12 months after intervention, especially if a drug-eluting stent has been implanted. Often, if the diagnostic process indicates that a patient requires urgent coronary artery bypass surgery, thienopyridine treatment is withheld because of the increased risk of bleeding associated with recent thienopyridine therapy. High-dose heparin is utilized to maintain blood flow during cardiopulmonary bypass. Chronic anticoagulation with warfarin after an acute coronary syndrome has also been explored.

Peripheral Vascular and Cerebrovascular Disease

The chronic management of peripheral vascular disease often includes antiplatelet therapy. The mainstay of therapy in this setting is aspirin. In some patients, the addition of a second agent, such as dipyridamole, clopidogrel, or cilostazol, may provide additional benefit. Patients usually receive systemic heparinization during vascular surgery procedures to maintain blood fluidity.

Extracorporeal Circulation

Circulation of blood through extracorporeal circuits, whether it be renal dialysis, cardiopulmonary bypass, continuous renal replacement therapy, or extracorporeal membrane oxygenation, leads to activation of the hemostatic system. Some degree of antithrombotic therapy is usually necessary to maintain blood in a fluid state and prevent occlusion of the extracorporeal circuit. Heparin has been the mainstay of anticoagulant strategies for extracorporeal circulation; however, with the increased recognition of heparin-induced thrombocytopenia as a serious clinical entity, the use of other anticoagulants has been explored in these settings.

A second issue, particularly with cardiopulmonary bypass, is reversal of the anticoagulant effect to permit completion of the surgical procedure and wound closure. In the case of heparin, this is usually accomplished by administration of protamine. There is no antidote for the direct thrombin inhibitors, and thus reversal of anticoagulation is more difficult with some of these agents. A drug with a short half-life has intrinsic benefits in this situation.

Implantation of Mechanical Heart Valves

Replacement heart valves used to treat valvular pathology are available that are composed of preserved allogeneic tissues or all-synthetic components. Although these prostheses are successful in correcting the valvular dysfunction, the prostheses, especially the all-synthetic valves, are associated with a high rate of thromboembolism. Postoperatively, patients are typically treated with an oral anticoagulant as a long-term treatment, and some clinicians advocate the use of early postoperative bridging heparin or low-molecular-weight heparin therapy. However, engineering advances that are associated with decreased thrombogenicity of these prostheses may require a re-evaluation of anticoagulant duration in the future.

Effects on Tests of Hemostasis

The effects of an antithrombotic drug on tests of hemostasis vary with the mechanism of action, the potency of the agent, its concentration in circulating blood, and the underlying condition of the patient. There has been a long-standing debate over whether a test used to monitor therapy should reflect the overall effect on hemostasis or more directly reflect the concentration of the drug in blood. Global tests of hemostasis—such as prothrombin time (PT), activated partial thromboplastin time (aPTT), activated clotting time (ACT), and bleeding time—provide useful information in many clinical settings. However, new tests have emerged that are more directly related to the drug concentration or effect. These tests include chromogenic factor Xa inhibition assays, ecarin clotting times, glycoprotein IIb/IIIa receptor assays, and platelet $P2Y_{12}$ receptor inhibition assays. The choice of assay should be based on the requirements for monitoring in a given clinical situation. In general, tests that more directly reflect the blood concentration or mechanism of action tend to perform better.

The presence of an antithrombotic agent may make it difficult to diagnose an underlying coagulation disorder. Heparin and direct thrombin inhibitors may interfere with tests used to make a diagnosis of a lupus anticoagulant. Warfarin therapy can make the diagnosis of protein C or protein S deficiency quite difficult. Diagnostic tests for hemostatic disorders need to be interpreted carefully and cautiously in the presence of antithrombotic agents.

Summary

The appropriate utilization of antithrombotic agents is essential for safe and effective therapy.

Chapters 24 and 25 discuss issues related to monitoring the currently available anticoagulant and antithrombotic agents.

Suggested Reading

Antithrombotic Therapy for Acute Venous Thromboembolism

Bates SM, Ginsberg JS. Clinical practice. Treatment of deep-vein thrombosis. *N Engl J Med.* 2004;351:268-277.

Kearon C. Long-term management of patients after venous thromboembolism. *Circulation.* 2004;110(9 suppl 1):I10-I18.

Kearon C, Ginsberg JS, Julian JA, et al. Comparison of fixed-dose weight-adjusted unfractionated heparin and low-molecular-weight heparin for acute treatment of venous thromboembolism. *JAMA.* 2006; 296:935-942.

Palareti G, Cosmi B, Legnani C, et al. D-dimer testing to determine the duration of anticoagulation therapy. *N Engl J Med.* 2006;355:1780-1789.

Qaseem A, Snow V, Barry P, et al. Current diagnosis of venous thromboembolism in primary care: a clinical practice guideline from the American Academy of Family Physicians and the American College of Physicians. *Ann Intern Med.* 2007;146:454-458.

Snow V, Qaseem A, Barry P, et al. Management of venous thromboembolism: a clinical practice guideline from the American College of Physicians and the American Academy of Family Physicians. *Ann Intern Med.* 2007;146:204-210.

Prophylaxis for Venous Thromboembolism

Cohen AT, Davidson BL, Gallus AS, et al. Efficacy and safety of fondaparinux for the prevention of venous thromboembolism in older acute medical patients: randomised placebo controlled trial. *BMJ.* 2006; 332:325-329.

Deheinzelin D, Braga AL, Martins LC, et al. Incorrect use of thromboprophylaxis for venous thromboembolism in medical and surgical patients: results of a multicentric, observational and cross-sectional study in Brazil. *J Thromb Haemost.* 2006;4:1266-1270.

Francis CW. Clinical practice: prophylaxis for thromboembolism in hospitalized medical patients. *N Engl J Med.* 2007;356:1438-1444.

Geerts WH, Bergqvist D, Pineo GF, et al. Prevention of venous thromboembolism: American College of Chest Physicians Evidence-Based Clinical Practice Guidelines (8th Edition) *Chest.* 2008;133(suppl 6): 381S-453S.

Turpie AG, Bauer KA, Eriksson BI, Lassen MR. Superiority of fondaparinux over enoxaparin in preventing venous thromboembolism in major orthopedic surgery using different efficacy end points. *Chest.* 2004;126:501-508.

Turpie AG, Norris TM. Thromboprophylaxis in medical patients: the role of low-molecular-weight heparin. *Thromb Haemost.* 2004;92:3-12.

Prophylaxis in Patients with Atrial Fibrillation

ACTIVE Writing Group of the ACTIVE Investigators, Connolly S, Pogue J, Hart R, et al. Clopidogrel plus aspirin versus oral anticoagulation for atrial fibrillation in the Atrial fibrillation Clopidogrel Trial with Irbesartan for prevention of Vascular Events (ACTIVE W): a randomised controlled trial. *Lancet.* 2006;367:1903-1912.

Glazer NL, Dublin S, Smith NL, et al. Newly detected atrial fibrillation and compliance with antithrombotic guidelines. *Arch Intern Med.* 2007;167:246-252.

Iqbal MB, Taneja AK, Lip GY, Flather M. Recent developments in atrial fibrillation. *BMJ.* 2005;330:238-243.

Singer DE, Albers GW, Dalen JE, et al. Antithrombotic therapy in atrial fibrillation: American College of Chest Physicians Evidence-Based Clinical Practice Guidelines (8th Edition). *Chest.* 2008;133(suppl 6): 546S-592S.

White HD, Gruber M, Feyzi J, et al. Comparison of outcomes among patients randomized to warfarin therapy according to anticoagulant control: results from SPORTIF III and V. *Arch Intern Med.* 2007;167:239-245.

Acute Coronary Syndromes

Braunwald E, Antman EM, Beasley JW, et al. ACC/AHA 2002 guideline update for the management of patients with unstable angina and non-ST-segment elevation myocardial infarction–summary article: a report of the American College of Cardiology/American Heart Association task force on practice guidelines (Committee on the Management of Patients With Unstable Angina). *J Am Coll Cardiol.* 2002;40:1366-1374.

Cannon CP. Evolving management of ST-segment elevation myocardial infarction: update on recent data. *Am J Cardiol.* 2006;98:10Q-21Q.

Curfman GD, Morrissey S, Jarcho JA, Drazen JM. Drug-eluting coronary stents: promise and uncertainty. *N Engl J Med.* 2007;356:1059-1060.

Kastrati A, Mehilli J, Pache J, et al. Analysis of 14 trials comparing sirolimus-eluting stents with bare-metal stents. *N Engl J Med.* 2007;356:1030-1039.

Peripheral Vascular and Cerebrovascular Disease

Watson K, Watson BD, Pater KS. Peripheral arterial disease: a review of disease awareness and management. *Am J Geriatr Pharmacother.* 2006;4:365-379.

Wilson JF, Laine C, Goldmann D. In the clinic: peripheral arterial disease. *Ann Intern Med.* 2007;146:ITC3-1-ITC3-16.

Extracorporeal Circulation

Hassell K. The management of patients with heparin-induced thrombocytopenia who require anticoagulant therapy. *Chest.* 2005;127:1S-8S.

Artificial Heart Valves

Kulik A, Rubens FD, Wells PS, et al. Early postoperative anticoagulation after mechanical valve replacement: a systematic review. *Ann Thorac Surg.* 2006;81:770-781

CHAPTER 24

Anticoagulant Agents

John T. Brandt, MD

Introduction

The two original anticoagulants, heparin and warfarin, continue to be the mainstay of anticoagulant therapy. The introduction of low-molecular-weight heparin into clinical practice represented a significant advance because a more predictable pharmacologic response permitted intermittent subcutaneous injection without laboratory monitoring and dose adjustment. As the safety and effectiveness of low-molecular-weight heparin became established, it began to be used on an outpatient basis, permitting therapy of uncomplicated deep vein thrombosis to move from the inpatient to the outpatient arena.

The increasing recognition of heparin-induced thrombocytopenia as a serious medical problem was associated with the need for alternative, immediate-acting anticoagulants. Novel direct thrombin inhibitors (DTIs), including argatroban and lepirudin, have entered the therapeutic arsenal to treat patients with heparin-induced thrombocytopenia during the acute period of high risk. The use of DTIs, which currently are all parenteral agents, has been explored in other clinical conditions, where they may offer advantages over heparin or low-molecular-weight heparin.

Clinical investigation of a number of novel anticoagulants is progressing at a rapid pace. Most of these agents are orally available anticoagulants that target either factor Xa or thrombin. They are designed to be administered by specified dose regimens, without laboratory monitoring. If these agents survive the rigors of clinical development, they may replace warfarin for chronic anticoagulant therapy. Although these agents may be the hope of the future, this chapter focuses on currently available drugs that inhibit fibrin formation, namely, the anticoagulants.

Unfractionated Heparin

Heparin is a naturally occurring polysaccharide that is isolated from animal tissues. It is composed of repeating disaccharide units consisting of a glucosamine and either D-glucuronic acid or L-iduronic acid. The saccharide units are extensively sulfated, giving heparin a relatively strong negative charge. Heparin is attached to cell membranes through a protein backbone with repeating serine and glycine residues, with nearly all serine residues attached to polysaccharide chains. Pharmaceutical heparin derived from biologic material is a heterogeneous mixture of polysaccharide chains of varying length and degree of sulfation, with molecular weights varying from 5000 to 30,000 Da and an average of 15,000 Da, corresponding to an average chain length of about 45 monosaccharide units.

Heparin interacts with antithrombin through a specific pentasaccharide sequence that is not present in all chains. When heparin binds to antithrombin, it induces a conformational change in antithrombin, resulting in increased ability of antithrombin to bind to the target protease and rapid formation of an inactive, covalent, antithrombin-protease complex. Heparin then dissociates from the antithrombin-protease complex and can catalyze a new antithrombin molecule, making heparin a uniquely "recyclable" anticoagulant (Figure 24-1). Inhibition of thrombin requires simultaneous binding to antithrombin and thrombin; this requires a minimum chain length of 18 monosaccharide units. In contrast, the inhibition of factor Xa only requires the pentasaccharide sequence of heparin to bind to antithrombin (Figure 24-2). The majority of polysaccharide chains in unfractionated heparin are longer than 18 monosaccharide units, and thus unfractionated heparin is effective at catalyzing the inhibition of both thrombin and factor Xa. To a lesser extent, heparin also catalyzes the antithrombin-mediated

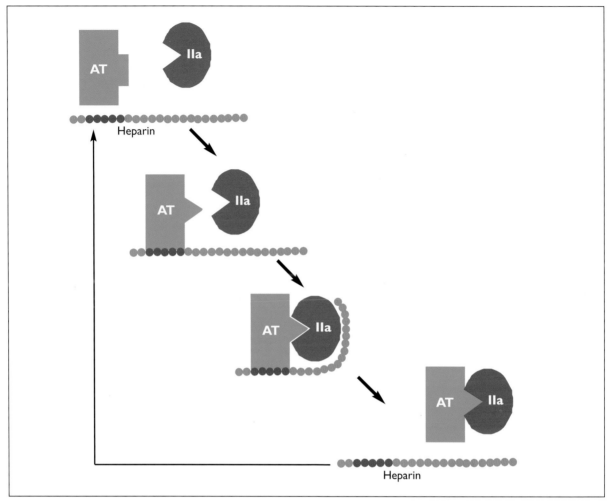

Figure 24-1. Heparin accelerates the inhibition of thrombin by antithrombin (AT). This process requires that heparin bind to antithrombin through a specific pentasaccharide sequence. Heparin interaction with antithrombin induces a change in the three-dimensional structure of antithrombin, facilitating interaction with the active site of thrombin. In addition, heparin must also bind to exosite 2 on thrombin, helping to bring the enzyme closer to its inhibitor. The simultaneous binding of antithrombin and thrombin requires a saccharide chain composed of at least 18 saccharides. Once antithrombin binds to the active site of thrombin, heparin is released and is able to further catalyze the inhibition of thrombin.

inhibition of other serine proteases, such as factors IXa, XIa, and XIIa.

In part because of its negative charge, unfractionated heparin interacts with a variety of proteins and cells, including endothelial cells. In particular, heparin interacts with many acute phase proteins, and this binding may alter the bioavailability of active heparin in patients with acute venous thromboembolism. Heparin also interacts with platelet factor 4 (PF4) to form a neo-epitope that is immunogenic in some patients. The resulting antibodies bind to the heparin-PF4 complex, forming an immune complex. This immune complex can engage and cluster Fc receptors on platelets, leading to platelet activation and aggregation, giving rise to the heparin-induced thrombocytopenia syndrome (see chapter 21). Heparin interacts with endothelial cells and can displace proteins bound to endothelial cell surfaces, such as PF4 and tissue factor pathway inhibitor. Consequently, administration of heparin may result in an increase in the plasma concentration of these proteins.

Heparin is one of the drugs most frequently administered to hospitalized patients. It is commonly used to maintain patent vascular access, which may lead to heparin contamination of samples drawn through those lines. Many indwelling catheters, intracoronary stents, and cardiopulmonary bypass membranes are available with an anticoagulant heparin coating. Heparin is commonly administered subcutaneously as a prophy-

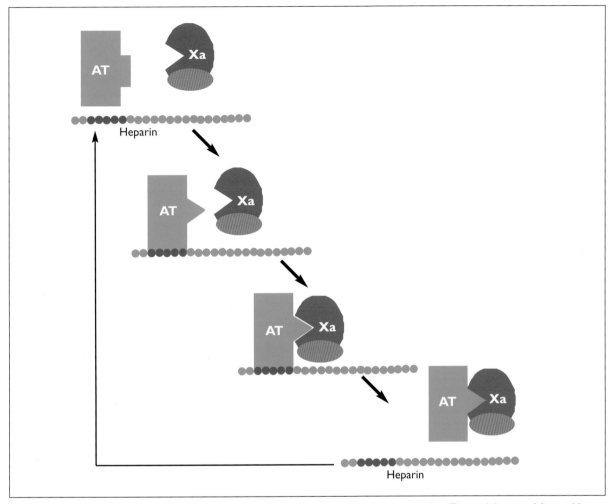

Figure 24-2. Heparin accelerates inhibition of factor Xa by antithrombin (AT). The pentasaccharide sequence of heparin binds to antithrombin, inducing a conformational change that facilitates interaction with factor Xa. This conformational change is sufficient to lead to formation of the inactive enzyme-inhibitor complex. Thus, inhibition of factor Xa can be mediated by shorter-length heparin fragments. Once the enzyme-inhibitor complex forms, heparin is released to further catalyze inhibition of factor Xa.

lactic agent for medical and surgical patients, although use in these settings is decreasing because of increased use of low-molecular-weight heparin. Continuous intravenous administration of heparin is still used in the initial treatment of patients with deep vein thrombosis and/or pulmonary embolism, although low-molecular-weight heparin allows initiation of anticoagulant therapy on an outpatient basis in many cases. Heparin infusions are also a routine component of the therapy of patients with acute coronary syndromes. Higher concentrations of heparin are used during renal dialysis, cardiopulmonary bypass, and percutaneous coronary intervention to prevent thrombotic complications, including occlusion of the vascular circuit.

Laboratory Monitoring of Unfractionated Heparin

Heparin therapy is generally not monitored when it is given to maintain catheter patency, used as a coating on vascular devices, or administered as a subcutaneous prophylactic anticoagulant. However, when heparin is used for treatment of venous thromboembolism, renal dialysis, cardiopulmonary bypass, or percutaneous coronary intervention, the response to therapy is usually monitored (Table 24-1). A variety of tests may be used to assess the patient's response to heparin, and the choice of test depends, in part, on the therapeutic concentration of heparin required for the clinical setting.

Table 24-1. Unfractionated Heparin

Mechanism of Action	Indirect, irreversible inhibitor of factor Xa and thrombin
Half-life	30 to 150 minutes in healthy individuals
Clinical Uses	Maintenance of vascular access Anticoagulant coating of catheters and vascular devices Prophylaxis in medical and surgical patients Treatment of patients with venous thromboembolism Maintenance of flow during renal dialysis Anticoagulation during percutaneous coronary intervention Maintenance of flow during cardiopulmonary bypass
Tests Used to Monitor	Activated partial thromboplastin time (aPTT) Factor Xa inhibition assay (chromogenic) Thrombin time Activated clotting time (ACT) Heptest
Therapeutic Range	Treatment of acute venous thromboembolism: 0.3 to 0.7 U/mL (factor Xa inhibition method) aPTT: establish reagent-specific, in-laboratory therapeutic ranges to correlate with a factor Xa inhibition level of 0.3 to 0.7 U/mL Dialysis: ACT of 120 to 200 seconds Percutaneous coronary intervention: ACT of 250 to 350 seconds (in absence of glycoprotein IIb/IIIa inhibitor) Cardiopulmonary bypass: ACT >400 seconds
Special Considerations	Associated with risk of heparin-induced thrombocytopenia; platelet count should be monitored at least every other day during therapy

Acute Venous Thromboembolism

The goals of heparin therapy in the setting of acute venous thromboembolism include rapid provision of anticoagulation to prevent extension of deep venous thrombosis and pulmonary embolism, and maintenance of adequate anticoagulation until therapeutic levels are achieved with vitamin K antagonists (VKAs). The initial treatment period should be at least 5 days, and heparin (or an alternate agent) should be continued until the patient is in the therapeutic range for VKA. The therapeutic window, defined by adequate anticoagulation without an excessive increase in the relative risk of bleeding, is relatively narrow for unfractionated heparin. Although the propensity to bleed with high concentrations of heparin may constrain initial dosing with heparin, patients who achieve therapeutic levels of heparin within the first 24 hours have better clinical outcomes than patients who take longer to reach the therapeutic range.

Clinical trials indicate that effective anticoagulation with unfractionated heparin in patients with acute venous thromboembolism is best achieved with a weight-based protocol. Such protocols include an initial bolus (eg, 80 U/kg) followed by a continuous infusion, initially at a rate of 16 to 18 U/kg (Figure 24-3). The heparin level is

then assessed about 6 hours after the initiation of therapy to allow the concentration of heparin to reach steady state. If checked too early, the heparin level may reflect residual effects of the bolus and give a false indication of adequate therapy. If the heparin level is within the therapeutic range, the infusion is maintained at the current rate; however, if the heparin level is below the therapeutic range, a second bolus of 40 U/kg may be given and the rate of infusion increased by 2 U/kg. The heparin level should be checked after another 6 hours and the cycle repeated until the patient is in the therapeutic range. If the heparin level is above the therapeutic range, the infusion should be decreased by 2 U/kg and the heparin level rechecked after about 6 hours. When the heparin level is at steady state within the therapeutic window, it may be monitored once daily. In addition to monitoring the heparin level, the platelet count should be determined at least every other day for evaluation of potential heparin-induced thrombocytopenia.

Several tests have been used to assess the heparin level, with the activated partial thromboplastin time (aPTT) still being the most common method. The therapeutic range for treatment of acute venous thromboembolism with heparin is

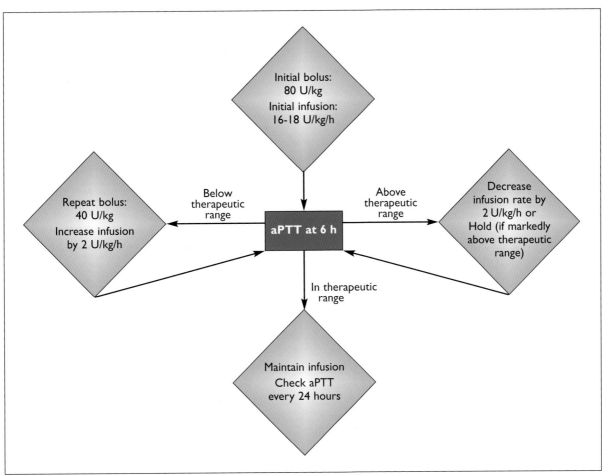

Figure 24-3. Example of a weight-based nomogram for heparin administration. Therapy for patients with venous thromboembolism is initiated with a bolus of unfractionated heparin followed by a continuous intravenous infusion. After 6 hours, the activated partial thromboplastin time (aPTT) is determined. If the aPTT is within the therapeutic range, the infusion is continued at the same rate. If the aPTT is above the therapeutic range, the infusion is decreased by 2 U/kg/h or held if markedly supra-therapeutic (>150 s). If the aPTT is below the therapeutic range, a second bolus is administered, and the infusion rate is increased by 2 U/kg/h. The aPTT should be checked 6 hours after a change in dose. Once the aPTT is within the therapeutic range, it may be checked once per day.

best defined in terms of concentration—not relative prolongation of the aPTT or some other global test of coagulation—and corresponds to a range of 0.3 to 0.7 U/mL when the concentration is determined by a factor Xa inhibition chromogenic method (this corresponds to a range of 0.2 to 0.5 U/mL when determined by a protamine titration method). The therapeutic range for any assay used to assess heparin therapy for venous thromboembolism should be defined by the test values corresponding to this range of heparin concentration.

The factor Xa inhibition assays employed for heparin monitoring utilize a chromogenic peptide substrate, typically a synthetic peptide that mimics a portion of prothrombin (the substrate of factor Xa), coupled to a chromogenic group, such as para-nitroaniline. The patient's plasma is added to the chromogenic substrate together with excess factor Xa, with or without exogenous added antithrombin. The heparin in the plasma catalyzes antithrombin-mediated inhibition of factor Xa. The residual factor Xa catalyzes cleavage of the peptide substrate, liberating the chromogenic group and developing a color product inversely proportional to the amount of heparin. Typically, the chromogenic factor Xa inhibition assays are able to measure heparin levels between 0 and 1.5 U/mL. Samples with higher levels of heparin must be diluted in heparin-free platelet-poor plasma before they are assayed.

There are a number of analytical and biologic issues associated with using the aPTT to monitor heparin therapy. Two key issues are the differences between reagents in the dose-response relation-

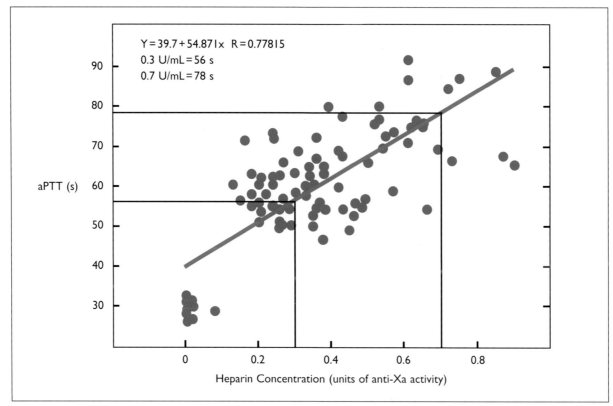

$$Y = 39.7 + 54.871x \quad R = 0.77815$$
$$0.3 \text{ U/mL} = 56 \text{ s}$$
$$0.7 \text{ U/mL} = 78 \text{ s}$$

Figure 24-4. Each laboratory should determine the therapeutic range for its own activated partial thromboplastin time (aPTT) reagent-instrument combination. An example of such a calculation is shown in this figure. The heparin concentration and aPTT are determined in samples from patients receiving heparin. The line of best fit is calculated, and the aPTT values corresponding to 0.3 and 0.7 U/mL are determined. These aPTT values then form the therapeutic range. In this case, the therapeutic range would be 56 to 78 seconds.

ship to heparin and the effect of increased levels of factor VIII on the dose-response characteristics of individual reagent-instrument combinations. Reagents vary significantly in the slope of the dose-response relationship, and thus expression as a ratio (eg, 1.5 times normal) is not appropriate. Consequently, each laboratory should determine the reagent- and lot-specific aPTT range for its method that corresponds to a concentration range of 0.3 to 0.7 U/mL. Because the dose-response relationship is different when heparin is added to plasma in vitro as compared with when heparin is administered in vivo, the correlation between a laboratory's aPTT and heparin concentration is best determined with samples from patients actually receiving heparin.

Both the aPTT and a heparin concentration should be measured in a series of patient samples. To avoid interference from the effect of oral anticoagulants, these samples should be obtained during the early phase of heparin administration. To avoid heparin neutralization by PF4 by activated platelets, it is best to test fresh samples rather than frozen-thawed samples. The results of the heparin assay are then plotted against the aPTT results. An example of data obtained from such a correlation is shown in Figure 24-4.

Factor VIII is an acute phase reactant, and the level of factor VIII is commonly increased in patients with acute venous thromboembolism. A high level of factor VIII lowers the dose-response relationship between heparin concentration and aPTT prolongation. Thus, the aPTT may be shorter than anticipated for a given heparin concentration in the presence of an elevated factor VIII. The effect of an increased factor VIII on the dose-response relationship between heparin and the aPTT is dependent on both the reagent and the instrument on which the aPTT is performed.

During acute venous thromboembolism, there is frequently an increase in acute phase reaction proteins. Many of these proteins bind heparin, neutralizing its anticoagulant effect. These acute phase proteins can essentially function to buffer the administered heparin. Consequently, increased doses of heparin may be needed to overcome the effect of these heparin-binding proteins. It can be difficult to distinguish between the effect of heparin-binding proteins and an increased factor VIII using just an aPTT. The most practical method for differentiating between these issues is a heparin factor Xa inhibition assay.

The problems associated with the use of the aPTT have led many laboratories to monitor heparin therapy with a factor Xa inhibition assay. The advent of automated chromogenic factor Xa inhibition assays has made this move practical for many moderate- to large-sized laboratories. Clinical studies have suggested that heparin therapy is improved by using a chromogenic heparin assay rather than the aPTT. Although the per-test cost may be higher, the overall cost of therapy appears to be lower because of the improved therapy, as measured by time to achieve a therapeutic level of heparin and time in the therapeutic range.

A modified thrombin time is used by some laboratories to monitor therapy. In this test, thrombin is added to platelet-poor plasma, and the time to clot formation is recorded. When used for monitoring heparin, the source of thrombin and the calcium concentration are important variables. A form of the assay in which human thrombin and calcium are added to plasma has been found to provide a reasonable dose-response relationship through the therapeutic range for heparin. Each laboratory must establish the therapeutic range for the thrombin time in terms of the clotting times that correspond to a heparin concentration of 0.3 to 0.7 U/mL. The thrombin time is not readily automated, and there are no interlaboratory proficiency programs to assess its performance.

The Heptest is a clot-based factor Xa inhibition assay that has been used to measure the response to heparin. In the presence of heparin, the time to clot formation is prolonged in a dose-dependent manner. It appears to correlate with factor Xa inhibition assays and is relatively easy to perform. Again, each laboratory needs to establish the therapeutic range in terms of clotting times that correspond to a concentration of 0.3 to 0.7 U/mL.

The thromboelastograph is a whole-blood assay that measures viscoelastic changes during blood coagulation and can be used to detect the effect of heparin anticoagulation. Typically, heparin and low-molecular-weight heparins cause prolongation of the coagulation time (R) and decrease the slope, or alpha-angle. The thromboelastograph can be performed with and without heparinase, and the difference between these conditions can be used to assess the effect of heparin on coagulation.

However, the therapeutic ranges for monitoring heparin therapy using the thromboelastograph are not well established.

Sample acquisition and processing may affect heparin monitoring. The sample should be drawn from an extremity different from the one used for heparin infusion to avoid a falsely increased concentration of heparin. Anything that induces platelet activation and release may increase the level of PF4 in the sample sufficiently to neutralize a portion of the heparin present. Thus, samples should be centrifuged within 1 hour of collection and tested within 4 hours or frozen at -70°C. If the sample is to be frozen for future testing, it should be truly platelet poor, with a platelet count of less than 10,000/μL, to prevent release of PF4 during the freeze-thaw cycle. Current recommendations are that samples for routine coagulation testing be collected in 109 mM (3.2%) citrate in a ratio of 9 parts blood to 1 part citrate. Other concentrations of citrate or incomplete draws may affect the results, particularly of the aPTT.

Acute Coronary Syndromes

Published guidelines for the treatment of patients with acute coronary syndromes, which include unstable angina, non–ST-elevation myocardial infarction and ST-elevation myocardial infarction, recommend that these patients be treated with heparin or another rapidly acting anticoagulant (eg, low-molecular-weight heparin, argatroban, bivalirudin). The dose for treatment of these patients is lower than that used for treatment of patients with venous thromboembolism. For unfractionated heparin, the recommendations are to start with a bolus of 60 U/kg (not to exceed a total of 4000 U) and an initial infusion of 12 U/kg/h (not to exceed 1000 U/h). The guidelines also recommend monitoring with the aPTT, with a designated therapeutic range of 50 to 70 seconds. These guidelines do not take into consideration the variability among aPTT test methods. However, clinical experience indicates that there is an increased risk of recurrent thrombotic events at subtherapeutic aPTTs and an increased risk of bleeding with supratherapeutic aPTTs.

Heparin is commonly administered in conjunction with fibrinolytic therapy in patients with acute myocardial infarction. The therapeutic window for heparin in this setting appears to be quite narrow, with higher heparin levels associated with an increased risk of intracerebral hemorrhage. The recommended dose for heparin depends on the fibrinolytic agent being used, and the package insert for the fibrinolytic should be consulted. However, clinical trials assessing the effectiveness of heparin in association with fibrinolytic therapy have usually used the aPTT to monitor therapy without regard to the responsiveness of the aPTT reagent.

Renal Dialysis, Percutaneous Coronary Intervention, and Cardiopulmonary Bypass

Higher concentrations of heparin are used for renal dialysis, percutaneous coronary intervention, and cardiopulmonary bypass. The aPTT is not used to monitor therapy in these conditions because the target concentration is above the evaluable dynamic range of the aPTT and the results are often needed in a shorter time frame than a routine aPTT provides.

The test most commonly used to monitor heparin in these settings is the activated clotting time (ACT). There are commercially available reagents and instruments for this assay. The classical ACT test uses a whole-blood sample without addition of an anticoagulant; the sample is collected into a glass tube containing an activator of the intrinsic coagulation pathway, such as diatomaceous earth. The ACT has been adapted to point-of-care platforms that are performed with single-use disposable cartridges. The therapeutic range for renal dialysis is an ACT of approximately 120 to 200 seconds, corresponding to a heparin concentration of 0.3 to 1.0 U/mL. For coronary angiography and percutaneous coronary intervention, the target therapeutic range is 250 to 350 seconds, corresponding to a heparin concentration of approximately 1.0 to 2.0 U/mL. If a glycoprotein IIb/IIIa inhibitor (see chapter 25) is used, then the target ACT is reduced to about 200 to 250 seconds. In addition, removal of the femoral sheath should be delayed until the ACT is less than 150 seconds. The highest concentration of heparin (about 4.0 to 5.0 U/mL) is used for cardiopulmonary bypass surgery. The target ACT in this setting is usually greater than 400 seconds.

Low-Molecular-Weight Heparins

Low-molecular-weight heparins are derived from unfractionated heparin through enzymatic or chemical depolymerization. The commercially available low-molecular-weight heparins consist of mixtures of polysaccharides ranging in size from 2000 to 10,000 Da, with an average molecular weight of 4000 to 5000 Da, corresponding to an average chain length of 15 monosaccharides. The pentasaccharide sequence necessary for binding antithrombin is present in approximately one-third of the polysaccharide chains, indicating that two-thirds of the material lacks anticoagulant activity. Low-molecular-weight heparin binds to antithrombin and accelerates its interaction with factor Xa, leading to irreversible inhibition of factor Xa by antithrombin. A portion of the polysaccharide chains in low-molecular-weight heparin is long enough to also mediate inhibition of thrombin by antithrombin. Thus, the anticoagulant activity of low-molecular-weight heparin is due to irreversible inhibition of both factor Xa and thrombin, with most of the activity due to inhibition of factor Xa.

Low-molecular-weight heparin has been utilized in a number of clinical settings, including prophylaxis in medical and surgical patients, treatment of patients with venous thromboembolism, treatment of patients with acute coronary syndromes, and anticoagulation during pregnancy. The dosing regimen depends on the clinical setting and the particular low-molecular-weight heparin being used; therefore, the package insert should be consulted to assure appropriate dosing for the setting and preparation of low-molecular-weight heparin. Administration of low-molecular-weight heparin to patients undergoing spinal or epidural anesthesia has been associated with an increased risk of spinal hematoma, which can be a devastating complication. Low-molecular-weight heparin should be used with extreme caution, if at all, in this setting.

Laboratory Monitoring of Low-Molecular-Weight Heparin

The pharmacokinetic and pharmacodynamic response to low-molecular-weight heparin is reasonably predictable after subcutaneous injection, in part because there is less binding of low-molecular-weight heparin to plasma proteins and the vessel wall as compared with the binding that occurs when unfractionated heparin is used. Consequently, therapy with low-molecular-weight heparin is not usually monitored with laboratory testing (Table 24-2). However, the clearance of low-molecular-weight heparin is reduced in patients with impaired renal function, and monitoring in these patients may be necessary to avoid overdosing. In addition, the pharmacokinetic response to low-molecular-weight heparin is different in newborns; therefore, the dose should be adjusted and therapy should be monitored in this

Table 24-2. Low-Molecular-Weight Heparin

Mechanism of Action	Indirect, irreversible inhibitor of factor Xa and thrombin Anticoagulant activity due to acceleration of the neutralization of factor Xa and thrombin (Xa greater than thrombin) by antithrombin
Half-life	3 to 4 hours in healthy individuals when administered subcutaneously
Clinical Uses	Prophylaxis in medical and surgical patients Treatment of patients with venous thromboembolism Treatment of patients with acute coronary syndrome Used in special populations, including neonates and pregnant women
Tests Used to Monitor	Routine monitoring not required Factor Xa inhibition assay in selected clinical situations
Therapeutic Range	For treatment of patients with venous thromboembolism, the target therapeutic range is 0.6 to 1.0 U/mL
Special Considerations	Clearance is altered in patients with impaired renal function; therapy may need to be monitored in patients with renal failure Clearance is altered in neonates; therapy should be monitored Associated with risk of heparin-induced thrombocytopenia, although risk is lower than for unfractionated heparin Risk of spinal hematoma when administered while patient is receiving epidural or spinal anesthesia or undergoing lumbar puncture

setting. Although the aPTT is affected by low-molecular-weight heparin, the relationship between concentration of low-molecular-weight heparin and prolongation of the aPTT is not sufficiently predictive to support use of the aPTT for monitoring therapy.

The assay of choice for monitoring low-molecular-weight heparin is a chromogenic factor Xa assay, which is similar to the assay described for unfractionated heparin. There are several commercially available assays designed to function on automated coagulation analyzers. Assays vary in terms of whether or not exogenous antithrombin is added to the reaction. It is unclear if addition of exogenous antithrombin improves the assay unless the patient's endogenous antithrombin is markedly reduced. The assay should be calibrated with a low-molecular-weight heparin preparation, preferably the same type of low-molecular-weight heparin that the patient is receiving.

Low-molecular-weight heparin is usually given by intermittent subcutaneous injection, and thus the blood level of low-molecular-weight heparin varies over time, raising the question of when to obtain a sample to assess the individual patient's response to therapy. The anticoagulant activity of low-molecular-weight heparin usually reaches a peak at 2 to 4 hours after injection and is at nadir just before the next dose. The nadir concentrations are quite low (~0.1 U/mL), near the lower limit of the analytical range of most factor Xa inhibition

assays, and often don't directly correlate with the dose or peak level. Clinical experience has shown that the benefit of therapy and risk of bleeding correlate better with the peak concentration. Consequently, the level of low-molecular-weight heparin in a sample obtained 4 hours after injection is most commonly used to guide or monitor therapy. For treatment of patients with venous thromboembolism who are dosed with low-molecular-weight heparin every 12 hours, the target therapeutic range is 0.6 to 1.0 U/mL when the assay is standardized against the appropriate low-molecular-weight heparin.

Administration of low-molecular-weight heparin is associated with a lower risk of heparin-induced thrombocytopenia as compared with the risk when unfractionated heparin is administered, but the risk is not zero. In addition, antibodies formed in response to unfractionated heparin may cross-react with low-molecular-weight heparin, leading to thrombocytopenia. Therefore, the platelet count should be monitored periodically in patients receiving low-molecular-weight heparin.

Fondaparinux

Fondaparinux is a synthetic heparin-mimetic pentasaccharide that binds to antithrombin and catalyzes the irreversible inhibition of factor Xa by antithrombin. Because of its small size, it is not capable of mediating the interaction between

Table 24-3. Fondaparinux

Mechanism of Action	Indirect, irreversible inhibitor of factor Xa
Half-life	~18 hours in healthy individuals when administered subcutaneously
Clinical Uses	Prophylaxis in medical and surgical patients Treatment of patients with venous thromboembolism
Tests Used to Monitor	Routine monitoring not required Factor Xa inhibition assay may be used to assess potential overdose
Therapeutic Range	Weight-based dosing without monitoring
Special Considerations	Clearance is altered in patients with impaired renal function; use in patients with severe renal impairment is contraindicated Prophylactic therapy contraindicated in patients weighing <50 kg Risk of spinal hematoma when administered while patient is receiving epidural or spinal anesthesia or undergoing lumbar puncture

antithrombin and thrombin. Thus, fondaparinux is a selective, indirect, irreversible inhibitor of factor Xa. Fondaparinux is usually administered by subcutaneous injection and has a half-life of approximately 18 hours when administered by this route. The pharmacokinetic and pharmacodynamic responses are quite predictable after subcutaneous administration and consequently therapy is usually not monitored. There is no known antidote for fondaparinux, and thus, treatment of a bleeding episode while on therapy can be problematic. Recombinant factor VIIa has been used in patients with serious, life-threatening bleeding.

The use of fondaparinux has been explored in surgical prophylaxis, medical prophylaxis, and treatment of patients with deep vein thrombosis, pulmonary embolism, heparin-induced thrombocytopenia, or acute coronary syndromes. Currently, fondaparinux is approved for initial treatment of patients with venous thromboembolism, including pulmonary embolism, and for prophylaxis of deep venous thrombosis in patients undergoing major orthopedic or abdominal surgery. The doses for these two indications differ; in venous thromboembolism, the recommended dose is 5 to 10 mg once daily (actual dose depends on the patient's weight), whereas for prophylaxis in patients undergoing orthopedic or abdominal surgery, the recommended dose is 2.5 mg once daily.

Fondaparinux is cleared by renal excretion and is contraindicated in patients with severe renal impairment (creatinine clearance less than 30 mL/min). In addition, prophylactic therapy in patients weighing less than 50 kg is contraindicated because of an increased risk of bleeding. As with low-molecular-weight heparin, there is a risk

of epidural or spinal hematoma when neuraxial anesthesia or spinal puncture is performed in patients receiving fondaparinux. Administration of fondaparinux has not been associated definitively with the development of heparin-induced thrombocytopenia. Clinical trials evaluating the use of fondaparinux in patients with heparin-induced thrombocytopenia suggest that it might be effective in this setting.

Laboratory Monitoring of Fondaparinux

The bioavailability of fondaparinux is quite predictable after subcutaneous injection, and therefore routine monitoring is not required (Table 24-3). Fondaparinux is given by set dosing regimens aligned by indication and body weight. The prothrombin time (PT) and aPTT are not sensitive to fondaparinux. Therefore, if the level of fondaparinux needs to be assessed (eg, washout before surgery), a factor Xa inhibition assay calibrated with fondaparinux diluted in pooled normal plasma, either with or without exogenous antithrombin, may be used.

Hirudins (Peptide Direct Thrombin Inhibitors)

There are currently three commercially available agents based on hirudin, a potent natural antithrombin peptide derived from the medicinal leech: lepirudin, desirudin, and bivalirudin. Because these compounds inhibit thrombin without the need for antithrombin, they are usually referred to as direct thrombin inhibitors (DTIs). Hirudin is a bivalent thrombin inhibitor that binds

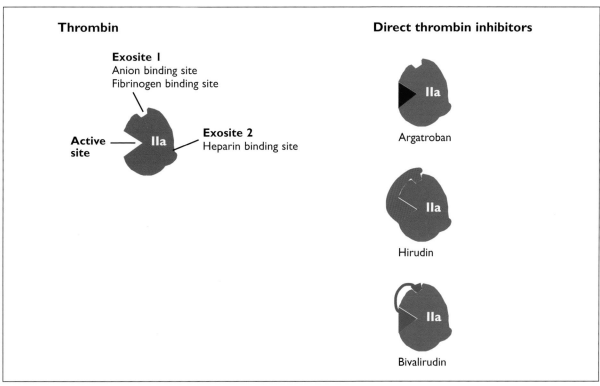

Thrombin

Exosite 1
Anion binding site
Fibrinogen binding site

Active
site
Exosite 2
Heparin binding site

Direct thrombin inhibitors

Argatroban

Hirudin

Bivalirudin

Figure 24-5. Direct thrombin inhibitors interact with the active site of thrombin, in contrast to heparins, which mediate antithrombin interaction with thrombin. In addition to the active site, two other sites are important in regulating thrombin activity. Exosite 1, the anion binding site, is involved in recognition of proteins, including fibrinogen and hirudin. Exosite 2 is the heparin binding site. Argatroban is a low-molecular-weight inhibitor that binds only to the active site of thrombin; its binding is reversible. Lepirudin is a recombinant form of hirudin that binds to the active site and to exosite 1, leading to irreversible inhibition of thrombin. Bivalirudin binds to the active site and to exosite 1. However, thrombin can cleave bivalirudin near the active site, releasing the domain that binds to the active site. Thus, the inhibitory activity of bivalirudin is slowly reversible.

to both the active site of thrombin and the fibrinogen/fibrin binding site (exosite 1); interaction with both sites is necessary for inhibition (Figure 24-5). Lepirudin is a recombinant form of hirudin in which the N-terminal isoleucine has been changed to leucine, and a sulfate group on tyrosine 63 has been deleted. Lepirudin binds tightly to thrombin through interaction with the active site and exosite 1. Thrombin does not cleave lepirudin, and consequently thrombin inhibition is essentially irreversible. The half-life of lepirudin is prolonged in patients with renal disease, and thus dosages should be adjusted in patients with renal failure. Lepirudin is highly immunogenic, with 40% to 70% of patients developing an antibody after treatment for heparin-induced thrombocytopenia. The antilepirudin antibodies may impede clearance, resulting in an enhanced pharmacodynamic effect and increased risk of bleeding. In addition, antibodies to lepirudin have been associated with fatal anaphylactic reactions.

Desirudin is also a recombinant form of hirudin. It differs structurally from lepirudin in that the N-terminal isoleucine has not been altered. Desirudin is also an irreversible inhibitor of thrombin. The half-life appears to be somewhat longer than for lepirudin. Desirudin clearance is also dependent on renal function; consequently, desirudin is contraindicated in patients with severe renal failure. As with lepirudin, desirudin is immunogenic.

Bivalirudin is a synthetic hirudin analogue composed of 20 amino acids. It has two functional domains linked by a spacer composed of four glycine residues, an active-site binding domain (a D-phenyl-prolyl-arginyl-prolyl sequence), and a 12 amino acid peptide that binds exosite 1. Once the thrombin-bivalirudin complex forms, thrombin can cleave the arginyl-prolyl bond in the active-site binding domain, leading to re-exposure of the thrombin active site. The carboxyl portion of bivalirudin may remain attached to exosite 1, but

Table 24-4. Lepirudin

Mechanism of Action	Bivalent, irreversible direct thrombin inhibitor (recombinant hirudin)
Half-life	1 to 2 hours in healthy individuals
Clinical Uses	Prophylaxis or treatment of thrombosis in patients with heparin-induced thrombocytopenia
Tests Used to Monitor	Activated partial thromboplastin time (aPTT)
Therapeutic Range	For treatment of heparin-induced thrombocytopenia, the target aPTT is 1.5 to 3.0 times the baseline value
Special Considerations	Clearance is altered in patients with impaired renal function; dose must be altered in patients with creatinine clearance <60 mL/min High rate of antibody formation; antibodies may increase anticoagulant activity or lead to anaphylaxis

other substrates such as fibrinogen can effectively compete for binding to exosite 1. Thus, bivalirudin is a slowly reversible thrombin inhibitor. The clearance of bivalirudin is dependent on renal function, and dose modification in patients with renal impairment is required. In contrast to lepirudin and desirudin, bivalirudin appears to be associated with a low immunogenic profile, and the antibodies detected have not been clinically significant.

The three hirudin analogues have been approved for use in different clinical settings. Desirudin is approved for use in prophylaxis of deep venous thrombosis in patients undergoing major orthopedic surgery. Lepirudin is approved for anticoagulation of patients with heparin-induced thrombocytopenia. Its use has also been explored in the maintenance of extracorporeal circulation in patients with heparin-induced thrombocytopenia who require cardiopulmonary bypass. Bivalirudin has been approved for use in patients with unstable angina who are undergoing percutaneous coronary intervention. Off-label use of these agents appears to occur commonly.

Laboratory Monitoring of Hirudin Analogues

Therapy with lepirudin requires laboratory monitoring, usually with the aPTT (Table 24-4). Lepirudin is commonly administered as an initial bolus of 0.4 mg/kg, followed by a continuous infusion of 0.15 mg/kg/h in patients with heparin-induced thrombocytopenia who require anticoagulation. Recent data, however, suggest that this maintenance dose may be too high, with the average requirement being closer to

0.11 mg/kg/h in patients with heparin-induced thrombocytopenia and thrombosis. The aPTT should be checked about 4 hours after initiation of therapy, with a target therapeutic range of 1.5 to 2.5 to 3 times the baseline value, depending on the aPTT reagent used. This target range corresponds to a lepirudin concentration of 0.15 to 1.5 g/mL. There are some published data that indicate meaningful interreagent variability in the response of aPTT reagents to hirudin-analogue DTIs. Reduced dosing should be used in patients with impaired renal function (creatinine clearance less than 60 mL/min). Because antibodies may form in 40% to 70% of patients during the course of therapy and alter the response to lepirudin, the aPTT should be monitored at least daily when the patient has been stabilized in the therapeutic range. Of note, antibodies to lepirudin may delay clearance without neutralizing the anticoagulant activity of lepirudin, leading to an enhanced pharmacodynamic response. This increased anticoagulant activity has been associated with increased bleeding.

The ecarin clotting time (ECT) can be used to detect the anticoagulant effect of hirudin-analogue DTIs in plasma. The assay uses an extract from the venom of the *Echis carinatus* snake that converts prothrombin to meizothrombin, an intermediate in the conversion to thrombin that is inhibited by DTIs, but only poorly by heparin. The ECT is not widely available commercially, and the therapeutic ranges for this assay have not been well established. Some ECT assay results may be affected by concomitant therapy with vitamin K antagonists.

Patients with heparin-induced thrombocytopenia who are being treated with a DTI such as lepirudin are often converted to long-term therapy with a VKA. Treatment with a VKA should not

Table 24-5. Bivalirudin

Mechanism of Action	Bivalent, reversible direct thrombin inhibitor (hirudin analogue)
Half-life	25 minutes in healthy individuals
Clinical Uses	Anticoagulation in patients with unstable angina undergoing percutaneous coronary intervention Patients with or at risk for heparin-induced thrombocytopenia undergoing percutaneous coronary intervention
Tests Used to Monitor	Activated clotting time (ACT)
Therapeutic Range	ACT >300 seconds in sample taken 5 minutes after bolus infusion
Special Considerations	Clearance is altered in patients with renal dysfunction; dose adjustment should be made based on renal function

begin until the platelet count has become normal (ie, a platelet count above ~150,000/µL), because of the risk of VKA-associated venous limb gangrene. Once therapy with a VKA is initiated, therapy with the DTI should be continued until the international normalized ratio (INR) is in the therapeutic range on two occasions 24 hours apart. Direct thrombin inhibitors may affect the PT, thus interfering with assessment of the response to VKAs. Of the various DTI drugs, lepirudin appears to have the smallest effect on the PT. It has been suggested that the infusion of lepirudin be decreased to bring the aPTT close to 1.5 times the baseline value when VKA therapy is initiated, with continued infusion of lepirudin until the INR is between 2.0 and 3.0 on two occasions 24 hours apart.

Desirudin is most commonly administered at a dose of 15 mg subcutaneously every 12 hours in patients undergoing major orthopedic surgery. In otherwise healthy individuals, therapy is usually not monitored with laboratory testing. In patients in whom reduced clearance may be an issue, such as patients with impaired renal function, the anticoagulant effect is usually assessed with the aPTT. Dosing should be interrupted for patients with an aPTT greater than twice the upper limit of normal or 85 seconds, whichever is lower, and should not be resumed until the aPTT returns to a value less than this cutoff.

Clinical pharmacology studies have demonstrated a predictable linear dose-response relationship for bivalirudin, and clinical trials have not shown a relationship between the activated clotting time and bleeding or thrombotic complications (Table 24-5). Consequently, bivalirudin is generally administered by a prespecified protocol based on the procedure and potential use of

platelet glycoprotein IIb/IIIa inhibitors. An ACT is usually performed 5 minutes after the initial bolus infusion. If the ACT is not greater than 300 seconds, administration of a second bolus should be considered. Although bivalirudin appears to be less immunogenic than lepirudin and desirudin, antibodies to the recombinant hirudins may cross-react with bivalirudin; therefore, caution is required when using bivalirudin in patients previously exposed to recombinant hirudin.

Argatroban

Argatroban is a synthetic, low-molecular-weight (~527 Da) direct thrombin inhibitor. It is a reversible competitive inhibitor of thrombin, binding at the active site but not at exosite 1 or 2 on thrombin (Figure 24-5). Consequently, argatroban can effectively inhibit both fluid-phase and clot-associated thrombin. It must be administered by intravenous infusion and has a half-life of approximately 40 to 50 minutes in individuals with normal hepatic function. Argatroban is metabolized by the liver, and the half-life of the drug is significantly prolonged in patients with hepatic dysfunction.

The use of argatroban as an anticoagulant has been explored in a number of clinical conditions. These studies have demonstrated that clinical outcomes are most often similar to those achieved with standard heparin therapy. Currently, indications for argatroban are limited to prophylaxis or treatment of thrombosis in patients with heparin-induced thrombocytopenia and to use as an anticoagulant in patients with, or at risk for, heparin-induced thrombocytopenia who are undergoing percutaneous coronary intervention.

Table 24-6. Argatroban

Mechanism of Action	Univalent, reversible direct thrombin inhibitor
Clinical Uses	Prophylaxis or treatment of thrombosis in patients with heparin-induced thrombocytopenia As an anticoagulant in patients with or at risk for heparin-induced thrombocytopenia undergoing percutaneous coronary intervention
Tests Used to Monitor	Activated partial thromboplastin time (aPTT) Activated clotting time (ACT)
Therapeutic Range	Prophylaxis or treatment: aPTT that is 1.5-3.0 times the initial baseline value (but <100 seconds) Percutaneous coronary intervention: ACT of 300 to 450 seconds
Special Considerations	Dose adjustment required for patients with liver disease

Laboratory Monitoring of Argatroban

Argatroban therapy should be monitored and the dose adjusted to target therapeutic ranges (Table 24-6). For the treatment of patients with heparin-induced thrombocytopenia, the aPTT is most commonly used to monitor therapy, with a target therapeutic range of 1.5 to 3 times the heparin-free baseline aPTT. The ECT is also sensitive to the effects of argatroban but is not widely available. A heparin-free baseline aPTT should be obtained before treatment is initiated. Argatroban therapy is then usually started at a dose of 2 g/kg/min, *without* an initial bolus. Because of a relatively high rate of supratherapeutic responses at this dose, Kiser et al (2005) have recommended starting argatroban at a dose of 1.00 to 1.25 g/kg/min. The aPTT should be checked after 2 hours of infusion and the dose adjusted to maintain the aPTT within the target therapeutic range; however, the dose should not exceed 10 g/kg/min. Francis and Hursting (2005) evaluated the response of 21 different aPTT reagents to increasing concentrations of argatroban. The behavior of the reagents was sufficiently similar that the target range of 1.5 to 3 times baseline is appropriate for essentially all currently available reagents. Patients with hepatic dysfunction are susceptible to overanticoagulation because of decreased clearance of argatroban. Consequently, the initial infusion rate should be reduced to 0.5 g/kg/min for patients with impaired hepatic function and subsequently adjusted to reach a target aPTT of 1.5 to 3 times the patient's baseline aPTT.

As with lepirudin, there should be an overlap between argatroban therapy and VKAs when transitioning to long-term oral therapy in patients with heparin-induced thrombocytopenia. Unlike lepirudin, argatroban can result in significant pro-

longation of the PT, making assessment of the response to VKA therapy difficult during the transition period. The package insert recommends that argatroban be continued until the INR is greater than or equal to 4.0, at which point the infusion of argatroban can be discontinued. The INR should then be re-evaluated 4 to 6 hours after cessation of argatroban to determine the argatroban-free INR; if the INR is below 2.0, argatroban should be restarted and this procedure repeated daily until the argatroban-free INR is greater than 2.0.

Arpino et al (2005) have suggested that a chromogenic factor X activity assay can be used to assess the response to oral anticoagulant therapy without cessation of argatroban. Because argatroban is specific for thrombin, it does not interfere with the chromogenic factor X assay. Arpino et al found that a factor X level of 45% or less by this method was an accurate predictor of an argatroban-free INR greater than 2.0. With this approach, argatroban therapy can be continued without interruption until the INR is likely to be greater than 2.0.

For anticoagulation during percutaneous coronary intervention, argatroban should be administered initially at a rate of 25 g/kg/min, accompanied by an initial bolus of 350 g/kg administered over 3 to 5 minutes. The ACT should be checked 5 to 10 minutes after completion of the bolus; if the ACT is less than 300 seconds, a second bolus of 150 g/kg should be administered and the infusion rate increased to 30 g/kg/min. The target range for the ACT is 300 to 450 seconds; if the ACT is greater than 450 seconds, the infusion rate should be decreased. The range should probably be modified if platelet glycoprotein IIb/IIIa inhibitors are to be used during the procedure; one group used an initial infusion rate of 15 g/kg/min to achieve a target ACT range of 275 to 325 seconds in

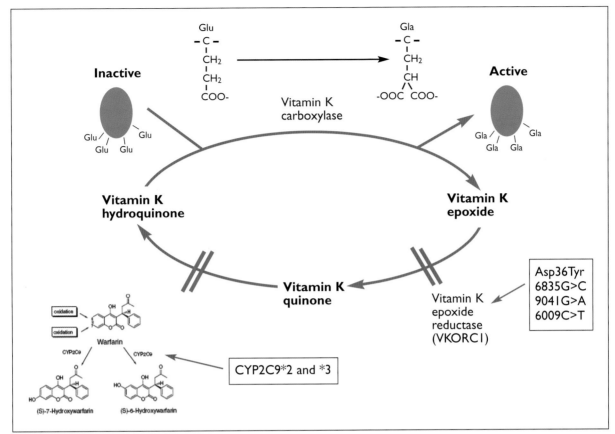

Figure 24-6. Vitamin K is necessary for conversion of glutamic acid residues (Glu) to γ carboxyglutamic acid residues (Gla) in vitamin K-dependent proteins. The dicarboxylic acid group in the Gla residues is essential for vitamin K-dependent proteins to interact effectively with phospholipid membranes where coagulation reactions take place. Vitamin K antagonists, such as warfarin, block the formation of Gla residues by trapping vitamin K in the inactive epoxide form. Reduction of vitamin K epoxide is dependent on vitamin K epoxide reductase complex I (VKORCI). Mutations in the gene for VKORCI (shown) have been shown to affect the response of individual patients to warfarin. Warfarin is metabolized by the cytochrome P450 (CYP) system, including CYP 2C9. Polymorphisms in *CYP 2C9* that affect CYP 2C9 activity alter the half-life of warfarin and thus affect the individual response to therapy with vitamin K antagonists.

patients also receiving a glycoprotein IIb/IIIa inhibitor.

Oral Anticoagulants

Currently, the only available oral anticoagulants are vitamin K agonists, which reduce the concentration of vitamin K-dependent proteins (factors II, VII, IX, and X; protein C; and protein S) in blood. The most commonly used agent is warfarin; two related drugs, acenocoumarol and phenprocoumon, are used in Europe. A fourth drug, fluindione, an indandione derivative, is used in France. The four VKAs differ in terms of pharmacokinetic parameters, particularly half-life, with acenocoumarol having the shortest half-life and phenprocoumon having the longest half-life.

Vitamin K antagonists block the reduction of vitamin K epoxide (Figure 24-6), trapping it in the inactive epoxide form, through inhibition of vitamin K epoxide reductase complex 1 (VKORC1). Consequently, vitamin K-catalyzed carboxylation of glutamic acid residues in the vitamin K-dependent proteins is diminished, leading to decreased synthesis and secretion of these proteins. In addition, the proteins that are secreted have decreased function relative to the mass of protein because of incomplete formation of γ-carboxyglutamic acid residues. The rate of decrease in the plasma concentration of the functional vitamin K-dependent factors is related to their half-life in the circulation. Protein C and factor VII decrease the most rapidly, whereas prothrombin decreases much more slowly. The anticoagulant activity appears to be dependent on adequate reduction of prothrombin and perhaps factor X. Thus, it may take 4 to 5 days to achieve true anticoagulation after initiation of VKA therapy, even though the PT, which is most sensitive to decreases in factor VII, may have become prolonged before this point.

The notable variation among individuals in their response to warfarin is due to a number of genetic, medical, and environmental parameters. Vitamin K antagonists are metabolized by the cytochrome P450 (CYP) family of enzymes. Of note, warfarin is a stereoisomer, and the S-enantiomer is more potent than the R-enantiomer. The S-enantiomer is inactivated through conversion to inactive metabolites, principally by CYP 2C9. A number of polymorphisms of CYP 2C9 that affect the function of this enzyme have been described. The CYP 2C9*2 and *3 alleles are relatively common and have been associated with decreased metabolism of S-warfarin, a requirement for lower doses of warfarin to maintain the therapeutic range, and a higher risk of bleeding.

Genetic variations in the gene for VKORC1 have also been found to be associated with the warfarin dose required to maintain the INR within the range of 2.0 to 3.0. These polymorphisms appear to account for 20% to 25% of the individual variability in response to warfarin. New treatment paradigms taking these genetic factors into account are being developed.

The response to warfarin appears to change with age, with lower doses being required in the elderly. Other attributes, such as body size, may also contribute to the response to warfarin. In addition, multiple drugs may interact with CYP 2C9 and alter its metabolism of warfarin by induction of higher levels of enzyme or inhibition of CYP 2C9 activity. Consequently, warfarin therapy is susceptible to multiple drug-drug interactions, and the effect of initiation or cessation of new drugs in patients on the response to VKAs must be monitored. Although there is some evidence that dietary intake of vitamin K may influence the response to warfarin, more recent data suggest that this may be less of a problem than previously thought.

Laboratory Monitoring of Vitamin K Antagonists

Because of the wide variation in individual responses to VKAs and the relatively narrow therapeutic range for VKAs, therapy must be closely monitored, especially during its initiation. The PT is the test most commonly used to monitor the response to VKAs. As with heparin, there are marked differences in the dose-response charac-

teristics of individual PT reagent-instrument test systems. The INR system was introduced in an attempt to bring some standardization to the monitoring of VKA therapy and has been widely adopted. The therapeutic range for most indications is a target INR of 2.5 (range 2.0 to 3.0). A higher therapeutic range may be required for patients with recurrent thrombosis while receiving adequate VKA therapy, higher risk of thrombosis, or mechanical heart valves, although the data supporting these higher ranges are not robust.

Recommendations for optimizing the performance of the PT used to monitor warfarin therapy have been provided. These include using a PT reagent with an international sensitivity index (ISI) between 0.9 and 1.7, verifying the ISI for the reagent on the instrument being used by the laboratory, using a reagent that is insensitive to the presence of heparin (usually accomplished by adding a heparin neutralizer), and using 109 mM (3.2%) sodium citrate as the anticoagulant. It is important for the laboratory to also verify that the INR is correctly calculated before reporting results, because reporting an erroneous INR may have dramatic consequences for the patient. As indicated by the following equation, the ratio of the patient's PT to the mean normal PT is raised to the power of the ISI:

$$INR = \left[\frac{PTpatient}{PTnormal} \right]^{ISI}$$

where:

PTpatient is the patient's PT in seconds.

PTnormal is the geometric mean PT determined from a group of normal samples.

Therapy should be initiated with the anticipated daily dose (which averages 4 to 5 mg/day); loading doses should be avoided. The PT should be monitored daily after 2 to 3 days of therapy until the patient is stably within the therapeutic range. The frequency of subsequent testing is often individualized, because some patients maintain a very stable level of anticoagulation and require less frequent monitoring (every 3 to 4 weeks), whereas other patients are less stable and require more frequent monitoring.

The goal of monitoring is to maintain the patient within the therapeutic range as much as possible. In well-controlled clinical studies, patients are in the therapeutic range for 60% to 80% of the time, and this is considered good therapy. In routine outpatient medical practices, patients may be in the therapeutic range only 35% to 40% of the time. Less time in the therapeutic range is associated with an increased risk of bleeding and thromboembolic events. Compliance with dosing instructions remains a significant contributing factor to quality of therapy with VKAs. Clinics that are devoted to management of anticoagulant therapy may achieve substantial improvement in the quality of therapy and clinical outcomes.

In patients who have a supratherapeutic INR without bleeding, a dose or two of VKA may need to be omitted and the dose adjusted downward to bring the patient back into the therapeutic range. This situation is also discussed in chapter 11 (see Table 11-3). If the INR is greater than 9.0 without significant bleeding, it is recommended that VKA therapy be temporarily discontinued and vitamin K administered. VKA therapy may be resumed once the INR is within the therapeutic range, but the response to therapy should be carefully monitored until it has stabilized. If the INR is elevated and the patient has serious but not life-threatening bleeding, intravenous vitamin K supplemented with fresh-frozen plasma, prothrombin complex concentrates, or recombinant factor VIIa is recommended. The recommendation for a patient with life-threatening bleeding is administration of either prothrombin complex concentrates or recombinant factor VIIa and supplemental intravenous vitamin K. Warfarin therapy should be discontinued in patients with serious or life-threatening bleeding until the clinical situation has stabilized.

Small instruments capable of performing a PT (INR) on whole blood obtained from a finger prick are available and may be used by individual patients to perform PT testing. A self-testing program may lead to improved therapy for some patients. With appropriate control and education, self-monitoring does not appear to be associated with an increased risk of thrombosis or bleeding.

Suggested Reading

Introduction

Arepally GM, Ortel TL. Clinical practice: heparin-induced thrombocytopenia. *N Engl J Med.* 2006;355: 809-817.

DiNisio M, Middeldorp S, Buller HR. Direct thrombin inhibitors. *N Engl J Med.* 2005;353:1028-1040.

Eriksson BI, Quinlan DJ. Oral anticoagulants in development: focus on thromboprophylaxis in patients undergoing orthopaedic surgery. *Drugs.* 2006;66: 1411-1429.

Unfractionated Heparin

Anand SS, Yusuf S, Pogue J, Ginsberg JS, Hirsh J. Relationship of activated partial thromboplastin time to coronary events and bleeding in patients with acute coronary syndromes who receive heparin. *Circulation.* 2003;107:2884-2888.

Becker RC, Meade TW, Berger PB, et al. The primary and secondary prevention of coronary artery disease: American College of Chest Physicians Evidence-Based Clinical Practice Guidelines (8th Edition). *Chest.* 2008;133(suppl 6):776S-814S.

Bowers J, Ferguson JJ. Use of the activated clotting time in anticoagulation monitoring of intravascular procedures. *Tex Heart Inst J.* 1993;20(4):258-263.

Braunwald E, Antman EM, Beasley JW, et al. ACC/AHA 2002 guideline update for the management of patients with unstable angina and non-ST-segment elevation myocardial infarction-summary article: a report of the American College of Cardiology/American Heart Association task force on practice guidelines (Committee on the Management of Patients With Unstable Angina). *J Am Coll Cardiol.* 2002;40:1366-1374.

Cannon CP. Evolving management of ST-segment elevation myocardial infarction: update on recent data. *Am J Cardiol.* 2006;98:10Q-21Q.

Coppell JA, Thalheimer U, Zambruni A, et al. The effects of unfractionated heparin, low molecular weight heparin and danaparoid on the thromboelastogram (TEG): an in-vitro comparison of standard and heparinase-modified TEGs with conventional coagulation assays. *Blood Coagul Fibrinolysis.* 2006;17:97-104.

Eikelboom JW, Hirsh J. Monitoring unfractionated heparin with the aPTT: time for a fresh look. *Thromb Haemost.* 2006;96:547-552.

Goodman SG, Menon V, Cannon CP, Steg G, Ohman EM, Harrington RA. Acute ST-segment elevation myocardial infarction: American College of Chest Physicians Evidence-Based Clinical Practice Guidelines (8th Edition). *Chest.* 2008; 133(suppl 6): 708S-775S.

Hirsh J, Bauer KA, Donati MB, Gould M, Samama MM, Weitz JI. Parenteral anticoagulants: American College of Chest Physicians Evidence-Based Clinical Practice Guidelines (8th Edition). *Chest.* 2008; 133(suppl 6):141S-159S.

Hirsh J, Raschke R. Heparin and low-molecular-weight heparin: the Seventh ACCP Conference on Antithrombotic and Thrombolytic Therapy. *Chest.* 2004;126:188S-203S.

Kearon C, Kahn SR, Agnelli G, Goldhaber S, Raskob GE, Comerota AJ. Antithrombotic therapy for venous thromboembolic disease: American College of Chest Physicians Evidence-Based Clinical Practice Guidelines (8th Edition). *Chest.* 2008;133(suppl 6): 454S-545S.

Lehman CM, Rettmann JA, Wilson LW, Markewitz BA. Comparative performance of three anti-factor Xa heparin assays in patients in a medical intensive care unit receiving intravenous, unfractionated heparin. *Am J Clin Pathol.* 2006;126:416-421.

Menon V, Harrington RA, Hochman JS, et al. Thrombolysis and adjunctive therapy in acute myocardial infarction: the Seventh ACCP Conference on Antithrombotic and Thrombolytic Therapy. *Chest.* 2004;126:549S-575S.

Olson JD, Arkin CF, Brandt JT, et al. College of American Pathologists Conference XXXI on laboratory monitoring of anticoagulant therapy: laboratory monitoring of unfractionated heparin therapy. *Arch Pathol Lab Med.* 1998;122:782-798.

Penner JA. Experience with a thrombin clotting time assay for measuring heparin activity. *Am J Clin Pathol.* 1974;61:645-653.

Popma JJ, Berger P, Ohman EM, Harrington RA, Grines C, Weitz JI. Antithrombotic therapy during percutaneous coronary intervention: the Seventh ACCP Conference on Antithrombotic and Thrombolytic Therapy. *Chest.* 2004;126:576S-599S.

Raschke RA, Reilly BM, Guidry JR, Fontana JR, Srinivas S. The weight-based heparin dosing nomogram compared with a "standard care" nomogram: a randomized controlled trial. *Ann Intern Med.* 1993;119:874-881.

Smuda K, Noefotistos D, Ts'ao CH. Effects of unfractionated heparin, low-molecular-weight heparin, and heparinoid on thromboelastographic assay of blood coagulation. *Am J Clin Pathol.* 2000;113:725-731.

Low-Molecular-Weight Heparin

Gosselin RC, King JH, Janatpour KA, Dager WE, Larkin EC, Owings JT. Variability of plasma anti-Xa activities with different lots of enoxaparin. *Ann Pharmacother.* 2004;38:563-568.

Harenberg J. Is laboratory monitoring of low-molecular-weight heparin therapy necessary?: yes. *J Thromb Haemost.* 2004;2:547-550.

Laposata M, Green D, Van Cott EM, Barrowcliffe TW, Goodnight SH, Sosolik RC. College of American Pathologists Conference XXXI on laboratory monitoring of anticoagulant therapy: the clinical use and laboratory monitoring of low-molecular-weight heparin, danaparoid, hirudin and related compounds, and argatroban. *Arch Pathol Lab Med.* 1998;122:799-807.

Fondaparinux

Bauer KA, Eriksson BI, Lassen MR, Turpie AG. Fondaparinux compared with enoxaparin for the prevention of venous thromboembolism after elective major knee surgery. *N Engl J Med.* 2001;345:1305-1310.

Buller HR, Davidson BL, Decousus H, et al. Fondaparinux or enoxaparin for the initial treatment of symptomatic deep venous thrombosis: a randomized trial. *Ann Intern Med.* 2004;140:867-873.

Buller HR, Davidson BL, Decousus H, et al. Subcutaneous fondaparinux versus intravenous unfractionated heparin in the initial treatment of pulmonary embolism. *N Engl J Med.* 2003;349:1695-1702.

Efird LE, Kockler DR. Fondaparinux for thromboembolic treatment and prophylaxis of heparin-induced thrombocytopenia. *Ann Pharmacother.* 2006;40:1383-1387.

Eriksson BI, Bauer KA, Lassen MR, Turpie AG. Fondaparinux compared with enoxaparin for the prevention of venous thromboembolism after hip-fracture surgery. *N Engl J Med.* 2001;345:1298-1304.

Klaeffling C, Piechottka G, Daemgen-von Brevern G, et al. Development and clinical evaluation of two chromogenic substrate methods for monitoring fondaparinux sodium. *Ther Drug Monit.* 2006;28:375-381.

Linkins LA, Julian JA, Rischke J, Hirsh J, Weitz JI. In vitro comparison of the effect of heparin, enoxaparin and fondaparinux on tests of coagulation. *Thromb Res.* 2002;107:241-244.

Turpie AG, Bauer KA, Eriksson BI, Lassen MR. Fondaparinux vs enoxaparin for the prevention of venous thromboembolism in major orthopedic surgery: a meta-analysis of 4 randomized double-blind studies. *Arch Intern Med.* 2002;162:1833-1840.

Turpie AG, Gallus AS, Hoek JA. A synthetic pentasaccharide for the prevention of deep-vein thrombosis after total hip replacement. *N Engl J Med.* 2001;344:619-625.

Warkentin TE, Maurer BT, Aster RH. Heparin-induced thrombocytopenia associated with fondaparinux. *N Engl J Med.* 2007;356:2653-2655.

Yusuf S, Mehta SR, Chrolavicius S, et al. Comparison of fondaparinux and enoxaparin in acute coronary syndromes. *N Engl J Med.* 2006;354:1464-1476.

Yusuf S, Mehta SR, Chrolavicius S, et al. Effects of fondaparinux on mortality and reinfarction in patients with acute ST-segment elevation myocardial infarction: the OASIS-6 randomized trial. *JAMA.* 2006;295:1519-1530.

Hirudins

Bates ER. Bivalirudin: an anticoagulant option for percutaneous coronary intervention. *Expert Rev Cardiovasc Ther.* 2004;2:153-162.

Eichler P, Lubenow N, Strobel U, Greinacher A. Antibodies against lepirudin are polyspecific and recognize epitopes on bivalirudin. *Blood.* 2004; 103:613-616.

Fenyvesi T, Harenberg J, Weiss C, Jorg I. Comparison of two different ecarin clotting time methods. *J Thromb Thrombolysis.* 2005;20:51-56.

Fenyvesi T, Jorg I, Harenberg J. Monitoring of anticoagulant effects of direct thrombin inhibitors. *Semin Thromb Hemost.* 2002;28:361-368.

Gosselin RC, Dager WE, King JH, et al. Effect of direct thrombin inhibitors, bivalirudin, lepirudin, and argatroban, on prothrombin time and INR values. *Am J Clin Pathol.* 2004;121:593-599.

Gosselin RC, King JH, Janatpour KA, Dager WE, Larkin EC, Owings JT. Comparing direct thrombin inhibitors using aPTT, ecarin clotting times, and thrombin inhibitor management testing. *Ann Pharmacother.* 2004; 38:1383-1388.

Greinacher A. Lepirudin: a bivalent direct thrombin inhibitor for anticoagulation therapy. *Expert Rev Cardiovasc Ther.* 2004;2:339-357.

Greinacher A. The use of direct thrombin inhibitors in cardiovascular surgery in patients with heparin-induced thrombocytopenia. *Semin Thromb Hemost.* 2004;30:315-327.

Greinacher A, Lubenow N, Eichler P. Anaphylactic and anaphylactoid reactions associated with lepirudin in patients with heparin-induced thrombocytopenia. *Circulation.* 2003;108:2062-2065.

Lubenow N, Eichler P, Lietz T, Greinacher A. Lepirudin in patients with heparin-induced thrombocytopenia: results of the third prospective study (HAT-3) and a combined analysis of HAT-1, HAT-2, and HAT-3. *J Thromb Haemost.* 2005;3:2428-2436.

Tardy B, Lecompte T, Boelhen F, et al. Predictive factors for thrombosis and major bleeding in an observational study in 181 patients with heparin-induced thrombocytopenia treated with lepirudin. *Blood.* 2006;108:1492-1496.

Warkentin TE, Greinacher A, Craven S, Dewar L, Sheppard JA, Ofosu FA. Differences in the clinically effective molar concentrations of four direct thrombin inhibitors explain their variable prothrombin time prolongation. *Thromb Haemost.* 2005;94:958-964.

Argatroban

Arpino PA, Demirjian Z, Van Cott EM. Use of the chromogenic factor X assay to predict the international normalized ratio in patients transitioning from argatroban to warfarin. *Pharmacotherapy.* 2005;25:157-164.

Francis JL, Hursting MJ. Effect of argatroban on the activated partial thromboplastin time: a comparison of 21 commercial reagents. *Blood Coagul Fibrinolysis.* 2005;16:251-257.

Jang IK, Lewis BE, Matthai WH Jr, Kleiman NS. Argatroban anticoagulation in conjunction with glycoprotein IIb/IIIa inhibition in patients undergoing percutaneous coronary intervention: an open-label, nonrandomized pilot study. *J Thromb Thrombolysis.* 2004;18:31-37.

Kiser TH, Jung R, MacLaren R, Fish DN. Evaluation of diagnostic tests and argatroban or lepirudin therapy in patients with suspected heparin-induced thrombocytopenia. *Pharmacotherapy.* 2005;25:1736-1745.

Yeh RW, Jang IK. Argatroban: update. *Am Heart J.* 2006;151:1131-1138.

Oral Anticoagulants

Aithal GP, Day CP, Kesteven PJ, Daly AK. Association of polymorphisms in the cytochrome P450 CYP2C9 with warfarin dose requirement and risk of bleeding complications. *Lancet.* 1999;353:717-719.

Ansell J, Hirsh J, Hylek E, Jacobson A, Crowther M, Palareti G. Pharmacology and management of the vitamin k antagonists: American College of Chest Physicians Evidence-Based Clinical Practice Guidelines (8th Edition). *Chest.* 2008;133(suppl 6): 160S-198S.

Ansell J, Jacobson A, Levy J, Voller H, Hasenkam JM. Guidelines for implementation of patient self-testing and patient self-management of oral anticoagulation: international consensus guidelines prepared by International Self-Monitoring Association for Oral Anticoagulation. *Int J Cardiol.* 2005;99:37-45.

Claes N, Buntinx F, Vijgen J, et al. The Belgian Improvement Study on Oral Anticoagulation Therapy: a randomized clinical trial. *Eur Heart J.* 2005;26:2159-2165.

Fairweather RB, Ansell J, van den Besselaar AM, et al. College of American Pathologists Conference XXXI on laboratory monitoring of anticoagulant therapy: laboratory monitoring of oral anticoagulant therapy. *Arch Pathol Lab Med.* 1998;122:768-781.

Fitzmaurice DA, Gardiner C, Kitchen S, Mackie I, Murray ET, Machin SJ. An evidence-based review and guidelines for patient self-testing and management of oral anticoagulation. *Br J Haematol.* 2005; 131:156-165.

Higashi MK, Veenstra DL, Kondo LM, et al. Association between CYP2C9 genetic variants and anticoagulation-related outcomes during warfarin therapy. *JAMA.* 2002;287:1690-1698.

Holbrook AM, Pereira JA, Labiris R, et al. Systematic overview of warfarin and its drug and food interactions. *Arch Intern Med.* 2005;165:1095-1106.

Horstkotte D, Piper C, Wiemer M. Optimal frequency of patient monitoring and intensity of oral anticoagulation therapy in valvular heart disease. *J Thromb Thrombolysis.* 1998;5(suppl 1):19-24.

Kimmel SE, Chen Z, Price M, et al. The influence of patient adherence on anticoagulation control with warfarin: results from the International Normalized Ratio Adherence and Genetics (IN-RANGE) Study. *Arch Intern Med.* 2007;167:229-235.

Margaglione M, Colaizzo D, D'Andrea G, et al. Genetic modulation of oral anticoagulation with warfarin. *Thromb Haemost.* 2000;84:775-778.

Olson JD, Brandt JT, Chandler WL, et al. Laboratory reporting of the International Normalized Ratio: progress and problems. *Arch Pathol Lab Med.* 2007;131(11):1641-1647.

Pengo V. Management of oral anticoagulant treatment in patients with venous thromboembolism. *Semin Thromb Hemost.* 2006;32:781-786.

Poller L. International normalized ratios (INR): the first 20 years. *J Thromb Haemost.* 2004;2:849-860.

Rieder MJ, Reiner AP, Gage BF, et al. Effect of VKORC1 haplotypes on transcriptional regulation and warfarin dose. *N Engl J Med.* 2005;352:2285-2293.

Schurgers LJ, Shearer MJ, Hamulyak K, Stocklin E, Vermeer C. Effect of vitamin K intake on the stability of oral anticoagulant treatment: dose-response relationships in healthy subjects. *Blood.* 2004;104:2682-2689.

White HD, Gruber M, Feyzi J, et al. Comparison of outcomes among patients randomized to warfarin therapy according to anticoagulant control: results from SPORTIF III and V. *Arch Intern Med.* 2007;167:239-245.

Antiplatelet Agents

Kandice Kottke-Marchant, MD, PhD

Introduction

Platelets play a major role in cardiovascular disease because platelet activation is a contributory mechanism in the pathogenesis of atherosclerosis, cerebrovascular disease, acute coronary syndromes, and peripheral vascular disease. With the formation of an atherosclerotic plaque, disruption of the plaque leads to exposure of extracellular matrix proteins and oxidized lipids, as well as endothelial activation. Platelets rapidly adhere to these sites of injury, leading to platelet activation and aggregation. This process ultimately leads to the development of a platelet thrombus and can result in arterial occlusion, with sequelae ranging from peripheral limb ischemia to myocardial infarction to stroke. Platelet-mediated vascular thrombosis, resulting from platelet adhesion, aggregation, and release of active granule constituents, is central to the development of thrombosis associated with implanted cardiovascular devices and also of arterial thrombosis associated with heparin-induced thrombocytopenia, thrombotic thrombocytopenia purpura, and antiphospholipid antibodies. Thus, pharmacologic inhibition of platelet function has become a mainstay of antithrombotic therapy in many of these disorders.

Platelets are activated through interaction with a number of different cell surface receptors and have many different parallel and overlapping activation pathways, as described in chapter 2. Pharmacologic inhibition of platelet function typically involves blockade of a cell surface receptor (P2Y$_{12}$ for thienopyridines), blockade of an activation pathway (cyclooxygenase for aspirin or adenylate cyclase for dipyridamole), or blockade of platelet aggregation through interference with fibrinogen binding to its receptor (glycoprotein IIb/IIIa [GPIIb/IIIa]) (Figure 25-1). However, with the exception of the GPIIb/IIIa antagonists, it should be kept in mind that most drugs that inhibit or block a single pathway have only a par-tial effect on platelet function because of the redundancy and multiplicity of platelet activation pathways.

Parenteral Agents

Glycoprotein IIb/IIIa Inhibitors

The final common pathway mediating platelet aggregation is activation of the glycoprotein IIb/IIIa receptor with subsequent fibrinogen binding that leads to aggregation of adjacent platelets through the fibrinogen bridge. Drugs that block the binding of GPIIb/IIIa ($\alpha_{IIb}\beta_3$) to fibrinogen, often called GPIIb/IIIa antagonists, disrupt this final common pathway and are specific platelet antagonists that are potent antithrombotic agents.

Mechanism of Action

The GPIIb/IIIa antagonist drugs vary in their design, but they are based on the knowledge that the activated GPIIb/IIIa receptor binds to fibrinogen through specific peptide sequences in the fibrinogen molecule. The binding results from the specific interactions between the arginine-glycine-aspartic acid (RGD) and the terminal dodecapeptide, KQAGDV, regions of the multivalent fibrinogen molecule with suitable docking regions for the receptor GPIIb/IIIa expressed on the activated platelet. Inhibition of binding of fibrinogen to the GPIIb/IIIa receptor by a monovalent antagonist drug should lead to decreased platelet aggregation and bleeding (Figure 25-2). This is a well-known consequence of Glanzmann thrombasthenia, an inherited deficiency of the GPIIb/IIIa receptor, as detailed in chapter 14. Likewise, bleeding is associated with envenomation by some types of snake venoms that contain linear peptide molecules, termed disintegrins, that include the RGD sequence and inhibit platelet aggregation through blockade of the GPIIb/IIIa receptor.

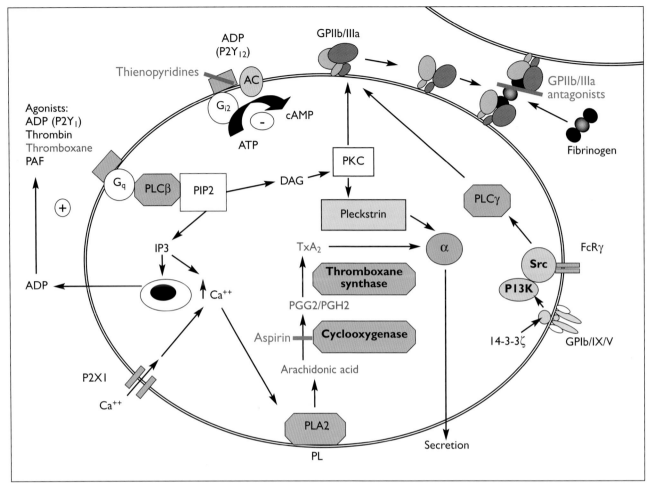

Figure 25-1. This diagram highlights the site of inhibition for common classes of antiplatelet drugs. The glycoprotein IIb/IIIa (GPIIb/IIIa) antagonists block the binding of fibrinogen to the GPIIb/IIIa receptor and are potent aggregation inhibitors. Aspirin and some other nonsteroidal anti-inflammatory drugs block thromboxane A2 (TxA_2) production by blocking cyclooxygenase activity. The thienopyridines (clopidogrel and ticlopidine) block the $P2Y_{12}$ adenosine diphosphate (ADP) receptor, leading to elevation of cyclic adenosine monophosphate (cAMP) and inhibition of platelet activation.

Three different classes of GPIIb/IIIa antagonists have been developed: (1) monoclonal antibody derived, (2) peptide-mimetic, and (3) nonpeptide based. The first GPIIb/IIIa antagonist drug developed, abciximab (c7E3 Fab, ReoPro) is a member of the first class of antagonists; it is a fragment antigen binding (Fab) fragment of mouse-human chimeric monoclonal antibody directed against the GPIIb/IIIa receptor, in which the immunoglobulin hypervariable regions are of mouse origin and the constant regions are of human origin. Abciximab became available for clinical use in the United States in 1995. In addition to binding to GPIIb/IIIa, abciximab is known to bind to two other integrins, $\alpha_V\beta_3$ (the vitronectin receptor) and $\alpha_M\beta_2$ (Mac-1), so abciximab may have additional function by binding to smooth muscle cells and leukocytes. It is used in an intravenous formulation and has a high affinity for the GPIIb/IIIa receptor (dissociation constant, Kd = 5 nmol/L). ReoPro has a biphasic plasma half-life, with an initial half-life of less than 10 minutes and a second-phase half-life of about 30 minutes. However, ReoPro has a biologic half-life of 12 to 24 hours as a result of its high affinity for the receptor.

Eptifibatide (Integrilin) is a member of the peptide-mimetic class of GPIIb/IIIa inhibitors. It is a cyclic heptapeptide that includes a lysine-glycine-aspartic acid (KGD) sequence and is derived from the linear snake venom disintegrin, barbourin. When compared with linear disintegrin peptides from snake venoms, the cyclic RGD-mimetic drugs, like eptifibatide, have a more constrained conformation and show higher specific activity for binding to the platelet receptor. Eptifibatide has a

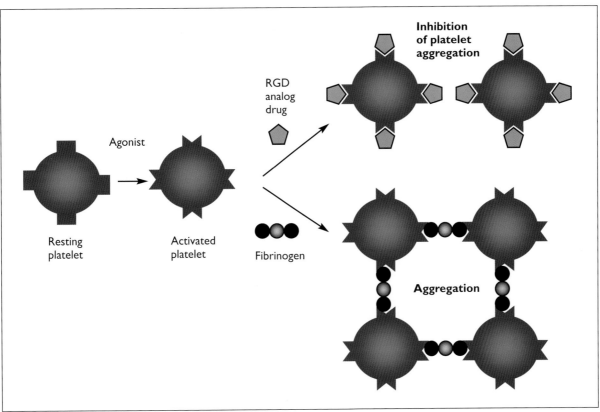

Figure 25-2. The mechanism for the activity of glycoprotein IIb/IIIa (GPIIb/IIIa) antagonist drugs is shown. In resting platelets, the GPIIb/IIIa fibrinogen receptor is in an inactive conformation. After agonist stimulation, the ligand binding site undergoes conformational activation to be able to bind the multiple arginine-glycine-aspartic acid (RGD) tripeptide sequences in fibrinogen, leading to platelet aggregation. However, a monovalent RGD analogue drug would bind to the activated receptor, preventing further binding to fibrinogen, causing inhibition of platelet aggregation.

lower affinity for the receptor than does abciximab (Kd = 120 nmol/L), is readily reversible, and has a plasma half-life between 60 and 150 minutes; clearance is predominantly through renal mechanisms.

An example of a drug from the third category (nonpeptide based) is tirofiban (Aggrastat), which is a synthetic nonpeptide tyrosine derivative. Tirofiban was designed to contain both COO^- and $NH3^+$ groups at a molecular distance equivalent to those in the native RGD peptide and thus function like a soluble RGD peptide in inhibition of fibrinogen binding to GPIIb/IIIa. It has an intermediate GPIIb/IIIa affinity (Kd = 15 nmol/L), with a short plasma half-life of 1.6 hours; clearance is through renal mechanisms.

Several orally active GPIIb/IIIa antagonists have been studied, but none have reached clinical use. Some of the tested compounds included xemilofiban, sibrafiban, orbofiban, lotrafiban, Klerval, lefradifiban, and foxifiban. These were typically prodrugs that were converted to active GPIIb/IIIa antagonists in vivo and had much longer half-lives—ranging from 4 to 24 hours—than the intravenous drug formulations. Phase III clinical trials with these agents in long-term secondary prevention of cardiovascular events showed little to no benefit of the oral compounds when compared with aspirin. In addition, increased bleeding complications were observed with the oral compounds at the higher drug doses. An evaluation of all of these phase III trials actually showed an unexpected excess mortality relative to aspirin, with an increased odds ratio of 1.31 (95% confidence interval [CI] 1.12-1.53). Studies showed interpatient variability in pharmacokinetics and platelet inhibition with most of the formulations. There was also considerable variability in the degree of platelet inhibition between the peak and trough drug levels, often resulting in inadequate platelet inhibition at the trough. Consequently, this class of drugs has not reached clinical approval and development is no longer being pursued.

Effect on Platelet Function Tests

The GPIIb/IIIa antagonist drugs show a profound inhibition of platelet aggregation by agonists such as adenosine diphosphate (ADP), epinephrine, collagen, arachidonic acid, and thrombin receptor-activating peptide (TRAP). They have little effect on ristocetin-induced aggregation. The binding of GPIIb/IIIa to fibrinogen is very calcium dependent, and accurate measurement of the inhibition of platelet aggregation by these drugs may be affected by the anticoagulant used to process the specimen. The original clinical study of eptifibatide (IMPACT II) used citrate anticoagulant to draw the aggregation specimens; the calcium chelation increased the binding of the drug to platelets and led to an overestimation of the inhibitory drug effect and a decreased clinical efficacy. PPACK (D-Phe-Pro-Arg chloromethylketone) anticoagulation, which does not alter the calcium concentration of blood, was used in later trials of eptifibatide (PURSUIT), with a more accurate correlation between drug concentration, inhibition of platelet function, and clinical efficacy. Because they block aggregation, this class of drugs will also give prolonged bleeding times, prolonged collagen/ADP (COL/ADP) and collagen/epinephrine (COL/EPI) closure times in the Platelet Function Analyzer-100 (PFA-100) system and inhibited platelet aggregation with the Plateletworks and the VerifyNow (Ultegra) system. The GPIIb/IIIa inhibitor drugs do not directly block platelet adhesion and do not interfere with assays of platelet adhesion. After cessation of drug infusion, platelet function recovers in about 2 to 4 hours for tirofiban and eptifibatide, and after 12 to 24 hours for abciximab.

Clinical Uses

The clinical utility of the GPIIb/IIIa antagonists to block platelet aggregation and decrease thrombotic outcomes has been demonstrated in several clinical trials of percutaneous coronary interventions (PCIs), such as angioplasty, stent placement, and atherectomy. These drugs are usually used in combination with aspirin and an anticoagulant. The initial clinical trials studied abciximab, with the EPIC study, involving 2099 patients, showing a 35% reduction in 30-day outcomes of death, myocardial infarction (MI), or recurrent ischemia with abciximab. However, there was a high rate of bleeding, and the subsequent EPILOG study showed that use of a lower dose of concomitant heparin could decrease the bleeding risk. The EPISTENT study showed abciximab to reduce

stent-related restenosis as well as thrombosis, and subsequent follow-up data showed efficacy out to 6 months. Subsequent clinical trials with eptifibatide and tirofiban also showed clinical efficacy in PCI management, although the initial IMPACT II study suggested lower efficacy with eptifibatide, probably because of inadequate dosing based on sample collection with the chelating anticoagulant sodium citrate. The subsequent ESPIRIT trial showed a similar benefit with eptifibatide as compared with abciximab, but at lower drug cost. In general, patients undergoing PCI who are considered at high risk for developing myocardial infarction (ie, those with ongoing ischemia) appear to derive the most benefit from therapy with GPIIb/IIIa antagonists. The efficacy of the GPIIb/IIIa antagonists in the medical management of unstable angina and MI is less clear, because several trials have shown conflicting results and modest clinical benefit. Recent studies have suggested that low-risk patients with PCI may benefit nearly as much from aspirin plus clopidogrel therapy as from the more expensive and potent GPIIb/IIIa antagonists.

Role of Monitoring

The effect of GPIIb/IIIa antagonist drugs can be followed by use of platelet aggregation studies. At clinically used doses, optical platelet aggregation with ADP typically shows inhibition of greater than 80% from baseline. In comparison, aggregation with thrombin receptor antagonist peptide (TRAP) shows less inhibition. Other rapid platelet function tests have been evaluated for assessing the platelet inhibition by the GPIIb/IIIa antagonists. The VerifyNow system, using a TRAP-based cartridge, measures platelet aggregation in whole blood by following the agglutination of fibrinogen-coated beads. It shows good correlation with GPIIb/IIIa antagonist-mediated inhibition of turbidimetric platelet aggregation and radiolabeled receptor-binding assays. The PFA-100 has been shown to detect the platelet inhibitory effect of GPIIb/IIIa antagonists; however, this instrument is not ideal for monitoring the drug effect at typical clinical doses, because the COL/ADP and COL/EPI closure times are typically maximal, greater than 300 seconds. Other tests, such as the Impact, the Plateletworks, and the thromboelastograph, can also detect the effect of GPIIb/IIIa antagonists, but they have not been studied in large trials, and recommendations for their use in this setting are not available. These tests have been described in more detail in chapter 7.

In the GOLD study using VerifyNow, Steinhubl et al (2001) documented patient-to-patient variability in the degree of platelet inhibition achieved by standard abciximab doses and some correlation with clinical outcome. There is limited clinical information correlating the degree of platelet inhibition with bleeding complications. The results of the large clinical trials validating the clinical efficacy of GPIIb/IIIa antagonists have not shown a requirement for platelet function monitoring in treated patients, and a role of routine platelet function monitoring has not been firmly established. Additional studies are required to determine whether monitoring platelet function with result-dependent dose adjustment is able to improve therapeutic efficacy and decrease adverse drug effects. Platelet function monitoring might have a role in select patients, especially those with high platelet counts, renal insufficiency, or unexpected bleeding or thrombosis despite GPIIb/IIIa therapy. It may also have a role in examining "washout" of the GPIIb/IIIa inhibitors and return to normal function in select patients, such as those requiring coronary artery bypass surgery or other surgical intervention.

Special Issues

Thrombocytopenia. Binding of fibrinogen and the GPIIb/IIIa antagonist drugs to the GPIIb/IIIa receptor induces a conformational change in the receptor, often referred to as a ligand-induced binding site. The induced conformational change can be detected by binding monoclonal antibodies to the receptor and can be evaluated by flow cytometry. One possible consequence of this conformational change in vivo could be the development of antibodies to the GPIIb/IIIa/drug complex, with resultant immune-mediated platelet clearance. Indeed, drug-dependent thrombocytopenia (platelet count <100,000/μL) has been observed in approximately 2.5% to 6.0% of patients during primary abciximab therapy, with severe thrombocytopenia (platelet count <50,000/μL) in 0.4% to 1.6% of patients. Drug-associated thrombocytopenia is reported in 1.2% to 6.8% of patients with eptifibatide and in 1.1% to 1.9% of patients with tirofiban. The thrombocytopenia can be detected by measuring platelet counts at approximately 2 hours and 24 hours after the start of GPIIb/IIIa therapy. The true thrombocytopenia should be distinguished from pseudothrombocytopenia, which occurs in approximately 2.1% of patients treated with GPIIb/IIIa antagonists because of in vitro platelet clumping from calcium chelation. There are no widely used laboratory assays to detect specific antibodies in these patients, and other causes of thrombocytopenia, such as heparin-induced thrombocytopenia (HIT), disseminated intravascular coagulation, and thrombotic thrombocytopenic purpura, should be excluded. Unlike patients with HIT, patients with GPIIb/IIIa-induced thrombocytopenia are thought to be at higher risk of bleeding, and discontinuation of antiplatelet agents may be necessary. Clinical bleeding or a profound thrombocytopenia (platelet count <20,000/μL) may necessitate platelet transfusions.

Oral Agents

Aspirin

Mechanism of Action

Aspirin is an ancient drug that has been used for a long time, primarily as a pain-relieving or antipyretic drug. The Assyrians first used willow leaves, which contain salicylic acid, to treat rheumatism. Salicylic acid was purified in 1838 and synthesized in 1859. In 1897, the Bayer Company developed the current formulation of aspirin, acetylsalicylic acid. The mechanism of aspirin's action was not determined until Vane et al described its inhibition of prostaglandin synthesis in 1971. This finding was followed by the determination of aspirin's ability to irreversibly inhibit cyclooxygenase by acetylation in 1975. The amino acid involved in that process was determined in 1994.

Figure 25-1 shows aspirin's role in the inhibition of platelet function. Aspirin irreversibly acetylates cyclooxygenase, which results in inhibition of the thromboxane pathway for the lifetime of the platelet (usually 7 to 10 days). In platelets, the principal isoform of cyclooxygenase is COX-1. Other nonsteroidal anti-inflammatory drugs also inhibit cyclooxygenase. However, unlike aspirin, these drugs cause reversible inhibition and only result in platelet inhibition while the drug is present in plasma.

Clinical Uses

Aspirin therapy has been firmly established as efficacious in secondary prevention of cardiovascular disease. In the Antiplatelet Trialists' Collaboration (a meta-analysis of 65 trials with more than 100,000 patients), aspirin was shown to

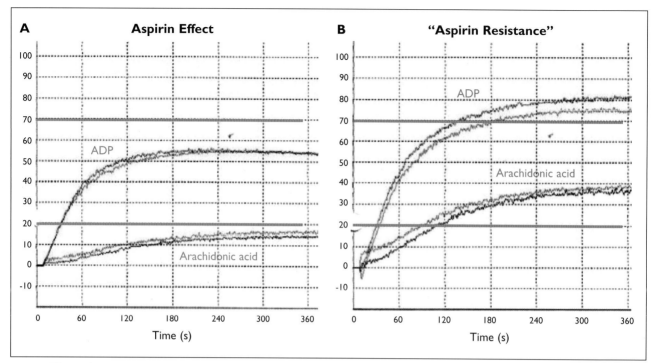

Figure 25-3. A. Optical platelet aggregation results typical of an aspirin effect, with inhibition of adenosine diphosphate (ADP) aggregation, typically less than 70% (upper red bar), and marked inhibition of arachidonic acid aggregation, typical-ly less than 20% (lower red bar). B. In patients with a sub-optimal aggregation response to aspirin (aspirin resistance), the respective aggregation responses with ADP and arachi-donic acid show less inhibition than in panel A.

result in a 25% decrease in death, myocardial infarction (MI), and stroke in patients with vascular disease. Aspirin was also shown to result in a 48% reduction in vascular occlusion during coronary artery bypass grafting, percutaneous coronary intervention (PCI), and arteriovenous fistulas. It also showed a reduction in venous thromboembolic events, with a 67% reduction in pulmonary embolism and 23% reduction in deep vein thrombosis.

Laboratory Tests Used to Assess Effect

Aspirin has a characteristic effect on platelet aggregation. Because aspirin inhibits cyclooxygenase, aspirin therapy shows a marked inhibition of arachidonic acid-induced aggregation. At some concentrations, aspirin causes inhibition of collagen and epinephrine because of cross-talk between receptors. Aspirin has a more minimal effect on ADP-induced aggregation. Stimulation of platelets with TRAP shows little to no inhibition by aspirin. Figure 25-3, A, shows a diagram of a typical platelet aggregation tracing in a patient taking aspirin therapy. The arachidonic acid aggregation is markedly inhibited. Adenosine diphosphate typically shows more preserved aggregation, but has significant disaggregation.

Notice, however, that even the ADP-induced aggregation is less than normal (normal is typically >70%).

Although platelet aggregation is one of the most widely used measures of platelet function, both in the study of aspirin effect and in the diagnosis of platelet disorders, the technique suffers from considerable variability in performance technique and in interpretation of results. A recent practice survey by the North American Specialized Coagulation Laboratory Association (NASCOLA) highlighted considerable interlaboratory variability in the way specimens are prepared for aggregation studies and in the types and concentrations of agonists used. For example, epinephrine is used as an agonist at concentrations ranging from 0.1 μM to 100,000 μM. There is considerable variation in precision in platelet testing among laboratories, and, until recently, external proficiency testing was not available for platelet aggregation.

The PFA-100 is a whole-blood platelet testing device that measures platelet adhesion and aggregation in a high-shear environment. The COL/EPI cartridge is typically sensitive to an aspirin effect, resulting in a prolonged closure time (>193 seconds). Because of a very high ADP concentration in the COL/ADP cartridge, this second cartridge is not considered sensitive to aspirin effects.

Laboratory Monitoring

Aspirin Resistance. Aspirin has been shown to be effective in large clinical studies; however, there is increasing evidence that some individuals may not achieve the full therapeutic effect of aspirin, a phenomenon often referred to as aspirin resistance (AR). Because cardiovascular thrombosis is multifactorial, resistance to the effect of aspirin needs to be defined. On the one hand, aspirin treatment failure is defined as development of a clinical event despite aspirin therapy. This is not necessarily due to lack of aspirin's antiplatelet effect, because the patient could have other reasons for the development of an adverse cardiovascular event, such as inflammation, abnormal lipids, or diabetes. On the other hand, the finding of suboptimal platelet inhibition despite aspirin therapy indicates a lack of pharmacodynamic response (Figure 25-3, B).

A lack of pharmacodynamic response to aspirin has been studied in many patient populations. These studies typically measure some aspect of platelet function, such as platelet aggregation, in patients receiving aspirin and look for a decreased therapeutic response or an increased risk of adverse clinical outcomes. The prevalence estimates vary widely, from 5% to 60%. However, this variability is likely due to the use of many different techniques to measure platelet reactivity to aspirin and a lack of consensus definition of a suboptimal response to aspirin. In addition, studies in the literature variably compare a patient's platelet function while on aspirin therapy to a population range or against a "cutoff" rather than the patient's own functional baseline.

In an early study by Helgason et al in 1993, AR for stroke prophylaxis was defined as the clinical dose of aspirin that resulted in any aggregation with arachidonic acid, when ADP-induced aggregation was normal or when aggregation with epinephrine or collagen was inhibited less than 70%. It was determined that many patients needed to have their aspirin doses increased to achieve optimal platelet inhibition. Overall, 3 of 113 patients continued to have only partial aggregation inhibition despite 1300 mg aspirin per day. In a study by Gum et al in 2001, AR was measured by optical platelet aggregation and PFA-100 in 326 individuals with stable coronary artery disease. They defined AR as maximal ADP aggregation greater than or equal to 70% and maximal arachidonic

acid aggregation greater than or equal to 20%. By these criteria, they identified 5.5% AR and 23.8% semiresponders. In a study by Sane et al in 2002, aspirin nonresponsiveness was defined when four of the following five parameters occurred: (1) collagen-induced aggregation greater than 70%, (2) ADP aggregation greater than 60%, (3) whole blood aggregation greater than 18 ohms, (4) expression of active GPIIb/IIIa greater than 220 log mean fluorescence units, and (5) P-selectin positivity greater than 8%. In patients meeting the definition of AR, their ADP aggregation was 68% ± 9% versus only 54% ± 13% for responders ($P = .007$). A recent study by Stejskal et al (2006) studied aspirin response by platelet aggregation, but with a combination of spontaneous aggregation (<5%) and propyl galate aggregation slope (<53%/min). They detected 55% AR.

The PFA-100 has been studied in the detection of AR. In the article by Gum et al (2001), the PFA-100 detected 9.5% AR in 326 patients with stable coronary artery disease, but it had a low concordance with platelet aggregation. A similar low correlation with platelet aggregation has been seen in other studies. In a study with a complex definition of AR based on five platelet function tests, the PFA-100 closure time was not significantly different in the aspirin-resistant and aspirin-sensitive groups. However, in a study by Grundmann et al in 2003, the PFA-100 closure time correlated with adverse events; the COL/EPI closure time was significantly shorter in aspirin-treated patients with recurrent stroke. There have been some data linking shortened PFA-100 closure times with increased von Willebrand factor levels rather than a suboptimal therapeutic response to aspirin.

Many other platelet function tests have been studied in the detection of AR. One device marketed specifically for AR is the VerifyNow aspirin cartridge, which is a whole-blood, near-patient test that measures platelet aggregation by the use of fibrinogen-coated beads and an agonist, arachidonic acid. It detects AR and correlates with an adverse outcome (myonecrosis) after PCI. The Diamed Impact-R is a cone and plate(let) analyzer that indirectly measures aspirin response by measuring platelet adhesion under shear. The urinary 11-dehydro thromboxane B2 is not a direct platelet function test; it measures the amount of a thromboxane A2 metabolite in urine. In a study by Eikelboom et al in 2002, urinary 11-dehydro thromboxane B2 levels in the upper quartile were shown to be associated with a 2-fold increased risk of MI and a 3.5-fold increased risk of cardiovascular death. Platelet flow cytometry has been used in many guises to measure platelet activation associated with AR, such as measurement of platelet P-selectin, microparticles, or activated GPIIb/IIIa receptor. Other tests under study for AR include the Plateletworks, which measures platelet function by differential platelet counting after agonist stimulation (ADP or collagen). A shear-induced collagen flow chamber is also being studied.

Platelet function tests have been shown to detect a certain proportion of patients with suboptimal pharmacodynamic response to aspirin. However, if there is no connection between the laboratory measurement of AR and outcome, the measurement of laboratory platelet function would just be an academic exercise. In 1993, Grotemeyer et al showed an 89% increased risk of a new cerebrovascular event in patients with AR. Subsequently, other studies also showed some correlation between AR and adverse outcome. In a study of 100 patients with peripheral vascular angioplasty, the 40% aspirin-resistant patients had an 87% increased risk of arterial reocclusion. In another study, preserved platelet function, as detected by increased urinary 11-dehydro thromboxane B2, was associated with an increased risk of MI and cardiovascular death. Several recent studies continue to support the finding of increased adverse outcomes in patients with AR. Aspirin resistance has been associated with myonecrosis during PCI, and a higher rate of cardiovascular events in patients undergoing PCI.

If AR is a real phenomenon and is associated with a risk of cardiovascular events, what is the potential mechanism? No unifying mechanism has been proven and the causes are likely to be multifactorial. Some lack of pharmacodynamic response may be linked to increased platelet activation despite inhibition of cyclooxygenase-2 (COX-2). Mechanisms extrinsic to the platelet, such as rapid platelet turnover or increased platelet production, also are likely. In addition, inadequate aspirin dosing or lack of compliance could certainly affect the therapeutic potential of aspirin. Drug competition by other nonsteroidal anti-inflammatory drugs may affect platelet response to aspirin. Intrinsic mechanisms postulated for AR include inducible COX-2 isoform in platelets, which is far less responsive to aspirin. There is some debate about the platelet's ability to express COX-2, but this has been demonstrated in patients with rapid platelet turnover, especially after cardiac surgery. In addition, uninhibited COX in nucleated cells that can synthesize more

COX despite aspirin therapy could lead to persistent COX activity. It is also possible that genetic variability in the enzymes in the cyclooxygenase pathway or in other platelet receptors could result in altered pharmacodynamic response to aspirin.

In 2005, a working group of the International Society on Thrombosis and Haemostasis (Michelson et al) met and published a position paper on AR. Although it was acknowledged that there is clinical evidence linking AR to adverse clinical outcomes, a clinically meaningful definition of aspirin "resistance" needs to be developed, based on data linking aspirin-dependent laboratory tests to clinical outcomes in patients. Because the correct treatment, if any, of AR is unknown and there are limited clinical data to address the clinical effectiveness of altering therapy of the basis of a laboratory finding of AR, it currently is not appropriate to test for aspirin resistance in most patients or to change therapy on the basis of such tests.

Thienopyridines

Mechanism of Action

Clopidogrel is a thienopyridine drug that inhibits platelet function by binding to $P2Y_{12}$, which is one of the ADP receptors on the platelet surface. Ticlopidine, another thienopyridine drug, acts by a similar mechanism. There are three ADP receptors on platelets: $P2X_1$, $P2Y_1$, and $P2Y_{12}$. The $P2Y_1$ receptor is the high-affinity receptor responsible for the initial platelet shape change, whereas the $P2Y_{12}$ receptor is a low-affinity receptor involved in amplification of platelet aggregation and thrombus stabilization. Stimulation of the receptors by ADP leads to platelet shape change and aggregation, functions that are blocked by clopidogrel therapy, as shown in Figure 25-4. The thienopyridines are specific for $P2Y_{12}$ and thus the activation of $P2Y_1$ still occurs. Thus, shape change can occur as well as a minor wave of reversible aggregation. One of the interesting findings to emerge over the last couple of years is the role of $P2Y_{12}$ in sustaining platelet aggregation, which requires ongoing signaling through $P2Y_{12}$. In the presence of a $P2Y_{12}$ inhibitor, aggregation is reversible.

When administered orally, the prodrug clopidogrel is converted into 2-oxoclopidogrel and further converted to an active metabolite in the liver by the cytochrome P450 (CYP) system. The CYP3A4 and CYP3A5 cytochromes, and to a lesser extent CYP1A2, CYP2C19, and CYP2B6, have

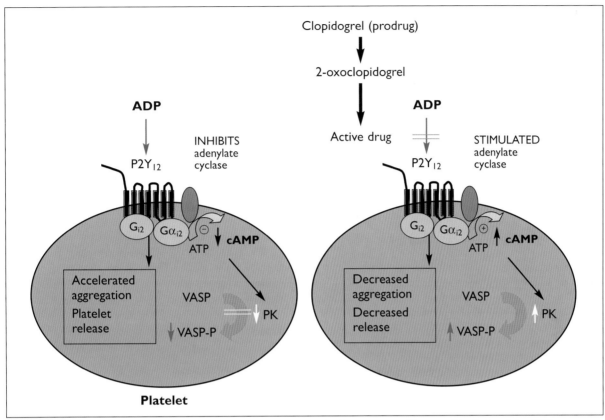

Figure 25-4. Clopidogrel is a prodrug that is converted to 2-oxoclopidogrel and then metabolized to the active drug in the liver cytochrome P450 system. The active drug binds irreversibly to the P2Y$_{12}$ adenosine diphosphate (ADP) receptor. The receptor is coupled to adenylate cyclase through G-protein G$_{i2}$. Typical binding of ADP to P2Y$_{12}$ leads to a downregulation of cyclic adenosine monophosphate (cAMP) and platelet activation. However, blockade of the P2Y$_{12}$ receptor by the active clopidogrel metabolite leads to increased cAMP production, with associated decrease in platelet aggregation and release. Vasodilator-stimulated phosphoprotein (VASP) is dephosphorylated after P2Y$_{12}$ stimulation and phosphorylated by P2Y$_{12}$ antagonists. Abbreviation: PK, cAMP-dependent protein kinase.

been shown to be responsible for clopidogrel metabolism. The active metabolite, which carries a free thiol group, reacts with the thiol of an amino acid of the platelet ADP receptor P2Y$_{12}$ to form a disulfide bond, thus blocking ADP from binding to the receptor. A recent article by Savi et al (2006) points to cysteine 97 (Cys97) as a target for clopidogrel. Cys97 forms a disulfide bond with Cys175; this bond is presumably disrupted by clopidogrel. The Cys97-Cys175 complex is critical for receptor conformation and function. Clopidogrel thus causes irreversible platelet inhibition. Ticlopidine is metabolized by a similar pathway, but the time course of action is slower and laboratory evidence of platelet inhibition may take up to 24 hours after an oral dose.

Clinical Uses

Clopidogrel was shown to be superior to aspirin in patients with atherothrombotic disease in the CAPRIE study in 1996. The CURE study validated this supposition by demonstrating a 20% relative risk reduction with clopidogrel plus aspirin versus placebo plus aspirin in patients presenting with non-ST-segment elevation acute coronary syndromes. The Clopidogrel to Reduce Events During Observation (CREDO) trial showed a 26.9% risk reduction by prolonged dual antiplatelet therapy in patients undergoing coronary stenting. In the recent CLARITY-TIMI 28 trial, patients with MI showed a 20% risk reduction with the addition of clopidogrel to aspirin plus thrombolytic drugs. However, in a lower-risk population, the CHARISMA study of nearly 16,000 patients showed that aspirin plus clopidogrel was not more effective than aspirin alone.

Although an effective antiplatelet drug, ticlopidine is now rarely used clinically because of its relatively long lag-time to development of a therapeutic effect and several reports of drug-associated thrombotic thrombocytopenic purpura. Ticlopidine is also associated with hematotoxicity (eg, leukopenia), which occurs in about 1% of patients receiving the drug.

Laboratory Monitoring

Clopidogrel has a marked inhibitory effect on platelet function, with a notable decrease in ADP-induced aggregation. The response to other agonists is decreased to a more variable extent. The VerifyNow whole-blood platelet device has a Food and Drug Administration-approved cartridge for measuring $P2Y_{12}$ inhibition. The PFA-100 is relatively insensitive to the effects of clopidogrel. The COL/EPI cartridge may show a prolonged closure time with clopidogrel, but the COL/ADP cartridge does not show a reproducible inhibition by clopidogrel, probably because of the very high ADP concentration in the cartridge. A research assay that has been used to monitor the effect of clopidogrel is the vasodilator-stimulated phosphoprotein (VASP) assay. VASP is dephosphorylated after $P2Y_{12}$ stimulation and phosphorylated by $P2Y_{12}$ antagonists. Flow cytometric measurement of the levels of the VASP phosphorylation and dephosphorylation have been shown to correlate with the results of ADP-induced platelet aggregation.

Clopidogrel Resistance

Not all patients respond to clopidogrel therapy with an appropriate therapeutic effect. Gurbel et al (2003) examined platelet function by ADP-induced aggregation and defined "clopidogrel resistance" as less than 10% inhibition of ADP-induced aggregation compared to the patient's own baseline. By 24 hours after dosing, 31% of patients did not achieve a therapeutic effect of clopidogrel; the percentage stayed the same for up to 5 days and then dropped to 15%.

There is evidence that the ADP agonist dose may affect the percentage of patients identified as clopidogrel resistant. Muller et al (2003) showed a difference between 5 and 20 uM ADP. Although there is considerable interpatient variability in the antiplatelet effect of clopidogrel, 90% of patients initially identified as clopidogrel responders continue to be identified as responders after 30 days of clopidogrel therapy. There is some evidence that clopidogrel dosing above 75 mg/day may further suppress platelet function. Like the studies of AR, the clinical studies of clopidogrel resistance are beginning to report an association between clopidogrel resistance and adverse cardiovascular events.

The postulated mechanisms of clopidogrel resistance are more complex than those for aspirin because of the metabolic conversion of clopidogrel to an active drug. In Figure 25-5, which depicts clopidogrel metabolism and activity, possible mechanisms for clopidogrel resistance are shown. Poor drug availability, either through inadequate dosing or absorption, could be a cause. Genetic polymorphisms of the cytochrome P450 system could affect the metabolic conversion to the active form of the drug. Genetic polymorphisms in the $P2Y_{12}$ or other ADP receptors could also affect the ability of clopidogrel to block ADP binding to the receptor. Because clopidogrel irreversibly inhibits $P2Y_{12}$ function for the platelet lifetime, increased platelet turnover could also modulate the effectiveness of the clopidogrel drug.

There is some evidence that genetic polymorphisms are associated with altered pharmacodynamic response to clopidogrel. A study by Angiolillo et al (2005) showed an 807T polymorphism of glycoprotein Ia (GPIa) to be associated with increased ADP aggregation response. A study by Fontana et al (2003) showed an H2 haplotype (i-139T, i-744C, +ins 801A, T52) to be associated with an increased aggregation response to ADP and showed some association with peripheral arterial disease. However, a later study by Ziegler et al (2005) did not show an association with peripheral arterial disease and showed a 34C to T and 53G to T mutation of $P2Y_{12}$ to be associated with increased risk of ischemic stroke. A crossover study between clopidogrel and a newer $P2Y_{12}$ antagonist, prasugrel, makes the receptor an unlikely major contributor to a poor response to clopidogrel. Recent studies have shown that polymorphisms of the cytochrome P450 3A5, 3A4, and 2C19 genes may be associated with pharmacodynamic response to clopidogrel and risk of atherothrombotic events in patients taking clopidogrel.

Dipyridamole

Mechanism of Action

Dipyridamole (Persantine) is a drug derived from the papaverine family. It inhibits the nucleoside transporter of cell membranes, decreasing the uptake of adenosine by erythrocytes and leading to increased plasma availability of adenosine. This process results in stimulation of platelet adenylate cyclase, resulting in the increase in cytoplasmic cyclic guanosine monophosphate (cGMP) and cyclic adenosine monophosphate (cAMP) levels that leads to a decrease in platelet activation.

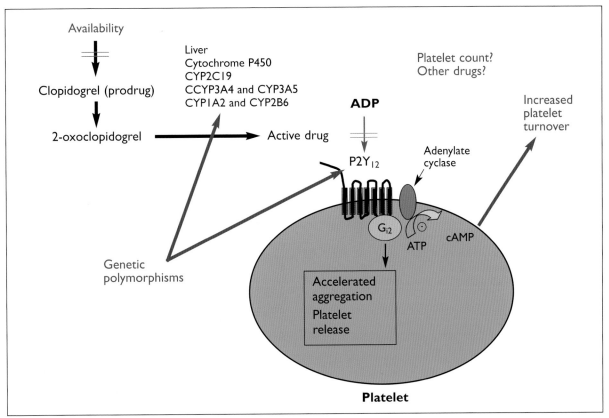

Figure 25-5. Postulated mechanisms of clopidogrel resistance include decreased availability, gastrointestinal absorption, genetic cytochrome P450 polymorphisms, decreased metabolism in the liver, genetic polymorphisms of the P2Y$_{12}$ receptor, and increased platelet turnover. Abbreviations: ADP, adenosine diphosphate; ATP, adenosine triphosphate; cAMP, cyclic adenosine monophosphate.

Dipyridamole also inhibits phosphodiesterase, to prevent the breakdown of cGMP. At the clinically used doses, however, dipyridamole does not prolong the bleeding time or inhibit platelet aggregation. Dipyridamole is also associated with vasodilation, but it was found to be associated with myocardial ischemia in some studies. Side effects may include headache, vertigo, and flushing.

Clinical Uses

Dipyridamole is rarely used alone as an antiplatelet agent. The intravenous formulation is used in stress-induced tests of myocardial ischemia in patients unable to exercise. Oral dipyridamole typically is used in conjunction with other antiplatelet agents, such as aspirin. A combination preparation, Aggrenox, is available that includes aspirin plus dipyridamole. The combination of aspirin and dipyridamole has been used in primary prevention of stroke and transient ischemic attacks. Clinical studies, such as the European Stroke Prevention Study (ESPS) have shown a 33% reduction in risk of stroke and death with combined aspirin plus dipyridamole therapy

as compared with placebo. Monotherapy with dipyridamole showed a reduction in stroke rate as compared with placebo in the ESPS-2, but the protective effect was less than that seen with aspirin alone. Some studies, such as the AICLA (1983) and the American-Canadian Cooperative Study Group (1985), have not shown an additive benefit of aspirin plus dipyridamole as compared with aspirin alone. The ESPS-2 study showed a combination of aspirin plus an extended release formulation of dipyridamole to be more effective at secondary stroke prevention (36% reduction) than aspirin alone (18% reduction), although there are concerns that a subtherapeutic dose of aspirin was used. Dipyridamole has been included in the antiplatelet protocol for thromboprophylaxis of patients with ventricular assist devices and total artificial hearts, but objective evidence for a therapeutic efficacy of dipyridamole in addition to aspirin is scant.

Monitoring Issues

Dipyridamole has been shown to decrease shear-induced platelet aggregation in whole blood but not in platelet-rich plasma. However, in vivo dipyridamole therapy alone has little, if any, effect on the laboratory measurement of platelet function. Therefore, dipyridamole is not typically monitored. Dipyridamole is typically used in conjunction with aspirin; however, few studies have addressed whether dipyridamole cotherapy markedly affects the platelet inhibition effects of aspirin.

Iloprost

Iloprost, a stable prostaglandin I2 analogue, increases cAMP levels and leads to inhibition of platelet function. It was tested as an alternative to heparin in patients with heparin-induced thrombocytopenia before development of danaparoid and direct thrombin inhibitors. Its use has not been continued because of vasoactive properties and hypotension.

Cilostazol

Mechanism of Action

Cilostazol (Pletal) is a phophodiesterase type-3 inhibitor that leads to platelet inhibition and vasodilation. It has also been reported to inhibit proliferation of vascular smooth muscle cells in vitro, to increase high-density lipoprotein cholesterol levels, and to decrease triglyceride, but the mechanisms are not clear.

Clinical Uses

Because of its antiplatelet and vasodilatory properties, cilostazol has been used in treatment of the intermittent claudication associated with peripheral arterial disease. It is one of two drugs approved for this application by the Food and Drug Administration in the United States, the other being pentoxifylline (Trental), a drug that increases erythrocyte and leukocyte flexibility but also decreases fibrinogen concentration and reduces blood viscosity. Cilostazol has been studied in several clinical trials for treatment of claudication. In a recent meta-analysis of several such trials, cilostazol increased maximal walking distance by 50% and pain-free walking distance by 67% as compared with placebo. In one study, cilostazol was shown to have a higher efficacy than pentoxifylline. Cilostazol is contraindicated in patients with congestive heart failure.

Monitoring Issues

In vitro addition of cilostazol has been reported to inhibit shear-induced platelet aggregation in platelet-rich plasma samples, but there are no current recommendations for the laboratory monitoring of patients being treated with cilostazol.

New Antiplatelet Drugs in Development

The clinically approved antiplatelet drugs target the ADP receptor $P2Y_{12}$, the cyclooxygenase pathway, and the GPIIb/IIIa fibrinogen receptor, as well as phosphodiesterase. However, given the many ways that platelets can be activated and the extensive cross-talk in platelet activation, it is reasonable to expect that other mechanisms of inhibiting platelet function could be successfully employed in the development of new antiplatelet agents. For example, inhibition of platelet adhesion through the glycoprotein Ib complex, inhibition of platelet activation through the protease-activated thrombin receptors, interference with signaling proteins and prevention of aggregation amplification through thromboxane receptors are all likely targets for development of antiplatelet agents. In addition, improvement in inhibition of the $P2Y_{12}$ receptor by a drug that is not affected by cytochrome P450 metabolism might result in more widespread $P2Y_{12}$ antagonist efficacy.

Inhibition of platelet adhesion to von Willebrand factor through the GPIb/IX/V complex is an attractive target for platelet inhibition. Reducing initial platelet adhesion probably would result in less overall platelet activation and diminished aggregation amplification. It is also reasonable to anticipate that such an adhesion-preventing drug would act in synergy with other available antiplatelet agents. Preclinical animal studies with von Willebrand factor protein fragments and monoclonal antibodies to GPIb have shown an ability to decrease thrombosis, but these compounds have not reached clinical trials. Aurin tricarboxylic acid, which binds to high-molecular-weight von Willebrand factor multimers, has been shown to inhibit platelet adhesion in vitro.

The protease-activated receptors (PAR-1 and PAR-4) are the primary thrombin receptors on the platelet surface; they are attractive targets for platelet inhibition. These receptors are unique in that they are activated by thrombin as a result of cleavage of a peptide, and the new N-terminal tethered ligand becomes the receptor agonist. Thus, PAR inhibitors would have to prevent the binding of the tethered ligands to the receptor. Several PAR-1 antagonists are in development.

Some of these are peptides based upon the sequence of the 14-amino acid tethered ligand (SFLLRNPNDKYEPF) with chemical modification providing antagonist rather than agonist properties. Nonpeptide thrombin receptor antagonists have been developed from pyrroloquinazoline analogues, benzimidazole derivatives, bicyclic amidine compounds, and himbacine-based compounds. These compounds have shown in vitro and ex vivo inhibition of platelet aggregation and decreased thrombosis in animal models. Clinical studies with orally active PAR-1 antagonists, such as SCH 530348, are underway; they are being studied as antithrombotic agents in acute coronary syndromes. Another approach to inhibition of the PARs is the use of pepducins, which are lipidated peptides that penetrate the platelet membrane and inhibit the intracellular PAR receptor/G-protein interaction.

One mechanism for the amplification of the platelet aggregation response is the production of thromboxane A2 through the cyclooxygenase pathway. Thromboxane A2 amplifies the aggregation response by acting as an agonist for other platelets by stimulating platelet activation through the thromboxane receptor. Inhibitors of the thromboxane receptor are being developed. These show inhibition of platelet aggregation in vitro and some antiplatelet activity in animal and clinical studies.

Another mechanism for platelet aggregation amplification is the release of ADP from dense granules, with the ADP binding to the $P2Y_{12}$ and $P2Y_1$ receptors leading to further aggregation. Local control of plasma ADP levels is regulated through the CD39/ADPase, an endothelial-based enzyme responsible for the degradation of ADP, thus decreasing the platelet-activating effect of ADP. Statin drugs used to lower cholesterol and lipid levels may have some antithrombotic effect due to preservation of the function of the CD39/ADPase. Animal studies of soluble recombinant CD39/ADPase have been shown to be protective against stroke.

Newer inhibitors of the ADP receptors are being developed that either have higher affinity for the $P2Y_{12}$ receptor, offer reversible binding, or don't require metabolism by the cytochrome P450 system. Prasugrel is one such compound that has higher affinity for $P2Y_{12}$ and is under clinical investigation. Two direct and reversible $P2Y_{12}$ antagonists, cangrelor and AZD6140, offer both rapid inhibition of the receptor and rapid reversal due to short half-life. The other ADP receptor,

$P2Y_1$, is also the target of pharmacologic development with inhibitor drugs in development.

Antiplatelet Effect of Other Drugs and Herbal Remedies

Many drugs can have an effect on platelet function, even though the primary action of the drug is not intended to be an antiplatelet agent. These include drugs such as antibiotics, antidepressants, and antipsychotic agents. A list is included in Table 25-1.

There are numerous herbal remedies commercially available and their use is increasing. Some herbal remedies contain warfarin-like compounds and have anticoagulant properties, whereas others can affect platelet function. The exact composition of many herbal preparations is not known, so assessing the effect of a particular preparation on platelet function in a given patient may be difficult. In addition, many patients do not relate their use of herbal preparations to their physicians. In the evaluation of platelet dysfunction or bleeding in patients not taking known antiplatelet medication, a thorough history of herbal medication use should be taken.

Some herbs contain salicylic acid and have aspirin-like activity; these include wintergreen leaf, sweet birch bark, and willow bark. In fact, as mentioned previously, the ancient Assyrians used willow as a treatment for rheumatism. Garlic is widely advertised to have cardioprotective effects, and many garlic preparations are available in health food stores. Extracts of garlic have been shown to inhibit platelet aggregation and platelet release in vitro. Garlic's antiplatelet effect is thought to be due to inhibition of thromboxane production or inhibition of the ADP receptor. Dietary supplementation with garlic extracts has been shown to inhibit ADP-induced platelet aggregation in humans. Indeed, limited clinical studies have indicated some promise in improving cardiovascular risk factors.

Other herbal remedies are thought to inhibit the cyclooxygenase pathway and have antiplatelet activity. These include ginger, used by pregnant women to relieve nausea, and feverfew, used to treat migraine headaches. Clove, a common kitchen spice, contains two compounds, eugenol and acetyl eugenol, that inhibit platelet thromboxane formation. Licorice has been shown to inhibit cyclooxygenase, lipoxygenase, and peroxidase activity in platelets; licorice also contains a

Table 25-1. Drugs that Affect Platelet Function

Antiplatelet drugs
Dipyridamole
Adenosine diphosphate receptor antagonists
  Ticlopidine
  Clopidogrel
Glycoprotein IIb/IIIa antagonists
  Abciximab
  Eptifibatide
  Tirofiban
 	
Anticoagulants*
Heparin
Coumadin
Direct thrombin inhibitors
 	
Thrombolytic agents*
Streptokinase
Urokinase
Tissue plasminogen activator
 	
Cardiovascular agents
Beta adrenergic blockers (eg, propranol)
Vasodilators (eg, nitroprusside, nitroglycerin)
Diuretics (eg, furosemide)
Calcium channel blockers
 	
Nonsteroidal anti-inflammatory drugs (NSAIDS)
Aspirin
Ibuprofen
Mefenamic acid
Indomethacin
COX-2 inhibitors
	Antimicrobials
	Penicillins
	Cephalosporin
	Nitrofurantoin
	Hydroxychloroquine
	Amphotericin
	
	Psychotropics and anesthetics
	Tricyclic antidepressants (eg, imipramine)
	Phenothiazines (eg, chlorpromazine)
	Local and general anesthesia (eg, halothane)
	
	Chemotherapeutic agents
	Mithramycin
	Daunorubicin
	BCNU
	
	Miscellaneous agents
	Dextrans
	Radiographic contrast
	Quinidine
	Ethanol
	
	Foods
	Caffeine
	Garlic
	Cumin
	Turmeric
	Dark chocolate
	
	* May increase bleeding risk in patients taking antiplatelet medication.

coumarin derivative and thus may have anticoagulant effects as well. Other herbs, such as ginseng and red pepper (capasaicin), can affect platelet function through mechanisms other than the cyclooxygenase pathway. Ginkgo biloba, an herb used for relief of chest complaints, has been associated with hemorrhage when used with other antiplatelet drugs; this effect has been thought to be associated with inhibition of platelet activation by platelet-activating factor. However, a recent study did not show a significant increase in platelet inhibition by Ginkgo biloba when coadministered with cilostazol or clopidogrel.

Summary

There are many antiplatelet drugs that target diverse pathways of platelet activation and have shown effectiveness as antithrombotic agents. Despite studies indicating that suboptimal response to these agents may be associated with adverse clinical outcomes, the role for laboratory monitoring of these drugs is not well established and requires further study. Promising new antiplatelet drugs that target complimentary path-

ways for platelet activation are in development. In assessing adverse bleeding in patients taking prescription antiplatelet drugs, the possible effect of other medications and herbal remedies should be carefully considered.

Suggested Reading

General

Bhatt DL, Topol EJ. Scientific and therapeutic advances in antiplatelet therapy. *Nat Rev Drug Discov.* 2003; 2:15-28.

Duffy B, Bhatt DL. Antiplatelet agents in patients undergoing percutaneous coronary intervention: how many and how much? *Am J Cardiovasc Drugs.* 2005;5:307-318.

Ferguson JJ III, Chronos NAF, Harrington RA, eds. *Antiplatelet Therapy in Clinical Practice.* London: Martin Dunitz; 2000.

Lange RA, Hillis LD. Antiplatelet therapy for ischemic heart disease. *N Engl J Med.* 2004;350:277-280.

Matzdorff A. Platelet function tests and flow cytometry to monitor antiplatelet therapy. *Semin Thromb Haemost.* 2005;31:393-399.

Patrono C, Bachmann F, Baigent C, et al. Expert consensus document on the use of antiplatelet agents. *Eur Heart J.* 2004;25:166-181.

Parenteral Agents

Glycoprotein IIb/IIIa Antagonists

Cierniewski CS, Byzova T, Papierak M, et al. Peptide ligands can bind to distinct sites in integrin $\alpha_{IIb}\beta_3$ and elicit different functional responses. *J Biol Chem.* 1999;274:16923-16932.

The EPIC Investigators. Use of a monoclonal antibody directed against the platelet glycoprotein IIb/IIIa receptor in high-risk coronary angioplasty. *N Engl J Med.* 1994;330:956-961.

The EPILOG Investigators. Platelet glycoprotein IIb/IIIa receptor blockade and low-dose heparin during percutaneous coronary revascularization. *N Engl J Med.* 1997;336:1689-1696.

EPISTENT Investigators. Randomized placebo-controlled and balloon-angioplasty-controlled trial to assess safety of coronary stenting with use of platelet glycoprotein IIb/IIIa blockade. *Lancet.* 1998;352:87-92.

ESPRIT Investigators: Enhanced Suppression of the Platelet IIb/IIIa Receptor with Integrilin Therapy. Novel dosing regimen of eptifibatide in planned coronary stent implantation (ESPRIT): a randomized, placebo-controlled trial. *Lancet.* 2000;356:2037-2044. Erratum in Lancet. 2001;357:1370.

Hartman GD, Egbertson MS, Halczenko W, et al. Nonpeptide fibrinogen receptor antagonists, 1: discovery and design of exosite inhibitors. *J Med Chem.* 1992;35:4640-4642.

Huxtable LM, Tafreshi MJ, Rakkar ANS. Frequency and management of thrombocytopenia with the glycoprotein IIb/IIIa receptor antagonists. *Am J Cardiol.* 2006;97:426-429.

Kastrati A, Mehilli J, Schuhlen H, et al; Intracoronary Stenting and Antithrombotic Regimen-Rapid Early Action for Coronary Treatment Study Investigators. A clinical trial of abciximab in elective percutaneous coronary intervention after pretreatment with clopidogrel. *N Engl J Med.* 2004;350:232-238.

Lincoff AM, ed. *Platelet Glycoprotein IIb/IIIa Inhibitors in Cardiovascular Disease.* 2nd ed. Totowa, NJ: Humana Press; 2003.

Newby LK, Califf RM, White HD, et al. The failure of orally administered glycoprotein IIb/IIIa inhibitors to prevent recurrent cardiac events. *Am J Med.* 2002;112:647-658.

Osende JI, Fuster V, Lev EI, et al. Testing platelet activation with a shear-dependent platelet function test versus aggregation-based tests: relevance for monitoring long-term glycoprotein IIb/IIIa inhibition. *Circulation.* 2001;103:1488-1491.

The PURSUIT Trial Investigators. Inhibition of platelet glycoprotein IIb/IIIa with eptifibatide in patients with acute coronary syndromes. Platelet glycoprotein IIb/IIIa in unstable angina: receptor suppression using integrelin therapy. *N Engl J Med.* 1998;339:436-443.

Scarborough RM, Naughton MA, Teng W, et al. Design of potent and specific integrin antagonists: peptide antagonists with high specificity for glycoprotein IIb/IIIa. *J Biol Chem.* 1993;268:1066-1073.

Steinhubl S, Talley J, Braden G, et al. Point-of-care measured platelet inhibition correlates with a reduced risk of an adverse cardiac event following percutaneous coronary intervention: Results of the GOLD (AU-Assessing Ultegra) multicenter study. *Circulation.* 2001;103:1403-1409.

Tcheng JE, Harrington RA, Kottke-Marchant K, et al. Multicenter, randomized, double-blind, placebo-controlled trial of the platelet integrin glycoprotein IIb/IIIa blocker Integrelin in elective coronary intervention. IMPACT Investigators. *Circulation.* 1995;91:2151-2157.

Oral Agents

Aspirin

Antithrombotic Trialists Collaboration. Collaborative meta-analysis of randomized trials of antiplatelet therapy for prevention of death, myocardial infarction, and stroke in high-risk patients. *BMJ.* 2002;324:71-86.

Chen WH, Lee PY, Ng W, Tse HF, Lau CP. Aspirin resistance is associated with a high incidence of myonecrosis after non-urgent percutaneous coronary intervention despite clopidogrel pretreatment. *J Am Coll Cardiol.* 2004;43:1122-1126.

Cuisset T, Frere C, Quilici J, et al. High post-treatment platelet reactivity identified low-responders to dual antiplatelet therapy at increased risk of recurrent cardiovascular events after stenting for acute coronary syndrome. *J Thromb Haemost.* 2006;4:542-549.

Eikelboom JW, Hirsh J, Weitz JI, Johnston M, Yi Q, Yusuf S. Aspirin-resistant thromboxane biosynthesis and the risk of myocardial infarction, stroke or cardiovascular death in patients at high risk for cardiovascular events. *Circulation.* 2002;105:1650-1655.

Gladding P, Webster M, Ormiston J, Olsen S, White H. Antiplatelet drug nonresponsiveness. *Am Heart J.* 2008;155:591-599.

Grotemeyer KH, Scharafinski HW, Husstedt IW. Two-year follow-up of aspirin responder and aspirin non responder: a pilot-study including 180 post-stroke patients. *Thromb Res.* 1993;71:397-403.

Grundmann K, Jaschonek K, Kleine B, Dichgans J, Topka H. Aspirin non-responder status in patients with recurrent cerebral ischemic attacks. *J Neurol.* 2003;250:63-66.

Gum PA, Kottke-Marchant K, Poggio ED, et al. Profile and prevalence of aspirin resistance in patients with cardiovascular disease. *Am J Cardiol.* 2001;88:230-235.

Haubelt H, Anders C, Hellstem P. Can platelet function tests predict the clinical efficacy of aspirin? *Semin Thromb Hemost.* 2005;31:404-410.

Helgason CM, Tortorice KL, Winkler SR, et al. Aspirin response and failure in cerebral infarction. *Stroke.* 1993;24:345-350.

Hezard N, Metz D, Nazeyrollas P, Droulle C, Potron G, Nguyen P. PFA-100 and flow cytometry: can they challenge aggregometry to assess antiplatelet agents, other than GPIIbIIIa blockers, in coronary angioplasty? *Thromb Res.* 2002;108:43-47.

Jack DB. One hundred years of aspirin. Lancet. 1997;350:437-439.

Kundu SK, Heilmann EJ, Sio R, Garci C, Davidson RM, Ostgaard RA. Description of an in vitro platelet function analyzer, PFA-100. *Semin Thromb Hemost.* 1995;21:106-112.

Matzdorff A. Platelet function tests and flow cytometry to monitor antiplatelet therapy. *Semin Thromb Hemost.* 2005;31:393-399.

Michelson A, Cattaneo M, Kunicki T, et al; Platelet Physiology Subcommittee of the Scientific and Standardization Committee of the International Society on Thrombosis and Haemostasis: Working Group on Aspirin Resistance. Aspirin resistance: position paper of the Working Group on Aspirin Resistance. *J Thromb Haemost.* 2005;3:1309-1311.

Michos ED, Ardehali R, Blumenthal RS, Lange RA, Ardehal H. Aspirin and clopidogrel resistance. *Mayo Clin Proc.* 2006;81:518-526.

Moffat KA, Ledford-Kraemer MR, Nichols WL, Hayward CPM. Variability in clinical laboratory practice in testing for disorders of platelet function. *Thromb Haemost.* 2005;93:549-553.

Mueller MR, Salat A, Stangl P, et al. Variable platelet response to low-dose ASA and the risk of limb deterioration in patients submitted to peripheral arterial angioplasty. *Thromb Haemost.* 1997;78:1003-1007.

Sane DC, McKee SA, Malinin AI, Serebruany VL. Frequency of aspirin resistance in patients with congestive heart failure treated with antecedent aspirin. *Am J Cardiol.* 2002;90:893-895.

Stejskal D, Vaclavik J, Lacnak B, Proskova J. Aspirin resistance measured by cationic propyl gallate platelet aggregometry and recurrent cardiovascular events during 4 years of follow-up. *Eur J Intern Med.* 2006;17:349-354.

Vane JR. Inhibition of prostaglandin synthesis as a mechanism of action for aspirin-like drugs. *Nat New Biol.* 1971;231:232-235.

Yusuf S, Zhao F, Mehta SR, et al; The Clopidogrel in Unstable Angina to Prevent Recurrent Events (CURE) Trial Investigators. Effects of clopidogrel in addition to aspirin in patients with acute coronary syndromes without ST-segment elevation. *N Engl J Med.* 2001;345(7):494-502.

Zhou L, Schmaier AH. Platelet aggregation testing in platelet-rich plasma: description of procedures with the aim to develop standards in the field. *Am J Clin Pathol.* 2005;123:172-183.

Thienopyridines

Angiolillo DJ, Fernandez-Ortiz A, Bernardo E, et al. Contribution of gene sequence variations of the hepatic cytochrome P450 3A4 enzyme to variability in individual responsiveness to clopidogrel. *Arterioscler Thromb Vasc Biol.* 2006;26:1895-1900.

Angiolillo DJ, Fernandez-Ortiz A, Bernardo E, et al. Variability in platelet aggregation following sustained aspirin and clopidogrel treatment in patients with coronary heart disease and influence of the 807 C/T polymorphism of the glycoprotein Ia gene. *Am J Cardiol.* 2005;96(8):1095-1099.

Barragan P, Bouvier J-L, Roquebert P-O, et al. Resistance to thienopyridines: clinical detection of coronary stent thrombosis by monitoring of vasodilator-stimulated phosphoprotein phosphorylation. *Catheter Cardiovasc Interv.* 2003;59:295-302.

Barsky AA, Arora RR. Clopidogrel resistance: myth or reality? J Cardiovasc Pharmacol Ther. 2006;11:47-53.

Bhatt DL, Fox KA, Hacke W, et al; CHARISMA Investigators. Clopidogrel and aspirin versus aspirin alone for the prevention of atherothrombotic events. *N Engl J Med.* 2006;354:1706-1717.

Brandt JT, Payne CD, Wiviott SD, et al. A comparison of prasugrel and clopidogrel loading doses on platelet function: magnitude of platelet inhibition is related to active metabolite formation. *Am Heart J.* 2007;153(1):66.e9-16.

CAPRIE Steering Committee. A randomised, blinded, trial of clopidogrel versus aspirin in patients at risk of ischaemic events (CAPRIE). *Lancet.* 1996; 348(9038):1329-1339.

Clarke TA, Waskell LA. The metabolism of clopidogrel is catalyzed by human cytochrome P450 3A and is inhibited by atorvastatin. *Drug Metab Dispos.* 2003;31:53-59.

Fontana P, Gaussem P, Aiach M, Fiessinger JN, Emmerich J, Reny JL. P2Y12 H2 haplotype is associated with peripheral arterial disease: a case-control study. *Circulation.* 2003;108:2971-2973.

Gurbel PA, Bliden KP. Durability of platelet inhibition by clopidogrel. *Am J Cardiol.* 2003;91:1123-1125.

Gurbel PA, Bliden KP, Hiatt BL, O'Connor CM. Clopidogrel for coronary stenting: response variability, drug resistance, and the effect of pretreatment platelet reactivity. *Circulation.* 2003;107:2908-2913.

Herbert JM, Savi P. P2Y12, a new platelet ADP receptor, target of clopidogrel. *Semin Vasc Med.* 2003;3:113-122.

Hulot JS, Bura A, Villard E, et al. Cytochrome P450 2C19 loss-of-function polymorphism is a major determinant of clopidogrel responsiveness in healthy subjects. *Blood.* 2006;108(7):2244-2247.

Kastrati A, von Beckerath N, Joost A, Pogatsa-Murray G, Gorchakova O, Schomig A. Loading with 600 mg clopidogrel in patients with coronary artery disease with and without chronic clopidogrel therapy. *Circulation.* 2004;110:1-4.

Matetzky S, Shenkman B, Guetta V, et al. Clopidogrel resistance is associated with increased risk of recurrent atherothrombotic events in patients with acute myocardial infarction. *Circulation.* 2004;109:3171-3175.

Muller I, Besta F, Schulz C, Massberg S, Schonig A, Gawaz M. Prevalence of clopidogrel non-responders among patients with stable angina pectoris scheduled for elective coronary stent placement. *Thromb Haemost.* 2003;89:783-787.

Pampuch A, Cerletti C, de Gaetano G. Comparison of VASP-phosphorylation assay to light-transmission aggregometry in assessing inhibition of the platelet ADP P2Y12 receptor. *Thromb Haemost.* 2006;96:767-773.

Sabatine MS, Cannon CP, Gibson CM, et al; CLARITY-TIMI 28 Investigators. Addition of clopidogrel to aspirin and fibrinolytic therapy for myocardial infarction with ST-segment elevation. *N Engl J Med.* 2005;352:1179-1189.

Savi P, Zachayus JL, Delesque-Touchard N, et al. The active metabolite of clopidogrel disrupts P2Y12 receptor oligomers and partitions them out of lipid rafts. *Proc Natl Acad Sci USA.* 2006;103(29):11069-11074.

Steinhubl SR, Berger PB, Mann JT 3rd, et al; CREDO Investigators. Clopidogrel for the Reduction of Events During Observation. Early and sustained dual oral antiplatelet therapy following percutaneous coronary intervention: a randomized controlled trial. *JAMA.* 2002;288:2411-2420.

Suh JW, Koo BK, Zhang SY, et al. Increased risk of atherothrombotic events associated with cytochrome P450 3A5 polymorphism in patients taking clopidogrel. *CMAJ.* 2006;174(12):1715-1722.

Van Werkum JW, Van der Stelt CAK, Seesing TH, Ten Berg M, Hackeng CM. The flow cytometric VASP assay can be used to determine the effectiveness of clopidogrel in patients treated with abciximab. *J Thromb Haemost.* 2007;5:881-883.

Wiviott SD, Antman EM. Clopidogrel resistance: a new chapter in a fast-moving story. *Circulation.* 2004;109:3064-3067.

Yusuf S, Zhao F, Mehta SR, et al. Effects of clopidogrel in addition to aspirin in patients with acute coronary syndromes without ST-segment elevation. *N Engl J Med.* 2001;345(7):494-502.

Ziegler S, Schillinger M, Funk M, et al. Association of a functional polymorphism in the clopidogrel target receptor gene, P2Y12, and the risk for ischemic cerebrovascular events in patients with peripheral artery disease. *Stroke.* 2005;36:1394-1399.

Dipyridamole

American-Canadian Cooperative Study Group. Persantine aspirin trial in cerebral ischemia, part II: endpoint results. *Stroke.* 1985;16:406-415.

Bousser MG, Eschwege E, Haguenau M, et al. "AICLA" controlled trial of aspirin and dipyridamole in the secondary prevention of athero-thrombotic cerebral ischemia. *Stroke.* 1983;14:5-14.

Copeland JG, Arabia FA, Smith RG, Sethi GK, Nolan PE, Banchy ME. Arizona experience with CardioWest Total Artificial Heart bridge to transplantation. *Ann Thorac Surg.* 1999 Aug;68(2):756-760.

Diener HC, Cunha L, Forbes C, Sivenius J, Smets P, Lowenthal A. European Stroke Prevention Study 2: dipyridamole and acetylsalicylic acid in the secondary prevention of stroke. *J Neurol Sci.* 1996;143:1-13.

European Stroke Prevention Study Group. The European Stroke Prevention Study (ESPS) principal end-points. *Lancet.* 1987;2:1351-1354.

Ling GS, Ling SM. Preventing ischemic stroke in the older adult. *Cleve Clin J Med.* 2005;72(suppl 3):S14-S25.

Nakamura T, Uchiyama S, Yamazaki M, Iwata M. Synergistic effect of cilostazol and dipyridamole mediated by adenosine on shear-induced platelet aggregation. *Thromb Res.* 2007;119(4):511-516. Epub 2006 Oct 12.

Schaper W. Dipyridamole, an underestimated vascular protective drug. *Cardiovasc Drugs Ther.* 2005;19:357-363.

Tran H, Anand SS. Oral antiplatelet therapy in cerebrovascular disease, coronary artery disease, and peripheral arterial disease. *JAMA.* 2004;292:1867-1874.

Weksler BB. Antiplatelet agents in stroke prevention: combination therapy: present and future. *Cerebrovasc Dis.* 2000;10(suppl 5):41-48.

Iloprost

Grant SM, Goa KL. Iloprost: a review of its pharmacodynamic and pharmacokinetic properties, and therapeutic potential in peripheral vascular disease, myocardial ischaemia and extracorporeal circulation. *Drugs.* 1992;43:889-924.

Cilostazol

Dawson DI, Cutler BS, Hiatt WR, et al. A comparison of cilostazol and pentoxifylline for treating intermittent claudication. *Am J Med.* 2000;109:523-530.

Rice TW, Lumsden AB. Optimal medical management of peripheral arterial disease. *Vasc Endovasc Surg.* 2006;40:312-327.

Thompson PD, Zimet R, Forbes WP, et al. Meta-analysis of results from eight randomized, placebo-controlled trials on the effect of cilostazol on patients with intermittent claudication. *Am J Cardiol.* 2002;90:1314-1319.

Zhang W, Ke H, Colman RW. Identification of interaction sites of cyclic nucleotide phosphodiesterase type 3A with milrinone and cilostazol using molecular modeling and site-directed mutagenesis. *Mol Pharmacol.* 2002;62:514-520.

New Antiplatelet Agents

Bonnefoy A, Vermylen J, Hoylaerts MF. Inhibition of von Willebrand factor-GPIb/IX/V interactions as a strategy to prevent arterial thrombosis. *Expert Rev Cardiovasc Ther.* 2003;1(2):257-269.

Brandt JT, Payne CD, Wiviott SD, et al. A comparison of prasugrel and clopidogrel loading doses on platelet function: magnitude of platelet inhibition is related to active metabolite formation. *Am Heart J.* 2007;153(1):66.e9-16

Cattaneo M. ADP receptors: inhibitory strategies for antiplatelet therapy. *Drug News Perspect.* 2006;19:253-259.

Chackalmannil S. Thrombin receptor (protease activated receptor-1) antagonists as potent antithrombotic agents with strong antiplatelet effects. *J Med Chem.* 2006;49:5389-5403.

David T, Ohlmann P, Eckly A, et al. Inhibition of adhesive and signaling functions of the platelet GPIb-V-IX complex by a cell penetrating GPIbalpha peptide. *J Thromb Haemost.* 2006;4(12):2645-2655.

Dogne JM, Hanson J, de Laval X, et al. From the design to the clinical application of thromboxane modulators. *Curr Pharm Des.* 2006;12(8):903-923.

Hanson J, Reynaud D, Qiao N, et al. Synthesis and pharmacological evaluation of novel nitrobenzenic thromboxane modulators as antiplatelet agents acting on both the alpha and beta isoforms of the human thromboxane receptor. *J Med Chem.* 2006;49:3701-3709.

Jneid H, Bhatt DL. Advances in antiplatelet therapy. *Expert Opin Emerg Drugs.* 2003;8:349-363.

Kaneider NC, Eger P, Dunzendorfer S, et al. Reversal of thrombin-induced deactivation of CD39/ATPDase in endothelial cells by HMG-CoA reductase inhibition: effects on Rho-GTPase and adenosine nucleotide metabolism. *Arterioscler Thromb Vasc Biol.* 2002;22(6):894-900.

Malinin AI, Ong S, Makarov M, Petukhova EY, Serebruany VL. Platelet inhibition beyond conventional antiplatelet agents: expanding role of angiotensin receptor blockers, statins and selective serotonoin reuptake inhibitors. *Int J Clin Pract.* 2006;60:993-1002.

McNicol A, Israels SJ. Platelets and anti-platelet therapy. *J Pharmacol Sci.* 2003;93:381-396.

Wiviott SD, Michelson AD, Berger PB, Lepor NE, Kereiakes DJ. Therapeutic goals for effective platelet inhibition: a consensus document. *Rev Cardiovasc Med.* 2006;7:214-225.

Wu CC, Teng CM. Comparison of the effects of PAR1 antagonists, PAR4 antagonists, and their combinations on thrombin-induced human platelet activation. *Eur J Pharmacol.* 2006;546(1-3):142-147.

Wu D, Meiring M, Kotze HF, Deckmyn H, Cawenbergs N. Inhibition of platelet glycoprotein Ib, glycoprotein IIb/IIIa, or both by monoclonal antibodies prevents arterial thrombosis in baboons. *Arterioscler Thromb Vasc Biol.* 2002;22(2):323-328.

Antiplatelet Properties of Other Drugs and Herbal Medications

Ackermann RT, Mulrow CD, Ramirez G, Gardner CD, Morbidoni L, Lawrence VAl. Garlic shows promise for improving some cardiovascular risk factors. *Arch Intern Med.* 2001;161:813-824.

Apitz-Castro R, Cabrera S, Cruz MR, Ledezma E, Jain MK. Effects of garlic extract and of three pure components isolated from it on human platelet aggregation, arachidonate metabolism, release reaction and platelet ultrastructure. *Thromb Res.* 1983;32:155-169.

Aruna D, Naidu MU. Pharmacodynamic interaction studies of Ginkgo biloba with cilostazol and clopidogrel in healthy human subjects. *Br J Clin Pharmacol.* 2007;63(3):333-338. Epub 2006 Sep 29.

Bordia A, Verma SK, Srivastava KC. Effect of ginger (Zingiber officinale Rose) and fenugreek (Trigonella foenumgracaecum L) on blood lipids, blood sugar and platelet aggregation in patients with coronary artery disease. *Prostaglandins Leukot Essent Fatty Acids.* 1997;56:379-384.

Cui J, Garle M, Eneroth P, Bjorkhem I. What do commercial ginseng preparations contain? *Lancet.* 1994; 344:134.

George JN, Shattil SJ. The clinical importance of acquired abnormalities of platelet function. *N Engl J Med.* 1991;324:27-39.

Mashour NH, Lin GI, Frishman WH. Herbal medicine for the treatment of cardiovascular disease. *Arch Intern Med.* 1998;158:2225-2234.

Rahman K, Billington D. Dietary supplementation with aged garlic extract inhibits ADP-induced platelet aggregation in humans. *J Nutr.* 2000;130:2662-2665.

Samuels N. Herbal remedies and anticoagulant therapy. *Thromb Haemost.* 2005;93:3-7.

Tawata M, Aida K, Noguchi T, et al. Anti-platelet action of isoliquirigenin, an aldose reductase inhibitor in licorice. *Eur J Pharmacol.* 1992;25:87-92.

Teng CM, Ko FN, Wang JP, et al. Antihaemostatic and antithrombotic effect of some antiplatelet agents isolated from Chinese herbs. *J Pharm Pharmacol.* 1991;43:667-669.

Section 6

Appendices and Index

Appendix A. Abbreviations

Appendix B. Example Reference Ranges

Index

Abbreviations

1,4,5-IP3	1,4,5-inositol 3 phosphate	CAP	College of American Pathologists
AA	arachidonic acid	CBC	complete blood count
AATP	acquired amegakaryocytic thrombocytopenic purpura	CCD	charge coupled display
		CEM	clot elastic modulus
ADAMTS-13	a disintegrin and metalloprotease with thrombospondin-1-like domains	CFU-MK	colony forming unit-megakaryocyte
		cGMP	cyclic guanosine monophosphate (GMP)
ACT	activated clotting time		
ADP	adenosine diphosphate	CML	chronic myelogenous leukemia
ADPase	an endothelial-based enzyme responsible for the degradation of ADP	COL/ADP	collagen/adenosine diphosphate
		COL/EPI	collagen/epinephrine
		COX	cyclooxygenase
AML	acute myelogenous leukemia	CPB	cardiopulmonary bypass
AMP	adenosine monophosphate	CRP	C-reactive protein
ANCA	antineutrophilic cytoplasmic antibody	CTAD	citrate, theophylline, adenosine, and dipyridamole
APC	activated protein C	CTRUS	congenital thrombocytopenia with radio-ulnar synostosis
APCr	activated protein C resistance		
apo(a)	apolipoprotein (a)	CV	coefficient of variation
aPTT	activated partial thromboplastin time	CYP	cytochrome P450
		DDAVP	desmopressin acetate
AR	aspirin resistance	DIC	disseminated intravascular coagulation
AT	antithrombin		
ATP	adenosine triphosphate	DMS	demarcation membrane system
β_2GPI	beta 2-glycoprotein I	DNA	deoxyribonucleic acid
BCNU	1,3-Bis(2-chloroethyl)-1-nitrosourea	dPT	dilute prothrombin time
		dRVVT	dilute Russell viper venom time
BU	Bethesda unit	DTI	direct thrombin inhibitor
calpain	calcium-activated neutral proteinase	DU	D-dimer units
		DVT	deep vein thrombosis
cAMP	cyclic adenosine monophosphate (cyclic AMP)	ECL	euglobulin clot lysis
		ECMO	extracorporeal membrane oxygenation
CAMT	congenital (or X-linked) amegakaryocytic thrombocytopenia		
		ECT	ecarin clotting time

EDTA	ethylenediaminetetraacetic acid		ITAM	immunoreceptor type activation motif
EIA	enzyme immunoassay		ITP	immune thrombocytopenic purpura
ELISA	enzyme-linked immunosorbent assay		JAK-2	Janus kinase 2
ESPS	European Stroke Prevention Study		KCT	kaolin clotting time
ET	essential thrombocythemia		LA	lupus anticoagulant
ETP	endogenous thrombin potential		LAMP	lysome associated membrane protein
FcRγ	Fc receptor γ		LDH	serum lactic dehydrogenase
FDA	Food and Drug Administration		LDL	low-density lipoprotein
FEU	fibrinogen equivalent units		LIBS	ligand-induced binding site
FGF	fibroblast growth factor		LMWH	low-molecular-weight heparin
FPA	fibrinopeptide A		Lp(a)	lipoprotein (a)
FPIA	fully automated fluorescence polarization immunoassay		LRR	leucine-rich repeat (protein family)
FRET	fluorescence resonance energy transfer		LYST	lysosomal trafficking regulator
GP	glycoprotein		MAIPA	monoclonal antibody immobilization of platelet antigens
GPIa/IIa	glycoprotein Ia/IIa complex		mRNA	messenger ribonucleic acid
GPIb	glycoprotein Ib		MI	myocardial infarction
GPIIb/IIIa	glycoprotein IIb/IIIa complex		MPV	mean platelet volume
GPIB/IX/V	glycoprotein Ib/IX/V complex		MTHFR	5,10-methylenetetrahydrofolate reductase
GPL	IgG phospholipid units		MYH9	myosin heavy chain gene
HAMA	human anti-mouse antibodies		NAIT	neonatal alloimmune thrombocytopenia
HAT	heparin-associated thrombocytopenia		NASCOLA	North American Specialized Coagulation Laboratory Association
HBS	heparin binding site			
HELLP	hemolysis elevated liver enzymes and low platelets (syndrome)		NHLBI	National Heart, Lung, and Blood Institute
HIPA	heparin-induced platelet aggregation		NMMHCA	nonmuscle myosin heavy chain-A
HIT	heparin-induced thrombocytopenia		NMMHC-IIA	nonmuscle myosin heavy chain IIA protein
HPLC	high performance (or pressure) liquid chromatography		NPP	normal pooled plasma
HRT	hormone replacement therapy		OCS	open-canalicular membrane system
HUS	hemolytic-uremic syndrome		PAI-1	plasminogen activator inhibitor-1
IABP	intra-aortic balloon counterpulsation		PAI-2	plasminogen activator inhibitor-2
IPF	immature platelet fraction		PAR	protease-activated receptor
INR	international normalized ratio		PCI	percutaneous coronary intervention
ISI	international sensitivity index			
ISTH	International Society on Thrombosis and Haemostasis		PCR	polymerase chain reaction

PDW	platelet distribution width		TAR	thrombocytopenia with absent radii (syndrome)
PE	pulmonary embolism		TAT	thrombin-antithrombin (complexes)
PEG	polyethylene glycol			
PF4	platelet factor 4		TCPT	Paris-Trousseau type thrombocytopenia
PFA-100	Platelet Function Analyzer-100			
PGM	prothrombin gene G20210A mutation		Tel-AML1	familial thrombocytopenia-leukemia
PI3K	phosphoinositide 3-kinase		TGF	transforming growth factor
PIVKA	proteins induced by vitamin K absence		TGT	thrombin generation time
			tHcy	total plasma homocysteine
PNH	paroxysmal nocturnal hemoglobinuria		TIA	transient ischemic attack
			tPA	tissue plasminogen activator
PNP	platelet neutralization procedure		TRAP	thrombin receptor-activating peptide
POCT	point-of-care test			
PPACK	D-Phe-Pro-Arg chloromethylketone		TT	thrombin time
			TTP	thrombotic thrombocytopenic purpura
PRP	platelet-rich plasma			
PT	prothrombin time		TxA$_2$	thromboxane A2
PTT	partial thromboplastin time		UFH	unfractionated heparin
RANK	receptor-activated NF-κB ligand		uPA	urokinase plasminogen activator
RCPA	Royal College of Pathologists of Australasia		VASP	vasodilator-stimulated phosphoprotein
RDW	red cell distribution width		VKA	vitamin K antagonist
RFL	recurrent fetal loss		VKORC1	vitamin K epoxide reductase complex 1
rFVIIa	recombinant factor VII			
RGD	arginine-glycine-aspartic acid		VTE	venous thromboembolism
RIPA	ristocetin-induced platelet aggregation		VW AgII	a 741-amino acid propeptide antigen cleaved from vWF
SAH	S-adenosyl-homocysteine		VWD	von Willebrand disease
SD	standard deviation		vWF	von Willebrand factor
SDF	stromal-derived factor		VWF	von Willebrand factor gene
SERM	selective estrogen receptor modulator		WAS	Wiskott-Aldrich syndrome
SERPIN	serine protease inhibitor		WASP	Wiskott-Aldrich syndrome protein gene
SLE	systemic lupus erythematosus			
SNARE	soluble N-ethylmaleimide-sensive factor attachment protein receptor		WHO	World Health Organization
			XLT	X-linked thrombocytopenia
SNP	single nucleotide polymorphism		XLTT	X-linked thrombocytopenia with thalassemia
SPD	storage pool disorders			
SRA	serotonin release assay			
TAFI	thrombin-activatable fibrinolysis inhibitor			

Example Reference Ranges

Note: These are example ranges only. These ranges will vary by test methodology, assay lot, and instrument. Each laboratory should establish its own reference ranges. For additional example adult and pediatric reference ranges, see chapter 6.

Test	Conventional Units	SI Units
Activated partial thromboplastin time (aPTT)	25–40 s	25–40 s
Alanine aminotransferase (ALT)	10–40 U/L	0.17–0.68 µkat/L
Antithrombin	21–30 mg/dL	210–300 mg/L
α_1-antitrypsin	78–200 mg/dL	14.5–36.5 µmol/L
Apolipoprotein A-1	80–151 mg/dL	0.8–1.5 g/L
Apolipoprotein B	50–123 mg/dL	0.5–1.2 g/L
Aspartate aminotransferase (AST)	10–30 U/L	0.17–0.51 µkat/L
Coagulation factor II (prothrombin)	70–130 %	.70–1.30 proportion of 1.0
Coagulation factor V, IX, X, XI, XII	70–130 %	.70–1.30 proportion of 1.0
Coagulation factor VII	60–140 %	.60–1.40 proportion of 1.0
Coagulation factor VIII	50–200 %	.50–2.00 proportion of 1.0
C-reactive protein	0.08–3.1 mg/L	0.76–28.5 nmol/L
D-dimer	<0.5 µg/mL	<3.0 nmol/L

Note: D-dimer may be reported in fibrinogen equivalent units (ng/mL FEU) or D-dimer units (DU); 1 DU = 2 FEU; normal range <500 ng/mL FEU

Test	Conventional Units	SI Units
Fibrin degradation products (FDP)	<10 µg/mL	<10 mg/L
Fibrinogen	200–400 mg/dL	5.8–11.8 µmol/L
Gamma glutamyltransferase (GGT)	2–30 U/L	0.03–0.51 µkat/L
Hematocrit	41–50 %	.41–.50 proportion of 1.0
Hemoglobin	14.0–17.5 g/dL	140–175 g/L
Homocysteine	0.68–2.02 mg/L	5–15 µmol/L
Immunoglobulin G (IgG)	650–1600 mg/dL	6.5–16.0 g/L
Immuoglobulin M (IgM)	54–300 mg/dL	550–3000 mg/L
Lipoprotein a (Lp[a])	10–30 mg/dL	0.35–1.0 µmol/L
Plasminogen (antigenic)	10–20 mg/dL	1.1–2.2 µmol/L
Plasminogen activator inhibitor (PAI)	4–40 ng/dL	75–750 pmol/L
Platelet aggregation	>65% aggregation	>0.65 aggregation
Platelet count	150–350 x10^3/µL	150–350 x10^9/L

Note: this text uses 150,000–350,000/µL

Test	Conventional Units	SI Units
Protein C	70%–140%	0.70–1.40
Protein S	70%–140%	0.70–1.40
Prothrombin time (PT)	10–13 s	10–13 s
Red blood cell count (RBC)	3.9–5.5 x10^6/µL	3.9–5.5 x10^{12}/L
Ristocetin cofactor	75%–125% mean of normal*	0.75–1.25 mean of normal*
Thrombin time (TT)	16–24 s	16–24 s
Vitamin K	0.13–1.19 ng/mL	0.29–2.64 nmol/L
von Willebrand factor	75%–125% mean of normal*	0.75–1.25 mean of normal*
Warfarin	1.0–10 µg/mL	3.2–32.4 µmol/L
White blood cell count (WBC)	4500–11,000 /µL	4.5–11.0 x10^9/L

* Dependent on blood type

Sources: AMA Manual of Style. A Guide for Authors and Editors. 10th ed. New York: Oxford University Press; 2007.
Laposata M. Laboratory Medicine. Chicago, Ill: ASCP Press; 2002.

Index

f = figure; *t* = table

A4070G mutation, 133–134

Abciximab, 334

Acquired amegakaryocytic thrombocytopenic purpura (AATP), 208

Actin, 23

Actin filaments, 15

Activated clotting time (ACT), 320

Activated partial thromboplastin time (aPTT), 3, 70–73, 136, 169, 316–317. *See also* Prothrombin time and activated partial thromboplastin time, normal

    abnormal, 143–151

        clinical etiologies, 143, 143*t*

            alloantibodies in hemophiliac patients, 146–147

            autoantibodies to factor VIII and other factors, 147

            coagulation factor deficiencies not associated with bleeding, 146

            coagulation factor deficiency associated with significant hemorrhage, 144

            factor IX deficiency, 145

            factor VIII deficiency, 144*f*, 145

            factor XI deficiency, 145

            lupus anticoagulant, 146

            medications, 143–144

            other inherited and acquired deficiencies, 146

            specific coagulation factor inhibitors, 146

            spurious causes of elevated, 147–148

            von Willebrand disease, 144–145

        laboratory testing, 148–150

    accelerated, 150–151

    algorithm for diagnosis of a bleeding etiology with normal, 172*f*

    algorithm for evaluation of prolonged, 148*f*, 149*f*, 163*f*

    clinical utility of, 71–72

    evaluation of unfractionated heparin sensitivity, 72

    method, 71, 71*f*

    pathways, 6*t*, 7, 8*f*, 9*f*

    principle, 70

    reagents, 71

    reference interval, 71

    spurious causes of elevated, 147–148

    steps to take in evaluation of prolonged, 72

    validation of heparin sensitivity of, using ex vivo heparin specimens, comparison with existing validated reagent, 72–73

Activated partial thromboplastin time and prothrombin time, normal. *See* Prothrombin time and activated partial thromboplastin time, normal

Activated protein C (APC), 5

Activated protein C resistance (APCr), 127*f*, 131–132, 260

    factor V Leiden in diagnosing inherited thrombophilia, 273–274, 273*f*

Activation peptides, 136–137

Acute coronary syndromes, 309–310, 319–320

Acute venous thromboembolism, 309, 316–319, 317*f*, 318*f*

ADAMTS-13

    assays, 110, 203

    cleavage site for, 225, 226*f*

    proteolysis of vWF, 227

    and thrombotic thrombocytopenic purpura (TTP), 205–207, 206*f*

Adenosine diphosphate (ADP), 336

    ADP-induced platelet aggregation, 17, 18*f*

    in platelet aggregation studies, 101–102, 102*f*, 103–104, 104*f*, 105, 105*f*

        asprin effect on, 338–339, 338*f*

Adenosine diphosphate (ADP) receptors (P2Y$_1$, P2Y$_{12}$, and P2X$_1$), 15, 17, 18*f*, 340-341, 341*f*, 342, 343*f*

    abnormalities, 191

Adenosine triphosphate (ATP), 17, 18*f*, 101

    in platelet aggregation studies, 105–106, 105*f*

Adenylate cyclase, 17, 341*f*

Adhesion assays, 106

    Impact, 106

    platelet adhesion chambers, 106

Adrenergic receptors, 15

Afibrinogenemia, 158, 222

Alloantibodies in hemophilic patients, 146–147

Alloimmune thrombocytopenias, 203–204

α-actinin, 20

Alpha chemokine receptor (CXCR4), 13

Alpha granule storage pool disorder (α-SPD), 191, 194*f*, 200–201

Alpha granules, 14, 21–22, 21*t*

Alphanate in treating von Willebrand disease, 233

Alpha 2-antiplasmin (α$_2$-antiplasmin), 32

    α$_2$-plasmin inibitor, 31*f*

    assays, 118

    binding to fibrin, 34

    excess in fibrinolytic thrombotic disorders, 280

α$_2$-macroglobulin, 32

American College of Chest Physicians Guidelines on management of warfarin, nontherapeutic INR, 156t

Amino acid propeptide antigen (VW AgII), 225, 226f

Amyloid-induced gastrointestinal malabsorption, 52

Amyloidosis, 50, 157, 218–219

Anemia, iron-deficiency, 51

Anesthetics, effect on platelet function, 94t

Angiodysplasia as underlying condition, 232

Angiotensin, 33

Angiotensin II, 33

Annexin V (annexin A5), 295

Antibiotics that affect platelet function, 94t

Anticardiolipin antibodies, 295

    and risk of venous or arterial thrombosis, 248, 251

    testing and other antiphospholipid antibody immunoassays, 301

Anticoagulant function of endothelium, 3–5, 4f

Anticoagulants, 313–329

    argatroban, 325–327, 326t

        laboratory monitoring of, 326–327

    blood collection tubes and, 40

    contamination of plasma samples by, 220t

    fondaparinux, 321–322, 322t

        laboratory monitoring of, 322, 322t

    hemorrhagic side effects of, 49, 52–53

    hirudins (peptide direct thrombin inhibitors), 322–325, 323f

        laboratory monitoring of, 324–325

    low-molecular-weight heparins, 320

    lupus, 42, 78, 133, 146, 162, 248, 251, 295

        anticoagulation monitoring in patients with, 303

        minimal sensitivity to, 66

        testing for, 297–300, 298f

    oral, 327–329, 327f

        laboratory monitoring of, 328–329

        point-of-care monitors for, 70

    treatment with, 49, 52–53

    unfractionated heparin, 313–320, 314f, 315f

        laboratory monitoring of, 315–320, 316t

Anticoagulants that affect platelet function, 94t

Anticoagulation, 4f

    function of endothelium, 3–5

Antifibrinolytic agents, in fibrinolytic thrombotic disorders, 282

Antifibrinolytic agents that affect platelet function, 94t

Antigen-capture assays, 109

Antihistamines that affect platelet function, 94t

Anti-PF4/heparin antibodies, rapid-turnaround-time assay for, 292

Antiphospholipid antibodies, 295–303, 333

    anticardiolipin antibody testing and other antiphospholipid antibody immunoassays, 301

    clinical features and diagnosis, 301–302, 302t

    laboratory testing for, 295

        lupus anticoagulant testing considerations and International Society on Thrombosis and Haemostasis (ISTH) criteria, 295–301, 298f

        monitoring anticoagulation in patients with lupus anticoagulants, 303

    and risk of venous or arterial thrombosis, 246, 248, 251

Antiphospholipid antibody syndrome, 54, 93, 207, 295

Antiplatelet agents, 333–346, 334f, 346t

    foods, 346t

    hemorrhagic side effects of, 49

    new drugs, 344–345

    oral

        aspirin, 337–340

        cilostazol, 344

        dipyridamole, 342–344

        Iloprost, 344

        thienopyridines, 340–342

    other drugs and herbal remedies, 345–346

    parenteral

        glycoprotein IIb/IIIa inhibitors, 333–337, 335f

Antithrombin (AT), 5, 130–131

    algorithm for testing, 272f

    antigen assay, 131

    deficiency, 244–245, 250–251, 253–254

        clinical history of, 55, 58

        type I, 271

        type II, 271

    D-helix of, 5

    in diagnosing inherited thrombophilia, 271–272, 272f

    functional assay, 131

Antithrombin (AT) heparin cofactor activity reference intervals, 131

Antithrombotic agents, 307–311, 307t, 308t

    common clinical settings where used

        acute coronary syndromes, 309–310

        acute venous thromboembolism, 309

        extracorporeal circulation, 310

        implantation of mechanical heart valves, 310

        peripheral vascular and cerebrovascular disease, 310

        prophylaxis for venous thromboembolism, 309

        prophylaxis in patients with atrial fibrillation, 309

    effects on tests of hemostasis, 310

    target therapeutic range and monitoring considerations, 308–309

APTT. *See* Activated partial thromboplastin time (aPTT)

Arachidonic acid, 25, 336

    in platelet aggregation studies, 101–103, 102f, 105f

    asprin effect on, 338–339, 338f

Argatroban, 160, 297, 313, 325–327, 326t

    effects of, 293

    laboratory monitoring of, 326–327

Arginine-glycine-aspartic acid (RGD) peptide, 225, 333

Arginine-glycine-aspartic acid (RGD) tripeptide, 19, 20f

Arterial thrombosis, 57, 93

    in adults, 239–254

        general considerations, 240–241

            assay calibration, reference range, and assay reliability, 240–241

            combined deficiency, 241

consent and counseling, 240
testing for thrombophilic risk in patients with, 241
acquired factors
antiphospholipid antibodies, 246
heparin-induced thrombocytopenia, 246, 247t
recommendations for testing of individual analytes, 244–246
factor V Leiden and prothrombin G20210A mutations, 245–246
hyperhomocysteinemia and high concentrations of factor VIII, 246
protein C, protein S, and antithrombin deficiency, 244–245
thrombophilic risk factor test selection in arterial thrombosis, 247–251
antiphospholipid antibodies (lupus anticoagulant and anticardiolipin antibodies), 248
coagulation factors, 251
C-reactive protein (CRP), 248
deficiencies of protein C, protein S, and antithrombin, 250–251
factor V Leiden (activated protein C resistance), 250
heparin-induced thrombocytopenia, 251
homocysteine, 249–250
lipoprotein (a), 248–249
prothrombin G20210A mutation, 250
thrombophilic risk factor test selection in neurovascular thrombosis, 251–254
antiphospholipid antibodies (lupus anticoagulant and anticardiolipin antibodies), 251
deficiencies of protein C, protein S, and antithrombin, 253–254
factor V Leiden (activated protein C resistance), 253
heparin-induced thrombocytopenia, 254
homocysteine, 252–253
lipoprotein (a), 251–252
prothrombin G20210A mutation, 253
thrombophilic risk factor test selection in provoked and spontaneous thromboembolism, 241–244, 241t
conclusions and recommendations, 252t
thrombophilic risk factor test selection in, 247–251
Arteriovenous malformation, 54
L-asparaginase therapy, 271, 272f
Aspirin, 337–340
clinical uses, 337–338
laboratory monitoring, 339–340
laboratory tests used to assess effect, 338–339, 338f
mechanism of action, 337
VerifyNow assay, 104–105
Aspirin resistance, 339–340
Assay of specific factors, 75–78
Atherosclerosis, 93, 249
Atrial fibrillation, prophylaxis in patients with, 309
Autoantibodies to factor VIII and other factors, 147

Autoimmune disorders, 52
as underlying condition, 232
Autoimmune thrombocytopenia, 49
Autologous red cell agglutination assay (SimpliRED™ D-dimer), 116
Automated latex immunoassays, 228
Automated nephelometry, 131

Bernard-Soulier syndrome, 96, 187f, 188t, 199–200, 200f
detection by ristocetin induced platelet aggregation, 102
genetic testing for, 109
β1-tubulin, 15
gene, 15
Beta 2-glycoprotein I (β2GPI), 248, 295, 297, 301, 302
anti-β2GPI antibodies, 248, 301, 302
Bethesada assay, 81–82
results of, 81–82
Bioimmunoassay, 120–121
Bivalirudin, 160, 297, 323–324
effects of, 293
Bleeding
after trauma or surgery, 225
coagulation factors deficiencies not associated with, 146
dental, 51, 225
due to organ system disorders, 51–54
due to vascular disorders, 50
duration of abnormality, 49
epistaxis and, 50
gastrointestinal, 225
gingival, 225
microvascular, 93
in neonates, 54
postpartum, 225
posttraumatic, 50
quantification of excessive, 47
spontaneous, 49, 50
from surgical procedures, 50–51
symptoms of, 225
type of, 49–50
Bleeding patients, clinical history in, 47–51
bleeding due to vascular disorders, 50
duration of bleeding abnormality, 49
epistaxis, 50
extent of bleeding, 47, 49
gingival/dental bleeding, 51
hemarthrosis and hematomas, 50
menstrual, 51
obstetric, 51
posttraumatic and spontaneous bleeding, 50
surgical procedures, 50–51
types of bleeding, 49–50
Bleeding time, 98
in von Willebrand disease, 230
Blood, clotting of, 3
Blood collection. See Sample collection and processing

Blood collection tubes, 39–40

Blood transfusion, bleeding sufficient to require, 47, 49

Blood tube draw order, 40

Bone marrow

    analysis, 96–97

        sample collection and processing considerations, 96–97, 96f

    disorders, acquired, 49

    interpretation, 97

Bovine topical thrombin, 51

Bradykinin, 32, 33

Brodifacoum, 53

Buerger disease, 57

C1-inhibitor, 32

C807T polymorphism, 210

Calcium ions (factor IV), 6t

Capillaries, dilated, 50

Cardiolipin, 301

Cardiopulmonary bypass (CPB), 210, 315, 320

    accelerated fibrinolysis during, 178

    as stimulator of tPA release, 177

Cardiovascular devices, 210

Cardiovascular drugs that affect platelet function, 94t

CD34, 13

CD36, 191

CD41, 13

Cerebral vein thrombosis, 58

Cerebrovascular disease, 58

Chang percentage, 80

Chediak-Higashi syndrome, 191, 192–193

Chemotherapeutic agents that affect platelet function, 94t

Childbirth, excessive bleeding after, 49

Children, thrombophilia in, 262–265, 263t, 264t

Christmas disease, 145

Chromogenic factor activity assays, 77

Chromogenic factor X assays, 76

Chronic myelogenous leukemia (CML), 97, 196

Chrono-log Platelet Aggregometer, 105–106, 292

Cilostazol, 52, 344, 346

Circumcision, excessive bleeding after, 49

Citrate, buffering with citric acid, 41

Citrate-theophylline-adenosine-dipyridamole (CTAD) mixture, 40

Citric acid, citrate buffering with, 41

C-Kit ligand, 13

Clauss fibrinogen assay, 74, 74f

Clinical Laboratory Standards Institute (CLSI)

    on evaluating coagulometers, 64

    on venipuncture technique, 37–38

Clopidogrel, 52, 340–342, 346

    clinical uses, 341–342

    laboratory monitoring of, 342

    mechanism of action, 340–341

Clopidogrel resistance, 342

Closure time, PFA-100 testing, 98–100, 99f, 336

    laboratory testing algorithm, 99f

    laboratory reporting algorithm, 100f

Clot-based tests, laboratory measurement of, 63

    activated partial thromboplastin time, 70–73

    analyzers, 63–64

    assay of specific factors, 75–78

    fibrinogen assay, 74–75, 74f

    mixing studies, 78–80, 79f

    prothrombin time (PT), 64–70

    thrombin time (TT) test, 73

Clot detection, 63–64

Clot elastic modulus (CEM), 107

Clot removal system, 10

Clove, 345

c-Mpl, 13

Coagulation analyzers, 63–64

    selecting and validating, 64

Coagulation cascade, factor VIII (antihemophilic factor) in, 225

Coagulation factor deficiencies, inherited, 54–55

Coagulation factors, 251

    in coagulation pathway, 7-10, 8f, 9f

    deficiencies not associated with bleeding, 146

    and hemostasis system, 3-6 4f, 6t

    inhibitors of, 11t

Coagulation pathways, fundamental operation of, 11, 11f

    extrinsic pathway, 6–7, 6f, 8f

    intrinsic pathway, 7, 8f

    newer coagulation model, 7–10, 9f

    PT and aPTT pathways, 7

Coagulation proteins, 3f, 5–6, 6t

    newer model of, 7–10, 9f

Coagulation testing, 63–89

    criteria for rejection of specimens, 45t

    effects of temperature and time on, 42–44, 43t

    establishing reference ranges and validating, 82–85

        preanalytic and postanalytic considerations

            accuracy, 82–83, 83f

            interfering conditions, 83–85

            linearity, 83

            precision, 83

            reference range, 82

    factors affecting specimen quality, 44–45, 44t, 45t

    laboratory measurement of clot-based, 63

        activated partial thromboplastin time, 70–73

        analyzers, 63–64

        assay of specific factors, 75–78

        fibrinogen assay, 74–75, 74f

        mixing studies, 78–80, 79f

        prothrombin time (PT), 64–70

        thrombin time (TT) test, 73

    prompt performance of, 42

    quality assurance in, 88, 89t

        external proficiency testing, 88–89

quality control in, 85–86, 85*f*

results reporting, 87–88

   turnaround time, 87

specific inhibitor studies, 80–82

   Bethesada assay, 80–81

susceptibility to error, 44–45

Cocaine, 59

Collagen, 23, 26

   in platelet aggregation studies, 101–103, 102*f*, 105*f*

Collagen/adenosine diphosphate (COL/ADP) closure time. *See* Closure time, PFA-100 testing

Collagen binding assays, 230

Collagen disorders, 51

Collagen/epinephrine (COL/EPI) closure time. *See* Closure time, PFA-100 testing

Collagen receptor disorders (GPIa/IIa and GPVI), 191

Collagen receptors (GPVI and GPla/IIa), 16–17, 17*f*

College of American Pathologists Concensus Conference XXXVI: Diagnostic Issues in Thrombophilia. *See* Thrombophilia testing

Colony forming unit-megakaryocyte (CFU-MK) defect, 208

Congenital amegakaryocytic thrombocytopenias (CAMT), 207

Congenital thrombocytopenia with radio-ulnar synostosis (CTRUS), 207

Congestive heart failure, 54

Connective tissue disorders, 172

Consumptive coagulopathies, 94, 151

Coomassie Brilliant Blue, 127

C-reactive protein (CRP), 248

Cushing syndrome, 49, 50, 173

CXCR4, 13

Cyclic adenosine monophosphate (cAMP), 17

   platelet function and, 94*t*

Cyclic thrombocytopenia, 208

Cyclooxygenase-2 (COX-2) inhibitors, 340

Cyclooxygenase pathway, 24*f*, 25

CYP 2C9, 328

Cys97-Cys175 complex, 341

Cytochrome P450 (CYP) system, 340–341, 341*f*

Cytotoxic drugs, 52

DDAVP. *See* Desmopressin acetate

D-dimer, 29

   role of, in excluding venous thromboembolism, 282–284, 283*t*

   role of, in fibrinolytic system, 30*f*

D-dimer assays, 115–117, 115*f*

   comparison of clinical utility of, 115*t*

   quantitive, use of, 181–182

   variations in types of, 283–284

D-dimer test, 155

Deadly quartet, 280

Decreased platelet production, 202*f*, 207

   hereditary thrombocytopenias with, 207–208

Deep vein thrombosis (DVT), 57, 93

Degradation peptides, 136–137

Demarcation membrane system (DMS), 14

Dense granule storage pool disorders (δ-SPD), 191–194, 194*f*

   and Chediak-Higashi syndrome, 191, 192–193

   diagnosis of, with flow cytometry, 193*f*

   diagnosis of, with lumiaggregation, 192*f*

   and Hermansky-Pudlak syndrome, 191, 193–194

Dense granules, 14, 21*t*, 22

Dental bleeding, 93, 177, 225

   excessive, after extractions, 49

   history of, 51

Desirudin, 323, 324, 325

Desmopressin acetate (DDAVP), 231

   Glanzmann thrombasthenia and, 190

   therapeutic approach to patients with Bernard-Soulier syndrome, 200

   in treating von Willebrand disease, 233

Diabetes mellitus, 54

1,2-diacylglycerol, 23

DIC. *See* Disseminated intravascular coagulation

Dilute Russell viper venom time (dRVVT), 296, 297, 298*f*, 299–300

Dilute prothrombin time, 297, 298*f*

Dilutional coagulopathies, 154–155, 169

Dilutional effect, 75

Dipyridamole, 52, 342–344

   clinical uses, 343

   mechanism of action, 342–343

   monitoring issues, 343–344

Direct thrombin inhibitors (DTIs), 160, 169, 218, 219, 297, 313

Disseminated intravascular coagulation (DIC), 52, 94, 178–182, 179*f*. *See also* Fibrinolytic bleeding disorders

   acquired purpuras with, 93

   bleeding, acquired etiology of, 49

   disorders associated with, 179*t*

   effect of, 52

   effect on PT and/or aPTT, 155, 169

   evaluation of, with D-dimer assay, 115, 115*t*, 155, 180*f*

   in fibrinolytic thrombotic disorders, 281–282

   gastric carcinoma, association with, 178

   and thrombin time, 221*f*, 222

   scoring systems, comparison of, 181*t*

   and thrombocytopenia, 204–205

DNA testing, 230

Dose-response relationship, 317–318

D-Phe-Pro-Arg chloromethylketone (PPACK), 40

Drug-induced thrombocytopenias, 204

Drugs that affect platelet function, 94*t*. *See also* Antiplatelet agents

Dyserythropoiesis, 208

Dysfibrinogenemia, 158, 220, 222

   acquired, 220

   diagnosis of, 220, 222

   in fibrinolytic thrombotic disorders, 281

   inherited, 220

Ecarin clotting time (ECT), 324

Ecchymosis, 50, 93

Eclampsia, 207

Ecto-ADPase, 23

EDTA (ethylenedaminetetraacetic acid), 94–95
    blood tube draw order and, 40
    in platelet aggregation studies, 103–104, 104f

Ehlers-Danlos syndrome, 50, 51, 173

Embolic disorders, 57

Endogenous thrombin potential, 135–136

Endoplasmic reticulum-Golgi intermediate complex (ERGIC-53), 157

Endothelial cell. See Endothelium

Endothelial cell protein C receptor (EPCR), 125

Endothelial protein thrombomodulin (TM), 127f

Endothelium, 3–5, 3f, 4f
    fibrinolysis testing, effect of, 118f, 119
    in fibrinolytic system, function of, 29–32, 30f, 31f
        migration in wound healing, 34
        regulation of fibrinolysis, 31, 32
    kallikrein binding to, 177–178

Enzyme-linked immunosorbent assays (ELISA), 106, 131, 227–228
    as immunologic fibrinogen assay, 74

Epinephrine, 26, 32
    in platelet aggregation studies, 101–103, 102f

Epistaxis, 93
    bleeding and, 50

Epsilon aminocaproic acid (Amicar), in treating von Willebrand disease, 233

Epstein syndrome, 197, 198

Eptifibatide (Integrilin), 334–335

Erythematous plaques, at site of heparin injections, 288

Essential thrombocythemia (ET), 97, 196

Estrogen therapy, 54

Euglobulin clot lysis time assay in fibrinolytic thrombotic disorders, 280

Exhausted platelets, 194f

Extracorporeal circulation, 310

Extracorporeal membrane oxygenation (ECMO), 210

Extravascular fibrinolysis, 29

Extrinsic pathway, prothrombin time (PT) and, 6–7, 6f, 8f

Fabry disease, 173

Factor assays, 75–78, 171, 175–178
    false-positives and false-negatives in, 77

Factor deficiencies, 52
    inherited, 169
    multiple, 52
        acquired causes of, 52

Factor II, 5, 6t
    deficiency, 155–156
    inhibitor, 161
    measuring, 76

Factor V, 6t, 26
    deficiency, 156–157
    inhibitors, 51, 161–162
    measuring, 76

Factor V and factor VIII deficiency, combined, 157

Factor V Cambridge, 133

Factor V Hong Kong, 133

Factor V HR2 haplotype, 133

Factor V Leiden, 55, 58, 59, 127f, 131, 133–134, 250, 253, 263
    activated protein C resistance (APCr) and, 273–274, 273f
    prothrombin G20210A mutations and, 133–134, 245–246
    polymerase chain reaction, use in identifying, 134

Factor Va, 5

Factor VII, 5, 6t
    deficiency, 157
    inhibitor, 162
    measuring, 76

Factor VIII, 4, 6t, 318–319
    activity, 132–133
    assay, 228
    in coagulation cascade, 225
    deficiency, 144f, 145
    in diagnosing inherited thrombophilia, 272–273
    and factor V deficiency, combined, 157
    inhibitor assays, 82
    measuring, 76–77, 76f

Factor VIIIa, 5

Factor IX, 5, 6t
    deficiency, 145
    measuring, 76–77

Factor IXa, 5

Factor X, 5, 6t
    deficiency, 157–158
    inhibitor, 162
    measuring, 76

Factor replacement therapy, monitoring, 77

Factor Xa, 5
    chromogenic assay, 132
    inhibition of, 317
    inhibition assays, 317, 319

Factor XI, 6t, 7
    deficiency of, 51, 145

Factor XIa, 5

Factor XII, 6, 6t, 7
    deficiency, 146
        in fibrinolytic thrombotic disorders, 281
    measuring, 76–77

Factor XIII, 6t, 9
    deficiency, 51, 171–172
    measuring, 77

Familial thrombocytopenia-leukemia (Tel-AML1), 207

Familial thrombophilia, clinical features of, 55, 55t

Family history, bleeding and, 53–54

Fc receptor γ (FcRγ), 17

Fechtner syndrome, 197–198
Fibrin, 9–10, 10*t*, 29, 30*f*
    degradation of, 115
    formation of, 64
Fibrin degradation fragment assays, 117
Fibrin degradation products (FDP), 220
    latex agglutination assays for, 181
Fibrinogen (factor I), 6*t*, 9, 10*f*, 25, 26, 30*f*
Fibrinogen abnormalities, 133, 220
    afibrinogenemia, 222
    dysfibrinogenemia, 220, 222
    hyperfibrinogenemia, 222
    hypofibrinogenemia, 222
Fibrinogen assay, 74–75, 74*f*
Fibrinogen deficiency, 158
Fibrinogen receptor (GPIIb/IIIa), 18–19, 20*f*
Fibrinolysis
    connection between lipoproteins and, 33
    extravascular, 29
    global tests of, 121–122
    intravascular, 29, 30*f*
    primary, 117
    regulation of, 32–33
    secondary, 117
    at site of thrombus, 33–34
Fibrinolysis pathway, 10
Fibrinolysis testing, 113–123
    alpha 2-antiplasmin assays, 118
    clinical history suggesting need for, 113, 113*f*
    D-dimer and fibrin split product assays, 115–117, 115*ft*
    global tests of fibrinolysis, 121–122
    lipoprotein (a), 122
    plasminogen activator inhibitor-I assays, 120–121
    plasminogen assays, 117
    research assays, 122–123
    sample collection and processing for, 113–115, 114*f*, 114*t*
    tissue plasminogen activator assays, 114*f*, 118–120, 118*f*
Fibrinolytic bleeding disorders, 175–182
    acquired causes of, 113, 175, 177–178
    alpha-2 antiplasmin deficiency, 177
    clinical evaluation of, 175, 176*f*
    disseminated intravascular coagulation, 178–182, 179*ft*, 180*f*, 181*t*
    plasminogen activator inhibitor-I deficiency, 175–177, 177*t*
Fibrinolytic system, 4, 5
    physiology, 29–34
        fibrinolysis at site of thrombus, 33–34
        history, 29
        proteins and function, 29–32, 30*f*, 31*f*
Fibrinolytic testing
    clinical history suggesting need for, 113, 113*f*
    D-dimer and fibrin split product assays, 115–117, 115*ft*
    sample collection and processing for, 113–115, 114*f*, 114*t*

Fibrinolytic bleeding disorders, 175–182
    acquired causes of, 177–178
    alpha 2-antiplasmin deficiency, 177
        disseminated intravascular coagulation, 178–182
    plasminogen activator inhibitor-1 deficiency, 175–177
        and tissue plasminogen activator (tPA), 176–177, 177*t*
    testing for, 176*f*
Fibrinolytic thrombotic disorders
    role of D-dimer in excluding venous thromboembolism, 282–284, 283*t*
    thrombosis and defects of, 279
        alpha 2-antiplasmin excess, 280
        antifibrinolytic medications in, 282
        disseminated intravascular coagulation in, 281–282
        dysfibrinogenemia, 281
        euglobulin clot lysis time assay, 280
        factor XII (Hageman factor) deficiency, 281
        lipoprotein (a) excess in, 281
        plasminogen activator inhibitor excess, 280
        plasminogen deficiency in, 279
        thrombin-activatable fibrinolysis inhibitor, 280–281
        tissue plasminogen activator deficiency, 279–280
Fibrinopeptide A, 9, 10*f*, 40, 117
Fibrinopeptide B, 9, 10*f*, 117
Fibrinopeptide release assays, 117
Fibroblast growth factor (FGF)-4, 15
Fibronectin, 23
Filamin, 20
Fitzgerald factor (HMWK), 6*t*, 76–77
Flaujeac factor (HMWK), 6*t*, 76–77
Fletcher factor (prekallikrein), 6*t*, 146
Flow cytometry, 107–109, 108*f*, 292
    circulating platelet analysis and, 210
    platelet turnover and, 109
Fluorescence polarization immunoassay (FPIA), 134–135
Fluorescence resonance energy transfer (FRET)-based assays, 110
FOG-1, 13
Fondaparinux, 160–161, 288–289, 292–293, 321–322, 322*t*
    laboratory monitoring of, 322, 322*t*
Foods that affect platelet function, 94*t*
Foxifiban, 335
Functional platelet disorders, 50

G proteins, 18, 18*f*, 19*f*
G20210A prothrombin mutation, 55, 57, 58, 133–134
Garlic, 345
Gastric carcinoma, disseminated intravascular coagulation (DIC) associated with, 178
Gastrointestinal bleeding, 225
Gastrointestinal telangiectasia as underlying condition, 232
GATA-1, 13, 208
GATA-2, 13, 208
Gaucher disease, 52

Genetic polymorphisms associated with thrombosis and cardiovascular disease, 209–210

Gingival bleeding, 93, 225
   history of, 51

Ginkgo biloba, 346

Ginseng, 346

Glanzmann thrombasthenia (GPIIb/IIIa deficiency), 50, 96, 106, 107, 109, 185, 186f, 188t, 189–190, 333

Glanzmann thrombasthenia-like phenotype, 203

Global tests of fibrinolysis, 121–122

D-glucuronic acid, 313

Glycoprotein disorders, 186f, 188t, 189–191
   adenosine diphosphate receptor abnormalities, 191
   collagen receptor disorders (GPIa/IIa and GPVI), 191
   Glanzmann thrombasthenia (GPIIb/IIIa deficiency), 185, 186f, 188t, 189–190, 333
   glycoprotein Ib/IX/V (Bernard-Soulier syndrome), 190
   glycoprotein IV (GPIV), 191
   platelet-type von Willebrand disease, 190

Glycoprotein Ib (GPIb), platelet adhesion and, 225

Glycoprotein Ib/IIa receptor antagonists that affect platelet function, 94t

Glycoprotein Ib/IX/V (Bernard-Soulier syndrome), 190

Glycoprotein Ib/IX/V complex, 26. See also GPIb/IX/V

Glycoprotein IIb/IIIa inhibitors, 333–337, 335f
   antagonist drugs, 333–335
      mechanism of action, 334f
   clinical uses, 336
   effect on platelet function tests, 336
   mechanisms of action, 333–335, 335f
   role of monitoring, 336–337
   special issues, 337

Glycoprotein IV (GPIV), 191

GMP33, 22

GPIa/IIa ($\alpha_2\beta_1$) integrin, 16–17, 17f

GPIb receptor, 210

GPIbα, 16, 16f

GPIbβ, 16, 16f

GPIb/IX/V, 13–14, 15–16, 16f, 17f, 22f, 23–24, 24f, 26

GPIIb/IIIa, 14f, 17f, 18–19, 20f, 22f, 23, 24f, 25

GPV, 16

GPVI/FcRγ, 23

Gray platelet syndrome, 188t, 191, 200–201

Hairy cell leukemia as underlying condition, 232

Heart valves, implantation of mechanical, 310

Hemangioma, 54

Hemarthrosis, 50, 177, 225

Hematin, 144

Hematocrit effect, 41, 41t

Hematomas, 50, 177

Hemodialysis, 210

Hemolysis, 84

Hemolysis elevated liver enzymes and low platelets syndrome (HELLP), 207

Hemolytic uremic syndrome (HUS), 205, 207

Hemophilia, 50, 225
   alloantibodies in, 146–147
   hereditary, 171
   severity of, 145t

Hemophilia A, 54, 76–77, 144–145, 227
   carriers of, 54
   family pedigree for, 144f

Hemophilia B, 54, 76–77, 145
   carriers of, 54

Hemostasis
   basic elements of, 3f
   clinical history in, 47–59, 48–49f
      bleeding due to organ system disorders, 51–54
         family history, 53–54
            medication and food habits, 52–53
            previous coagulation test results, 54
      in bleeding patients, 47–51
      features of familial thrombophilia, 55, 55t
      in patients with thrombosis, 54–59, 55t, 56f
      prevalence of hereditary thrombophilia, 55
   effects on tests of, 310
   and thrombosis pathway, operation of, 11, 11f

Hemostasis Analysis System, 97, 106–107

Hemostasis testing, 63
   activated partial thromboplastin time (aPTT), 70–73
   clot-based, 63–64
   fibinogen assay, 74–75
   prothrombin time (PT), 64–70
   mixing studies, 78–80, 79f
   specific factor assays, 75–78
   specific inhibitor studies, 80–82
   thrombin time (TT), 73

Hemostatic system, constituents of, 3–11, 3f
   coagulation proteins, 3f, 5–6, 6t
   endothelium, 3–5, 3f, 4f
   extrinsic pathway and prothrombin time (PT), 6–7, 6f
   intrinsic pathway and activated partial thromboplastin time (aPTT), 7
   platelets, 3f, 5
   prothrombin time (PT) and activated partial thromboplastin time (aPTT) pathways, 6t, 7, 8f, 9f
   regulatory mechanisms, 10–11, 11t

Heparin, 160–161, 169. See also Unfractionated heparin
   binding site deficiency, 271–272
   in drug-induced thrombocytopenias, 204
   as glycosaminoglycan, 287
   interacting with antithrombin, 313
   low sensitivity, 66
   unfractionated, 219–220, 220t, 288

Heparin associated thrombocytopenia (HAT), 287

Heparin cofactor activity assay, 131

Heparin cofactor II, 131

Heparin-induced platelet aggregation (HIPA) assay, 291–292

Heparin-induced thrombocytopenia (HIT), 54, 59, 93, 203, 204, 207, 209, 246, 247t, 251, 254, 287–293, 313, 324–325, 333
    clinical laboratory diagnosis and algorithm, 289–292, 289t, 290f, 291f
    clinical presentation, 287–289
    pathophysiology of, 288f, 289
    treatment of, 292–293
Heparin-induced thrombocytopenia syndrome, 314
Heparin sensitivity, validation of
    of activated partial thromboplastin time (aPTT), 72
    using ex vivo heparin specimens, 72–73
Heparin sulfate, 5
Heparin therapy, 59
Hepatic disease, 49
Heptest, 319
Herbal remedies, antiplatelet effect, 345–346, 346t
Herbal supplements
    effect on platelet function, 94t
    interaction with warfarin, 155
Hereditary hemorrhagic telangiectasia, 50, 173
Hermansky-Pudlak syndrome, 191, 193–194
High-molecular-weight kininogen (HMWK), 6t, 146
High-molecular-weight kininogen and prekallikrein (Fitzgerald, Flaujeac, or William factor), measuring, 76–77
Hippocrates, 29
Hirudins (peptide direct thrombin inhibitors), 322–325, 323f
    laboratory monitoring of, 324–325
Homocysteine, 249–250, 252–253
    in diagnosing inherited thrombophilia, 274
    elevated levels of, 59
        association with cardiovascular disease, 249
        MTHFR C677T mutation, 249–250
        risk factor for venous and arterial thrombosis, 249–250
Homocystinuria, 50, 134–135
Hormone therapy
    bleeding and, 58–59
    venous thromboembolism and, 258–259, 259t
HOXA11, mutation in, 207
Humate-P in treating von Willebrand disease, 233
Hypercoagulability, global tests of
    activated partial thromboplastin time, 136
    activation and degradation peptides, 136–137
    endogenous thrombin potential, 135–136
    undiluted activated protein C resistance, 136
Hypercoagulable risk factors, 58
Hyperfibrinogenemia, 222
Hyperhomocysteinemia, 134–135, 260, 261, 263
    high concentrations of factor VIII and, 246
    risk factors for acquired, 134
Hyperlipidemia, 54
Hypertensive disorders of pregnancy with thrombocytopenia, 207
Hyperviscosity syndromes, 54
Hypofibrinogenemia, 158, 222
    acquired, 222
    inherited, 222

Hypoprothrombinemia, 302
Hypothyroidism, 52
    as underlying condition, 232

Icteric samples, 84–85
Idiopathic thrombocytopenic purpura (ITP), 54
L-iduronic acid, 313
IgG anti-UFH/PF4 antibody, 289
Iloprost, 344
Immune destructive thrombocytopenia, 202–204
Immune thrombocytopenia, 54
Immune thrombocytopenic purpura (ITP), 202–203, 204
Immunoreceptor type activation motif (ITAM), 17
Impact-R, 97, 106
Implantation of mechanical heart valves, 310
Indwelling lines and catheters, blood collection from, 39
Infants, thrombophilia in, 262–265, 263t, 264t
Infections, 93
Inhibitor studies, 80–82
1,4,5-inositol 3 phosphate (1,4,5-IP3), 23
Insulin resistance syndrome, 280
Interleukin-3 (IL-3), 13
Interleukin-6 (IL-6), 13
Interleukin-11 (IL-11), 13
International normalized ratio (INR), 66–70
    calculation of, 66–69
    checklist actions for, 68t
    validation and calibration of, 67
    warfarin management when nontherapeutic, 156t
International reference preparation (IRP), 65–66
International sensitivity index (ISI), 40, 65–67, 171, 328
International Society on Thrombosis and Haemostasis (ISTH)
    criteria for lupus anticoagulant testing, 295–297
    position paper on aspirin resistance (AR), 340
    scoring system for DIC, 181t, 182
    Subcommittee on Fibrinogen, 281
    Subcommittee on von Willebrand Factor, 231
Intra-aortic balloon counterpulsation (IABP), 210
Intravascular fibrinolysis, 29, 30f
Intrinsic pathway, activated partial thromboplastin time (aPTT) and, 7, 8f
In vitro platelet clumping, 95
Iron-deficiency anemia, 51

Jacobsen syndrome, 208
Janus kinase 2 (JAK-2) mutations, 195, 196
Joints, repeated bleeding into, 50

Kallikrein, 30–31
    binding to endothelium, 177–178
    formation of, 30
Kaolin clotting time (KCT), 296, 300
Klerval, 335
Kozak dimorphism (-5T/C), 210

Laminin, 23

Latex agglutination assays
    for fibrin degradation products (FDP), 181
    in measuring D-dimer, 115–116

Laurell electroimmunodiffusion, 131

Lefradifiban, 335

Lepirudin, 160, 293, 297, 313, 323, 324, 324*t*
    half-life of, 323

Leucine-rich repeat (LRR) protein family, 15–16, 16*f*

Leukemias, 94
    as underlying condition, 232

Levey-Jennings charts, 85, 85*f*

Licorice, 345–346

Lipemia, 83–84

Lipoprotein (a), 122, 248–249, 251–252
    connection between fibrinolysis and, 33
    excess in fibrinolytic thrombotic disorders, 281

Livedo reticularis, 58

Liver disease, 49, 52, 169
    factor deficiency due to, 159
    and prolonged thrombin time, 221*f*, 222

LMAN1 (lectin, mannose-binding 1) gene, 157

Lotrafiban, 335

Low-molecular-weight heparins, 320

Lumiaggregation, 105–106

Lumiaggregometry, 292

Lupus, 50

Lupus anticoagulants, 42, 78, 133, 146, 162, 248, 251, 295
    algorithm for diagnosis of presence of, 298*f*
    anticoagulation monitoring in patients with, 303
    minimal sensitivity to, 66
    testing for, 295–301, 298*f*
        activated partial thromboplastin time, 296, 297–299, 298*f*
        dilute prothrombin time (dPT), 297, 298*f*
        dilute Russell viper venom time (dRVVT), 296, 297, 298*f*, 299–300
        International Society on Thrombosis and Haemostasis (ISTH) criteria, 295–297
        kaolin clotting time (KCT), 296, 300

Lymphoma as underlying condition, 232

Lymphoproliferative disorders, 52

Lyonization, 54

Lysine-glycine-aspartic acid (KGD) sequence, 334

Lysosomal trafficking regulator (LYST), 192–193

Lysosome associated membrane protein (CD63) (LAMP), 25

Macrothrombocytopenia, 96
    with neutrophilic inclusions, 197–199
    true congenital, 96

Malabsorption of vitamin K, 52

Malignancy, 54
    bleeding and, 58

Malignant paraprotein disorders, 50

Malpighi, 29

Manual clot detection, by tilt-tube method, 63–64

Marfan syndrome, 50, 173

May-Hegglin anomaly, 197, 198, 199

MCFD2 (multiple combined factor deficiency 2) gene, 157

Mean platelet volume (MPV), 196

Medications, risk of thrombosis from, 59

Megakaryocytes, 5, 13–15
    development and maturation of, 13–15, 13*f*, 14*f*
    platelet production from, 14*f*, 15

Menorrhagia, 51, 225
    treating, 233

Menstrual history, bleeding and, 51

Messenger ribonucleic acid (mRNA) detection, 96

Metabolic syndrome, 33, 280

5,10-methylenetetrahydrofolate reductase. See *MTHFR* gene mutations

Microvascular bleeding, 93

Mitral valve prolapse as underlying condition, 232

Mixing studies, 78–80, 148
    1:1, 80
    4:1, 80
    criteria for interpretation of, 80
    differential diagnosis of corrected, 153–154, 153*t*
        dilutional coagulopathy, 154–155
        disseminated intravascular coagulation, 155
        factor II deficiency, 155–156
        factor V and factor VIII deficiency, combined, 157
        factor V deficiency, 156–157
        factor VII deficiency, 157
        factor X deficiency, 157–158
        fibrinogen deficiency, 158
        vitamin K-dependent coagulation factor deficiency, hereditary, 158
        warfarin, 155, 156*t*
    differential diagnosis of noncorrected, 160–162
        direct thrombin inhibitors, 160
        factor II inhibitor, 161
        factor VII inhibitor, 162
        factor V inhibitor, 161–162
        factor X inhibitor, 162
        heparin, 160–161
        lupus anticoagulant, 162
    guidelines for interpretation of, 81*t*
    ristocetin cofactor, 233
    value of, 149

Modified thrombin time, 319

Monoclonal antibody immobilization of platelet antigens (MAIPA), 109

Monoclonal antibody inhibitors, 52

Monoclonal gammopathies, 52
    as underlying condition, 232

Morawitz, Paul, 6

MTHFR gene mutations, 135
    MTHFR C677T variant
        genetic testing for, 274
        and elevated homocysteine, 249–250

and risk of fetal loss, 58

MTHFR T/T genotype, 260

Multifactorial acquired bleeding disorders, 52

Multimers, 225

analysis of, 228, 228*t*, 229*f*

Multiple myeloma, thalidomide therapy for, 59

Munchausen syndrome, 54

Myelodysplastic disorders, 94

Myelodysplastic syndromes, 52

Myelofibrosis, 97, 196

Myeloma as underlying condition, 232

Myeloproliferative disorders, 52, 54, 57, 94, 96

associated with thrombocytosis, 189*t*, 196

as underlying condition, 232

Myosin heavy chain gene defects (MYH9), 96

Myosin II polymerization, 15

Needle selection for venipuncture, 38

Neonatal alloimmune thrombocytopenia (NAIT), 54, 204

Neonates

bleeding in, 54

thrombophilia in, 262–265, 263*t*, 264*t*

Nephrotic syndrome, 54

Neurovascular thrombosis

conclusion and recommendations, 254*t*

thrombophilic risk factor test selection in, 251–254

NF-E2, transcription factor, regulation of, 15

Nijmegen modification, 81

Nitrous oxide, 23

Nonimmune destructive thrombocytopenias, 204–207, 206*f*

Nonmuscle myosin heavy chain-A (NMMHCA), 198–199

Nonsteroidal anti-inflammatory drugs (NSAIDS) that affect platelet function, 94*t*

North American Specialized Coagulation Laboratory Association (NASCOLA), 230, 338

study of platelet aggregation testing, 103

Obesity, association between venous thrombosis and, 57

Obesity dyslipidemia syndrome, 280

Obstetrical therapy, bleeding and, 58–59

Obstetric history, bleeding and, 51

Open-canalicular membrane system (OCS), 21, 21*f*

Operational Process Specification (OPspecs) charts, 85–86, 85*f*

Oral agents

aspirin, 337–340

cilostazol, 344

dipyridamole, 342–344

lloprost, 344

thienopyridines, 340–342

Oral anticoagulants, 327–329, 327*f*

laboratory monitoring of, 328–329

point-of-care monitors for, 70

Oral contraceptives

menstrual bleeding and, 51

risk of venous thromboembolism and, 257–258

Orbofiban, 335

Organ system disorders, bleeding due to, 51–54

Osteogenesis imperfecta, 50, 173

Osteonectin, 22

P2X$_1$ receptor, 340

P2Y$_1$ receptor, 17, 340

P2Y$_{12}$ receptor, 17, 340

PAR-1, 17–18, 24–25, 24*f*

PAR-4, 17–18, 24–25, 24*f*

Parenteral agents, glycoprotein IIb/IIIa inhibitors, 333–337, 335*f*

Paris-Trousseau type thrombocytopenia (TCPT), 208

Paroxysmal nocturnal hemoglobinuria (PNH), 54, 57–58, 207

Partial thromboplastin time (PTT), 7

Partial thromboplastin time, activated. *See* Activated partial thromboplastin time (aPTT)

Penicillin in drug-induced thrombocytopenias, 204

Peptides, activation and degradation, 136–137

Percutaneous coronary intervention, 315, 320

Perinatal thrombosis, risk of, 259

Peripheral platelet destruction, 202–207

immune destructive thrombocytopenias, 202–204

Peripheral vascular and cerebrovascular disease, 310

Peripheral venipuncture, sample collection technique for, 37–38

Petechiae, 49, 93

PFA-100, 97, 98–101, 99*f*, 230, 336, 338–339

laboratory testing algorithm, 99*f*

laboratory reporting algorithm, 100*f*

Phospholipase pathway, 24–25

Phospholipids, 7, 8*f*, 9*f*

platelet membranes, 26

specimen handling and lupus anticoagulants, 42

Placental insufficiency, 58

Plasma homocysteine concentration, 134–135, 135*f*

Plasmin-antiplasmin (PAP) complexes, 31

enzyme immunoassays for measuring, 123

Plasminogen, 29, 32

Plasminogen activator inhibitor, 4

Plasminogen activator inhibitor-1 (PAI-1), 17–18, 30, 31–34

assays, 120–121

binding to fibrin, 34

deficiency, 175–177, 177*t*

laboratory assessment of, 176

secretion of, by endothelial cells, 32

tPA–PAI-1 complex, 29–30, 30*f*, 31, 118, 118*f*, 120–121

uPA–PAI-1 complex, 30, 31, 120

Plasminogen activator inhibitor-2 (PAI-2), 32

Plasminogen activator inhibitor excess in fibrinolytic thrombotic disorders, 280

Plasminogen activators, 220

types of, 29–30

Plasminogen assays, 117

Plasminogen deficiency in fibrinolytic thrombotic disorders, 279

Plasminogen kringle-4, 33

Platelet activating factor, 26

Platelet activation, 22–23, 24f, 209

Platelet activation and granule release, 24–25
    cyclooxygenase pathway, 24f, 25
    phospholipase pathway, 24–25

Platelet adhesion
    cytoskeletal changes, 23–24
    inhibition of, to von Willebrand factor (vWF) through GPIb/IX/V complex, 344
    role of collagen receptors, 16–17, 23, 24f
    role of GPIb/IX/V and von Willebrand factor, 16f, 23

Platelet adhesion chambers, 106

Platelet agglutination, speed of, 227

Platelet aggregation, 20f, 25–26, 101–105
    performance and interpretation, 101–103, 102f
    Plateletworks, 103–104, 104f
    quality control for, 102–103
    in releasing thrombogenic platelet microparticles, 289
    results of, in platelet disorders, 188t
    ristocetin-induced, 227, 229f, 231
    specimen considerations, 101
    VerifyNow, 104–105

Platelet aggregation studies, 196

Platelet cytoskeleton, 19–21, 23–24
    changes during platelet activation, 23–24

Platelet disorders, 185–211
    associated with thrombosis or platelet activation, 208–209
        cardiovascular devices, 210
        genetic polymorphisms associated with thrombosis and cardiovascular disease, 209–210
        platelet activation, 209
    dysfunction with normal platelet count, 185, 187f, 188t
        glycoprotein disorders, 186f, 188t, 189–191
            adenosine diphosphate receptor abnormalities, 191
            collagen receptor disorders (GPIa/IIa and GPVI), 191
            Glanzmann thrombasthenia (GPIIb/IIIa deficiency), 185, 186f, 188t, 189–190
            glycoprotein Ib/IX/V (Bernard-Soulier syndrome), 190
            glycoprotein IV (GPIV), 191
            platelet-type von Willebrand disease, 188t, 190
        platelet release defects
            disorders of platelet procoagulant activity (Scott syndrome), 195
            signal transduction disorders, 194–195
            storage pool (alpha and dense granule disorders), 191–194, 192f, 193f
    with thrombocytopenia, 196–197, 197f
        decreased platelet production, 202f, 207
            hereditary thrombocytopenias with, 207–208
        with decreased platelet size, Wiskott-Aldrich syndrome and X-linked thrombocytopenia, 201
        idiopathic thrombocytopenic purpura (ITP), 202–203, 204
        with increased platelet size, 197

Bernard-Soulier syndrome, 187f, 188t, 199–200, 200f
    gray platelet syndrome (alpha granule storage pool disorder), 188t, 191, 194f, 200–201
    macrothrombocytopenias with neutrophilic inclusions (MYH9 disorders), 197–199
    nonimmune destructive thrombocytopenias, 204–207, 206f
    with normal platelet size, 201–202, 202f
    peripheral platelet destruction, 202–203
        immune destructive thrombocytopenias, 202–203
  with thrombocytosis, 195
    myeloproliferative disorders associated with, 189t, 196
    reactive, 195–196

Platelet disorders associated with thrombosis or platelet activation, 208–209
    cardiovascular devices, 210
    genetic polymorphisms associated with thrombosis and cardiovascular disease, 209–210
    platelet activation, 209

Platelet disorders with thrombocytopenia, 196–197, 197f
    decreased platelet production, 202f, 207
        hereditary thrombocytopenias with, 207–208
    with decreased platelet size, Wiskott-Aldrich syndrome and X-linked thrombocytopenia, 201
    idiopathic thrombocytopenic purpura (ITP), 202–204
    with increased platelet size, 197
        Bernard-Soulier syndrome, 187f, 188t, 199–200, 200f
        gray platelet syndrome (alpha granule storage pool disorder), 189t, 191, 194f, 200–201
        macrothrombocytopenias with neutrophilic inclusions (MYH9 disorders), 197–199
    nonimmune destructive thrombocytopenias, 204–207, 206f
    with normal platelet size, 201–202, 202f
    peripheral platelet destruction, 202–203
        immune destructive thrombocytopenias, 202–203

Platelet disorders with thrombocytosis, 195
    myeloproliferative disorders associated with, 189t, 196
    reactive, 195–196

Platelet distribution width (PDW), 96

Platelet dysfunction, 52
    associated with other illnesses, 208
    with normal platelet count (qualitative platelet disorders), 185, 187f, 188t
        glycoprotein disorders, 186f, 188t, 189–190
            adenosine diphosphate receptor abnormalities, 191
            collagen receptor disorders (GPIa/IIa and GPVI), 191
            Glanzmann thrombasthenia (GPIIb/IIIa deficiency), 185, 186f, 188t, 189–190
            glycoprotein Ib/IX/V (Bernard-Soulier syndrome), 190
            glycoprotein IV (GPIV), 191
            platelet-type von Willebrand disease, 190, 226, 231
        platelet release defects
            disorders of platelet procoagulant activity (Scott syndrome), 195
            signal transduction disorders, 194–195

storage pool (alpha and dense granule disorders), 191–194, 192*f*, 193*f*, 194*f*

Platelet factor 4 (PF4), 289, 314

β-thromboglobulin, 210

Platelet flow cytometry, 107–109, 209, 340

Platelet Function Analyzer-100. *See* PFA-100.

Platelet function, drugs that affect, 94*t*. *See also* Antiplatelet agents

Platelet function testing, 97–100, 340. *See also* PFA-100.

bleeding time, 98

Platelet genetic testing, 109–110

Platelet granules, 21–22, 21*f*

alpha, 21–22, 21*t*

dense, 21*t*, 22

release of, 25

Platelet mapping, 107

Platelet mechanical assays, 106–107

Platelet-mediated bleeding disorders, 93

Platelet-mediated vascular thrombosis, 333

Platelet membrane phospholipids, 26

Platelet membrane systems, 21, 21*f*

Platelet neutralization procedure (PNP), 298–299

Platelet-poor plasma (PPP)

coagulation tests on, 42

preparation of, 42

Platelet procoagulant activity, 26

disorders of (Scott syndrome), 195

Platelet production

megakaryocyte development and maturation, 13–15, 13*f*, 14*f*

from megakaryocytes, 14*f*, 15

Platelet P-selectin, 209

Platelet receptors, 15–19, 16*f*

Platelet release defects

disorders of platelet procoagulant activity (Scott syndrome), 195

signal transduction disorders, 194–195

storage pool (alpha and dense granule disorders), 191–194, 192*f*, 193*f*, 194*f*

Platelet-rich plasma (PRP), 101

platelet count in, 101

Platelets, 3*f*, 5, 13–26

cytoskeleton, 19–21

changes during activation, 23–24

granules, 21–22, 21*f*

alpha granules, 21–22, 21*f*

dense granules, 21*t*, 22

release of, 25

production, 13–15, 13f, 14*f*

megakaryocytic development and maturation, 13–15, 13*f*

from megakaryocytes, 14*f*, 15

receptors, 15–19, 16*f*

role in cardiovascular disease, 333

Platelet storage pool disorder, 102

Platelet structure

adenosine diphosphate receptors, 17, 18*f*

collagen receptors (GPVI and GPla/IIa), 16–17, 17*f*

fibrinogen receptor (GPIIb/IIIa), 18–19, 20*f*

surface glycoproteins and plasma membrane, 15

thrombin receptors (PAR-1 and PAR-4), 17–18, 19*f*

thromboxane receptor and prostaglandin receptors, 18

von Willebrand receptor (GPIb/IX/V), 15–16, 16*f*

Platelet testing, 93–110. *See also* Platelet function testing; PFA-100

ADAMTS-13 assays, 110

adhesion assays, 106

platelet adhesion chambers, 106

Impact, 106

aggregation, 101–105,

performance and interpretation, 101–103, 102*f*

Plateletworks, 103–104, 104*f*

specimen considerations, 101

VerifyNow, 104–105

bone marrow analysis, 96–97

sample collection and processing considerations, 96–97, 96*f*

bone marrow interpretation, 97

count, indices, and morphology, 94

performance and interpretation, 95–96, 95*f*

sample collection and processing considerations, 94–95

electron microscopy, 109

flow cytometry, 107–109, 108*f*

genetic, 109–110

mechanical assays, 106–107

patient history, 93–94

clinical

for patients with bleeding diathesis, 93, 94*t*

for patients with thrombosis, 93–94

platelet function testing, 97–100

bleeding time, 98

PFA-100, 98–100, 99*f*

release assays, 105

lumiaggregation, 105–106

markers, 106

P-selectin, 106

turnover (platelet reticulocyte analysis), 109

whole blood viscoelastometry, 107

Platelet turnover (platelet reticulocyte analysis), 109

Platelet-type von Willebrand disease, 190, 226, 231

treatment of, 234

Plateletworks, 97, 103–104, 104*f*, 336

Polycythemia vera, 97, 196

Polymerase chain reaction (PCR), 134

Polymorphisms in PAI-1 promoter region, 280

Postpartum bleeding, 225

Posttraumatic bleeding, 50

PPACK (protease inhibitor D-Phe-Pro-Arg chloromethylketone), 114, 118, 336

Preeclampsia, 54, 207

Pregnancy, hypertensive disorders of, with thrombocytopenia, 207

Pregnancy and hormonal therapy
    hormone replacement therapy and venous thromboembolism, 258–259, 259t
    oral contraceptives and risk of venous thromboembolism, 257–258
    pregnancy-related venous thromboembolism, 259
    thrombophilia and late gestational vascular complications, 260–261
    thrombophilia and recurrent fetal loss, 259–260
    treatment of pregnant women with thrombophilias, 261–262, 261t

Pregnancy-related venous thromboembolism, 259

Prekallikrein, 6t, 7, 8t
    deficiency of, 146
    formation of kallikrein from, 30

Primary fibrinolysis, 117

Procoagulation, 4f
    function of endothelium, 3–4

Prophylaxis
    in patients with atrial fibrillation, 309
    for venous thromboembolism, 309

Prostacyclin (PGI2), 4t, 5

Prostaglandin I2, 23

Protamine sulfate, 161

Protease-activated receptors (PAR-1 and PAR-4), 344–345

Protease-activated receptors (PARs), 17–18, 19f

Protease inhibitor D-Phe-Pro-Arg chloromethylketone. See PPACK

Protein C, 4, 125–128
    antigen assays, 127–128
    in diagnosing inherited thrombophilia, 268–270, 269f
    deficiencies of, 55, 58, 244–245, 250–251, 253–254
        hereditary, 59
    functional assays, 125–127

Protein phosphoinositide 3-kinase (PI3K), 16, 16f, 17, 18f, 24f, 25

Protein S, 128–130, 128f
    activity assays, 129
    antigenic assays, 129–130
    in diagnosing inherited thrombophilia, 270–271, 270f
    deficiencies of, 55, 58, 244–245, 250–251, 253–254

Prothrombin (factor II), 6t, 9f
    conversion of, to thrombin, 8–9

Prothrombinase, formation of, 77

Prothrombin F1.2, 40

Prothrombin G20210A mutations, 55, 58, 133–134, 250, 253, 263
    in diagnosing inherited thrombophilia, 274
    factor V Leiden and, 133–134
    polymerase chain reaction, use in identifying, 134

Prothrombin time (PT), 3, 64–70, 169. See also Prothrombin time and activated partial thromboplastin time, normal
    clinical utility of, 70
    extrinsic pathway and, 6–7, 6f, 8f

    prolonged, 153–163
        algorithm, 162, 163f
        differential diagnosis of corrected mixing study, 153–154, 153t
            dilutional coagulopathy, 154–155
            disseminated intravascular coagulation, 155
            factor II deficiency, 155–156
            factor V and factor VIII deficiency, combined, 157
            factor V deficiency, 156–157
            factor VII deficiency, 157
            factor X deficiency, 157–158
            fibrinogen deficiency, 158
            vitamin K-dependent coagulation factor deficiency, hereditary, 158
            warfarin, 155, 156t
        differential diagnosis of noncorrected mixing study, 160–162
            direct thrombin inhibitors, 160
            factor II inhibitor, 161
            factor VII inhibitor, 162
            factor V inhibitor, 161–162
            factor X inhibitor, 162
            heparin, 160–161
            lupus anticoagulant, 162
        liver disease, 159
        superwarfarin poisoning, 159
        vitamin K deficiency, 159–160
            in adults, 160
            in newborns, 159–160
    reagent considerations, 65–66
    reference ranges for, 67–68
    reporting of, 66
    testing, purposes for performing, 65t

Prothrombin time and activated partial thromboplastin time, normal, 169–173
    additional testing, 171–173
        fibrinolysis defects, 172
        primary hemostasis defects, 171
        secondary hemostasis defects, 171–172
            factor assays, 171
            factor XIII deficiency, 171–172
        vascular disorders, 172–173
    algorithm, 172f, 173
    clinical evaluation, 169
    false-negative, 169–171
        analytic causes, 170–171
        patient-specific causes, 170
        postanalytic causes, 171
        preanalytic causes, 170

Protoplatelets, 15

P-selectin, 22, 106

Pseudothrombocytopenia, 95, 204
    detection of, 95, 95f

Pseudo von Willebrand disease, 190

Pseudoxanthoma elasticum, 50

Psychotropic drugs that affect platelet function, 94t

PT. *See* Prothrombin time

Pulmonary embolism, 57, 93

Purpura, 49, 93

Purpura fulminans, 263

Quality assurance in coagulation testing, 88–89, 89*t*
    external proficiency testing, 88–89

Quality control (QC)
    in coagulation testing, 85–86, 85*f*
    frequency of testing, 86
    PFA-100 and, 98
    for platelet aggregation, 102–103
    prothrombin time (PT) and, 68

Quebec syndrome, 201

Quick, Armand, 6–7, 64

Quinidine, 204

Radial immunodiffusion, 131

Radiocontrast agents, 222

Raloxifene, venous thromboembolism and, 258

Rapid plasma reagin (RPR) test, 301

Rapid-turnaround-time assay for anti-PF4/heparin antibodies, 292

Reactive thrombocytosis, 195–196

Receptor-activated NF-κB (RANK) ligand, 15

Red pepper (capasaicin), 346

Reference ranges, establishing and validating, 82–85
    preanalytic and postanalytic considerations, 82–85
    accuracy, 82–83, 83*f*
    interfering conditions, 83–85
    linearity, 83
    precision, 83

Regulatory mechanisms, 10–11, 11*t*

Release assays, 105–106
    lumiaggregation, 105–106, 105*f*
    markers, 106
    P-selectin, 106

Renal dialysis, 315, 320

Renal disease, 49

Renal transplantation, tissue plasminogen activator release during, 177

Renin-angiotensin system, 33

Research assays, 122–123

Retroperitoneal hematomas, 50

Rheumatoid arthritis, 50

Rheumatologic disorders, 58

Ristocetin, 102

Ristocetin cofactor, 228*t*, 231
    assay, 227, 229*f*
    mixing studies, 233

Ristocetin-induced platelet aggregation (RIPA), 102, 103*f*, 227, 228*t*, 229*f*, 231

Rodenticides, 53

Rosner index, 80

Russell viper venom (RVV)-activated clotting assays, 132

Russell viper venom (RVV) screening method
    advantages of, for factor V Leiden (FVL), 132

Russell viper venom (RVV) time, 157

S-adenosyl-homocysteine (SAH), 134

Salicylic acid, 337, 345. *See also* Aspirin

Sample collection and processing, 37–45
    blood collection tubes, 39–40
        anticoagulant, 40
        draw order, 40
        hematocrit effect, 41, 41*t*
        volume effect, 40–41
    from indwelling lines and catheters, 39
    for peripheral venipuncture, 37–38
        collection sets, 38
        needle selection for, 38
        specimen labeling, 38
        tourniquet effect, 38
    sample processing, 42–45
        effects of temperature and time, 42–44, 43*t*
        factors affecting specimen quality, 44–45, 44*t*, 45*t*
        preparation of platelet-poor plasma, 42

Sapporo criteria, revised, for diagnosis of antiphospholipid antibody syndrome, 302, 302*t*

SCAT-1 tube, 40

Scleroderma, 50

Scott syndrome, 195

Sebastian syndrome, 197, 198, 199

Secondary fibrinolysis, 117

Secondary hemostasis, defects in, 169

Selective estrogen receptor molecules (SERMs), venous thromboembolism and, 258

Senile purpura, 50, 172

Serine protease inhibitor (SERPIN), 31–32, 280

Serotonin release assay, 291

Severe aortic stenosis, 52

Sibrafiban, 335

Signal transduction disorders, 194–195

Single nucleotide polymorphism (SNP) analysis, 109

Skin necrosis, at site of heparin injections, 288

Sodium citrate, 37, 40

Soluble fibrin monomer assays, 182

SNARE (soluble N-ethyl-maleimide-sensitive factor attachment protein receptor) complex, 25

Specific coagulation factor inhibitors, 146

Specific inhibitor studies, 80–82
    Bethesada assay, 80–81
    Nijmegen modification, 81

Specimen quality, factors affecting, 44–45, 44*t*, 45*t*

Spectrin, 20

Split sample testing, 89

Spontaneous bleeding, 49, 50

Staphylokinase, 29

Steroid-induced purpura, 172

Storage pool disorders (alpha and dense granule disorders), 191–194, 192*f*, 193*f*, 194*f*

Streptokinase, 29

Stroke, thrombotic, 58

Stromal-derived factor (SDF)-1 chemokine, 15

Sulfite precipitation as immunologic fibrinogen assay, 74

Sulfonamide drugs in drug-induced thrombocytopenias, 204

Superficial thrombophlebitis, 57

Superwarfarin poisoning, 159

Suramin, 144

Surface glycoproteins and plasma membrane, 15

Surgery, bleeding after trauma or, 177, 225

Surgical operations, excessive bleeding after, 49

Surgical procedures, bleeding from, 50–51

Syndrome X, 280

Systemic lupus erythematosus, 58

Takayasu disease, 57

Tamoxifen, venous thromboembolism and, 258

Taularidine, 144

T-cell lymphoproliferative disorders, 52

Telangiectasia, dilated, 50

Thalidomide therapy for multiple myeloma, 59

Thienopyridines, 340–342

    clinical uses, 341–342

    clopidogrel resistance, 342

    laboratory monitoring, 342

    mechanism of action, 340–341

Thienopyridines that effect platelet function, 94t

Thr145Met dimorphism (Br or HPA-2), 210

Thrombin, 4–5, 4f, 6f, 8f, 9, 9f, 10f, 26. See also Thrombin time (TT), abnormal; Thrombin time test

    activation of platelets, 19f

    in coagulation pathway, newer model, 9f

    conversion of prothrombin (factor II) to, 8–9

    inhibition of, 313

    bovine, 219, 221f

Thrombin-activatable fibrinolysis inhibitor (TAFI), 4, 4f, 5, 31–32

    in fibrinolytic thrombotic disorders, 280–281

    measuring, 123

Thrombin generation, regulation of, 10–11, 11t

Thrombin generation time (TGT), 107

Thrombin receptor-activating peptide (TRAP), 18, 101, 336

Thrombin receptors (PAR-1 and PAR-4), 15, 17–18, 19f

Thrombin time (TT), abnormal, 217–222. See also Thrombin; Thrombin time (TT) test

    algorithm for evaluation of, 221f, 222

    amyloidosis, 218–219

    autoantibodies, 219

    clinical indications, 217–218

    differential diagnosis, 218, 218t

    direct thrombin inhibitors, 219

    disseminated intravascular coagulation, 221f, 222

    drugs, 219–220

    fibrin degradation products, 220

    fibrinogen abnormalities, 220, 221f, 222

    plasminogen activators, 220

    radiocontrast agents, 222

    unfractionated heparin, 219–220, 220t

    volume expanders, 220

Thrombin time (TT) test, 73. See also Thrombin; Thrombin time (TT), abnormal

    clinical utility of, 73

    global test of hypercoagulability

        activated partial thromboplastin time, 136

        activation and degradation peptides, 136–137

        endogenous thrombin potential, 135–136

        undiluted activated protein C resistance, 136

    method, 73

    reagents, 73

    reference ranges, 73

Thrombocythemia, essential, 196

Thrombocytopenia, 16, 49, 50, 52, 337

    with absent radii (TAR syndrome), 191, 208

    with decreased platelet production, 207–208

        acquired, 208

            acquired amegakaryocytic thrombocytopenic purpura (AATP), 208

            cyclic thrombocytopenia, 208

        hereditary, 207–208

            congenital amegakaryocytic thrombocytopenia (CAMT), 207

            congenital thrombocytopenia with radioulnar synostosis (CTRUS), 207

            familial thrombocytopenia-leukemia (Tel-AML1), 207

            thrombocytopenia with absent radii (TAR), 208

    with decreased platelet size, 201

        Wiskott-Aldrich syndrome and X-linked thrombocytopenia, 201

    drug-induced, 204

    heparin-induced, 93, 204, 207, 246, 247t, 251

    hypertensive disorders of pregnancy with, 207

    immune destructive thrombocytopenias, 202–204

        alloimmune, 203–204

            neonatal alloimmune thrombocytopenia (NAIT), 204

        drug-induced, 204

        heparin-induced thrombocytopenia (HIT), 204

        immune thrombocytopenic purpura (ITP), 202–203

    with increased platelet size, 197

        Bernard-Soulier syndrome, 187f, 188t, 199–200, 200f

        gray platelet syndrome (alpha granule storage pool disorder), 189t, 191, 194f, 200–201

        macrothrombocytopenias with neutrophilic inclusions (MYH9 disorders), 197–199

    nonimmune destructive thrombocytopenias, 204–207

        disseminated intravascular coagulation (DIC), 204–205

        thrombotic thrombocytopenic purpura (TTP), 205–207

            role of ADAMTS-13, 206f

        hypertensive disorders of pregnancy with thrombocytopenia, 207

    with normal platelet size, 201–202, 202f

Thrombocytosis
    myeloproliferative disorders associated with, 196
    reactive, 195–196
Thromboelastograph, 97, 107, 319
Thromboelastometry, 178
Thromboembolism, 58
    venous, 59
Thrombokinase, 6, 7
Thrombomodulin, 4
Thrombophilia. *See also* Thrombophilia testing; Thrombophilic risk factors; Thrombophilic risk factor test selection
    inherited thrombophilia, 55
        laboratory diagnosis of, 267–276
            activated protein C resistance and factor V Leiden, 273–274, 273*f*
            algorithmic approach to risk factors, 267–268, 267*f*, 269*f*, 270*f*, 272*f*
            antithrombin, 271–272, 272*f*
            factor VIII in, 272–273
            homocysteine, 274
            protein C, 268–270, 269*f*
            protein S, 270–271, 270*f*
            prothrombin G20210A mutation, 274
        laboratory testing, duration of anticoagulation therapy, and screening first-degree relatives, 274–276, 276*t*
    late gestational vascular complications and, 260–261
    in neonates infants, and children, 262–265, 263*t*, 264*t*
    recurrent fetal loss and, 259–260
    risk factors for, 243*t*
        prevalence of inherited, 243*t*
    testing for, in children, 263–265
    in women and children, 257–265
        in neonates, infants, and children, 262–265, 263*t*, 264*t*
        pregnancy and hormonal therapy
            hormone replacement therapy and venous thromboembolism, 258–259, 259*t*
            oral contraceptives and risk of venous thromboembolism, 257–258
            pregnancy-related venous thromboembolism, 259
            thrombophilia and late gestational vascular complications, 260–261
            thrombophilia and recurrent fetal loss, 259–260
            treatment of pregnant women with thrombophilias, 261–262, 261*t*
Thrombophilia testing, 125–137
    activated protein C resistance, 127*f*, 131–132
        assays, 131–132
    antithrombin, 130–131
        assays, 131
    College of American Pathologists Concensus Conference XXXVI: Diagnostic Issues in Thrombophilia, 239–254
        criteria for level of evidence assigned, 240*t*
        general recommendations, 240–241, 241*t*
        thrombotic risk in patients with venous or arterial thrombosis, testing for, 241–254
            acquired thrombophilic risk factors, 246
            arterial thrombosis, conclusions and recommendations, 252*t*
            arterial thrombosis, thrombophilic risk factor test selection in, 247–251
            individual analytes, recommendations for testing of, 244–246
            neurovascular thrombosis, thrombophilic risk factor test selection in, 251–254
            neurovascular thrombosis, conclusions and recommendations, 254*t*
            venous thromboembolism, conclusions and recommendations, 247*t*
            venous thromboembolism, thrombophilic risk factor test selection in provoked and spontaneous, 241
    duration of anticoagulation therapy, and screening first-degree relatives, 274–276, 276*t*
    inherited thrombophilia testing algorithm, 244*f*
    factor VIII activity, 132–133
    factor V Leiden and prothrombin G20210A mutations, 133–134
    fibrinogen abnormalities, 133
    general recommendations for, 241*t*
    lupus anticoagulant, 133
    plasma homocysteine concentration, 134–135, 135*f*
    protein C, 125–128
        antigen assays, 127–128
        functional assays, 125–127
    protein S, 128–130, 128*f*
        activity assays, 129
        antigenic assays, 129–130
    timing in performing, 244
Thrombophilic risk factors
    acquired, 57–59, 246
        antiphospholipid antibodies, 246
        heparin-induced thrombocytopenia, 246, 247*t*
        malignancy, 58
        medical history, 57–58
        obstetrical history and hormone therapy, 58–59
            elevated levels of homocysteine, 59
            medical history, 59
            surgery and trauma, 59
    algorithmic approach to laboratory diagnosis, 267–268, 267*f*, 269*f*, 270*f*, 272*f*, 273*f*
    inherited, 243*t*
    prevalence of, 243*t*
Thrombophilic risk factor test selection
    in arterial thrombosis, 247–251
        antiphospholipid antibodies (lupus anticoagulant and anticardiolipin antibodies), 248
        coagulation factors, 251
        C-reactive protein (CRP), 248
        deficiencies of protein C, protein S, and antithrombin, 250–251
        factor V Leiden (activated protein C resistance), 250
        heparin-induced thrombocytopenia, 251
        homocysteine, 249–250

lipoprotein (a), 248–249

prothrombin G20210A mutation, 250

in neurovascular thrombosis, 251–254

antiphospholipid antibodies (lupus anticoagulant and anticardiolipin antibodies), 251

deficiencies of protein C, protein S, and antithrombin, 253–254

factor V Leiden (activated protein C resistance), 253

heparin-induced thrombocytopenia, 254

homocysteine, 252–253

lipoprotein (a), 251–252

prothrombin G20210A mutation, 253

in provoked and spontaneous thromboembolism, 241–244, 241*t*

Thrombophlebitis, superficial, 57

Thrombopoietin receptor (c-Mpl), 13

Thrombosis. *See also* Venous thrombosis

arterial, 57

cerebral vein, 58

clinical history in patients with, 54–59, 55*t*, 56*f*, 93–94

age of onset, 55

ethnicity, 55

location and types of, 57

recurrences, 57

neurovascular, 251–254

pathway, operation of, 11, 11*f*

venous, 57, 58, 239–254

Thrombospondin, 23

Thrombotic disorders, 185

stroke, 58

Thrombotic thrombocytopenic purpura (TTP), 54, 203, 205–207, 206*f*

Thrombo-Wellco Test, 117

Thromboxane A2, 24*f*, 25

Thromboxane B2, 210

Thromboxane receptors, 15, 18

prostaglandin receptors and, 18

Thrombus, fibrinolysis at site of, 33–34

Thyroid disease, 49

Ticlopidine, 52, 340, 341–342

Tilt-tube method, manual clot detection by, 63–64

Tirofiban (Aggrastat), 335

Tissue factor, 3–4, 6*t*, 8*t*, 9*t*

Tissue factor pathway inhibitor (TFPI), 5, 8, 161

Tissue plasminogen activator (tPA), 5, 10, 29–34, 30*f*, 31*f*, 114*t*

assays, 114*f*, 118–120, 118*f*

deficiency in fibrinolytic thrombotic disorders, 279–280

intravascular fibrinolysis in initiating, 29

reduced hepatic clearance of, 178

release during renal transplantation, 177

release of, from endothelial cells, 32–33

tPA–PAI-1 complex, 29–30, 30*f*, 31, 118, 118*f*, 120–121

Tonsillectomy, excessive bleeding after, 49

Topical bovine thrombin, 219

Total thrombin-clottable fibrinogen as immunologic fibrinogen assay, 74

Tourniquet effect, 38

Transforming growth factor, beta 1 (TGFβ1), 15

Transglutaminase enzyme factor XIIIa, 9–10

Transient thromboembolic events, 58

Trisodium citrate as recommended anticoagulant, 40

True congenital macrothrombocytopenias, 96

TTP. *See* Thrombotic thrombocytopenic purpura (TTP)

Turbidimetric method, functional fibrinogen assays as, 74

Turbidimetric platelet aggregation assay, 101

Turbidometric coagulation assays, 85

Turnaround time, 87

UFH/PF4 complex, 289

Ultegra assay, 105

Ultracentrifugation, 84

Undiluted activated protein C resistance, 136

Unfractionated heparin, 219–220, 220*t*, 288, 313–320, 314*f*, 315*f*

laboratory monitoring of, 315–320, 316*t*

UFH/PF4 complex, 289

Upshaw-Schulman syndrome, 205

Urokinase-plasminogen activator (uPA), 29, 30, 31*f*

single-chain zymogen (scuPA), 30

uPA–PAI-1 complex, 30, 31, 120

Vacutainers, 39

Valproic acid, use of, as underlying condition, 232

Variable number of tandem repeats (VNTR) polymorphism, 210

Vascular disorders, bleeding due to, 50

Vascular inflammation, 93

Vascular malformations, 172

Vasculitis, 93, 172

Vasodilator-stimulated phosphoprotein (VASP) assay, 342

Vasopressin, 32

Venereal Disease Research Laboratory (VDRL) test, 301

Venipuncture, needle selection for, 38

Venous thromboembolism, 59, 315

hormone replacement therapy and, 258–259, 259*t*

oral contraceptives and risk of, 257–258

pregnancy-related, 259

prophylaxis for, 309

role of D-dimer in excluding, 282–284, 283*t*

Venous thrombosis, 57, 58, 93. *See also* Thrombosis

in adults, 239–254

general considerations, 240–241

assay calibration, reference range, and assay reliability, 240–241

combined deficiency, 241

consent and counseling, 240

testing for thrombophilic risk in patients with, 241

acquired factors

antiphospholipid antibodies, 246

heparin-induced thrombocytopenia, 246, 247*t*

recommendations for testing of individual analytes, 244–246

factor V Leiden and prothrombin G20210A mutations, 245–246

hyperhomocysteinemia and high concentrations of factor VIII, 246

protein C, protein S, and antithrombin deficiency, 244–245

thrombophilic risk factor test selection in arterial thrombosis, 247–251

antiphospholipid antibodies (lupus anticoagulant and anticardiolipin antibodies), 248

coagulation factors, 251

C-reactive protein (CRP), 248

deficiencies of protein C, protein S, and antithrombin, 250–251

factor V Leiden (activated protein C resistance), 250

heparin-induced thrombocytopenia, 251

homocysteine, 249–250

lipoprotein (a), 248–249

prothrombin G20210A mutation, 250

thrombophilic risk factor test selection in neurovascular thrombosis, 251–254

antiphospholipid antibodies (lupus anticoagulant and anticardiolipin antibodies), 251

deficiencies of protein C, protein S, and antithrombin, 253–254

factor V Leiden (activated protein C resistance), 253

heparin-induced thrombocytopenia, 254

homocysteine, 252–253

lipoprotein (a), 251–252

prothrombin G20210A mutation, 253

thrombophilic risk factor test selection in provoked and spontaneous thromboembolism, 241–244, 241t

association between obesity and, 57

Ventricular septal defects as underlying condition, 232

VerifyNow (Ultegra) system, 97, 104–105, 336

Vitamin C deficiency, 50, 173

Vitamin K, 5, 327f

malabsorption of, 52

Vitamin K antagonists (VKAs), 316, 327–328

laboratory monitoring of, 328–329

Vitamin K deficiency, 49, 52, 54, 159–160, 169

in adults, 160

evaluating, 52

in newborns, 159–160

Vitamin K-dependent coagulation factor deficiency, hereditary, 158

Vitamin K-dependent coagulation proteins, 5

Vitamin K epoxide reductase complex 1 (VKORC1), 158, 327, 327f

Vitronectin, 23

Vitronectin receptor ($\alpha_v\beta_3$), 18

Volume effect, 40–41

Volume expanders, 220

Volume expanders that affect platelet function, 94t

Von Willebrand disease, 50, 51, 54, 93, 144–145, 225–234

acquired, 52, 232–233

clinical features of, 225–227

classification of, 226–227

factor VIII and, 225, 226f

inheritance of, 227

laboratory testing of, 227–228, 228t, 229f

findings in type 1, 230

findings in type 2, 230–231

findings in type 3, 231

molecular basis of, 231–232, 232f

Normandy (type 2N *VWF* mutations), 227

platelet-type or pseudo-, 190, 226, 231

pseudo, 190

tests for, 185, 187, 187f

treatment, 233–234

type 1, 100, 225–226, 230, 233

type 2, 225–226, 228, 230–231

type 2A, 226, 230–231, 232, 233

type 2B, 226, 228, 230–231, 232, 233

type 2M, 227, 231, 232, 233

type 2N, 100, 227, 231, 232, 233

type 3, 226, 228, 231, 233

types of, 225–227

Von Willebrand factor (vWF), 4, 4f, 15, 16f, 21, 22t, 23, 24f, 26, 49, 93

A1 domain binding site, 16f, 229f

domain structure and function, 226f

measuring, 78

platelet adhesion to, 205, 210

quantitative or qualitative deficiencies of, 225

ristocetin-induced conformational change, 229f

von Willebrand disease and, 225, 226f

Von Willebrand factor (vWF) antigen assay, 227–228, 228t

Von Willebrand factor (vWF) multimer analysis, 228, 229f

Von Willebrand receptor (GPIb/IX/V), 15–16, 16f, 23, 24f

VWF (von Willebrand gene), 226–227, 230

type 2N mutations (von Willebrand disease Normandy), 227

Waldenstrom macroglobulinemia as underlying condition, 232

Warfarin over-anticoagulation, 155

Warfarins, 5, 155, 169

as cause of prolonged prothrombin time (PT), 155

management of, 156t

Waterhouse-Friderichsen syndrome, 263

Westgard multirule sets, 86

Whole blood clot lysis times, 121–122

Whole blood platelet function assays, 97

Whole blood viscoelastometry, 107

William factor (HMWK), 6t, 76–77

Wilms tumor, 52

as underlying condition, 232

Wiskott-Aldrich syndrome, 54, 96, 191, 201

World Health Organization (WHO), 65–66

Xemilofiban, 335

X-linked thrombocytopenia, 201

X-linked thrombocytopenia with thalassemia (XLTT), 208